Translations

of

Mathematical Monographs

Volume 39

Introduction to Spectral Theory: Selfadjoint Ordinary Differential Operators

by

B. M. Levitan and I. S. Sargsjan

American Mathematical Society
Providence, Rhode Island
1975

ВВЕДЕНИЕ В СПЕКТРАЛЬНУЮ ТЕОРИЮ:
САМОСОПРЯЖЕННЫЕ ОБЫКНОВЕННЫЕ
ДИФФЕРЕНЦИАЛЬНЫЕ ОПЕРАТОРЫ

Б. М. ЛЕВИТАН и И. С. САРГСЯН

Издательство "Наука"
Главная Редакция
Физико-Математической Литературы
Москва 1970

Translated from the Russian by Amiel Feinstein

AMS (MOS) subject classifications (1970).
Primary 34B25, 47E05; Secondary 35L05, 47A10, 81A06.

Library of Congress Cataloging in Publication Data CIP

Levitan, Boris Moiseevich.
Introduction to spectral theory.

(Translations of mathematical monographs ; v. 39)
Translation of Vvedenie v spektral'nui͡u teorii͡u.
Bibliography: p.
1. Boundary value problems. 2. Spectral theory (Mathematics) 3. Selfadjoint operators. 4. Differential operators. I. Sargsi͡an, Ishkhan Saribekovich, joint author. II. Title. III. Series.
QA379.L4813 515'.7222 75-15565
ISBN 0-8218-1589-X

Table of Contents

PREFACE

The spectral theory of linear operators has as its basic origins, on the one hand, linear algebra—more precise theorems on the reduction of quadratic forms to sums of squares—and on the other hand, problems in the theory of oscillations (vibrating strings, membranes, etc.).

The analogy between problems of linear algebra and the theory of oscillations was, apparently, noted a long time ago. However, it was only D. Hilbert who began to make systematic use of this analogy, in his fundamental work on the theory of integral equations. A result of this work was the appearance of the Hilbert space l_2, and then of general Hilbert space.

The introduction in mathematics of the space l_2 and of abstract Hilbert space led to the intense development of the spectral theory of linear selfadjoint operators in Hilbert space. During recent years, by the efforts of a number of outstanding mathematicians, this theory has been brought to a high degree of completeness and, perhaps, even been completed.

It appeared, at the beginning, that abstract spectral theory completely covers the various special cases and gives an answer, in principle, in all problems.

However, it gradually became clear that this was far from the case. There are many very important questions to which abstract spectral theory either gives no answers at all, or gives very inadequate answers. For example, such is the case concerning questions on the asymptotic behaviour of eigenvalues and eigenfunctions and other spectral quantities. Moreover, the theorem on the spectral decomposition of an operator, which is one of the basic theorems of the entire theory, describes this decomposition, in abstract terms, by means of a so-called resolution of the identity. For concrete operators (differential or finite-difference operators) the spectral decomposition can usually be described in terms of the solution of an appropriate equation. In questions concerning the specific description of the resolution of the identity, general spectral theory is of little help. This, presumably, explains the paradoxical fact that a few years after the actual completion of the abstract spectral theory of selfadjoint operators there began, almost simultaneously in various countries, intensive work on the spectral theory of selfadjoint differential operators. This work has continued to the present day, in particular on partial differential operators.

The present book is devoted to the systematic account of a number of basic topics in the modern spectral theory of selfadjoint ordinary differential equations, principally of one second-order equation and of a system of two first-order equations (of which so-called Dirac systems are a special case). We also consider some topics in the spectral theory of nth order differential operators, however in substantially less detail than for second-order operators. Thus, those parts of this book which concern nth order operators can serve as simply an introduction to this domain, which at the present time has already had time to become very broad.

One is led to distinguish between regular and singular differential operators. A differential operator is said to be regular if the domain of its variable is finite and its coefficients are continuous. If the domain is infinite or the coefficients (all or some) are not summable (or both), then the differential operator is said to be singular.

The basic spectral theory of regular second-order operators consists of the so-called Sturm-Liouville theory, named in honor of two outstanding mathematicians whose work initiated the deep study of these questions. Regular systems of two first-order equations began to be studied substantially later.

Although rather much space is devoted in this book to regular problems, they do not form its principal content, which is the spectral theory of singular operators. This theory was founded in papers by H. Weyl. These papers, together with the classical moment problem, played an outstanding role in the construction, through the efforts of F. Riesz, J. von Neumann and other mathematicians, of a general spectral theory of symmetric and selfadjoint operators. We note that Weyl's remarkable theorem on the limit circle and limit point gives a complete description, for a symmetric second-order differential operator, of all its selfadjoint extensions. The general problem of describing all selfadjoint extensions of a symmetric operator was solved substantially later by von Neumann.

A large role in drawing the attention of mathematicians to the spectral theory of differential operators was played by the monographs of E. C. Titchmarsh [**1,2**], in which he gave a new approach to the theory of singular second-order operators, and stated (influenced in part by problems from quantum mechanics) and solved an entire series of new problems.

The first discussion, in a monograph published in this country, of spectral theory for singular second-order operators was given by one of the authors (B. M. Levitan [1]), who also presented a new method for establishing this theory. The idea behind this method is very simple, and consists in obtaining the basic spectral relations for a singular operator from the corresponding relations in the regular case by passage to the limit. This method, and the results of the aforementioned monograph, are widely used in the present book. However, this book can in no way be regarded as a revision of the other. The changes and additions are so substantial that this book is definitely a new one.

Much space has also been allotted in this book to singular systems of two first-order equations, which were absent from the 1950 monograph. We remark that these results are comparatively new and, at any rate, were on the whole obtained after 1950. From the point of view of general spectral theory, many fundamental theorems in this book are special cases of some very general theorems.

For the convenience of the reader who is not familiar with abstract spectral theory, we have inserted a chapter (Chapter 13) in which we discuss this theory, concisely and in the main without proofs, and indicate various connections with the spectral theory of differential operators.

At the request of the authors, V. A. Sadovničiĭ wrote §17 of Chapter 1, and A. G. Kostjučenko wrote §8, Chapter 7, and §8, Chapter 12. A. G. Kostjučenko and M. V. Fedorjuk carefully read the entire manuscript and gave us much valuable advice, which we have made use of.

We wish to express our gratitude to all of those named.

In addition, we warmly thank Dr. Feinstein, who in the course of translation introduced a number of improvements in the original text.

February, 1969 *B. Levitan, I. Sargsjan*
Moscow

Chapter 1

Expansion in a Finite Interval

§1. Sturm-Liouville operators

1. Let L be a linear operator defined on some set of elements. An element $y \neq 0$ is called an eigenvector of L if $Ly = \lambda y$; the number λ is called an eigenvalue of L.

One of the simplest operators which is frequently encountered in applications is an operator of the form

$$L \equiv -\frac{d^2}{dx^2} + q(x),$$

where the function $q(x)$ will be assumed real and, to begin with, continuous on some interval $[a, b]$. For this operator the set of elements (functions) $y(x)$ mentioned above is determined by the obvious differentiability condition and also by certain conditions on the boundary of the interval $[a, b]$.

The most impostant boundary conditions for the operator L are the following:

$$\text{I.}\quad y(a)\cos\alpha + y'(a)\sin\alpha = 0,$$
$$y(b)\cos\beta + y'(b)\sin\beta = 0.$$
$$\text{II.}\quad y(a) = y(b), \quad y'(a) = y'(b).$$

One of the basic objectives of this book is the study of the following boundary-value problem:

$$Ly(x) = -\frac{d^2y}{dx^2} + q(x)\,y = \lambda y, \tag{1.1}$$

$$\begin{aligned} y(a)\cos\alpha + y'(a)\sin\alpha &= 0, \\ y(b)\cos\beta + y'(b)\sin\beta &= 0. \end{aligned} \tag{1.2}$$

The boundary-value problem (1.1) – (1.2) is known in the literature as the Sturm-Liouville problem.

We remark that the more general second-order equation

$$-\frac{d}{dx}\left\{p(x)\frac{dy}{dx}\right\} + l(x)y = \lambda r(x)y \qquad (a \leqslant x \leqslant b), \tag{1.3}$$

where the functions $p(x)$ and $r(x)$ are positive in the interval $[a, b]$, can be reduced to the form (1.1). If we assume that $p(x)$ has a continuous first derivative, and $p(x)r(x)$ has a continuous second derivative, then by means of the substitutions

$$z = \frac{1}{c}\int_a^x \left(\frac{r(x)}{p(x)}\right)^{1/2} dx, \quad u = (r(x)\,p(x))^{1/4} y, \quad \mu = c\lambda,$$

where c is given by

$$c = \frac{1}{\pi}\int_a^b \left(\frac{r(x)}{p(x)}\right)^{1/2} dx,$$

equation (1.3) assumes the form (1.1) with λ replaced by μ, where

$$q(z) = \frac{\theta''(z)}{\theta(z)} - c^2\frac{l(x)}{r(x)}, \qquad \theta(z) = (r(x)\,p(x))^{1/4}.$$

As a result of this change the interval $[a, b]$ is transformed to the interval $[0, \pi]$, while the boundary conditions do not change their form.

2. Thus we shall consider the boundary-value problem (1.1) – (1.2). Without loss of generality we may suppose that $a = 0$ and $b = \pi$. Indeed, the substitution $t = \pi(x - a)/(b - a)$ transforms the interval $[a, b]$ into the interval $[0, \pi]$ without changing the form of the boundary-value problem (1.1) – (1.2).

Let us assume that for some λ_1 the boundary-value problem under consideration has a nontrivial solution $y(x, \lambda_1) \not\equiv 0$. In this case, according to the definition given at the beginning of the section, the number λ_1 is called an eigenvalue, and the function $y(x, \lambda_1)$ – an eigenfunction of the boundary-value problem (1.1) – (1.2).

LEMMA 1.1. *Eigenfunctions $y(x, \lambda_1)$ and $y(x, \lambda_2)$, corresponding to distinct eigenvalues, are orthogonal, i.e.*

$$\int_0^\pi y(x, \lambda_1)\, y(x, \lambda_2)\, dx = 0 \qquad (\lambda_1 \neq \lambda_2).$$

PROOF. Let $f(x)$ and $g(x)$ be continuous and twice-differentiable functions. Put

$$Lf = f''(x) - q(x)f(x).$$

Integrating by parts twice, we obtain the identity

$$\int_0^\pi Lf \cdot g(x)\,dx = W_\pi\{f, g\} - W_0\{f, g\} + \int_0^\pi f(x) \cdot Lg\,dx, \tag{1.4}$$

where

$$W_x\{f, g\} = \begin{vmatrix} f(x) & g(x) \\ f'(x) & g'(x) \end{vmatrix}.$$

Let $f(x) = y(x, \lambda_1)$ and $g(x) = y(x, \lambda_2)$. It follows easily from the boundary conditions (1.2) that $W_0\{f, g\} = W_\pi\{f, g\} = 0$. Therefore it follows from the identity (1.4) that

$$(\lambda_1 - \lambda_2) \int_0^\pi y(x, \lambda_1)\, y(x, \lambda_2)\,dx = 0,$$

and since $\lambda_1 \neq \lambda_2$, the lemma is proved.

Lemma 1.2. *Eigenvalues of the boundary-value problem* (1.1)–(1.2) *are real.*

Proof. Let $\lambda_1 = u + iv$ be a complex eigenvalue. Since $q(x)$ is a real function, and the numbers α and β are also real, the number $\lambda_2 = \overline{\lambda}_1 = u - iv$ is also an eigenvalue, with eigenfunction $\bar{y}(x, \lambda_1)$. Then on the basis of the preceding lemma we have

$$\int_0^\pi |y(x, \lambda_1)|^2\,dx = 0,$$

from which it follows that $y(x, \lambda_1) \equiv 0$.

Theorem 1.1. *If $q(x)$ is a continuous function on the interval $[a, b]$, then for every α there exists a unique solution $\varphi(x, \lambda)$, $a \leqq x \leqq b$, of equation* (1.1) *such that*

$$\varphi(a, \lambda) = \sin\alpha, \quad \varphi'_x(a, \lambda) = -\cos\alpha.$$

For every fixed $x \in [a, b]$ the function $\varphi(x, \lambda)$ is an entire function of λ.

Proof. Put

$$\varphi_0(x, \lambda) = \sin\alpha - (x - a)\cos\alpha,$$

and for $n > 0$ let

$$\varphi_n(x, \lambda) = \varphi_0(x, \lambda) + \int_a^x \{q(t) - \lambda\}\, \varphi_{n-1}(t, \lambda)(x - t)\,dt.$$

Since q is continuous, we have $|q(x)| < M$ for $a \leqq x \leqq b$. Let $|\lambda| \leqq N$. Then $|\varphi_0(x, \lambda)| \leqq K$ for $a \leqq x \leqq b$. Therefore

$$|\varphi_1(x,\lambda)-\varphi_0(x,\lambda)|\leqslant\int_a^x (M+N)K(x-t)\,dt=\frac{1}{2}K(M+N)(x-a)^2.$$

For $n \geqq 2$ we have

$$\varphi_n(x,\lambda)-\varphi_{n-1}(x,\lambda)=\int_a^x \{q(t)-\lambda\}\cdot\{\varphi_{n-1}(t,\lambda)-\varphi_{n-2}(t,\lambda)\}(x-t)\,dt,$$

$$|\varphi_n(x,\lambda)-\varphi_{n-1}(x,\lambda)|\leqslant(M+N)(b-a)\int_a^x|\varphi_{n-1}(t,\lambda)-\varphi_{n-2}(t,\lambda)|\,dt.$$

Hence

$$|\varphi_2(x,\lambda)-\varphi_1(x,\lambda)|\leqslant\frac{K(M+N)^2(b-a)}{2}\int_a^x (t-a)^2\,dx$$

$$=\frac{K(M+N)^2(b-a)(x-a)^3}{3!}$$

and in general

$$|\varphi_n(x,\lambda)-\varphi_{n-1}(x,\lambda)|\leqslant\frac{K(M+N)^n(b-a)^{n-1}(x-a)^{n+1}}{(n+1)!}.$$

Therefore the series

$$\varphi(x,\lambda)=\varphi_0(x,\lambda)+\sum_{n=1}^{\infty}\{\varphi_n(x,\lambda)-\varphi_{n-1}(x,\lambda)\} \tag{1.5}$$

converges uniformly in λ for $|\lambda| \leqq N$ and uniformly in x for $a \leqq x \leqq b$. Since for $n \geqq 2$

$$\varphi_n'(x,\lambda)-\varphi_{n-1}'(x,\lambda)=\int_a^x\{q(t)-\lambda\}\{\varphi_{n-1}(t,\lambda)-\varphi_{n-2}(t,\lambda)\}\,dt,$$

$$\varphi_n''(x,\lambda)-\varphi_{n-1}''(x,\lambda)=\{q(x)-\lambda\}\{\varphi_{n-1}(x,\lambda)-\varphi_{n-2}(x,\lambda)\},$$

the series obtained by differentiating the series (1.5) once or twice also converge uniformly in x. Consequently

$$\begin{aligned}\varphi''(x,\lambda)&=\sum_{n=1}^{\infty}\{\varphi_n''(x,\lambda)-\varphi_{n-1}''(x,\lambda)\}\\&=\varphi_1''(x,\lambda)-\varphi_0''(x,\lambda)+\sum_{n=2}^{\infty}\{\varphi_n''(x,\lambda)-\varphi_{n-1}''(x,\lambda)\}\\&=\{q(x)-\lambda\}\cdot\left\{\varphi_0(x,\lambda)+\sum_{n=2}^{\infty}[\varphi_{n-1}(x,\lambda)-\varphi_{n-2}(x,\lambda)]\right\}\\&=\{q(x)-\lambda\}\,\varphi(x,\lambda),\end{aligned}$$

and so $\varphi(x,\lambda)$ satisfies equation (1.1). It is also clear that it satisfies the stated initial conditions. That $\varphi(x,\lambda)$ is an entire function of the variable λ follows from the uniform convergence of the series (1.5) and the structure of the functions $\varphi_n(x,\lambda)$.

§2. Asymptotic formulas for the eigenvalues and eigenfunctions

1. Put $\operatorname{ctg}\alpha = -h$ and $\operatorname{ctg}\beta = H$.[1] Then the boundary conditions (1.2) can be rewritten in the form

$$\begin{aligned} y'(0) - hy(0) &= 0, \\ y'(\pi) + Hy(\pi) &= 0. \end{aligned} \tag{2.1}$$

Let us denote by $\varphi(x,\lambda)$ the solution of equation (1.1) satisfying the initial conditions

$$\varphi(0,\lambda) = 1, \qquad \varphi'(0,\lambda) = h, \tag{2.2}$$

and by $\psi(x,\lambda)$ the solution of the same equation, satisfying the initial conditions

$$\psi(0,\lambda) = 0, \qquad \psi'(0,\lambda) = 1. \tag{2.3}$$

LEMMA 2.1. *Let $\lambda = s^2$. Then*

$$\varphi(x,\lambda) = \cos sx + \frac{h}{s}\sin sx + \frac{1}{s}\int_0^x \sin\{s(x-\tau)\}\, q(\tau)\,\varphi(\tau,\lambda)\,d\tau, \tag{2.4}$$

$$\psi(x,\lambda) = \frac{\sin sx}{s} + \frac{1}{s}\int_0^x \sin\{s(x-\tau)\}\, q(\tau)\,\psi(\tau,\lambda)\,d\tau. \tag{2.5}$$

PROOF. We shall prove the relation (2.4). Since $\varphi(x,\lambda)$ satisfies equation (1.1), we have

$$\int_0^x \sin\{s(x-\tau)\}\, q(\tau)\,\varphi(\tau,\lambda)\,d\tau$$

$$= \int_0^x \sin\{s(x-\tau)\}\,\varphi''(\tau,\lambda)\,d\tau + s^2\int_0^x \sin\{s(x-\tau)\}\,\varphi(\tau,\lambda)\,d\tau.$$

Integrating the first integral on the right side by parts twice and taking the conditions (2.2) into account, we find that

[1] We shall assume for the present that neither h nor H equals ∞. The case in which one or both of these numbers equals ∞ will be considered below.

$$\int_0^x \sin\{s(x-\tau)\}\, q(\tau)\varphi(\tau, \lambda)\, d\tau = -h \sin sx + s\varphi(x, \lambda) - s \cos sx,$$

which is (2.4). The relation (2.5) can be proved similarly.

Lemma 2.2. *Let $s = \sigma + it$. Then there exists $s_0 > 0$ such that for $|s| > s_0$ one has the estimates*

(2.6) $$\varphi(x, \lambda) = O(e^{|t|x}), \quad \psi(x, \lambda) = O(|s|^{-1}e^{|t|x}),$$

and more precisely

(2.7) $$\varphi(x, \lambda) = \cos sx + O(|s|^{-1}e^{|t|x}),$$

(2.8) $$\psi(x, \lambda) = \frac{\sin sx}{s} + O\left(\frac{e^{|t|x}}{|s|^2}\right).$$

All of these estimates hold uniformly in x for $0 \leqq x \leqq \pi$.

Proof. Put $\varphi(x, \lambda) = e^{|t|x}F(x)$. Then from (2.4) we obtain

$$F(x) = \left\{\cos sx + \frac{h}{s} \sin sx\right\} e^{-|t|x}$$
$$+ \frac{1}{s} \int_0^x \sin\{s(x-\tau)\}\, e^{-|t|(x-\tau)} q(\tau) F(\tau)\, d\tau.$$

Let $\mu = \max_{0 \leqq x \leqq \pi} |F(x)|$. Then it follows from the last relation that

$$\mu \leqslant 1 + \frac{|h|}{|s|} + \frac{\mu}{|s|} \int_0^\pi |q(\tau)|\, d\tau,$$

and so

$$\mu \leqslant \frac{1 + \frac{|h|}{|s|}}{1 - \frac{1}{|s|}\int_0^\pi |q(\tau)|\, d\tau}$$

under the assumption that the denominator is positive. This is certainly the case if $|s| > \int_0^\pi |q(\tau)|\, d\tau$, and thus the estimate (2.6) is proved for the function $\varphi(x, \lambda)$. The proof of the estimate (2.6) for the function $\psi(x, \lambda)$, based on the relation (2.5), is similar. The estimates (2.7) and (2.8) can be obtained by substituting the estimates (2.6) into the integrals on the right sides of (2.4) and (2.5) respectively.

It is clear that general asymptotic expansions for $\varphi(x, \lambda)$ and $\psi(x, \lambda)$ as functions of s could be obtained by repeating this process.

2. We will now consider the derivation of asymptotic formulas for the eigenvalues and eigenfunctions. From these formulas, in particular, will follow the existence of an infinite set of eigenvalues.

As before we assume, to begin with, that $h \neq \infty$ and $H \neq \infty$. For any λ the function $\varphi(x, \lambda)$ obviously satisfies the first of the boundary conditions (2.1). Therefore we will determine an eigenvalue if we insert $\varphi(x, \lambda)$ in the second boundary condition.

According to Lemma 1.2 the eigenvalues are real. The number of negative eigenvalues is finite (this follows, for example, from the Sturmian theory; see the next section). For positive λ we have $\operatorname{Im} s = 0$, and the estimate (2.7) assumes the form

$$\varphi(x, \lambda) = \cos sx + O(1/s). \tag{2.7'}$$

Furthermore, differentiating the relation (2.4) with respect to x and using the estimate (2.7′), it is not difficult to obtain the estimate

$$\varphi_x'(x, \lambda) = -s \sin sx + h \cos sx + O(1/s). \tag{2.9}$$

Now inserting the values of the functions $\varphi(x, \lambda)$ and $\varphi_x'(x, \lambda)$ from the estimates (2.7) and (2.9) into the second of the boundary conditions in (2.1), we obtain the following equation for the determination of the eigenvalues:

$$-s \sin s\pi + (h + H) \cos s\pi + O(1/s) = 0. \tag{2.10}$$

For large s equation (2.10) obviously has solutions which lie close to integers. From this already follows the existence of an infinite set of eigenvalues.

We will show that, starting with some sufficiently large integer n, there exists only one root of (2.10) near every n. To this end we differentiate the left side of (2.10) with respect to s, which is possible by virtue of (2.4) (note that the expression $O(1/s)$ in the left side of (2.10) is in fact, by Theorem 1.1, an analytic function of λ). We obtain

$$-\pi s \cos s\pi + O(1).$$

It is not hard to see that this expression cannot equal zero for values of s near large integers.

Let s_n be the nth root of equation (2.10). From Sturm's theorem (cf. §3) and asymptotic formulas for the eigenfunctions it follows that s_n has to lie near n, and not some other integer. Another proof of this assertion, not based upon Sturmian theory, can be obtained in the following way.

As was already noted above, the eigenvalues are the roots of the equation[2)]

$$\varphi(\pi,\lambda)+H\varphi_x'(\pi,\lambda)\equiv\omega(\lambda)=0.$$

Put $\lambda=s^2$. Then $\omega(\lambda)=\omega_1(s)$. It follows from (2.4) that $\omega_1(s)$ is an entire function of s. Furthermore, it follows from the asymptotic formulas (2.7′) and (2.9) that for $\sin s\pi\neq 0$

$$\omega_1(s)=-Hs\sin s\pi\{1+O(1/|s|)\}. \tag{2.10′}$$

Let D_R be the disc in the s-plane with center at the origin and radius $R=N+\frac{1}{2}$, where N is a positive integer. By Rouché's theorem and the asymptotic formula (2.10′), the number of zeros of $\omega_1(s)$ inside the disc D_R is equal to the number of zeros of the function $s\sin s\pi$ inside D_R, i.e. equals $2n+2$. The function $\omega_1(s)$ is even, and we can therefore consider only its positive zeros. To every positive zero of $\omega_1(s)$ there corresponds an eigenvalue, i.e. the number of s_k which are less than $N+\frac{1}{2}$ is $N+1$. It follows that in the asymptotic formula for s_n we have

$$s_n=n+o(1). \tag{2.11}$$

In fact, let $s_n=m_n+o(1)$, $m_n\neq n$. Then on the one hand, the number of s_k which are less than s_n is equal to n $(k=0,1,\cdots,n-1)$. On the other hand, in view of the foregoing there must be $2m_n+2$ zeros of the function $\omega_1(s)$ in the disc with center at the origin and radius $m_n+\frac{1}{2}$, i.e. the number of s_k which are less than s_n would be equal to $m_n\neq n$. This contradiction proves the validity of formula (2.11).

Let $s_n=n+\delta_n$. Then equation (2.10) assumes the form

$$n\sin\delta_n\pi+O(1)=0.$$

It follows that $\sin\delta_n\pi=O(1/n)$, i.e. $\delta_n=O(1/n)$. Thus for large n the roots of equation (2.10) have the form

$$s_n=n+O(1/n). \tag{2.11′}$$

The asymptotic formula (2.11′) can be improved substantially if we assume that the coefficient $q(x)$ in equation (1.1) has a bounded derivative. In fact, differentiating (2.4) with respect to x and then inserting the values of $\varphi(x,\lambda)$ and $\varphi_x'(x,\lambda)$ in the second of the boundary conditions (2.1), we obtain after simple transformations

$$(-s+B)\sin s\pi+A\cos s\pi=0, \tag{2.12}$$

where

[2)] That every root of this equation has multiplicity 1 can be shown by an argument similar to the proof of Lemma 11.2 of this chapter.

$$A = h + H + \int_0^\pi \left\{\cos s\tau + \frac{H}{s}\sin s\tau\right\} q(\tau)\,\varphi(\tau, \lambda)\,d\tau,$$

$$B = \frac{hH}{s} + \int_0^\pi \left\{\sin s\tau + \frac{H}{s}\right\} q(\tau)\,\varphi(\tau, \lambda)\,d\tau.$$

By virtue of the estimate (2.7′) the expressions for A and B assume the form

$$A = h + H + \frac{1}{2}\int_0^\pi q(\tau)\,d\tau + \frac{1}{2}\int_0^\pi q(\tau)\cos 2s\tau d\tau + O\left(\frac{1}{s}\right),$$

$$B = \frac{1}{2}\int_0^\pi q(\tau)\sin 2s\tau d\tau + O\left(\frac{1}{s}\right).$$

Since by hypothesis $q(x)$ has a bounded derivative, integrating by parts yields

$$\int_0^\pi q(\tau)\cos 2s\tau d\tau = O\left(\frac{1}{s}\right), \quad \int_0^\pi q(\tau)\sin 2s\tau d\tau = O\left(\frac{1}{s}\right).$$

Consequently we obtain for A and B the expressions

$$A = h + H + h_1 + O\left(\frac{1}{s}\right), \qquad h_1 = \frac{1}{2}\int_0^\pi q(\tau)\,d\tau,$$

$$B = O\left(\frac{1}{s}\right).$$

Therefore equation (2.12) can be written in the form

$$\operatorname{tg} s\pi = \frac{h + H + h_1 + O\left(\frac{1}{s}\right)}{s + O\left(\frac{1}{s}\right)}.$$

Putting once again $s_n = n + \delta_n$, we obtain

$$\operatorname{tg} \pi\delta_n = \frac{h + H + h_1}{n} + O\left(\frac{1}{n^2}\right),$$

whence

$$\delta_n = \frac{h + H + h_1}{\pi n} + O\left(\frac{1}{n^2}\right),$$

and so

$$s_n = n + \frac{c}{n} + O\left(\frac{1}{n^2}\right), \tag{2.13}$$

where

$$c = \frac{1}{\pi}\left(h + H + \frac{1}{2}\int_0^\pi q(\tau)\,d\tau\right).$$

Assuming that $q(x) \in C^2(0, \pi)$, one can prove a more precise asymptotic formula, namely:

$$s_n = n + \frac{c}{n} + \frac{c_1}{n^3} + O\left(\frac{1}{n^4}\right), \tag{2.13'}$$

where c_1 is a constant.

Using formula (2.13), we will obtain an asymptotic formula for the eigenfunctions $\varphi(x, \lambda_n) = \varphi_n(x)$. Substituting into (2.4) the expression (2.7′) for $\varphi(x, \lambda)$ and using the differentiability of $q(x)$, we obtain

$$\varphi(x, \lambda) = \cos sx + \frac{h}{s}\sin sx$$

$$+ \frac{1}{s}\int_0^x \sin\{s(x-\tau)\}\cos s\tau \cdot q(\tau)\,d\tau + O\left(\frac{1}{s^2}\right)$$

$$= \cos sx + \frac{h}{s}\sin sx + \frac{\sin sx}{2s}\int_0^x q(\tau)\,d\tau + O\left(\frac{1}{s^2}\right).$$

Inserting s_n everywhere for s, in view of formula (2.13) we obtain

$$\varphi(x, \lambda_n) = \varphi_n(x) = \cos nx - \frac{cx}{n}\sin nx + \frac{h}{n}\sin nx$$

$$+ \frac{\sin nx}{2n}\int_0^x q(\tau)\,d\tau + O\left(\frac{1}{n^2}\right) = \cos nx + \frac{\beta(x)}{n}\sin nx + O\left(\frac{1}{n^2}\right),$$

where

$$\beta(x) = -cx + h + \frac{1}{2}\int_0^x q(\tau)\,d\tau.$$

To obtain an asymptotic expansion for the normalized eigenfunctions $\varphi_n(x)$, let us consider the integral

$$\alpha_n^2 = \int_0^\pi \varphi_n^2(x)\,dx = \int_0^\pi \cos^2 nx\,dx + \frac{1}{n}\int_0^\pi \beta(x)\sin 2nx\,dx + O\left(\frac{1}{n^2}\right).$$

By virtue of the differentiability of the function $\beta(x)$

$$\int_0^\pi \beta(x)\sin 2nx\,dx = O\left(\frac{1}{n}\right).$$

Therefore $\alpha_n^2 = \pi/2 + O(1/n^2)$, from which

$$\frac{1}{a_n}=\sqrt{\frac{2}{\pi}}\left\{1+O\left(\frac{1}{n^2}\right)\right\}=\sqrt{\frac{2}{\pi}}+O\left(\frac{1}{n^2}\right).$$

Consequently the following asymptotic formula holds for the normalized eigenfunctions:

$$v_n(x)=\frac{1}{a_n}\varphi_n(x)=\sqrt{\frac{2}{\pi}}\left\{\cos nx+\frac{\beta(x)}{n}\sin nx\right\}+O\left(\frac{1}{n^2}\right). \tag{2.14}$$

3. Let us now examine the case $h=\infty$, $H\neq\infty$ (the case $h\neq\infty$, $H=\infty$ can be reduced to this case by the substitution $t=\pi-x$). The first boundary condition in (2.1) assumes the form

$$y(0)=0. \tag{2.3'}$$

The function $\psi(x,\lambda)$ from §2.1 satisfies condition (2.3′). Therefore for the case under consideration we can determine the eigenvalues by inserting the function $\psi(x,\lambda)$ in the second of the boundary conditions (2.1). Differentiating (2.5) with respect to x, we obtain

$$\psi'_x(x,\lambda)=\cos sx+\int_0^x \cos\{s(x-\tau)\}\,q(\tau)\,\psi(\tau,\lambda)\,d\tau.$$

Therefore it follows from the second boundary condition in (2.1) that

$$\begin{aligned}&\cos s\pi+\int_0^\pi \cos\{s(\pi-\tau)\}\,q(\tau)\,\psi(\tau,\lambda)\,d\tau\\ &+H\left\{\frac{\sin s\pi}{s}+\frac{1}{s}\int_0^\pi \sin\{s(\pi-\tau)\}\,q(\tau)\,\psi(\tau,\lambda)\,d\tau\right\}=0.\end{aligned} \tag{2.15}$$

By virtue of the estimate (2.8) we obtain from (2.15)

$$\cos s\pi+\frac{1}{s}\int_0^\pi \cos\{s(\pi-\tau)\}\,q(\tau)\sin s\tau\,d\tau+\frac{\sin s\pi}{s}+O\left(\frac{1}{s^2}\right)=0. \tag{2.15'}$$

Assuming once again that $q(x)$ has a bounded derivative, we obtain

$$\begin{aligned}&\int_0^\pi q(\tau)\cos\{s(\pi-\tau)\}\sin s\tau\,d\tau\\ &=\cos s\pi\int_0^\pi q(\tau)\cos s\tau\cdot\sin s\tau\,d\tau+\sin s\pi\int_0^\pi q(\tau)\sin^2 s\tau\,d\tau\\ &=\frac{\sin s\pi}{2}\int_0^\pi q(\tau)\,d\tau+O\left(\frac{1}{s}\right).\end{aligned}$$

Therefore equation (2.15′) assumes the form

$$\cos s\pi + \frac{\sin s\pi}{s}\left\{H + \frac{1}{2}\int_0^\pi q(\tau)\,d\tau\right\} + O\left(\frac{1}{s^2}\right) \tag{2.16}$$
$$= \cos s\pi + H_1\frac{\sin s\pi}{s} + O\left(\frac{1}{s^2}\right) = 0.$$

It is evident from this equation that for large s its roots must be close to the numbers of the form $n + \frac{1}{2}$, where n is an integer. Moreover, just as before one can prove that, starting with some sufficiently large integer n, equation (2.16) has only one root close to each number $n + \frac{1}{2}$.

Put $s_n = n + \frac{1}{2} + \delta_n$. Then it follows from equation (2.16) that

$$\operatorname{ctg}\left(n + \frac{1}{2} + \delta_n\right)\pi = -\operatorname{tg}\delta_n\pi = -\frac{H_1}{n + \frac{1}{2}} + O\left(\frac{1}{n^2}\right),$$

whence

$$\delta_n = \frac{H_1}{\pi\left(n + \frac{1}{2}\right)} + O\left(\frac{1}{n^2}\right),$$

and so

$$s_n = n + \frac{1}{2} + \frac{H_1}{\pi\left(n + \frac{1}{2}\right)} + O\left(\frac{1}{n^2}\right),$$

where $H_1 = H + \frac{1}{2}\int_0^\pi q(\tau)\,d\tau$.

Now, inserting the value of s_n in (2.5), we obtain the following asymptotic formula for the eigenfunctions $\psi(x, \lambda_n) = \psi_n(x)$:

$$\psi_n(x) = \frac{1}{n + \frac{1}{2}}\sin\left(n + \frac{1}{2}\right)x + O\left(\frac{1}{n^2}\right).$$

For the normalizing coefficients α_n^{-1} we obtain the formula

$$\frac{1}{\alpha_n} = \sqrt{\frac{2}{\pi}}\left(n + \frac{1}{2}\right)\left\{1 + O\left(\frac{1}{n}\right)\right\}.$$

Therefore for the case under consideration the normalized eigenfunctions $v_n(x) = (1/\alpha_n)\psi_n(x)$ have the form

$$v_n(x) = \sqrt{\frac{2}{\pi}}\sin\left(n + \frac{1}{2}\right)x + O\left(\frac{1}{n}\right). \tag{2.17}$$

4. Finally, let us examine the case $h = \infty$ and $H = \infty$. This says that the boundary conditions (2.1) assume the form $y(0) = y(\pi) = 0$, and

therefore the function $\psi(x, \lambda)$ of the previous subsection must also satisfy the condition $\psi(\pi, \lambda) = 0$. It follows from (2.5) that

$$\sin s\pi + \int_0^\pi \sin\{s(\pi - \tau)\} q(\tau) \psi(\tau, \lambda)\, d\tau = 0$$

or

$$\sin s\pi \left\{1 + \int_0^\pi \cos s\tau q(\tau) \psi(\tau, \lambda)\, d\tau\right\} - \cos s\pi \int_0^\pi \sin s\tau \cdot q(\tau) \psi(\tau, \lambda)\, d\tau = 0.$$

By virtue of the estimate (2.8) it follows from this equation (we are assuming that $q(x)$ has a bounded derivative) that

$$\sin s\pi - \frac{1}{2s} \cos s\pi \int_0^\pi q(\tau)\, d\tau + O\left(\frac{1}{s^2}\right)$$

(2.18)

$$= \sin s\pi - \frac{a}{s} \cos s\pi + O\left(\frac{1}{s^2}\right) = 0.$$

This equation is of the same form as equation (2.10). Therefore proceeding as in §2.2, we can establish that the roots of equation (2.18) lie close to integers n, and that starting with some sufficiently large integer n, only one root lies near such n. Therefore the roots s_n of equation (2.18) have the form

$$s_n = n + \frac{a_1}{n} + O\left(\frac{1}{n^2}\right), \tag{2.19}$$

where $\alpha_1 = (1/2\pi) \int_0^\pi q(\tau)\, d\tau$.

Inserting the value of s_n in (2.5), we obtain the asymptotic formula

$$\psi_n(x) = \frac{\sin nx}{n} + O\left(\frac{1}{n^2}\right)$$

for the eigenfunctions $\psi(x, \lambda_n) = \psi_n(x)$. For the normalizing coefficients α_n^{-1} we obtain the formula

$$\frac{1}{\alpha_n} = \sqrt{\frac{2}{\pi}}\, n \left\{1 + O\left(\frac{1}{n}\right)\right\}.$$

Therefore the normalized eigenfunctions $v_n(x) = (1/\alpha_n) \psi_n(x)$ have the form

$$v_n(x) = \sqrt{\frac{2}{\pi}} \sin nx + O\left(\frac{1}{n}\right). \tag{2.20}$$

§3. Zeros of eigenfunctions

A deep study of the distribution of the zeros of eigenfunctions led Sturm to another proof of the existence of an infinite set of eigenvalues for the boundary-value problem (1.1) – (1.2).

Let us consider, as a guide to the results of this section, the simple boundary-value problem

$$y'' + \lambda y = 0, \qquad y'(0) = y'(\pi) = 0.$$

Here the eigenfunctions are $\varphi_0(x) = 1$, $\varphi_1(x) = \cos x$, $\varphi_2(x) = \cos 2x, \cdots$, $\varphi_n(x) = \cos nx, \cdots$. The corresponding eigenvalues are $\lambda_0 = 0$, $\lambda_1 = 1^2$, $\lambda_2 = 2^2, \cdots, \lambda_n = n^2, \cdots$. We have arranged the eigenfunctions in the order of increasing eigenvalues, starting the count from zero. It is at once evident that the zeros of the eigenfunctions have the following two properties:

1) the nth eigenfunction has exactly n zeros in the interval $(0, \pi)$;

2) the zeros of the nth and $(n+1)$st eigenfunctions interlace, i.e. between any two successive zeros of the nth eigenfunction there lies a zero of the $(n+1)$st eigenfunction.

It turns out that these properties of the eigenfunctions hold in the general case.

The following fundamental theorem of Sturm is basic in this circle of topics.

Theorem 3.1. *Suppose that we are given two equations*

$$u'' + g(x)u = 0, \tag{3.1}$$

$$v'' + h(x)v = 0. \tag{3.2}$$

If $g(x) < h(x)$ over the entire interval $[a, b]$, then between every two zeros of any nontrivial solution of the first equation there is at least one zero of every solution of the second equation.

Proof. Multiplying equation (3.1) by v, equation (3.2) by u, and subtracting, we obtain

$$u''v - v''u = \frac{d}{dx}\{u'v - v'u\} = \{h(x) - g(x)\}\, uv. \tag{3.3}$$

Let us denote two successive zeros of u by x_1 and x_2. Then integrating the identity (3.3) from x_1 to x_2, we obtain

$$\{u'v - v'u\}_{x_1}^{x_2} = u'(x_2)v(x_2) - u'(x_1)v(x_1) = \int_{x_1}^{x_2} \{h(x) - g(x)\}\, u(x)v(x)\, dx.$$

Let us assume that v does not equal zero anywhere in the interval (x_1, x_2). We can assume without loss of generality that $u > 0$ and $v > 0$ in the interval (x_1, x_2). Consequently the right side of the preceding equality is positive. Since by assumption $u(x) \geqq 0$, the function u is increasing at the point x_1. Therefore[3] $u'(x_1) > 0$. By a similar argument $u'(x_2) < 0$. Therefore

$$u'(x_2)v(x_2) - u'(x_1)v(x_1) \leqq 0,$$

and we arrive at a contradiction. The theorem is proved.

COROLLARY. *No solution of the equation*

$$y'' + g(x)y = 0 \qquad (-\infty \leqq a \leqq x \leqq b \leqq \infty) \tag{3.4}$$

with $g(x) < -m^2 < 0$ *can have more than one zero.*

In fact, the equation $y'' - m^2y = 0$ has the solution e^{mx} which is nowhere zero. Therefore on the basis of the preceding theorem no solution of (3.4) can have more than one zero in any finite interval.

THEOREM 3.2 (THE COMPARISON THEOREM). *Let* $u(x)$ *be the solution of equation* (3.1) *satisfying the initial conditions*

$$u(a) = \sin\alpha, \qquad u'(a) = -\cos\alpha, \tag{3.5}$$

and $v(x)$ *the solution of equation* (3.2) *satisfying the same initial conditions. Moreover, suppose that* $g(x) < h(x)$ *over the entire interval* $[a, b]$.

If $u)x)$ *has* m *zeros in the interval* $a < x \leqq b$, *then* $v(x)$ *has not less than* m *zeros in the same interval, and the* k*th zero of* $v(x)$ *is less than the* k*th zero of* $u(x)$.

PROOF. Let x_1 denote the zero of $u(x)$ closest to (but different from) the point a.[4] On the basis of the preceding theorem it suffices to prove that $v(x)$ has at least one zero in the interval (a, x_1). Assume the contrary. Without loss of generality we may suppose that $v(x) > 0$ and $u(x) > 0$ in the interval (a, x_1). Since $u(x_1) = 0$, the function $u(x)$ is decreasing

---

[3] $u'(x_1)$ cannot equal zero, since it would then follow from the uniqueness theorem for solutions of a differential equation that in this case $u(x) \equiv 0$, which we have excluded.

[4] A smallest zero exists, since a nontrivial solution $y(x)$ of a linear differential equation of second order has only isolated zeros. Indeed, suppose that $x_0 \neq \pm\infty$ is a limit point of zeros $x_1, x_2, \cdots$ different from x_0 of some solution $y(x)$ of the differential equation. Then $y(x_0) = 0$ by virtue of the continuity of $y(x)$. Furthermore, $[y(x_0) - y(x_n)]/(x_0 - x_n) = 0$. Passing to the limit $(n \to \infty)$, we obtain $y'(x_0) = 0$, i.e. $y(x) \equiv 0$.

in a neighborhood of the point x_1. Hence $u'(x_1) \leqq 0$. Integrating the identity (3.3) from a to x_1, we obtain

$$u'(x_1)v(x_1) = \int_a^{x_1} \{h(x) - g(x)\} u(x) v(x)\, dx.$$

Since by assumption $v(x) > 0$, $u(x) > 0$ and $h(x) > g(x)$ in the interval (a, x_1), the right side in the equality above is positive. But the left side is $\leqq 0$, and we have obtained a contradiction, which proves the theorem.

Let $\varphi(x, \lambda)$ be the function introduced in §2. Consider the equation $(a \leqq x \leqq b)$

$$\varphi(x, \lambda) = 0.$$

The zeros of this equation are obviously functions of λ. We will prove that these functions are continuous with respect to λ, namely:

Lemma 3.1. *If x_0 $(a < x_0 < b)$ is a zero of the function $\varphi(x, \lambda_0)$, then for any sufficiently small $\epsilon > 0$ there exists $\delta > 0$ such that for $|\lambda - \lambda_0| < \delta$ the function $\varphi(x, \lambda)$ has exactly one zero in the interval $|x - x_0| < \epsilon$.*

Proof. A zero x_0 of a solution $\varphi(x, \lambda_0)$ of the differential equation (1.1) is a simple zero, since if we had $\varphi_x'(x_0, \lambda_0) = 0$, it would follow from the uniqueness theorem for the equation (1.1) that $\varphi(x, \lambda_0) \equiv 0$. Hence $\varphi_x'(x_0, \lambda_0) \neq 0$, and for definiteness let us assume that $\varphi_x'(x_0, \lambda_0) > 0$. Let ϵ be so small that $\varphi_x'(x, \lambda_0) > 0$ for $|x - x_0| \leqq \epsilon$. Then $\varphi(x - \epsilon, \lambda_0) < 0$ and $\varphi(x + \epsilon, \lambda_0) > 0$. Moreover, in view of the continuity of $\varphi_x'(x, \lambda)$ with respect to λ[5)] there exists δ such that for $|\lambda - \lambda_0| \leqq \delta$ the function $\varphi_x'(x, \lambda)$ also remains positive over the entire interval $|x - x_0| \leqq \epsilon$. Therefore the monotonically increasing function $\varphi(x, \lambda)$ obviously cannot have two zeros in this interval. If, moreover, we choose δ so small that for $|\lambda - \lambda_0| < \delta$ the function $\varphi(x - \epsilon, \lambda)$ of λ remains negative, while $\varphi(x + \epsilon, \lambda)$ remains positive (by virtue of the continuity of $\varphi(x, \lambda)$ with respect to λ, this is possible), we will have proved the assertion of the lemma: the solution $\varphi(x, \lambda)$ has, for $|\lambda - \lambda_0| < \delta$, exactly one zero in the interval $[x - \epsilon, x + \epsilon]$.

From this lemma there follows an important corollary.

Corollary. *As λ varies, the solution $\varphi(x, \lambda)$ can lose zeros or gain zeros only by these zeros leaving or entering the interval $[a, b]$ through its endpoints a and b.*

[5)] By Theorem 1.1, $\varphi(x, \lambda)$ is an entire function of λ.

The following theorem of Sturm proves the existence of an infinite set of eigenvalues.

THEOREM 3.3 (OSCILLATION THEOREM). *There exists an unbounded increasing sequence of eigenvalues* $\lambda_0, \lambda_1, \lambda_2, \cdots$ *of the boundary-value problem* (1.1) – (1.2). *An eigenfunction corresponding to the eigenvalue* λ_m *has exactly* m *zeros in the interval* $a < x < b$.

PROOF. Let $\varphi(x, \lambda)$ be the solution of equation (1.1) satisfying the initial conditions (3.5).

By Theorem 3.2, as λ increases the number of zeros of $\varphi(x, \lambda)$ does not decrease. Let $|q(x)| < c$ for $a \leqq x \leqq b$. We will compare equation (1.1) with the equation

$$y'' + (\lambda + c)y = 0.$$

The solution of this equation which satisfies the initial conditions (3.5) is

$$y = \sin\alpha \cdot \cosh\left\{(-\lambda - c)^{1/2}(x - a)\right\} - \frac{\cos\alpha}{(-\lambda - c)^{1/2}} \sinh\left\{(-\lambda - c)^{1/2}(x - a)\right\}.$$

For negative values of λ with sufficiently large absolute value this function is obviously nowhere zero (in the interval $a < x \leqq b$). Therefore, using Theorem 3.2 again, we see that for negative values of λ with sufficiently large absolute value $\varphi(x, \lambda)$ does not equal zero for $a < x \leqq b$.

Choosing as the comparison equation $y'' + (\lambda - c)y = 0$, we see that for λ positive and increasing unboundedly, the number of zeros of the solution $\varphi(x, \lambda)$ lying in the interval $[a, b]$ increases without limit.

Let us consider the equation $\varphi(x, \lambda) = 0$.

By Lemma 3.1 the roots of this equation are continuous functions of λ. On the other hand, by Theorem 3.2 every zero of $\varphi(x, \lambda)$ moves to the left as λ increases, and no zero can pass through the point a, since the number of zeros does not decrease. By the corollary to Lemma 3.1 new zeros enter through the point b. Let μ_0 be the first value of the parameter λ for which $\varphi(b, \lambda) = 0$. Such a value obviously exists. Let μ_1 be the second value of the parameter λ for which $\varphi(b, \lambda) = 0$, and so on. The sequence of numbers $\mu_0, \mu_1, \cdots$ has the property that the function $\varphi(x, \mu_m)$ has exactly m zeros in the interval (a, b), and $\varphi(b, \mu_m) = 0$. If $\sin\beta = 0$, then the second of the boundary conditions (1.2) will be satisfied,[6] and consequently the numbers μ_m are the eigenvalues. Thus

[6] Since $\varphi(x, \lambda)$ satisfies the initial conditions (3.5), it will obviously satisfy the first of the boundary conditions (1.2).

in this case the theorem is proved.

Suppose now that $\sin\beta \neq 0$, and let $u(x)$ and $v(x)$ be the functions considered in Theorem 3.2. Then

$$\frac{d}{dx}\left\{u^2\left(\frac{u'}{u}-\frac{v'}{v}\right)\right\} = 2uu'\left(\frac{u'}{u}-\frac{v'}{v}\right)+u^2\left(\frac{u''}{u}-\frac{v''}{v}\right)-u^2\left(\frac{u'^2}{u^2}-\frac{v'^2}{v^2}\right)$$

(3.6)
$$= \frac{(u'v-v'u)^2}{v^2}+u^2\{h(x)-g(x)\} > 0.$$

Therefore the function $u^2(u'/u - v'/v)$ is monotonically increasing on every interval on which v is nowhere zero. Let us assume that $u(x)$ and $v(x)$ have the same number of zeros in the interval (a,b).

Denote by x_ν that zero of $u(x)$ which is closest to the point b. We will show that $v(x)$ cannot have any zero in the interval $x_\nu \leqq x \leqq b$. In fact, by Theorem 3.2 the function $v(x)$ has at least ν zeros lying between a and x_ν. If $v(x)$ equalled zero somewhere in the interval $x_\nu \leqq x < b$, then it would have more zeros in the interval (a,b) than does $u(x)$, contrary to hypothesis.

Integrating the relation (3.6) between x_ν and b, we obtain

$$u^2(b)\left\{\frac{u'(b)}{u(b)}-\frac{v'(b)}{v(b)}\right\} > u^2(x_\nu)\left\{\frac{u'(x_\nu)}{u(x_\nu)}-\frac{v'(x_\nu)}{v(x_\nu)}\right\} = 0.$$

Hence

(3.7)
$$\frac{u'(b)}{u(b)} > \frac{v'(b)}{v(b)}.$$

Take $\varphi(x,\lambda')$ for $u(x)$ and $\varphi(x,\lambda'')$ for $v(x)$, where $\mu_m < \lambda' < \lambda'' < \mu_{m+1}$. On the basis of inequality (3.7) the function $\varphi'(b,\lambda)/\varphi(b,\lambda)$ decreases monotonically in the interval (μ_m,μ_{m+1}). Since $\varphi(b,\mu_m) = \varphi(b,\mu_{m+1}) = 0$, it has to decrease from $+\infty$ to $-\infty$. Therefore there exists one point λ_{m+1} in the interval (μ_m,μ_{m+1}) for which

$$\varphi'(b,\lambda_{m+1})/\varphi(b,\lambda_{m+1}) = -\operatorname{ctg}\beta,$$

i.e. for which the second of the conditions (1.2) is satisfied. Thus λ_{m+1} is an eigenvalue, and $\varphi(x,\lambda_{m+1})$ has as many zeros in (a,b) as does $\varphi(x,\mu_{m+1})$, i.e. $m+1$.

§4. Periodic solutions

We will consider the problem

(4.1)
$$d^2y/dx^2 + \{\lambda - q(x)\}y = 0,$$

(4.2)
$$y(0) = y(\pi), \quad y'(0) = y'(\pi).$$

Let us assume that $q(0)=q(\pi)$. Then $q(x)$ can be defined for all values of x as a continuous periodic function with period π. In this case solutions of equation (4.1) also become defined for all values of x. If, then, we are required to find all periodic solutions $y(x)$ with period π, we are led to the boundary-value problem (4.1) – (4.2).

In order to prove for this case the existence of an infinite set of eigenvalues, we proceed in the following way.

We will consider two problems simultaneously:

$$d^2y/dx^2+\{\lambda-q(x)\}y=0, \tag{4.1'}$$

$$y(0)=y(\pi)=0, \tag{4.2'}$$

and

$$d^2z/dx^2+\{\lambda-q(x)\}z=0, \tag{4.1''}$$

$$z'(0)=z'(\pi)=0. \tag{4.2''}$$

The problem (4.1′) – (4.2′) corresponds to the case $h=\infty$ and $H=\infty$ for the problem (1.1) – (1.2), which we considered in §2.4, and the problem (4.1″) – (4.2″), to the case $h=H=0$, which we considered in §2.2. Therefore both problems have sequences of eigenvalues and eigenvectors:

$$\mu_0,\ \mu_1,\ \mu_2,\ \dots,\ \mu_n,\ \dots;\quad y_0(x),\ y_1(x),\ y_2(x),\ \dots,\ y_n(x),\ \dots,$$
$$\nu_0,\ \nu_1,\ \nu_2,\ \dots,\ \nu_n,\ \dots;\quad z_0(x),\ z_1(x),\ z_2(x),\ \dots,\ z_n(x),\ \dots$$

By Theorem 3.3, the functions $y_k(x)$ and $z_k(x)$ each have k zeros in the interval $(0,\pi)$.

Let $\varphi(x,\lambda)$ and $\psi(x,\lambda)$ be linearly independent solutions of equation (4.1).

LEMMA 4.1. *If $\varphi(x,\lambda)$ has at least two zeros in the interval $[0,\pi]$, and if x_1 and x_2 are two successive zeros lying in this interval, then $\psi(x,\lambda)$ has exactly one zero in (x_1,x_2), i.e. the zeros of two linearly independent solutions separate one another.*

PROOF. Let us assume that $\psi(x,\lambda)$ has no zeros in the interval (x_1,x_2). Then the function $f(x)=\varphi(x,\lambda)/\psi(x,\lambda)$ is continuous on the entire interval (x_1,x_2) and equals zero at its endpoints.[7] Consequently,

[7] Clearly $\psi(x,\lambda)$ cannot equal zero at x_1 or x_2, since otherwise $\varphi(x,\lambda)$ and $\psi(x,\lambda)$ would be linearly dependent.

at some intermediate point ξ we must have $f'(\xi)=0$. But this is not possible if the solutions $\varphi(x,\lambda)$ and $\psi(x,\lambda)$ form a fundamental system, since

$$f'(x)=\frac{\varphi'(x,\lambda)\psi(x,\lambda)-\varphi(x,\lambda)\psi'(x,\lambda)}{\psi^2(x,\lambda)}$$

where the numerator is the Wronskian determinant $W\{\varphi,\psi\}\neq 0$. Hence $\psi(x,\lambda)$ has at least one zero in (x_1,x_2). It cannot have more than one in this interval because, as was just proved, between any two zeros of the solution $\psi(x,\lambda)$ there lies at least one zero of the solution $\varphi(x,\lambda)$, contrary to hypothesis. This proves the lemma.

Let the solutions $\varphi(x,\lambda)$ and $\psi(x,\lambda)$ form a fundamental system of solutions of equation (4.1), determined by the initial conditions

$$\varphi(0,\lambda)=1,\quad \varphi'(0,\lambda)=0, \tag{4.3}$$

$$\psi(0,\lambda)=0,\quad \psi'(0,\lambda)=1. \tag{4.4}$$

In order that the function $y(x,\lambda)=C_1\varphi(x,\lambda)+C_2\psi(x,\lambda)$ be an eigenfunction of the problem (4.1) – (4.2), the constants C_1 and C_2 have to satisfy the system of linear equations

$$C_1\varphi(0,\lambda)+C_2\psi(0,\lambda)=C_1\varphi(\pi,\lambda)+C_2\psi(\pi,\lambda),$$
$$C_1\varphi'(0,\lambda)+C_2\psi'(0,\lambda)=C_1\varphi'(\pi,\lambda)+C_2\psi'(\pi,\lambda)$$

with determinant

$$F(\lambda)=\begin{vmatrix}1-\varphi(\pi,\lambda) & -\psi(\pi,\lambda)\\ -\varphi'(\pi,\lambda) & 1-\psi'(\pi,\lambda)\end{vmatrix}. \tag{4.5}$$

Multiplying out, we obtain

$$F(\lambda)=\varphi(\pi,\lambda)+\psi'(\pi,\lambda)-2, \tag{4.5$'$}$$

since the Wronskian determinant is $\varphi(x,\lambda)\psi'(x,\lambda)-\psi(x,\lambda)\varphi'(x,\lambda)\equiv 1$, which one can see by putting $x=0$. Only when the determinant (4.5) equals zero will a solution $C_1\varphi+C_2\psi$ be an eigenfunction. Consequently the eigenvalues of the problem (4.1) – (4.2) are the roots of the equation

$$F(\lambda)=\varphi(\pi,\lambda)+\psi'(\pi,\lambda)-2=0. \tag{4.6}$$

If, in particular, for one of the eigenvalues λ^0 all the elements of the determinant (4.5) equal zero, i.e. if

$$\varphi(\pi,\lambda^0)=1,\quad \varphi'(\pi,\lambda^0)=0, \tag{4.3$'$}$$

$$\psi(\pi,\lambda^0)=0,\quad \psi'(\pi,\lambda^0)=1, \tag{4.4$'$}$$

which is quite possible, then C_1 and C_2 are arbitrary. In this case every solution of the equation

$$d^2y/dx^2 + \{\lambda^0 - q(x)\} y = 0$$

is a periodic function.

From conditions (4.3) – (4.3′) and (4.4) – (4.4′) it follows that for $\lambda = \lambda^0$ the functions $\varphi(x, \lambda^0)$ and $\psi(x, \lambda^0)$ satisfy the boundary conditions (4.2) and are therefore eigenfunctions corresponding to one and the same eigenvalue λ^0. Such a λ^0 is called an eigenvalue of multiplicity two.

Theorem 4.1. *Between the larger of the numbers μ_{k-1}, ν_k and the smaller of the numbers μ_k, ν_{k+1} there lies exactly one zero of the function $F(\lambda)$, i.e. an eigenvalue λ_k, $k = 0, 1, 2, \cdots$ of the boundary-value problem* (4.1 – (4.2). *These eigenvalues comprise all the eigenvalues of the problem.*

Proof. For $\lambda = \nu_0$ the solution $z_0(x)$ does not have any zeros in the interval $(0, \pi)$. It follows that $\nu_0 < \mu_0$, since if $\mu_0 < \nu_0$ then by Theorem 3.1 the function $z_0(x)$ would have at least one zero between the zeros 0 and π of the function $y_0(x)$, contrary to hypothesis. A similar argument proves the inequality $\nu_k < \mu_k$, $k = 1, 2, \cdots$. Moreover, since $z_{k+1}(x)$ has $k+1$ zeros, it follows from Theorem 3.1 that for $\lambda > \nu_{k+1}$ every solution of (4.1) has not less than k zeros in $(0, \pi)$. Since $y_{k-1}(x)$ has only $k-1$ zeros in $(0, \pi)$, we conclude that $\mu_{k-1} \leqq \nu_{k+1}$, and finally that $\mu_{k-1} < \nu_{k+1}$, since $\mu_{k-1} = \nu_{k+1}$ is ruled out by Lemma 4.1. We will now determine what sign the function $F(\lambda)$ has for the values μ_k and ν_k. For $\lambda = \mu_k$ the function $\varphi(x, \mu_k)$ has exactly $k+1$ zeros in the interval $[0, \pi]$, one between each adjacent pair of the $k+2$ zeros (including the endpoints)[8] of the function $y_k(x)$; consequently $\varphi(x, \mu_k)$ changes sign $k+1$ times and therefore $\varphi(\pi, \mu_k)$ is positive if k is odd, and negative if k is even. Furthermore we have $\psi(\pi, \mu_k) = 0$, as $\psi(x, \mu_k)$ can be taken for the eigenfunction $y_k(x)$. Consequently (by virtue of the fact that we have for the Wronskian determinant $W(\pi) = \varphi(\pi, \mu_k)\psi'(\pi, \mu_k) = 1$) $\psi'(\pi, \mu_k) > 0$ for odd k and < 0 for even k. But then for odd k the function

$$F(\mu_k) = \varphi(\pi, \mu_k) + \frac{1}{\varphi(\pi, \mu_k)} - 2 = \frac{[\varphi(\pi, \mu_k) - 1]^2}{\varphi(\pi, \mu_k)}$$

is nonnegative. It vanishes only when

$$\varphi(\pi, \mu_k) = \psi'(\pi, \mu_k) = 1, \tag{4.7}$$

---

[8] By virtue of Lemma 4.1, since $y_k(x)$ and $\varphi(x, \mu_k)$ are linearly independent solutions.

but we will not consider this case for the present. In the same way, for even k we have that $F(\mu_k)$ is negative. It follows that $F(\lambda)$ has at least one zero between μ_k and μ_{k+1}, which is an eigenvalue λ_{k+1} of the problem (4.1) – (4.2).

For $\lambda = \nu_k$, as is not hard to see, $\varphi(x, \nu_k)$ can be taken for the eigenfunction $z_k(x)$. Hence again $\varphi(\pi, \nu_k)\psi'(\pi, \nu_k) = 1$. Since the function $z_k(x) = \varphi(x, \nu_k)$ has exactly k zeros in the interval $(0, \pi)$, it follows that $\varphi(\pi, \nu_k) > 0$ for even k and < 0 for odd k. But then it follows that $F(\nu_k) > 0$ for even k and $F(\nu_k) < 0$ for odd k. Thus the function $F(\lambda)$ has at least one zero between ν_k and ν_{k+1}. We again exclude from consideration the case (4.7). Thus one can finally assert that $F(\lambda)$ has at least one zero between $\max(\mu_{k-1}, \nu_k)$ and $\min(\mu_k, \nu_{k+1})$. This proves the existence of at least one eigenvalue in the specified interval. It remains to prove uniqueness. To do this it obviously suffices to prove that the derivative $F'(\lambda)$ has constant sign in this interval.

From (4.5′) we have

(4.8) $$F'(\lambda) = \partial\varphi(\pi, \lambda)/\partial\lambda + \partial\psi'_x(\pi, \lambda)/\partial\lambda.$$

The existence of the derivatives $\partial\varphi(\pi, \lambda)/\partial\lambda$ and $\partial\psi'_x(\pi, \lambda)/\partial\lambda$ follows from Theorem 1.1. For both $\varphi(x, \lambda)$ and $\psi(x, \lambda)$ we have the relation

(4.9) $$\frac{d^2}{dx^2}\left(\frac{\partial y}{\partial \lambda}\right) + \{\lambda - q(x)\}\frac{\partial y}{\partial \lambda} = -y,$$

which is obtained by differentiating the differential equation (4.1) with respect to λ. Furthermore, for both $\varphi(x, \lambda)$ and $\psi(x, \lambda)$ we have

(4.10) $$\frac{\partial y(0, \lambda)}{\partial \lambda} = 0, \quad \frac{\partial y'_x(0, \lambda)}{\partial \lambda} = 0,$$

since the initial conditions for $\varphi(x, \lambda)$ and $\psi(x, \lambda)$ do not depend upon λ. We now determine $\partial y/\partial\lambda$ as a solution of the inhomogeneous equation (4.9) with the initial conditions (4.10). The method of variation of parameters leads to the formulas

(4.11) $$\frac{\partial\varphi(x, \lambda)}{\partial\lambda} = \int_0^x \varphi(t, \lambda)\{\varphi(x, \lambda)\psi(t, \lambda) - \varphi(t, \lambda)\psi(x, \lambda)\}\,dt,$$

(4.12) $$\frac{\partial\psi(x, \lambda)}{\partial\lambda} = \int_0^x \psi(t, \lambda)\{\varphi(x, \lambda)\psi(t, \lambda) - \varphi(t, \lambda)\psi(x, \lambda)\}\,dt.$$

Differentiating (4.12) with respect to x, we find that

(4.13) $$\frac{\partial\psi'_x(x, \lambda)}{\partial\lambda} = \int_0^x \psi(t, \lambda)\{\varphi'_x(x, \lambda)\psi(t, \lambda) - \varphi(t, \lambda)\psi'_x(x, \lambda)\}\,dt.$$

Putting $x = \pi$ in (4.11) and (4.13) and adding, by virtue of (4.8) we obtain

$$F'(\lambda) = \int_0^\pi \{-\psi(\pi, \lambda)\varphi^2(t, \lambda) + [\varphi(\pi, \lambda) - \psi'_x(\pi, \lambda)] \times \varphi(t, \lambda)\psi(t, \lambda) + \varphi'_x(\pi, \lambda)\psi^2(t, \lambda)\}\, dt.$$

The integrand

$$a\varphi^2(t,\lambda) + 2b\varphi(t,\lambda)\psi(t,\lambda) + c\psi^2(t,\lambda), \tag{4.14}$$

where $a = -\psi(\pi,\lambda)$, $b = \frac{1}{2}\{\varphi(\pi,\lambda) - \psi'_x(\pi,\lambda)\}$ and $c = \varphi'_x(\pi,\lambda)$, maintains one sign if $b^2 - ac \leqq 0$, since

$$a\varphi^2 + 2b\varphi\psi + c\psi^2 = \frac{1}{a}(a\varphi + b\psi)^2 + \frac{1}{a}(ac - b^2)\psi^2.$$

In our case

$$b^2 - ac = \frac{1}{4}\{\varphi(\pi, \lambda) + \psi'_x(\pi, \lambda)\}^2 - 1 = \frac{1}{4}\{F(\lambda) + 2\}^2 - 1.$$

If the point λ is a zero of $F(\lambda)$, then at this point the integrand (4.14) will be negative for example, for even k (we have $\psi(\pi,\lambda) > 0$ and $\varphi'_x(\pi,\lambda) < 0$) and can be represented in the form

$$-\{\sqrt{\psi(\pi, \lambda)}\varphi(t, \lambda) + \sqrt{-\varphi'_x(\pi, \lambda)}\,\psi(t, \lambda)\}^2.$$

The derivative $F'(\lambda)$ will also be negative at every zero lying between μ_{k-1} and μ_k; the same will be true for the interval (ν_k, ν_{k+1}). But then the function $F(\lambda)$ can vanish only once in each of these intervals, since the derivative $F'(\lambda)$ cannot have the same sign at two successive zeros of $F(\lambda)$. This proves the theorem.

Let $v_k(x)$ be an eigenfunction corresponding to the eigenvalue λ_k. Then by the preceding theorem $v_k(x)$ can have only k or $k+1$ zeros in $[0,\pi)$. But every eigenfunction always has an even number of zeros in this interval, i.e. points at which it changes sign, since by the boundary conditions (4.2) $v_k(0) = v_k(\pi)$ and $v'_k(0) = v'_k(\pi)$.

We thus have:

THEOREM 4.2. *The boundary-value problem* (4.1)–(4.2) *is solvable for an infinite sequence of eigenvalues* λ_k. *A corresponding eigenfunction* $v_k(x)$ *vanishes* k *times in the interval* $[0,\pi)$ *if* k *is even, and* $k+1$ *times if* k *is odd.*

Thus $v_0(x)$ has no zeros, $v_1(x)$ and $v_2(x)$ have two zeros each, $v_3(x)$ and $v_4(x)$ have four zeros each, etc., in $[0,\pi)$.

It remains to consider the case (4.7) which we excluded earlier, i.e.

when $\varphi(\pi,\mu_k)=\psi'_x(\pi,\mu_k)=1$. In this case μ_k is an eigenvalue of (4.1) – (4.2). Note first that k must be odd. Indeed, for k even $\psi(x,\mu_k)\equiv y_k(x)$ would have k zeros in $(0,\pi)$. Since $\psi(0,\mu_k)=\psi(\pi,\mu_k)=0$, Lemma 4.1 tells us that $\varphi(x,\mu_k)$ has $k+1$ zeros in $(0,\pi)$, and therefore $\varphi(\pi,\mu_k)<0$. It is now clear that no linear combination of $\psi(x,\mu_k)$ and $\varphi(x,\mu_k)$ can satisfy the first condition in (4.2), let alone both of them.

If μ_k is an eigenvalue of multiplicity 1 of (4.1)–(4.2), then the previous arguments apply. If it is an eigenvalue of multiplicity 2, this means that all solutions of (4.1) with $\lambda=\mu_k$ satisfy the boundary conditions (4.2). In particular, for some m we will have $\nu_m=\mu_k$. Using Lemma 4.1, it is not hard to see that $m=k+1$, i.e., $\nu_{k+1}=\mu_k$, so that $\min(\mu_k,\nu_{k+1})=\max(\mu_k,\nu_{k+1})$. This eigenvalue is to be counted twice, i.e., we put $\lambda_k=\lambda_{k+1}=\mu_k$. As two corresponding linearly independent eigenfunctions we can take

$$v_k(x)=\varphi(x,\mu_k),\qquad v_{k+1}(x)=\psi(x,\mu_k).$$

It remains to show that if μ_k is an eigenvalue of multiplicity 2, then it is the only eigenvalue in the intervals $[\max(\mu_{k-1},\nu_k),\mu_k]$ and $[\mu_k,\min(\mu_{k+1},\nu_{k+2})]$. Note first that in this case the coefficients a, b, c in (4.14) all vanish for $\lambda=\mu_k$, so that $F'(\mu_k)=0$. We assert that $F''(\mu_k)<0$. Since, as we saw above, F' is positive at any zero of F lying between μ_{k-1} and μ_k, and negative at any zero of F lying between μ_k and μ_{k+1}, the relations $F(\mu_k)=F'(\mu_k)=0$, $F''(\mu_k)<0$ show that no such zeros can exist. Similarly, F has no zeros between ν_k and ν_{k+1} or between ν_{k+1} and ν_{k+2}.

To show, finally, that $F''(\mu_k)<0$, we observe by differentiating the expression for $F'(\lambda)$ that $F''(\mu_k)$ is an integral whose integrand is

$$A\varphi^2(t,\mu_k)+2B\varphi(t,\mu_k)\psi(t,\mu_k)+c\psi^2(t,\mu_k), \tag{4.15}$$

where

$$A=-\frac{\partial\psi}{\partial\lambda}(\pi,\mu_k),\qquad B=\frac{1}{2}\left\{\frac{\partial\varphi}{\partial\lambda}(\pi,\mu_k)-\frac{\partial}{\partial\lambda}\psi'_x(\pi,\mu_k)\right\},$$

$$C=\frac{\partial}{\partial\lambda}\varphi'_x(\pi,\mu_k).$$

By (4.8), $F'(\mu_k)=0$ implies

$$\frac{\partial\varphi}{\partial\lambda}(\pi,\mu_k)+\frac{\partial}{\partial\lambda}\psi'_x(\pi,\mu_k)=0.$$

Therefore

$$B^2 - AC = \left[\frac{\partial\varphi}{\partial\lambda}(\pi, \mu_k)\right]^2 + \frac{\partial\psi}{\partial\lambda}(\pi, \mu_k)\frac{\partial}{\partial\lambda}\varphi'_x(\pi, \mu_k).$$

Using (4.11), (4.12) and an expression for $(\partial/\partial\lambda)\varphi'_x(\pi, \lambda)$ analogous to (4.13), we find that

$$\frac{\partial\varphi}{\partial\lambda}(\pi, \mu_k) = \int_0^\pi \varphi(t, \mu_k)\psi(t, \mu_k)\,dt$$

$$\frac{\partial\psi}{\partial\lambda}(\pi, \mu_k) = \int_0^\pi [\psi(t, \mu_k)]^2 dt$$

$$\frac{\partial}{\partial\lambda}\varphi'_x(\pi, \mu_k) = -\int_0^\pi [\varphi(t, \mu_k)]^2 dt.$$

Therefore $A < 0$, and it follows from the Cauchy-Bunjakovskiĭ inequality that $B^2 - AC \leqq 0$. Hence the expression (4.15) is nonpositive and not identically zero. We conclude that $F''(\mu_k) < 0$.

Theorem 4.1 enables us to write down asymptotic formulas for the eigenvalues λ_k. These formulas easily follow from formulas (2.13) and (2.19).

§5. Proof of the expansion theorem by the method of finite differences

1. In §§2 and 3 we proved the existence of an infinite set of eigenvalues by two different methods. The method of §2 goes back to Liouville, and that of §3 to Sturm. For this reason the boundary-value problem

(5.1) $$y'' + \{\lambda - q(x)\}y = 0,$$

(5.2) $$y(0)\cos\alpha + y'(0)\sin\alpha = 0,$$

(5.3) $$y(\pi)\cos\beta + y'(\pi)\sin\beta = 0$$

is called the Sturm-Liouville problem. However neither Sturm nor Liouville was able to prove the completeness of the eigenfunctions. This was done by V. A. Steklov.

At the present time there exist various methods for proving the completeness of the eigenfunctions, among which the most important are the method of integral equations (or the Green's function method), the method of contour integration and the method of finite differences. In this section we will discuss the method of finite differences. The other methods will be considered in the following two sections.

2. Thus we will consider the boundary-value problem (5.1)–(5.3). If $y = y(x, \lambda)$ is the solution of equation (5.1) satisfying the initial conditions

(5.4) $$y(0) = \sin\alpha,$$

(5.5) $$y'(0) = -\cos\alpha,$$

then $y(x, \lambda)$ obviously satisfies the boundary condition (5.2).

Let n be an arbitrary positive integer, and let $h = \pi/n$. Further, put $x_\nu = \nu h$, $q_\nu = q(\nu h)$, and let y_ν, $\nu = 0, 1, \cdots, n-1$ be the components of a vector which will be defined below. We will replace the first and second derivatives by the difference ratios:

(5.6) $$y' \sim \frac{y_\nu - y_{\nu-1}}{h}, \qquad y'' \sim \frac{y_{\nu+1} - 2y_\nu + y_{\nu-1}}{h^2} \qquad (\nu = 0, 1, \ldots, n-1).$$

From the initial condition (5.4) we obtain for y_0

(5.7) $$y_0 = \sin\alpha,$$

while y_{-1} is determined by means of the initial condition (5.5):

(5.8) $$y_{-1} = \sin\alpha + h\cos\alpha.$$

Replacing the second derivative in equation (5.1) by the expression (5.6), we obtain a finite-difference equation

(5.9) $$\frac{y_{\nu+1} - 2y_\nu + y_{\nu-1}}{h^2} + (\lambda - q_\nu)\, y_\nu = 0 \qquad (\nu = 0, 1, 2, \ldots, n-1).$$

From equation (5.9) and the initial conditions (5.7) and (5.8) one can determine the sequence $y_1, y_2, \cdots, y_n$. As is known, for $n \to \infty$ and $\nu h \to x$ we have $y_\nu \to y(x)$, where $y(x)$ is the solution of the problem (5.1) + (5.4) + (5.5), and indeed the convergence is uniform in any region $(0 \leqq x \leqq \pi,\ 0 \leqq |\lambda| \leqq R < \infty)$.

Let us now adjoin to the problem (5.9) + (5.7) + (5.8) the boundary condition

(5.10) $$y_{n-1}\cos\beta + \frac{y_n - y_{n-1}}{h}\sin\beta = 0,$$

which one obtains from the boundary condition (5.3) by replacing the derivative by the difference ratio. The problem thus obtained for the finite-difference equation is obviously an algebraic problem concerning eigenvalues in an n-dimensional vector space. From the boundary conditions (5.7), (5.8) and (5.10) it follows that

$$y_{-1} = y_0(1 + h\operatorname{ctg}\alpha), \tag{5.11}$$

$$y_n = y_{n-1}(1 - h\operatorname{ctg}\beta).^{9)} \tag{5.12}$$

Inserting these values in the system of equations (5.9), we obtain a system of n homogeneous equations in the n unknowns $y_0, y_1, \cdots, y_{n-1}$. It is easily seen that the matrix of this system is symmetric and is a Jacobi matrix. Therefore, as is well known from algebra, the eigenvalues of the system (5.9) + (5.7) + (5.8) + (5.10) are real, and the eigenvectors corresponding to different eigenvalues are orthogonal (the scalar product of two vectors $y = (y_0, y_1, \cdots, y_{n-1})$ and $z = (z_0, z_1, \cdots, z_{n-1})$ is defined by $(y, z) = \sum_{\nu=0}^{n-1} hy_\nu z_\nu$).

We will show that all the eigenvalues are simple. Let us assume that λ is an eigenvalue of multiplicity greater than 1. Then since the system is symmetric, there must be at least two linearly independent eigenvectors: $y = (y_0, y_1, \cdots, y_{n-1})$ and $z = (z_0, z_1, \cdots, z_{n-1})$. We can assume that $y_0 \neq 0$ and $z_0 \neq 0$ (if, for example, $y_0 = 0$, then by (5.11) $y_{-1} = 0$, and therefore from (5.9) we obtain $y_\nu = 0$ for $\nu = 1, 2, \cdots, n$). Moreover, since equation (5.9) is homogeneous, we may assume without loss of generality that $y_0 = z_0$. But then from (5.11) we obtain $y_{-1} = z_{-1}$, and so it follows from (5.9) that $y_\nu = z_\nu$ $(\nu = 1, 2, \cdots, n-1)$.

We denote the eigenvalues of the problem (5.9) + (5.7) + (5.8) + (5.10), in increasing order, by $\lambda_{k,h}$ $(k = 1, 2, \cdots, n)$ and denote corresponding eigenvectors by $y_k^{(h)} = (y_{0,k}^{(h)}, \cdots, y_{n-1,k}^{(h)})$. Further, let $\alpha_{k,h}^2$ denote the norm squared of $y_k^{(h)}$:

$$\alpha_{k,h}^2 = \| y_k^{(h)} \|^2 = (y_k^{(h)}, y_k^{(h)}) = \sum_{\nu=0}^{n-1} h\,[y_{\nu,k}^{(h)}]^2.$$

The vectors $y_1^{(h)}, y_2^{(h)}, \cdots, y_n^{(h)}$ form an orthogonal basis in the n-dimensional vector space R^n. Therefore for any vector $f = (f_0, f_1, \cdots, f_{n-1})$ we have the Parseval equality

$$(f, f) = \sum_{\nu=0}^{n-1} hf_\nu^2 = \sum_{k=1}^{n} \frac{1}{\alpha_{k,h}^2} (f, y_k^{(h)})^2. \tag{5.13}$$

Let $y_\nu(\lambda)$ denote the solution of equation (5.9) satisfying the initial conditions (5.7) and (5.8). If $h \to 0$ and $\nu h \to x$, then $y_\nu(\lambda) \to y(x, \lambda)$ uniformly in every bounded region in the λ-plane. It follows that the

---

9) We will assume, from this point on, that $\operatorname{ctg}\alpha \neq \infty$ and $\operatorname{ctg}\beta \neq \infty$. If $\operatorname{ctg}\alpha = \infty$ one has to put $y_{-1} = 0$; if $\operatorname{ctg}\beta = \infty$, then we put $y_n = 0$.

roots of equation (5.12) (i.e., the equation $y_n(\lambda) = y_{n-1}(\lambda)(1 - h\operatorname{ctg}\beta)$), which we have denoted by $\lambda_{1,h}, \lambda_{2,h}, \cdots, \lambda_{n,h}$, tend to the successive roots of equation (5.3) (i.e., the equation $y(\pi,\lambda)\cos\beta + y_x(\pi,\lambda)\sin\beta = 0$). In particular, if $h \to 0$, the number of roots $\lambda_{k,h}$ in each finite interval $(-N, N)$ is bounded, uniformly with respect to h.[10]

We now show that the Parseval equality for the problem (5.1) + (5.2) + (5.3) can be obtained from (5.13) by letting $h \to 0$.

Suppose that the function $f(x)$ $(0 \leq x \leq \pi)$ has a continuous second derivative and vanishes identically in some neighborhoods of the points $x = 0$ and $x = \pi$. We put $f = \{f(\nu h)\} = \{f_\nu\}$ $(\nu = 0, 1, \cdots, n-1)$ and write the equality (5.13) for f:

$$(f, f) = \sum_{\nu=0}^{n-1} hf^2(\nu h) = \sum_{k=1}^{n} \frac{1}{\alpha_{k,h}^2}(f, y_k^{(h)})^2$$

(5.14)

$$= \sum_{|\lambda_{k,h}| \leq N} \frac{1}{\alpha_{k,h}^2}(f, y_k^{(h)})^2 + \sum_{|\lambda_{k,h}| > N} \frac{1}{\alpha_{k,h}^2}(f, y_k^{(h)})^2 = I_1 + I_2,$$

where N denotes an arbitrary positive number. Let us estimate I_2. To this end we will transform the Fourier coefficients $(f, y_k^{(h)})$. It follows from equation (5.9) that

$$(f, y_k^{(h)}) = \sum_{\nu=0}^{n-1} hf_\nu y_{k,\nu}^{(h)}$$

$$= -\frac{1}{\lambda_{k,h}} \sum_{\nu=0}^{n-1} hf_\nu \left[\frac{y_{k,\nu+1}^{(h)} - 2y_{k,\nu}^{(h)} + y_{k,\nu-1}^{(h)}}{h^2} - q_\nu y_{k,\nu}^{(h)}\right]$$

(5.15)

$$= -\frac{1}{\lambda_{k,h}} \Bigg\{[-f_{-1}y_{k,0}^{(h)} + f_{n-1}y_{k,n}^{(h)} + f_0 y_{k,-1}^{(h)} - f_n y_{k,n-1}^{(h)}]h^{-1}$$

$$+ \sum_{\nu=0}^{n-1} hy_{k,\nu}^{(h)}\left[\frac{f_{\nu+1} - 2f_\nu + f_{\nu-1}}{h^2} - q_\nu f_\nu\right]\Bigg\}.$$

If h is sufficiently small, then $f_0 = 0$, $f_{n-1} = 0$ and $f_n = 0$. We also put $f_{-1} = 0$. Then from (5.15) we obtain

$$(f, y_k^{(h)}) = -\frac{1}{\lambda_{k,h}} \sum_{\nu=0}^{n-1} hy_{k,\nu}^{(h)}\left[\frac{f_{\nu+1} - 2f_\nu + f_{\nu-1}}{h^2} - q_\nu f_\nu\right]$$

$$= -\frac{1}{\lambda_{k,h}}\left(y_k^{(h)}, \frac{\Delta^2 f}{h^2} - qf\right).$$

[10] For another proof of this see the Remark on p. 00.

Therefore by Bessel's inequality

$$
\begin{aligned}
I_2 &\leqslant \frac{1}{N^2} \sum_{|\lambda_{k,h}|>N} \frac{1}{\alpha_{k,h}^2}\left(y_k^{(h)}, \frac{\Delta^2 f}{h^2} - qf\right)^2 \\
&< \frac{1}{N^2} \sum_{k=1}^{n} \frac{1}{\alpha_{k,h}^2}\left(y_k^{(h)}, \frac{\Delta^2 f}{h^2} - qf\right)^2 = \frac{1}{N^2}\left\| \frac{\Delta^2 f}{h^2} - qf \right\|^2.
\end{aligned} \tag{5.16}
$$

If $h \to 0$, then

$$\left\| \frac{\Delta^2 f}{h^2} - qf \right\|^2 \to \int_0^\pi \{f''(x) - q(x) f(x)\}^2 \, dx.$$

It therefore follows from (5.16) that there exists a constant $A > 0$ such that, uniformly in $h \to 0$,

$$I_2 < A/N^2.$$

Hence from (5.14) we obtain

$$\left| \sum_{\nu=0}^{n-1} h f^2(\nu h) - \sum_{|\lambda_{k,h}|\leqslant N} \frac{1}{\alpha_{k,h}^2} (f, y_k^{(h)})^2 \right| < \frac{A}{N^2}. \tag{5.17}$$

Since as $h \to 0$

$$\sum_{\nu=0}^{n-1} h f^2(\nu h) \to \int_0^\pi f^2(x)\, dx;$$

$$\alpha_{k,h}^2 \to \int_0^\pi y^2(x, \lambda_k)\, dx = \alpha_k^2; \qquad (f, y_k^{(h)}) \to \int_0^\pi f(x)\, y(x, \lambda_k)\, dx,$$

where λ_k is the kth (in increasing order) zero of equation (5.3), and since the number of roots $\lambda_{k,h}$ in a fixed interval $(-N, N)$ is bounded uniformly in h, it follows that by passing to the limit $h \to 0$ in (5.17) we obtain

$$\left| \int_0^\pi f^2(x)\, dx - \sum_{|\lambda_k|\leqslant N} \frac{1}{\alpha_k^2} \left\{ \int_0^\pi f(x)\, y(x, \lambda_k)\, dx \right\}^2 \right| < \frac{A}{N^2}.$$

Letting $N \to \infty$ in this inequality, it follows that

$$\int_0^\pi f^2(x)\, dx = \sum_{k=1}^{\infty} \frac{1}{\alpha_k^2} \left\{ \int_0^\pi f(x)\, y(x, \lambda_k)\, dx \right\}^2, \tag{5.18}$$

which is the Parseval equality for functions satisfying the conditions indicated above.

3. Let us denote by $v_k(x)$ normalized eigenfunctions, i.e., we put

$$v_k(x) = (1/\alpha_k)\, y(x, \lambda_k).$$

Then the Parseval equality (5.18) takes the form

$$\int_0^\pi f^2(x)\,dx = \sum_{k=1}^\infty a_k^2, \tag{5.19}$$

where the a_k are the Fourier coefficients with respect to the normalized eigenfunctions, i.e.

$$a_k = \int_0^\pi f(x)\, v_k(x)\,dx.$$

Let $f(x)$ be an arbitrary square integrable function. As is known, one can find a sequence of twice continuously differentiable (even infinitely differentiable) functions $f_k(x)$ which converges to $f(x)$ in quadratic mean. One can also assume that the functions $f_k(x)$ vanish identically in neighborhoods of the points $x = 0$ and $x = \pi$. Therefore by virtue of (5.19) we have

$$\int_0^\pi \{f_k(x) - f_l(x)\}^2\,dx = \sum_{n=0}^\infty \{a_n^{(k)} - a_n^{(l)}\}^2, \tag{5.20}$$

where

$$a_n^{(k)} = \int_0^\pi f_k(x)\, v_n(x)\,dx.$$

If $k, l \to \infty$, then the left side of (5.20) tends to zero. Hence the right side tends to zero.

It follows from the Cauchy-Bunjakovskiĭ inequality that

$$|a_n - a_n^{(k)}| \leqslant \left\{\int_0^\pi [f(x) - f_k(x)]^2\,dx\right\}^{1/2}.$$

Therefore the convergence in the mean of $f_k(x)$ to $f(x)$ implies that $\lim_{k\to\infty} a_n^{(k)} = a_n$ $(n = 0, 1, \cdots)$. Let N denote a fixed positive integer. It follows from (5.20) that

$$\sum_{n=0}^N \{a_n^{(k)} - a_n^{(l)}\}^2 \leqslant \int_0^\pi \{f_k(x) - f_l(x)\}^2\,dx.$$

Letting $k \to \infty$, we obtain

$$\sum_{n=0}^N \{a_n - a_n^{(l)}\}^2 \leqslant \int_0^\pi \{f(x) - f_l(x)\}^2\,dx.$$

Now letting $N \to \infty$, we obtain

$$\sum_{n=0}^{\infty} \{a_n - a_n^{(l)}\}^2 \leqslant \int_0^{\pi} \{f(x) - f_l(x)\}^2 \, dx.$$

From this inequality there follows, in particular (by means of Minkowski's inequality), the convergence of the series $\sum_{n=0}^{\infty} a_n^2$. Since

$$\left| \sum_{n=0}^{\infty} a_n^2 - \sum_{n=0}^{\infty} \{a_n^{(l)}\}^2 \right| = \left| \sum_{n=0}^{\infty} \{a_n - a_n^{(l)}\} \{a_n + a_n^{(l)}\} \right|$$

$$\leqslant \left(\sum_{n=0}^{\infty} | a_n - a_n^{(l)} |^2 \right)^{1/2} \left(\sum_{n=0}^{\infty} | a_n + a_n^{(l)} |^2 \right)^{1/2},$$

it follows from the foregoing that as $l \to \infty$

$$\sum_{n=0}^{\infty} \{a_n^{(l)}\}^2 \to \sum_{n=0}^{\infty} a_n^2.$$

On the other hand, the convergence in quadratic mean of $f_l(x)$ to $f(x)$ implies that

$$\int_0^{\pi} f_l^2(x) \, dx \to \int_0^{\pi} f^2(x) \, dx.$$

Therefore, passing to the limit $l \to \infty$ in the equality

$$\int_0^{\pi} f_l^2(x) \, dx = \sum_{n=0}^{\infty} \{a_n^{(l)}\}^2$$

we obtain

$$\int_0^{\pi} f^2(x) \, dx = \sum_{n=0}^{\infty} a_n^2.$$

We have thus proved:

THEOREM 5.1. *For every square integrable function $f(x)$ on the interval $[0, \pi]$ one has the Parseval equality*

$$\int_0^{\pi} f^2(x) \, dx = \sum_{n=0}^{\infty} a_n^2, \tag{5.21}$$

where

$$a_n = \int_0^{\pi} f(x) v_n(x) \, dx. \tag{5.22}$$

REMARK. The boundedness of the number of eigenvalues $\lambda_{k,h}$ in a fixed interval $(-N, N)$ can be proved without using the theory of functions of a complex variable.

Let δ denote an arbitrary positive number $< \pi$, and let

$$f_\delta(x) = \begin{cases} 1/\delta & \text{for } 0 \leqq x \leqq \delta, \\ 0 & \text{for } \delta < x \leqq \pi. \end{cases}$$

Further, denote by $\bar{f}_\delta$ the vector with components $f_{\delta\nu} = f_\delta(\nu h)$ $(\nu = 0, 1, \cdots, n-1)$. Applying the equality (5.13) to the vector $\bar{f}_\delta$, we find that

$$(\bar{f}_\delta, \bar{f}_\delta) = \sum_{\nu=0}^{n-1} h f_{\delta\nu}^2 = \frac{h}{\delta^2}\left[\frac{\delta}{h}\right] = \sum_{k=1}^{n} \frac{1}{\alpha_{k,h}^2} (\bar{f}_\delta, y_k^{(h)})^2.$$

Hence for any $N > 0$ we obtain the inequality

$$\sum_{|\lambda_{k,h}| \leqslant N} \frac{1}{\alpha_{k,h}^2} (\bar{f}_\delta, y_k^{(h)})^2 \leqslant \frac{1}{\delta}. \tag{5.23}$$

From the convergence, as $h \to 0$ and $\nu h \to x$, of $y_\nu(\lambda)$ to $y(x, \lambda)$ it follows that for sufficiently small h the quantity $(\bar{f}_\delta, y_k^{(h)})$ differs arbitrarily little from the integral

$$\frac{1}{\delta} \int_0^\delta y(x, \lambda_{k,h})\, dx$$

and indeed uniformly for $|\lambda_{k,h}| \leqq N$. Therefore if $\sin \alpha \neq 0$, then for sufficiently small h and sufficiently small δ we have

$$(\bar{f}_\delta, y_k^{(h)})^2 > \tfrac{1}{2} \sin^2 \alpha.$$

Hence, according to (5.23)

$$\sum_{|\lambda_{k,h}| \leqslant N} \frac{1}{\alpha_{k,h}^2} < \frac{2}{\delta \sin^2 \alpha}. \tag{5.24}$$

Since as $h \to 0$ the quantities $\alpha_{k,h}^2$ are uniformly bounded, it follows from the bound (5.24) that the number of eigenvalues $\lambda_{k,h}$ in the interval $(-N, N)$ is bounded, uniformly in h. Therefore we can choose a sequence $h_1, h_2, \cdots \to 0$ such that the limits

$$\lim_{p \to \infty} \lambda_{k, h_p} = \lambda_k \qquad (k = 1, 2, \ldots; \ |\lambda_{k, h_p}| \leqslant N)$$

exist. Obviously the numbers λ_k are the roots of the equation (5.3), and therefore there is in fact no need to choose a sequence $\{h_p\}$, i.e. the limits $\lim_{h \to 0} \lambda_{k,h} = \lambda_k$ exist. If $\sin \alpha = 0$, then we obtain the same result upon defining the function $f_\delta(x)$ by

$$f_\delta(x) = \begin{cases} 1/\delta^2 & \text{for } 0 \leqq x \leqq \delta, \\ 0 & \text{for } \delta < x \leqq \pi. \end{cases}$$

THEOREM 5.2 (THE EXPANSION THEOREM). *If, for a continuous function $f(x)$, the series*

$$\sum_{n=0}^{\infty} a_n v_n(x), \qquad a_n = \int_0^{\pi} f(x) v_n(x)\,dx,$$

converges uniformly in the interval $[0, \pi]$, then this series represents the function $f(x)$:

$$f(x) = \sum_{n=0}^{\infty} a_n v_n(x).$$

PROOF. We put

$$h(x) = f(x) - \sum_{n=0}^{\infty} a_n v_n(x). \tag{5.25}$$

By the hypothesis of the theorem, $h(x)$ is a continuous function on $[0, \pi]$. It follows from (5.25) that

$$b_n = \int_0^{\pi} h(x) v_n(x)\,dx = 0$$

$$(n = 0, 1, 2, \ldots).$$

Applying the Parseval equality (5.21) to $h(x)$, we obtain

$$\int_0^{\pi} h^2(x)\,dx = 0.$$

Since $h(x)$ is continuous, it follows that $h(x) \equiv 0$. Therefore the theorem follows from (5.25).

§6. Proof of the expansion theorem by the method of integral equations

1. We will consider the problem

$$y'' + \{\lambda - q(x)\} y = f(x), \tag{6.1}$$

$$y(0)\cos\alpha + y'(0)\sin\alpha = 0, \qquad y(\pi)\cos\beta + y'(0)\sin\beta = 0, \tag{6.2}$$

where $f(x) \not\equiv 0$ is a continuous function.

Let λ be a fixed complex number. We denote by $u(x, \lambda)$ the solution of equation (5.1) which satisfies the initial conditions

$$u(0, \lambda) = \sin\alpha, \qquad u'(0, \lambda) = -\cos\alpha,$$

and by $v(x, \lambda)$ the solution of the same equation which satisfies the initial conditions

$$v(\pi, \lambda) = \sin\beta, \qquad v'(\pi, \lambda) = -\cos\beta.$$

If $u(x,\lambda)$ and $v(x,\lambda)$ are linearly independent, i.e. if $u(x,\lambda)$ is not an eigenfunction of the problem (5.1) – (5.3),[11] then the Wronskian determinant is nonzero:

$$W\{u, v\} = \begin{vmatrix} u & v \\ u' & v' \end{vmatrix} \neq 0.$$

Conversely, if for some λ the Wronskian determinant equals zero, then $u = Cv$ and so u is an eigenfunction. Thus the eigenvalues of the problem (5.1) – (5.3) coincide with the zeros of the Wronskian determinant. Since in our case the coefficient of the first derivative equals zero, by Liouville's well-known formula W does not depend upon x: $W\{u,v\} = \omega(\lambda)$.

Let us introduce the function

$$G(x, t; \lambda) = \begin{cases} \frac{1}{\omega(\lambda)} u(x, \lambda) v(t, \lambda), & x \leqslant t, \\ \frac{1}{\omega(\lambda)} u(t, \lambda) v(x, \lambda), & x \geqslant t. \end{cases}$$

$G(x,t;\lambda)$ is called the Green's function. It is symmetric with respect to x and t and is real for real λ. We will show that the function

$$y(x, \lambda) = \int_0^\pi G(x, t; \lambda) f(t)\, dt \tag{6.3}$$

is the solution of the boundary-value problem (6.1) – (6.2). In fact,

$$y(x, \lambda) = \frac{1}{\omega(\lambda)} \left\{ v(x, \lambda) \int_0^x u(t, \lambda) f(t)\, dt + u(x, \lambda) \int_x^\pi v(t, \lambda) f(t)\, dt \right\}. \tag{6.3'}$$

Therefore

$$y''(x, \lambda) = \frac{1}{\omega(\lambda)} \left\{ v''(x, \lambda) \int_0^x u(t, \lambda) f(t)\, dt + u''(x, \lambda) \int_x^\pi v(t, \lambda) f(t)\, dt \right.$$
$$\left. + v'(x, \lambda) u(x, \lambda) f(x) - u'(x, \lambda) v(x, \lambda) f(x) \right\}$$
$$= \frac{1}{\omega(\lambda)} [q(x) - \lambda] \left\{ v(x, \lambda) \int_0^x u(t, \lambda) f(t)\, dt + u(x, \lambda) \int_x^\pi v(t, \lambda) f(t)\, dt \right\} + f(x)$$
$$= [q(x) - \lambda]\, y(x, \lambda) + f(x),$$

i.e. $y'' + \{\lambda - q(x)\} y = f(x)$.

[11] If $u = Cv$, then u satisfies the boundary conditions (6.2), and is consequently an eigenfunction of the problem (5.1) – (5.3).

One can verify directly that the function $y(x, \lambda)$ satisfies the boundary conditions (6.2).

Thus, *if* λ *is not an eigenvalue of the homogeneous problem* (5.1) – (5.3), *then the inhomogeneous system* (6.1) – (6.2) *is solvable for any function* $f(x)$. *Its solution is given by* (6.3). Conversely, *if* λ *is an eigenvalue of the homogeneous problem, then the inhomogeneous system* (6.1) – (6.2) *is, generally speaking, insoluble.*

If λ is not an eigenvalue of the homogeneous problem, then the system (6.1) – (6.2) has a unique solution. In fact, the difference of two solutions of the inhomogeneous problem is obviously an eigenfunction of the homogeneous problem, and by virtue of our assumption will be identically zero.

We can assume that $\lambda = 0$ is not an eigenvalue. Otherwise we choose a fixed number η and consider the boundary-value problem

$$y'' + \{ (\lambda + \eta) - q(x) \} y = 0,$$

$$y(0) \cos \alpha + y'(0) \sin \alpha = 0, \qquad y(\pi) \cos \beta + y'(\pi) \sin \beta = 0.$$

The eigenfunctions of this system are the same as those of the system (5.1) – (5.3). All eigenvalues are shifted to the left by η. We can obviously choose η in such a way that zero will not be an eigenvalue of the new problem. Put $G(x, t; 0) = G(x, t)$. Then the function

$$y(x) = \int_0^\pi G(x, t) f(t)\, dt \tag{6.3''}$$

is the solution of the equation $y'' - q(x) y = f(x)$ which satisfies the boundary conditions (6.2). Let us rewrite equation (6.1) in the form

$$y'' - q(x) y = f(x) - \lambda y.$$

On the basis of the foregoing we can assert that the system (6.1) – (6.2) is equivalent to the integral equation

$$y + \lambda \int_0^\pi G(x, t) y(t)\, dt = \int_0^\pi G(x, t) f(t)\, dt \equiv g(x).$$

In particular, the homogeneous problem $[f(x) \equiv 0]$ is equivalent to the integral equation

$$y(x) + \lambda \int_0^\pi G(x, t) y(t)\, dt = 0. \tag{6.4}$$

2. Let us denote by $\lambda_0, \lambda_1, \lambda_2, \cdots$ the set of all eigenvalues of the problem (5.1) – (5.3), and by $v_0(x), v_1(x), \cdots$ corresponding normalized eigenfunctions. Consider the kernel

$$H(x, \xi) = \sum_{n=0}^{\infty} \frac{v_n(x) v_n(\xi)}{\lambda_n}.$$

By virtue of the asymptotic formulas obtained[12] in §2, the series for $H(x,\xi)$ converges absolutely and uniformly, and hence the kernel $H(x,\xi)$ is continuous. Let us now consider the kernel

$$Q(x, \xi) = G(x, \xi) + H(x, \xi) = G(x, \xi) + \sum_{n=0}^{\infty} \frac{v_n(x) v_n(\xi)}{\lambda_n}.$$

This kernel is obviously continuous and symmetric.

By a well-known theorem of the theory of integral equations every symmetric kernel $Q(x,\xi)$, not identically zero, has at least one eigenfunction,[13] i.e. there exists a number λ_0 and function $u(x) \not\equiv 0$ satisfying the equation

$$u(x) + \lambda_0 \int_0^{\pi} Q(x, \xi) u(\xi) \, d\xi = 0. \tag{6.5}$$

Thus if we show that the kernel $Q(x,\xi)$ has no eigenfunctions, we can conclude that $Q(x,\xi) \equiv 0$, i.e.

$$G(x, \xi) = -\sum_{n=0}^{\infty} \frac{v_n(x) v_n(\xi)}{\lambda_n}. \tag{6.6}$$

From this expansion it is then easy to obtain the completeness of the eigenfunctions.

It follows from the equation (6.4) that

$$\int_0^{\pi} G(x, \xi) v_n(\xi) \, d\xi = -\frac{1}{\lambda_n} v_n(x).$$

Therefore

$$\int_0^{\pi} Q(x, \xi) v_n(\xi) \, d\xi = 0,$$

[12] For our purposes it is obviously sufficient to use formula (2.11), and consequently it is not necessary to require the differentiability of $q(x)$. A similar remark holds for the cases h or $H = \infty$ and $h = H = \infty$. The corresponding (less precise) asymptotic formulas can be obtained by the reader without difficulty.

[13] See Petrovskiĭ [1], p. 68 (transl. p. 51).

i.e. the kernel $Q(x,\xi)$ is orthogonal to all eigenfunctions of the boundary-value problem (5.1) – (5.3).

Let $u(x)$ be a solution of the integral equation (6.5). We will show that $u(x)$ is orthogonal to all of the $v_n(x)$. In fact, it follows from (6.5) that

$$0=\int_0^\pi u(x)v_n(x)\,dx+\lambda_0\int_0^\pi v_n(x)\left\{\int_0^\pi Q(x,\xi)u(\xi)\,d\xi\right\}dx$$

$$=\int_0^\pi u(x)v_n(x)\,dx+\lambda_0\int_0^\pi u(\xi)\left\{\int_0^\pi Q(x,\xi)v_n(x)\,dx\right\}d\xi=\int_0^\pi u(x)v_n(x)\,dx.$$

Hence it follows that

$$0=u(x)+\lambda_0\int_0^\pi Q(x,\xi)u(\xi)\,d\xi=u(x)+\lambda_0\int_0^\pi G(x,\xi)u(\xi)\,d\xi,$$

i.e. $u(x)$ is an eigenfunction of the boundary-value problem (5.1) – (5.3). But since $u(x)$ is orthogonal to all the $v_n(x)$, we have $u(x)\equiv 0$; consequently $Q(x,\xi)\equiv 0$. This proves (6.6).

THEOREM 6.1 (EXPANSION THEOREM). *If $f(x)$ has a continuous second derivative and satisfies the boundary conditions* (5.2) – (5.3), *then $f(x)$ can be expanded in an absolutely and uniformly convergent Fourier series with respect to the eigenfunctions of the boundary-value problem* (5.1) – (5.3):

$$f(x)=\sum_{n=0}^{\infty}a_n v_n(x),\qquad a_n=\int_0^\pi f(x)v_n(x)\,dx. \tag{6.7}$$

PROOF. We put $f''(x)-q(x)f(x)=h(x)$. Then by virtue of (6.3) and (6.6)

$$f(x)=\int_0^\pi G(x,\xi)h(\xi)\,d\xi=-\sum_{n=0}^{\infty}v_n(x)\frac{1}{\lambda_n}\int_0^\pi v_n(\xi)h(\xi)\,d\xi\equiv\sum_{n=0}^{\infty}a_n v_n(x).$$

From the orthonormality of the functions $v_n(x)$ it follows that

$$a_n=\int_0^\pi f(x)v_n(x)\,dx.$$

THEOREM 6.2. *For every function $f(x)$ which is square integrable over the interval $[0,\pi]$ we have the Parseval equality:*

$$\int_0^\pi f^2(x)\,dx = \sum_{n=0}^{\infty} a_n^2. \tag{6.8}$$

Proof. If $f(x)$ satisfies the hypothesis of the preceding theorem, then (6.8) follows at once from the uniform convergence of the series (6.7). In fact,

$$\int_0^\pi f^2(x)\,dx = \sum_{n=0}^{\infty} a_n \int_0^\pi f(x)\,v_n(x)\,dx = \sum_{n=0}^{\infty} a_n^2.$$

Extending the Parseval equality to an arbitrary square integrable function can be carried out by the methods used in §5 for proving Theorem 5.1.

3. Let us return to formula (6.3). The right side of this formula is called the *resolvent*. We know that the resolvent exists for all λ which are not eigenvalues of the boundary-value problem (5.1)–(5.3).

We will now show how to obtain the Fourier expansion of the resolvent if one knows the expansion of the function $f(x)$.

The function $y(x,\lambda)$ defined by (6.3) satisfies the boundary conditions (6.2). Therefore, integrating by parts, we obtain

$$\begin{aligned}
&\int_0^\pi \{y''(x,\lambda) - q(x)\,y(x,\lambda)\}\,v_n(x)\,dx \\
&= \int_0^\pi \{v_n''(x) - q(x)\,v_n(x)\}\,y(x,\lambda)\,dx \\
&= -\lambda_n \int_0^\pi y(x,\lambda)\,v_n(x)\,dx = -\lambda_n d_n(\lambda).
\end{aligned} \tag{6.9}$$

Let

$$y(x,\lambda) = \sum_{n=0}^{\infty} d_n(\lambda)\,v_n(x), \qquad a_n = \int_0^\pi f(x)\,v_n(x)\,dx.$$

Then it follows from (6.1) and (6.9) that

$$a_n = \int_0^\pi \{y''(x,\lambda) + [\lambda - q(x)]\,y(x,\lambda)\}\,v_n(x)\,dx = -\lambda_n d_n(\lambda) + \lambda d_n(\lambda).$$

Hence

$$d_n(\lambda) = a_n/(\lambda - \lambda_n).$$

Consequently the expansion of the resolvent has the form

(6.10) $$y(x, \lambda) = \sum_{n=0}^{\infty} \frac{a_n}{\lambda - \lambda_n} v_n(x).$$

§7. Proof of the expansion theorem in the periodic case

For the proof that functions satisfying periodic boundary conditions can be expanded in a series with respect to the eigenfunctions of the problem (4.1) – (4.2), i.e. the problem

(7.1) $$y'' + \{\lambda - q(x)\} y = 0,$$

(7.2) $$y(0) = y(\pi), \qquad y'(0) = y'(\pi),$$

we will apply the method of integral equations (the method of finite differences is also applicable to this case).

To begin with we will prove the uniform and absolute convergence of the series

(7.3) $$\sum_{n=0}^{\infty} \frac{v_n(x)\, v_n(\xi)}{\lambda_n}.$$

By virtue of Theorem 4.1 and the asymptotic formulas (2.13) and (2.19) we obtain information concerning the eigenvalues λ_n of the problem (7.1) – (7.2), namely: $\lambda_n / n^2 \to 1$. Therefore to prove the uniform and absolute convergence of the series (7.3) it suffices to prove that the eigenfunctions $v_n(x)$ are uniformly bounded.

To simplify the proof we will exclude the case of an eigenvalue of multiplicity two. We put $\lambda = s^2$ and rewrite equation (7.1) in the form $y'' + s^2 y = q(x) y$.

Applying the method of variation of constants, we obtain

$$y(x, \lambda) = c_1 \sin sx + c_2 \cos sx + \frac{1}{s} \int_0^x y(t, \lambda)\, q(t) \sin s(x - t)\, dt.$$

The constants c_1 and c_2 are determined up to a constant factor by the relation

$$c_2 = c_1 \sin s\pi + c_2 \cos s\pi + \frac{1}{s} \int_0^\pi y(t, \lambda)\, q(t) \sin s(\pi - t)\, dt.$$

We choose c_1 and c_2 in such a way that the function $y(x, \lambda)$ is normalized: $\int_0^\pi y^2(x, \lambda)\, dx = 1$. Then $y(x, \lambda)$, for $s^2 = \lambda_n$, will be a normalized eigenfunction. In fact, since we have excluded a root of multiplicity two, for no n will λ_n equal an eigenvalue μ_n (cf. §4); therefore no solution of the equation $y'' + [\lambda_n - q(x)]y = 0$ (except the trivial solution $y \equiv 0$) can vanish at the endpoints of the interval. Hence we conclude that for every solution y, different from the $v_n(x)$, for which $y(0) = y(\pi)$, one can find c such that $v_n(0) - cy(0) = v_n(\pi) - cy(\pi) = 0$, and thus $v_n(x) - cy(x) \equiv 0$. Consequently, for $s_n^2 = \lambda_n$ the normalized function $y(x, \lambda)$ is identical with a normalized eigenfunction.

Now applying the Cauchy-Bunjakovskiĭ inequality, we obtain

$$\left(\int_0^x y(t, \lambda) q(t) \sin s(x-t)\, dt\right)^2 \leqslant \int_0^\pi y^2(t, \lambda)\, dt \int_0^\pi q^2(t)\, dt \leqslant C.$$

Hence

$$y(x, \lambda) = c_1(s) \sin sx + c_2(s) \cos sx + O(1/s).$$

Moving $O(1/s)$ to the left side, squaring and then integrating, we obtain

$$\int_0^\pi \left\{y(x, \lambda) - O\left(\frac{1}{s}\right)\right\}^2 dx = 1 - 2\int_0^\pi y(x, \lambda) O\left(\frac{1}{s}\right) dx + \int_0^\pi O^2\left(\frac{1}{s}\right) dx$$

$$(7.4) \qquad c_1^2\left[\frac{\pi}{2} + O\left(\frac{1}{s}\right)\right] + c_2^2\left[\frac{\pi}{2} + O\left(\frac{1}{s}\right)\right] + 2c_1c_2 O\left(\frac{1}{s}\right).$$

Further,

$$\left(\int_0 y(x, \lambda) O\left(\frac{1}{s}\right) dx\right)^2 \leqslant \int_0^\pi y^2(x, \lambda)\, dx \int_0^\pi O^2\left(\frac{1}{s}\right) ds = O\left(\frac{1}{s^2}\right).$$

Since the numbers c_1 and c_2 enter symmetrically in the solution (7.4), we can suppose, for example, that $|c_1| \geqq |c_2|$. Putting then $\eta = (c_2/c_1)^2$, $\eta \leqq 1$, we will have

$$c_1^2\left(\frac{\pi}{2} + \eta\frac{\pi}{2} + O\left(\frac{1}{s}\right)\right) \leqslant 1 + O\left(\frac{1}{s^2}\right).$$

This inequality shows that c_1 (and c_2 along with it) is bounded. But this proves the uniform boundedness of the eigenfunctions. We have thus proved that the series (7.3) converges uniformly and absolutely. All the arguments of the preceding section now remain valid (the Green's function for periodic conditions is constructed, for example, in Chapter 7, §2, of the book by Coddington and Levinson). In particular, the formula

$$G(x, \xi) = -\sum_{n=0}^{\infty} \frac{v_n(x)\, v_n(\xi)}{\lambda_n}$$

holds.

THEOREM 7.1. *Every continuous function $f(x)$ which has a continuous second derivative can be expanded in a series with respect to the eigenfunctions of the problem* (7.1) – (7.2), *if it satisfies the boundary conditions* (7.2).

The proof coincides word for word with the proof of Theorem 6.1.

§8. Proof of the expansion theorem by the method of contour integration

We will first present a formal argument.

Let us assume that the eigenfunction expansion of the form (6.7) has already been justified. Then, as was shown in §6, the resolvent $y(x, \lambda)$ defined by formula (6.3) has the eigenfunction expansion (6.10) and satisfies the equation (6.1). Solving this equation, we can, as follows from (6.10), determine the terms in the expansion of a function $f(x)$ by finding the residues at the poles of the function $y(x, \lambda)$. As was shown in §6.1, the zeros of the function $\omega(\lambda)$ coincide with the eigenvalues of the problem (5.1) – (5.3) and are therefore simple. Let $v(x, \lambda_n) = k_n u(x, \lambda_n)$. It follows from the boundary conditions that the constant k_n is finite and different from zero. Consequently, the function $y(x, \lambda)$ has the residue

$$\frac{k_n}{\omega'(\lambda_n)} u(x, \lambda_n) \int_0^\pi u(t, \lambda_n) f(t)\, dt$$

at the point $\lambda = \lambda_n$.

These formal arguments show that we must have an expansion of the form

$$f(x) = \sum_{n=0}^{\infty} \frac{k_n}{\omega'(\lambda_n)} u(x, \lambda_n) \int_0^\pi u(t, \lambda_n) f(t)\, dt.$$

The general method which we shall use below consists in the following: the function $y(x, \lambda)$ is defined by (6.3); integrating it along expanding contours in the complex λ-plane, we obtain $f(x)$ in the limit; contracting the contour of integration onto the real axis, on which the singularities of $y(x, \lambda)$ lie, we obtain the eigenfunction expansion of $f(x)$.

THEOREM 8.1. *Suppose that $f(x)$ belongs to $\mathscr{L}(0, \pi)$. Then for $0 < x < \pi$ the expansion of $f(x)$ with respect to the eigenfunctions of the problem* (5.1) – (5.3) *behaves, as concerns convergence, in the same way as the* *ordinary trigonometric Fourier series of* $f(x)$. *In particular, this expansion converges to* $\frac{1}{2}[f(x+0)+f(x-0)]$, *if* $f(x)$ *is of bounded variation in a neighborhood of* x.

Proof. We consider the integral

$$\frac{1}{2\pi i}\int_C y(x, \lambda)\, d\lambda \tag{8.1}$$

($y(x,\lambda)$ is defined by (6.3′)), which is taken over a closed contour in the λ-plane, described in the following way. Let $\lambda = s^2$; then the upper half of the contour is the path described by λ as s traverses the line segments in the s-plane ($s=\sigma+it$)

$$\sigma = n+\tfrac{1}{2}, \qquad 0 \leqq t \leqq n+\tfrac{1}{2},$$

$$t = n+\tfrac{1}{2}, \qquad 0 \leqq \sigma \leqq n+\tfrac{1}{2}.$$

The lower half of the contour is the reflection of the upper half about the real axis. It is easily seen that the integral (8.1) is equal to a partial sum of the eigenfunction expansion. Indeed, there are only a finite number of eigenvalues, i.e. a finite number of poles of the function $y(x,\lambda)$, inside the path of integration. Then by a familiar theorem of residue theory the integral (8.1) is equal to the sum of the residues at the poles of $y(x,\lambda)$, and these residues are the terms in the expansion of the function $f(x)$.

Let us consider the case $\sin\alpha \neq 0$, $\sin\beta \neq 0$. By Lemma 2.2 we have, on the above-mentioned line segments,

$$u(y, \lambda) = \cos ys \cdot \sin\alpha + O\{|s|^{-1} e^{ty}\},$$

$$v(x, \lambda) = \cos\{s(\pi - x)\}\sin\beta + O\{|s|^{-1} e^{t(\pi-x)}\},$$

$$u'(y, \lambda) = -s\sin sy \cdot \sin\alpha + O\{e^{ty}\},$$

$$v'(x, \lambda) = s\sin\{s(\pi-x)\}\sin\beta + O\{e^{t(\pi-x)}\}.$$

Then

$$\omega(\lambda) = s\sin s\pi\cdot\sin\alpha\cdot\sin\beta + O\{e^{t\pi}\}.$$

On these same line segments $|\sin s\pi| > Ae^{t\pi}$, and consequently

$$\frac{1}{\omega(\lambda)} = \frac{1}{s\sin s\pi\cdot\sin\alpha\cdot\sin\beta}\left\{1+O\left(\frac{1}{|s|}\right)\right\}.$$

Hence

$$\frac{v(x,\lambda)\,u(y,\lambda)}{\omega(\lambda)} = \frac{\cos\{s(\pi-x)\}\cos sy}{s\sin s\pi} + O\left(\frac{e^{t(y-x)}}{|s|^2}\right)$$

and

$$\frac{v(x,\lambda)}{\omega(\lambda)}\int_0^x u(y,\lambda)f(y)\,dy$$

$$=\int_0^x \frac{\cos\{s(\pi-x)\}\cos sy}{s\sin s\pi}f(y)\,dy+O\left\{\frac{1}{|s|^2}\int_0^x e^{t(y-x)}|f(y)|\,dy\right\}.$$

Let $0<\delta<x$; then the last term can be transformed to the form

$$O\left(\frac{e^{-\delta t}}{|s|^2}\right)+O\left(\frac{1}{|s|^2}\int_{x-\delta}^x |f(y)|\,dy\right).$$

In view of the fact that $d\lambda/ds=2s$, this term leads to the appearance in the integral (8.1) of the term

$$\int_C\left\{e^{-\delta t}+\int_{x-\delta}^x |f(y)|\,dy\right\}O\left(\frac{1}{|s|}\right)|ds|=O\left\{\int_C e^{-\delta t}\left|\frac{ds}{s}\right|\right\}+O\left\{\int_{x-\delta}^x |f(y)|\,dy\right\}.$$

The second term standing on the right side can be made arbitrarily small by a suitable choice of δ, while for fixed δ the first term tends to zero as $n\to\infty$, as it has the form

$$O\left\{\frac{1}{n}\int_0^{n+\frac{1}{2}} e^{-\delta t}dt\right\}+O\left\{\frac{1}{n}\int_0^{n+\frac{1}{2}} e^{-\delta\left(n+\frac{1}{2}\right)}d\sigma\right\}=O\left(\frac{1}{n\delta}\right)+O\left(e^{-\delta\left(n+\frac{1}{2}\right)}\right).$$

A similar examination can be made of the other term in $y(x,\lambda)$, in which $x\leqq y\leqq\pi$. We obtain, as a consequence, that the integral (8.1) is equal to

$$\frac{1}{2\pi i}\int_C\left\{\int_0^x \frac{\cos\{s(\pi-x)\}\cos y}{s\sin s\pi}f(y)\,dy\right.$$
$$\left.+\int_x^\pi \frac{\cos\{s(\pi-y)\}\cos sx}{s\sin s\pi}f(y)\,dy\right\}d\lambda+o(1). \tag{8.2}$$

The first term in this formula coincides exactly with the expression we would have obtained in the problem of expanding $f(x)$ in a Fourier cosine series, and is therefore equal to the partial sum of the Fourier cosine series. This fact, of course, can easily be verified directly by means of residue theory. The theorem on the connection between eigenfunction expansions and Fourier cosine series which we are considering now follows. If $\sin\alpha = 0$ or $\sin\beta = 0$, the proof of the theorem is similar.

For the case in which $f(x)$ is of bounded variation in a neighborhood of the point x, the result established above can easily be obtained directly without drawing upon the theory of Fourier series. First of all we have

$$\frac{\cos\{s(\pi - x)\}\cos sy}{\sin s\pi} = O\left\{\frac{e^{t(\pi-x)+ty}}{e^{t\pi}}\right\} = O\{e^{-t(x-y)}\},$$

and consequently that part of the expression (8.2) where $0 \leqq y \leqq x - \delta$ tends to zero as before. For $x - \delta \leqq y \leqq x$ we have

$$\frac{\cos\{s(\pi - x)\}\cos sy}{\sin s\pi} = \frac{e^{-is(\pi-x)}\{1 + O(e^{-2t(\pi-x)})\}\, e^{-isy}\{1 + O(e^{-2ty})\}}{2ie^{-is\pi}\{1 + O(e^{-2t\pi})\}}$$

$$= -\frac{1}{2} ie^{is(x-y)}\{1 + O(e^{-2\delta t})\},$$

if $\delta < \pi - x$. The term which contains $O(e^{-2\delta t})$ leads, as before, to a term which tends to zero. The principal term in (8.2) is given by the integral

$$-\frac{i}{2s}\int_{x-\delta}^{x} e^{is(x-y)} f(y)\,dy = -\frac{i}{2s} f(x-0)\int_{x-\delta}^{x} e^{is(x-y)}dy$$

$$+\frac{i}{2s}\int_{x-\delta}^{x} e^{is(x-y)}\{f(x-0) - f(y)\}\,dy$$

$$= f(x-0)\frac{1 - e^{is\delta}}{2s^2} + \frac{i}{2s}\int_{x-\delta}^{x} e^{is(x-y)}\{f(x-0) - f(y)\}\,dy.$$

The first term $f(x-0)/2\lambda$ yields, after integration, $f(x-0)/2$, as required. The integral of the second term gives $O(e^{-\delta t}|\lambda|^{-1})$, and consequently tends to zero. Further, since $f(x)$ is of bounded variation, we can put $f(x-0) - f(y) = g(y) - h(y)$, where $g(y)$ and $h(y)$ are positive monotonically nonincreasing functions which tend to zero as $y \to x$. By the second theorem of the mean we have

$$\int_{x-\delta}^{x} \operatorname{Re}\{e^{is(x-y)}\}\, g(y)\,dy = g(x-\delta)\int_{x-\delta}^{\xi} \operatorname{Re}\{e^{is(x-y)}\}\,dy$$

$$= g(x-\delta)\operatorname{Re}\int_{x-\delta}^{\xi} e^{is(x-y)}dy = O\left\{\frac{g(x-\delta)}{|s|}\right\}.$$

This term gives rise in (8.1) to the term

$$\int O\left\{\frac{g(x-\delta)}{|\lambda|}\right\} d|\lambda| = O\{g(x-\delta)\},$$

which can be made arbitrarily small by suitably choosing δ. The term containing $h(y)$ can be studied in similar fashion, as can the terms in which $y > x$. This completes the proof of the theorem.

If $x = 0$, then under the condition $\sin\alpha \neq 0$ a similar assertion concerning convergence holds. If $\sin\alpha = 0$, then $y(x,\lambda) \equiv 0$, and our assertion no longer holds, with the exception only of the case for which $f(+0) = 0$. For $x = \pi$ the situation is analogous.

§9. Sharpening of the expansion theorem

The theorem on Fourier series expansions with respect to eigenfunctions was obtained in §§5 and 6 under very restrictive conditions.

The asymptotic formulas for the eigenfunctions derived in §2 enabled us to prove in §8 that an eigenfunction expansion converges under the same conditions as does a Fourier cosine series.

In this section we will give a different, more elementary proof of this fact.

For definiteness we will consider the case $h \neq \infty$ and $H \neq \infty$.

We will also assume that $q(x)$ has a bounded derivative, so that the asymptotic formula (2.14) holds.

Let $f(x)$ be an absolutely (Lebesgue) integrable function on the interval $[0,\pi]$. We put

$$\sigma_{n,f}(x) = \int_0^\pi f(t)\left\{\frac{1}{\pi} + \frac{2}{\pi}\sum_{k=1}^n \cos kx \cdot \cos kt\right\} dt,$$

$$s_{n,f}(x) = \int_0^\pi f(t) \sum_{k=0}^n v_k(x) v_k(t)\, dt.$$

If

$$\Phi_n(x,t) \doteq \sum_{k=0}^n v_k(x) v_k(t) - \left\{\frac{1}{\pi} + \frac{2}{\pi}\sum_{k=1}^n \cos kx \cdot \cos kt\right\},$$

then

$$s_{n,f}(x) - \sigma_{n,f}(x) = \int_0^\pi \Phi_n(x,t) f(t)\, dt.$$

LEMMA 9.1. *There exists a constant M such that for all n*

$$|\Phi_n(x,t)| < M.$$

PROOF. It follows from the asymptotic formula (2.14) that

$$v_k(x)v_k(t)-\frac{2}{\pi}\cos kx\cdot\cos kt$$

$$\begin{aligned}(9.1)\qquad &=\sqrt{\frac{2}{\pi}}\{\beta(t)\cos kx\cdot\sin kt+\beta(x)\cos kt\sin kx\}+O\left(\frac{1}{k^2}\right)\\ &=\frac{1}{\sqrt{2\pi}}\{\beta(x)+\beta(t)\}\frac{\sin k(x+t)}{k}\\ &+\frac{1}{\sqrt{2\pi}}\{\beta(x)-\beta(t)\}\frac{\sin k(x-t)}{k}+O\left(\frac{1}{k^2}\right).\end{aligned}$$

Obviously the function $\beta(x)$ is bounded in the interval $[0,\pi]$. We will prove that the sum $\sum_{k=1}^n \sin kx/k$ is uniformly bounded in the interval $[0,\pi]$. We have[14]

$$\tau_n(x)=\sum_{k=1}^n\frac{\sin kx}{k}=\int_0^x\left(\sum_{k=1}^n\cos kt\right)dt$$

$$=\int_0^x\frac{\sin\left(n+\frac{1}{2}\right)t-\sin\frac{t}{2}}{2\sin\frac{t}{2}}dt=\int_0^x\frac{1}{t}\sin\left(n+\frac{1}{2}\right)t\,dt$$

$$+\int_0^x\left(\frac{1}{2\sin\frac{t}{2}}-\frac{1}{t}\right)\sin\left(n+\frac{1}{2}\right)t\,dt-\frac{1}{2}x$$

$$=\int_0^{\left(n+\frac{1}{2}\right)x}\frac{\sin u}{u}du+\int_0^x\left(\frac{1}{2\sin\frac{t}{2}}-\frac{1}{t}\right)\sin\left(n+\frac{1}{2}\right)t\,dt-\frac{1}{2}x.$$

But the integral $\int_0^N(\sin u/u)\,du$ is bounded and has a maximum at $N=\pi$. Therefore for $0\leq x\leq\pi$

$$|\tau_n(x)|\leqslant\int_0^\pi\frac{\sin t}{t}dt+\int_0^\pi\left|\frac{1}{2\sin\frac{t}{2}}-\frac{1}{t}\right|dt+\frac{1}{2}\pi,$$

---

[14] By virtue of the well-known identity

$$\sum_{k=1}^n\cos kt=\frac{\sin\left(n+\frac{1}{2}\right)t-\sin\frac{t}{2}}{2\sin\frac{t}{2}}$$

which proves our assertion concerning the uniform boundedness of $\tau_n(x)$. The lemma now follows at once from (9.1) by virtue of the uniform boundedness of $\beta(x)$ and $\tau_n(x)$ in the interval $[0,\pi]$.

LEMMA 9.2. *If the function $\varphi(x)$ defined in the interval $[0,\pi]$ has a continuous second derivative and satisfies the boundary conditions*

$$\varphi(0)\cos\alpha + \varphi'(0)\sin\alpha = 0, \quad \varphi(\pi)\cos\beta + \varphi'(\pi)\sin\beta = 0, \tag{9.2}$$

then as $n \to \infty$

$$\int_0^\pi \Phi_n(x, t)\,\varphi(t)\,dt \to 0.$$

PROOF. In fact, under our hypotheses it follows from Theorem 6.1 and a familiar theorem of the theory of Fourier series that $s_{n,\varphi}(x)$ and $\sigma_{n,\varphi}(x)$ tend uniformly to $\varphi(x)$. Hence their difference

$$s_{n,\varphi}(x) - \sigma_{n,\varphi}(x) = \int_0^x \Phi_n(x, t)\,\varphi(t)\,dt$$

tends to zero uniformly.

THEOREM 9.1. *For any function $f(x)$ which is absolutely Lebesgue integrable over the interval $[0,\pi]$ we have, uniformly over $[0,\pi]$,*

$$s_{n,f}(x) - \sigma_{n,f}(x) \to 0$$

as $n \to \infty$.

PROOF. Let $f(x)$ be an arbitrary absolutely integrable function. As is known, for any $\epsilon > 0$ we can find a function $\varphi_\epsilon(x)$ which is twice continuously differentiable and satisfies the boundary conditions (9.2), such that

$$\int_0^\pi |f(x) - \varphi_\varepsilon(x)|\,dx < \frac{\varepsilon}{2M},$$

where M is as in Lemma 9.1.

We further have

$$s_{n,f}(x) - \sigma_{n,f}(x) = \int_0^\pi \Phi_n(x, t)\{f(t) - \varphi_\varepsilon(t)\}\,dt + \int_0^\pi \Phi_n(x, t)\,\varphi_\varepsilon(t)\,dt. \tag{9.3}$$

On the basis of Lemma 9.1

$$\left|\int_0^\pi \Phi_n(x, t)\{f(t) - \varphi_\varepsilon(t)\}\,dt\right| \leqslant M\int_0^\pi |f(t) - \varphi_\varepsilon(t)|\,dt < M\frac{\varepsilon}{2M} = \frac{\varepsilon}{2}. \tag{9.4}$$

On the other hand, by virtue of Lemma 9.2 we can, for fixed ϵ, take n so large that we will have, uniformly over $[0, \pi]$, the inequality

$$\left| \int_0^\pi \Phi_n(x, t) \varphi_s(t)\, dt \right| < \frac{\varepsilon}{2}. \tag{9.5}$$

By virtue of (9.4) and (9.5) it follows from (9.3) that

$$|s_{n,f}(x) - \sigma_{n,f}(x)| < \epsilon,$$

and since ϵ is arbitrary, the theorem is proved.

From this theorem it follows, in particular, that the expansion of a summable function in the eigenfunctions of a Sturm-Liouville operator converges or diverges at any point of the interval $[0, \pi]$ according as its cosine series converges or diverges at this point.

For the case $h = \infty$ or $H = \infty$ the Sturm-Liouville expansion has to be compared with the Fourier series in the functions $\sin(n + \frac{1}{2})x$, $n = 0, 1, \cdots$, and for the case $h = \infty$ and $H = \infty$, with a series in the functions $\sin nx$.

§10. One-dimensional Dirac systems

1. Let L denote a matrix operator

$$L \equiv \begin{pmatrix} p_{11}(x) & p_{12}(x) \\ p_{21}(x) & p_{22}(x) \end{pmatrix}, \qquad p_{12}(x) \equiv p_{21}(x), \tag{10.1}$$

where the $p_{ik}(x)$ $(i, k = 1, 2)$ are real functions which are defined and continuous on the interval $[0, \pi]$. Further, let $y(x)$ denote a two-component vector-function

$$y(x) = \begin{pmatrix} y_1(x) \\ y_2(x) \end{pmatrix}.$$

Then the equation

$$\left(B \frac{d}{dx} + L - \lambda I\right) y = 0, \tag{10.2}$$

where λ is a parameter, and

$$B = \begin{pmatrix} 0 & 1 \\ -1 & 0 \end{pmatrix}, \qquad I = \begin{pmatrix} 1 & 0 \\ 0 & 1 \end{pmatrix},$$

is equivalent to a system of two simultaneous first-order ordinary differential equations

$$\begin{aligned} \frac{dy_2}{dx} + p_{11}(x) y_1 + p_{12}(x) y_2 &= \lambda y_1, \\ -\frac{dy_1}{dx} + p_{21}(x) y_1 + p_{22}(x) y_2 &= \lambda y_2. \end{aligned} \tag{10.2'}$$

For the case in which $p_{12}(x) = p_{21}(x) \equiv 0$, $p_{11}(x) = V(x) + m$, $p_{22}(x) = V(x) - m$, where $V(x)$ is a potential function, and m is the mass of a particle, the system (10.2′) is known in relativistic quantum theory as a stationary one-dimensional Dirac system. Let $H = H(x)$ denote a smooth orthogonal transformation in two-dimensional space. As is known, every orthogonal transformation in two-dimensional space has, in a fixed orthonormal basis, a matrix of the form

$$H(x) = \begin{pmatrix} \cos\varphi(x) & -\sin\varphi(x) \\ \sin\varphi(x) & \cos\varphi(x) \end{pmatrix}.$$

It is easily seen that the matrices B and H commute, i.e. $BH = HB$. Let us put $y = Hz$. Then inserting this value for y in (10.2) and multiplying both sides of the resulting equation from the left by H^{-1}, we obtain

$$H^{-1}B\frac{d}{dx}(Hz) + H^{-1}LHz = H^{-1}\lambda Hz$$

or

$$B\frac{dz}{dx} + \left(H^{-1}B\frac{d}{dx}H + H^{-1}LH\right)z = \lambda z. \tag{10.3}$$

Let us compute the matrix

$$Q = H^{-1}B\frac{d}{dx}H + H^{-1}LH.$$

We have

$$H^{-1}B\frac{d}{dx}H = \begin{pmatrix} \varphi'(x) & 0 \\ 0 & \varphi'(x) \end{pmatrix},$$

$$H^{-1}LH = \begin{pmatrix} p_{11}\cos^2\varphi + p_{12}\sin 2\varphi + p_{22}\sin^2\varphi, & p_{12}\cos 2\varphi + \frac{1}{2}(p_{22} - p_{11})\sin 2\varphi, \\ p_{12}\cos 2\varphi + \frac{1}{2}(p_{22} - p_{11})\sin 2\varphi, & p_{11}\sin^2\varphi - p_{12}\sin 2\varphi + p_{22}\cos^2\varphi \end{pmatrix}.$$

Therefore the matrix Q has the form

$$Q \equiv \begin{pmatrix} q_{11} & q_{12} \\ q_{12} & q_{22} \end{pmatrix} =$$

$$= \begin{pmatrix} \varphi' + p_{11}\cos^2\varphi + p_{12}\sin 2\varphi + p_{22}\sin^2\varphi, & p_{12}\cos 2\varphi + \frac{1}{2}(p_{22} - p_{11})\sin 2\varphi \\ p_{12}\cos 2\varphi + \frac{1}{2}(p_{22} - p_{11})\sin 2\varphi, & \varphi' + p_{11}\sin^2\varphi - p_{12}\sin 2\varphi + p_{22}\cos^2\varphi \end{pmatrix}.$$

We choose the function $\varphi(x)$ so that $q_{12}(x) \equiv 0$. Then

$$p_{12}(x)\cos 2\varphi(x) + \tfrac{1}{2}\{p_{22}(x) - p_{11}(x)\}\sin 2\varphi(x) = 0.$$

Hence

$$\varphi(x) = \frac{1}{2} \operatorname{arctg} \frac{2p_{12}(x)}{p_{11}(x) - p_{22}(x)},$$

and the matrix $Q(x)$ assumes the form

$$Q(x) = \begin{pmatrix} q_{11}(x) & 0 \\ 0 & q_{22}(x) \end{pmatrix} \equiv \begin{pmatrix} p(x) & 0 \\ 0 & r(x) \end{pmatrix}.$$

Therefore the equation (10.3) can be rewritten in the form

(10.4) $$\begin{pmatrix} 0 & 1 \\ -1 & 0 \end{pmatrix} \frac{dz}{dx} + \begin{pmatrix} p(x) & 0 \\ 0 & r(x) \end{pmatrix} z = \lambda z.$$

Now choose the function $\varphi(x)$ so that $\operatorname{tr} Q(x) \equiv q_{11}(x) + q_{22}(x) = 0$, i.e. $2\varphi'(x) + p_{11}(x) + p_{22}(x) = 0$. Hence

$$\varphi(x) = -\frac{1}{2} \int_0^x \{p_{11}(\tau) + p_{22}(\tau)\}\, d\tau.$$

Then equation (10.3) assumes the form

(10.5) $$\begin{pmatrix} 0 & 1 \\ -1 & 0 \end{pmatrix} \frac{dz}{dx} + \begin{pmatrix} p(x) & q(x) \\ q(x) & -p(x) \end{pmatrix} z = \lambda z.$$

We will refer to equations (10.4) and (10.5) as canonical forms of the system (10.2). In studying various questions of the spectral theory of the equation (10.2), it is convenient to use one or another of these canonical forms. Thus, for example, in studying the asymptotic behaviour of the eigenvalues and eigenfunctions, and also questions concerning the expansion of an arbitrary vector-function with respect to the eigenfunctions of equation (10.2) (in the presence of certain homogeneous boundary conditions at the points 0 and π) it is convenient to use the canonical equation (10.4); while in studying such questions, for example, as the asymptotic behaviour of the number of eigenvalues of equation (10.2), defined on an infinite interval, and also inverse problems, it is convenient to use the canonical equation (10.5).

2. Thus, let us consider the boundary-value problem for the system (10.2′), first reducing it to the canonical form (10.4):

(10.6) $$y_2' - \{\lambda + p(x)\} y_1 = 0, \quad y_1' + \{\lambda + r(x)\} y_2 = 0,$$

(10.7) $$y_2(0)\cos\alpha + y_1(0)\sin\alpha = 0,$$

(10.8) $$y_2(\pi)\cos\beta + y_1(\pi)\sin\beta = 0.$$

We will assume that the functions $p(x)$ and $r(x)$ are continuous on the interval $[0, \pi]$. Suppose that for some λ_1 the boundary-value problem under consideration has a nontrivial solution

$$y(x, \lambda_1) = \begin{pmatrix} y_1(x, \lambda_1) \\ y_2(x, \lambda_1) \end{pmatrix}.$$

In this case the number λ_1 is called an eigenvalue, and the corresponding solution $y(x, \lambda_1)$ is called an eigen-vector-function.

Lemma 10.1. *Eigen-vector-functions $y(x, \lambda_1)$ and $z(x, \lambda_2)$ corresponding to distinct eigenvalues $\lambda_1 \neq \lambda_2$ are orthogonal, i.e.*

$$\int_0^\pi \{y_1(x, \lambda_1) z_1(x, \lambda_2) + y_2(x, \lambda_1) z_2(x, \lambda_2)\} \, dx = 0.$$

Proof. Since the functions $y(x, \lambda_1)$ and $z(x, \lambda_2)$ are solutions of the system (10.6), we have

$$\begin{aligned} y_2'(x, \lambda_1) - \{\lambda_1 + p(x)\} y_1(x, \lambda_1) &= 0, \\ y_1'(x, \lambda_1) + \{\lambda_1 + r(x)\} y_2(x, \lambda_1) &= 0, \\ z_2'(x, \lambda_2) - \{\lambda_2 + p(x)\} z_1(x, \lambda_2) &= 0, \\ z_1'(x, \lambda_1) + \{\lambda_2 + r(x)\} z_2(x, \lambda_2) &= 0. \end{aligned}$$

Multiplying these equations by $z_1(x, \lambda_2)$, $-z_2(x, \lambda_2)$, $-y_1(x, \lambda_1)$ and $y_2(x, \lambda_1)$ respectively, and adding, we obtain

$$\frac{d}{dx}\{y_1(x, \lambda_1) z_2(x, \lambda_2) - y_2(x, \lambda_1) z_1(x, \lambda_2)\}$$
$$= (\lambda_1 - \lambda_2)\{y_1(x, \lambda_1) z_1(x, \lambda_2) + y_2(x, \lambda_1) z_2(x, \lambda_2)\}.$$

Integrating this relation from 0 to π, we find that

$$(\lambda_1 - \lambda_2) \int_0^\pi \{y_1(x, \lambda_1) z_1(x, \lambda_2) + y_2(x, \lambda_1) z_2(x, \lambda_2)\} \, dx$$
$$= \{y_1(x, \lambda_1) z_2(x, \lambda_2) - y_2(x, \lambda_1) z_1(x, \lambda_2)\}_0^\pi.$$

The right side vanishes by virtue of the boundary conditions (10.7) and (10.8). We therefore have

$$(\lambda_1 - \lambda_2) \int_0^\pi \{y_1(x, \lambda_1) z_1(x, \lambda_2) + y_2(x, \lambda_1) z_2(x, \lambda_2)\} \, dx = 0,$$

which proves the lemma since, by hypothesis, $\lambda_1 \neq \lambda_2$.

Lemma 10.2. *The eigenvalues of the boundary-value problem* (10.6) – (10.8) *are real.*

We leave the proof to the reader (cf. the proof of Lemma 1.2).

§11. Asymptotic formulas for the eigenvalues and eigen-vector-functions

1. We now take up the question of deriving asymptotic formulas for the eigenvalues and eigen-vector-functions of the boundary-value problem (10.6) – (10.8). From these formulas, as was the case for the Sturm-Liouville problem, will follow in particular the existence of an infinite set of eigenvalues.

In deriving asymptotic formulas for the eigenvalues and eigenfunctions of the Sturm-Liouville problem in §2, the use of the equalities (2.4) and (2.5) was essential. For the case of a Dirac system one can also write down a formula which is analogous to (2.4) and (2.5) and carry our further arguments along the same plan. Here, however, we will present a different method. In studying the asymptotic behaviour of the eigenvalues and eigen-vector-functions of a Dirac system it is very convenient to make use of so-called operator-matrix transformations, which are introduced along with some of their applications in Chapter 9. Here we will simply define these operators and obtain their final form.

Let A and B be two linear differential operators and $\mathscr{E}_1$, $\mathscr{E}_2$ two linear function spaces.

DEFINITION. A continuous linear operator X which maps $\mathscr{E}_1$ into $\mathscr{E}_2$ is called an *operator transformation* if it satisfies the following two conditions:

1°. $AX = XB$.

2°. A continuous inverse X^{-1} exists.

Let

$$A \equiv \begin{pmatrix} p_1(x) & \frac{d}{dx} \\ -\frac{d}{dx} & r_1(x) \end{pmatrix}, \qquad B \equiv \begin{pmatrix} p_2(x) & \frac{d}{dx} \\ -\frac{d}{dx} & r_2(x) \end{pmatrix}, \tag{11.1}$$

where $p_1(x)$, $p_2(x)$, $r_1(x)$ and $r_2(x)$ are real and continuous functions on the interval $[0, \pi]$.

Let $\mathscr{E}_1$ and $\mathscr{E}_2$ be the set of all continuously differentiable vector functions

$$f(x) = \begin{pmatrix} f_1(x) \\ f_2(x) \end{pmatrix} \text{ and } g(x) = \begin{pmatrix} g_1(x) \\ g_2(x) \end{pmatrix}$$

satisfying the respective boundary conditions

$$\begin{aligned} f_2(0)\cos\gamma + f_1(0)\sin\gamma &= 0, \\ g_2(0)\cos\delta + g_1(0)\sin\delta &= 0, \end{aligned} \tag{11.2}$$

where γ and δ are arbitrary real numbers.

As is proved in Chapter 9, §3, in this case the operator-matrix transformation X can be given in the form $g(x) = X[f(x)]$, i.e.

$$
\begin{aligned}
g_1(x) &= \alpha(x) f_1(x) + \beta(x) f_2(x) \\
&\quad + \int_0^x \{P(x, s) f_1(s) + R(x, s) f_2(s)\}\, ds, \\
g_2(x) &= -\beta(x) f_1(x) + \alpha(x) f_2(x) + \\
&\quad + \int_0^x \{Q(x, s) f_1(s) + H(x, s) f_2(s)\}\, dx,
\end{aligned}
\tag{11.3}
$$

where

$$
\begin{aligned}
\alpha(x) &= \frac{1}{\varkappa} \sin \left\{ \frac{1}{2} \int_0^x [p_1(\tau) + r_1(\tau) - \right. \\
&\qquad \left. - p_2(\tau) - r_2(\tau)]\, d\tau + \arcsin \frac{1}{\varkappa} \right\}, \\
\beta(x) &= \frac{1}{\varkappa} \cos \left\{ \frac{1}{2} \int_0^x [p_1(\tau) + r_1(\tau) - \right. \\
&\qquad \left. - p_2(\tau) - r_2(\tau)]\, d\tau + \arcsin \frac{1}{\varkappa} \right\}
\end{aligned}
\tag{11.4}
$$

($\kappa = \sec(\delta - \gamma)$), and the functions $P(x,s)$, $R(x,s)$, $Q(x,s)$ and $H(x,s)$ are continuously differentiable functions of their arguments.

2. Let us denote by

$$\varphi(x, \lambda) = \begin{pmatrix} \varphi_1(x, \lambda) \\ \varphi_2(x, \lambda) \end{pmatrix}$$

the solution of the system (10.6) satisfying the initial conditions

$$\varphi_1(0, \lambda) = \cos\alpha, \qquad \varphi_2(0, \lambda) = -\sin\alpha. \tag{11.5}$$

The function $\varphi(x, \lambda)$ obviously satisfies the boundary condition (10.7). Let us consider the problem (10.6) + (11.5) for $p(x) = r(x) \equiv 0$. As is not difficult to see, in this case

$$\varphi_1(x, \lambda) = \cos(\lambda x - \alpha), \qquad \varphi_2(x, \lambda) = \sin(\lambda x - \alpha). \tag{11.6}$$

We now apply an operator-transformation to the solution of the problem (10.6) + (10.7). The system (10.6) has the form

$$A_1 y = \begin{pmatrix} -p(x) & \dfrac{d}{dx} \\ -\dfrac{d}{dx} & -r(x) \end{pmatrix} y = \lambda y. \tag{11.7}$$

The function (11.6), i.e. the vector-function

$$\begin{pmatrix} \cos(\lambda x - \alpha) \\ \sin(\lambda x - \alpha) \end{pmatrix},$$

is a solution of the equation

$$B_1 y = \begin{pmatrix} 0 & \frac{d}{dx} \\ -\frac{d}{dx} & 0 \end{pmatrix} y = \lambda y. \tag{11.8}$$

If $\psi(x, \lambda)$ is a solution of the equation $B_1\psi = \lambda\psi$, then by virtue of the definition of an operator-transformation X we have

$$A_1 X[\psi] = XB_1[\psi] = X[\lambda\psi] = \lambda X[\psi],$$

i.e. $\varphi = X[\psi]$ is a solution of the equation $A_1[\varphi] = \lambda\varphi$. Thus, if

$$\psi(x, \lambda) = \begin{pmatrix} \cos(\lambda x - \alpha) \\ \sin(\lambda x - \alpha) \end{pmatrix}$$

is a solution of the equation (11.8), then the vector-function $\varphi(x, \lambda) = X[\psi(x, \lambda)]$ is a solution of the equation (11.7). In our case $p_1(x) = -p(x)$, $r_1(x) = -r(x)$, $p_2(x) = r_2(x) \equiv 0$. Further, since the boundary conditions (11.2) are now replaced by the boundary condition (10.7), we have $\gamma = \delta$, so that $\kappa = 1$. Therefore by (11.4) we obtain for the functions $\alpha(x)$ and $\beta(x)$ the expressions

$$\begin{aligned} \alpha(x) &= \cos\left\{\frac{1}{2}\int_0^x [p(\tau) + r(\tau)]\, d\tau\right\}, \\ \beta(x) &= \sin\left\{\frac{1}{2}\int_0^x [p(\tau) + r(\tau)]\, d\tau\right\}. \end{aligned} \tag{11.9}$$

From formula (11.3) for the solution

$$\varphi(x, \lambda) = \begin{pmatrix} \varphi_1(x, \lambda) \\ \varphi_2(x, \lambda) \end{pmatrix}$$

of the problem (10.6) + (11.5) we obtain the following formulas:

$$\begin{aligned} \varphi_1(x, \lambda) = {} & \alpha(x)\cos(\lambda x - \alpha) + \beta(x)\sin(\lambda x - \alpha) \\ & + \int_0^x \{P(x, s)\cos(\lambda s - \alpha) + R(x, s)\sin(\lambda s - \alpha)\}\, ds \end{aligned}$$

and

$$\varphi_2(x, \lambda) = \alpha(x) \sin(\lambda x - \alpha) - \beta(x) \cos(\lambda x - \alpha)$$
$$+ \int_0^x \{Q(x, s) \cos(\lambda x - \alpha) + H(x, s) \sin(\lambda s - \alpha)\}\, ds$$

or, inserting the values for $\alpha(x)$ and $\beta(x)$ from (11.9),

$$\varphi_1(x, \lambda) = \cos\{\xi(x, \lambda) - \alpha\} + \int_0^x \{P(x, s) \cos(\lambda s - \alpha) + R(x, s) \sin(\lambda s - \alpha)\}\, ds, \tag{11.10}$$

$$\varphi_2(x, \lambda) = \sin\{\xi(x, \lambda) - \alpha\} + \int_0^x \{Q(x, s) \cos(\lambda s - \alpha) + H(x, s) \sin(\lambda s - \alpha)\}\, ds, \tag{11.11}$$

where

$$\xi(x, \lambda) = \lambda x - \frac{1}{2} \int_0^x [p(\tau) + r(\tau)]\, d\tau. \tag{11.12}$$

LEMMA 11.1. *For $|\lambda| \to \infty$ the following estimates hold uniformly with respect to x, $0 \leqq x \leqq \pi$:*

$$\begin{aligned} \varphi_1(x, \lambda) &= \cos\{\xi(x, \lambda) - \alpha\} + O\left(\frac{1}{\lambda}\right), \\ \varphi_2(x, \lambda) &= \sin\{\xi(x, \lambda) - \alpha\} + O\left(\frac{1}{\lambda}\right), \end{aligned} \tag{11.13}$$

$$\begin{aligned} \frac{\partial \varphi_1}{\partial \lambda} &= -x \sin\{\xi(x, \lambda) - \alpha\} + O(1), \\ \frac{\partial \varphi_2}{\partial \lambda} &= x \cos\{\xi(x, \lambda) - \alpha\} + O(1). \end{aligned} \tag{11.14}$$

PROOF. To obtain the estimates (11.13) it suffices to integrate by parts the integrals in the right sides of (11.10) and (11.11), which is possible by virtue of the differentiability of the functions $P(x,s)$, $R(x,s)$, $Q(x,s)$ and $H(x,s)$. The estimates (11.14) can be proved in a similar way, if one first differentiates (11.10) and (11.11) with respect to λ. This proves the lemma.

LEMMA 11.2. *The eigenvalues of the boundary-value problem* (10.6) + (10.7) + (10.8) *are just the zeros of the function* $\varphi_2(\pi, \lambda) \cos\beta + \varphi_1(\pi, \lambda) \sin\beta$. *Moreover, every zero of this function has multiplicity* 1.

PROOF. Since the functions $\varphi_1(x, \lambda)$ and $\varphi_2(x, \lambda)$ satisfy the boundary condition (10.7), to find the eigenvalues of the boundary-value problem (10.6) – (10.8) we have to insert the functions $\varphi_1(x, \lambda)$ and $\varphi_2(x, \lambda)$ in the boundary condition (10.8) and find the roots of this equation. Put

$$D(\lambda) = \varphi_2(\pi, \lambda)\cos\beta + \varphi_1(\pi, \lambda)\sin\beta;$$

then

$$\frac{dD(\lambda)}{d\lambda} = \frac{\partial\varphi_2(\pi, \lambda)}{\partial\lambda}\cos\beta + \frac{\partial\varphi_1(\pi, \lambda)}{\partial\lambda}\sin\beta.$$

Suppose that λ_1 is a root of multiplicity two, and let

$$\begin{pmatrix}\varphi_1(x, \lambda_1)\\ \varphi_2(x, \lambda_1)\end{pmatrix}$$

be one of the corresponding eigen-vector-functions. Then the conditions $D(\lambda_1) = 0$ and $dD_1(\lambda_1)/d\lambda = 0$ must hold simultaneously, i.e.

$$\varphi_2(\pi, \lambda_1)\cos\beta + \varphi_1(\pi, \lambda_1)\sin\beta = 0,$$
$$\frac{\partial\varphi_2(\pi, \lambda_1)}{\partial\lambda}\cos\beta + \frac{\partial\varphi_1(\pi, \lambda_1)}{\partial\lambda}\sin\beta = 0.$$

These equalities imply that

$$\varphi_2(\pi, \lambda_1)\frac{\partial\varphi_1(\pi, \lambda_1)}{\partial\lambda} - \varphi_1(\pi, \lambda_1)\frac{\partial\varphi_2(\pi, \lambda_1)}{\partial\lambda} = 0. \tag{11.15}$$

Differentiating the system (10.6) with respect to λ, we obtain

$$\begin{aligned}\left(\frac{\partial y_2}{\partial\lambda}\right)'_x - \{\lambda + p(x)\}\frac{\partial y_1}{\partial\lambda} &= y_1,\\ \left(\frac{\partial y_1}{\partial\lambda}\right)'_x + \{\lambda + r(x)\}\frac{\partial y_2}{\partial\lambda} &= -y_2\end{aligned} \tag{11.16}$$

Now multiplying the equations (10.6) and (11.16) by $\partial y_1/\partial\lambda$, $-\partial y_2/\partial\lambda$, $-y_1$ and y_2 respectively, adding and then integrating with respect to x from 0 to π, we obtain

$$\left\{y_2(x, \lambda)\frac{\partial y_1}{\partial\lambda} - y_1(x, \lambda)\frac{\partial y_2}{\partial\lambda}\right\}_0^\pi = -\int_0^\pi \{y_1^2(x, \lambda) + y_2^2(x, \lambda)\}\,dx.$$

Putting $\lambda = \lambda_1$, taking into account that by virtue of (11.5)

$$\left.\frac{\partial\varphi_1(x, \lambda_1)}{\partial\lambda}\right|_{x=0} = \left.\frac{\partial\varphi_2(x, \lambda_1)}{\partial\lambda}\right|_{x=0} = 0,$$

and using (11.15), we obtain the relation

$$-\int_0^\pi \{\varphi_1^2(x, \lambda_1) + \varphi_2^2(x, \lambda_1)\}\,dx =$$
$$= \varphi_2(\pi, \lambda_1)\frac{\partial\varphi_1(\pi, \lambda_1)}{\partial\lambda} - \varphi_1(\pi, \lambda_1)\frac{\partial\varphi_2(\pi, \lambda_1)}{\partial\lambda} = 0.$$

It follows that $\varphi_1(x,\lambda_1)=\varphi_2(x,\lambda_1)\equiv 0$, which is impossible. This proves the lemma.

As was already mentioned above, the eigenvalues of the boundary-value problem (10.6)–(10.8) coincide with the roots of the equation

$$\varphi_2(\pi,\lambda)\cos\beta+\varphi_1(\pi,\lambda)\sin\beta=0.$$

Inserting now the values of the functions $\varphi_1(\pi,\lambda)$ and $\varphi_2(\pi,\lambda)$ from the estimates (11.13), we obtain

$$\sin(\lambda\pi-\vartheta)+O(1/\lambda)=0, \tag{11.16'}$$

where by virtue of (11.12)

$$\vartheta=\beta-\alpha-\frac{1}{2}\int_0^\pi\{p(\tau)+r(\tau)\}\,d\tau. \tag{11.17}$$

For large $|\lambda|$ equation (11.16′) obviously has solutions which, as is not hard to see, have the form[15)]

$$\pi\lambda_n-\vartheta=n\pi+\delta_n \qquad (n=0,\pm1,\pm2,\cdots).$$

Inserting these values in (11.16′), we find that $\sin\delta_n=O(1/n)$, i.e. $\delta_n=O(1/n)$. Therefore we obtain the following asymptotic formula for the eigenvalues:

$$\lambda_n=n+\frac{\vartheta}{\pi}+O\left(\frac{1}{n}\right) \qquad (n=0,\ \pm1,\ \pm2,\ldots), \tag{11.18}$$

where ϑ is defined by (11.17).

Using the formula (11.18), we obtain an asymptotic formula for the eigen-vector-functions $\varphi_1(x,\lambda_n)=u_n(x)$, $\varphi_2(x,\lambda_n)=v_n(x)$, namely:

$$u_n(x)=\cos(\xi_n-\alpha)+O\left(\frac{1}{n}\right), \tag{11.19}$$

$$v_n(x)=\sin(\xi_n-\alpha)+O\left(\frac{1}{n}\right), \tag{11.20}$$

where

$$\xi_n=\xi(x,\ \lambda_n)=\lambda_n x-\frac{1}{2}\int_0^x\{p(\tau)+r(\tau)\}\,d\tau.$$

To obtain an asymptotic decomposition of the normalized eigen-

15) This fact can be proved in the same way as for the case of a Sturm-Liouville operator (cf. §2).

vector-functions, we consider the integral

$$\alpha_n^2 = \int_0^\pi \{u_n^2(x) + v_n^2(x)\}\, dx = \pi + O\left(\frac{1}{n}\right).$$

Consequently the normalized eigen-vector-functions have the form

$$u_n(x) = \frac{1}{\sqrt{\pi}} \cos(\xi_n - \alpha) + O\left(\frac{1}{n}\right),$$

$$v_n(x) = \frac{1}{\sqrt{\pi}} \sin(\xi_n - \alpha) + O\left(\frac{1}{n}\right).$$

§12. Proof of the expansion theorem by the method of finite differences

We will consider the boundary-value problem (10.6) + (10.7) + (10.8). As before, we assume that the functions $p(x)$ and $r(x)$ are real and continuous over the interval $[0, \pi]$, while α and β are arbitrary real numbers.

Let us denote by n an arbitrary positive integer and put $h = \pi/n$. Further, we put $x_\nu = \nu h$, $p_\nu = p(\nu h)$, $r_\nu = r(\nu h)$, and let $y_{1,\nu}$ and $y_{2,\nu+1}$ $(\nu = 0, 1, \cdots, n-1)$ be the components of vectors which will be defined below. We replace the derivatives of the functions $y_1(x)$ and $y_2(x)$ by difference ratios:

$$y_1'(x) \sim \frac{y_{1,\nu} - y_{1,\nu-1}}{h}, \qquad y_2'(x) \sim \frac{y_{2,\nu+1} - y_{2,\nu}}{h} \tag{12.1}$$

$$(\nu = 0,\ 1, \ldots, n-1).$$

If $y_1 = y_1(x, \lambda)$ and $y_2 = y_2(x, \lambda)$ form a solution of the system (10.6) satisfying the initial conditions

$$y_1(0) = \cos\alpha, \tag{12.2}$$

$$y_2(0) = -\sin\alpha, \tag{12.3}$$

then the functions $y_1(x, \lambda)$ and $y_2(x, \lambda)$ obviously satisfy the boundary condition (10.7). Therefore from condition (12.3) we obtain for $y_{2,0}$

$$y_{2,0} = -\sin\alpha. \tag{12.4}$$

Put

$$y_{1,-1} = \cos\alpha. \tag{12.5}$$

Replacing the derivatives in the equations of the system (10.6) by the expressions (12.1), we obtain a system of finite-difference equations:

$$\begin{aligned}&\frac{1}{h}(y_{2,\nu+1}-y_{2,\nu})-(\lambda+p_\nu)\,y_{1,\nu}=0,\\ (12.6)\qquad &\frac{1}{h}(y_{1,\nu}-y_{1,\nu-1})+(\lambda+r_\nu)\,y_{2,\nu}=0\\ &(\nu=0,\ 1,\ \ldots,\ n-1).\end{aligned}$$

From the equations (12.6) and the initial conditions (12.4) – (12.5) we can successively determine $y_{1,0}, y_{1,1}, \cdots, y_{1,n-1}$ and $y_{2,1}, y_{2,2}, \cdots, y_{2,n}$. As is known, for $n \to \infty$ and $\nu h \to x$ the quantities $y_{1,\nu}$ and $y_{2,\nu}$ tend to the solution of the problem (10.6) + (12.4) + (12.5), and indeed uniformly in every region $(0 \leqq x \leqq \pi;\ 0 \leqq |\lambda| \leqq R < \infty)$. Let us now adjoin to the problem (12.6) + (12.4) + (12.5) the boundary condition

$$(12.7)\qquad y_{2,n}\cos\beta + y_{1,n-1}\sin\beta = 0,$$

which is obtained from the boundary condition (10.8). The problem thereby obtained for the finite-difference equations is obviously an algebraic eigenvalue problem in $2n$-dimensional vector space. It follows from the conditions (12.4), (12.5) and (12.7) that[16]

$$(12.8)\qquad y_{1,-1} = -y_{2,0}\operatorname{ctg}\alpha,$$

$$(12.9)\qquad y_{1,n-1} = -y_{2,n}\operatorname{ctg}\beta.$$

Inserting these values in the system (12.6), we obtain a homogeneous system of equations with $2n$ unknowns: $y_{1,0}, y_{1,1}, \cdots, y_{1,n-1};\ y_{2,1}, y_{2,2}, \cdots, y_{2,n}$.

LEMMA 12.1. *The eigenvalues of the problem* (12.4) – (12.7) *are simple.*

PROOF. Suppose that λ is an eigenvalue of multiplicity greater than 1. Then there must exist at least two corresponding linearly independent eigenvectors:[17] $y = \{y_{1,0}, y_{1,1}, \cdots, y_{1,n-1};\ y_{2,1}, y_{2,2}, \cdots, y_{2,n}\}$ and $z = \{z_{1,0}, z_{1,1}, \cdots, z_{1,n-1};\ z_{2,1}, z_{2,2}, \cdots, z_{2,n}\}$. We can assume that $y_{1,-1} \neq 0$, $z_{1,-1} \neq 0$, since if, say, $y_{1,-1} = 0$, then by (12.8) we have $y_{2,0} = 0$ and therefore we obtain from the system (12.6) $y_{1,\nu-1} = y_{2,\nu} = 0$ for $\nu = 0, 1, \cdots, n$. Moreover, since the system (12.6) is homogeneous, we may assume without loss of generality that $y_{1,-1} = z_{1,-1}$. But then from (12.8) we obtain $y_{2,0} = z_{2,0}$, and so it follows from the system (12.6) that $y_{1,\nu-1} = z_{1,\nu-1}$, $y_{2,\nu} = z_{2,\nu}$ $(\nu = 0, 1, \cdots, n)$. This proves the lemma.

We define the scalar product of two vectors y and z by

[16] If $\operatorname{ctg}\alpha = \infty$, then one has to put $y_{2,0} = 0$; if $\operatorname{ctg}\beta = \infty$, we put $y_{2,n} = 0$.

[17] The algebraic eigenvalue problem under consideration is selfadjoint (cf. the Remark on p. 61).

$$(y, z) = \sum_{\nu=0}^{n} (y_{1,\nu} z_{1,\nu} + y_{2,\nu} z_{2,\nu})\, h. \tag{12.10}$$

Lemma 12.2. *Eigenvectors corresponding to distinct eigenvalues are orthogonal.*

Proof. According to (12.6) we have, on the one hand,

$$y_{2,\nu+1} - y_{2,\nu} - p_\nu h y_{1,\nu} = \lambda h y_{1,\nu}, \tag{12.11}$$

$$-y_{1,\nu} + y_{1,\nu-1} - r_\nu h y_{2,\nu} = \lambda h y_{2,\nu}, \tag{12.12}$$

and on the other hand

$$z_{2,\nu+1} - z_{2,\nu} - p_\nu h z_{1,\nu} = \mu h z_{1,\nu}, \tag{12.13}$$

$$-z_{1,\nu} + z_{1,\nu-1} - r_\nu h z_{2,\nu} = \mu h z_{2,\nu}. \tag{12.14}$$

Multiplying equation (12.11) by $z_{1,\nu}$ and equation (12.13) by $y_{1,\nu}$ and then subtracting, we obtain

$$z_{1,\nu} y_{2,\nu+1} - z_{1,\nu} y_{2,\nu} - y_{1,\nu} z_{2,\nu+1} + y_{1,\nu} z_{2,\nu} = h(\lambda - \mu)\, y_{1,\nu} z_{1,\nu}. \tag{12.15}$$

Further, multiplying equation (12.12) by $z_{2,\nu}$ and equation (12.14) by $y_{2,\nu}$ and then subtracting, we obtain

$$-z_{2,\nu} y_{1,\nu} + z_{2,\nu} y_{1,\nu-1} + y_{2,\nu} z_{1,\nu} - y_{2,\nu} z_{1,\nu-1} = h(\lambda - \mu)\, y_{2,\nu} z_{2,\nu}. \tag{12.16}$$

Now letting ν in equations (12.15) and (12.16) take the values $0, 1, \cdots, n-1$ and adding the $2n$ equations thereby obtained, we find that

$$\begin{aligned} &(\lambda - \mu) \sum_{\nu=0}^{n-1} h\,(y_{1,\nu} z_{1,\nu} + y_{2,\nu} z_{2,\nu}) \\ &\quad = \sum_{\nu=0}^{n-1} (z_{1,\nu} y_{2,\nu+1} - y_{1,\nu} z_{2,\nu+1} + z_{2,\nu} y_{1,\nu-1} - y_{2,\nu} z_{1,\nu-1}). \end{aligned} \tag{12.17}$$

By virtue of (12.10) the equality (12.17) can be rewritten in the form

$$\begin{aligned} (\lambda - \mu)(y, z) &= \sum_{\nu=0}^{n-1} (z_{1,\nu} y_{2,\nu+1} - y_{1,\nu} z_{2,\nu+1}) \\ &\quad + \sum_{\nu=0}^{n-1} (z_{2,\nu} y_{1,\nu-1} - y_{2,\nu} z_{1,\nu-1}). \end{aligned} \tag{12.18}$$

We transform the right side of this equality:

$$\begin{aligned} \sum_{\nu=0}^{n-1} (z_{1,\nu} y_{2,\nu+1} - y_{1,\nu} z_{2,\nu+1}) &= z_{1,n-1} y_{2,n} - z_{2,n} y_{1,n-1} \\ - \sum_{\nu=0}^{n-2} (y_{1,\nu} z_{2,\nu+1} - z_{1,\nu} y_{2,\nu+1}) &= z_{1,n-1} y_{2,n} - z_{2,n} y_{1,n-1} \\ &\quad - \sum_{\nu=1}^{n-1} (y_{1,\nu-1} z_{2,\nu} - y_{2,\nu} z_{1,\nu-1}). \end{aligned} \tag{12.19}$$

We now transform the second sum in (12.18):

$$\sum_{\nu=0}^{n-1} (z_{2,\nu}y_{1,\nu-1} - y_{2,\nu}z_{1,\nu-1}) = z_{2,0}y_{1,-1} - y_{2,0}z_{1,-1} + \sum_{\nu=1}^{n-1} (y_{1,\nu-1}z_{2,\nu} - y_{2,\nu}z_{1,\nu-1}). \tag{12.20}$$

By virtue of (12.19) and (12.20), (12.18) assumes the form

$$(\lambda-\mu)(y-z) = z_{2,0}y_{1,-1} - y_{2,0}z_{1,-1} + z_{1,n-1}y_{2,n} - z_{2,n}y_{1,n-1}.$$

Now using the conditions (12.8) and (12.9), we obtain $(\lambda-\mu)(y,z) = 0$, which proves the lemma since $\lambda \neq \mu$.

REMARK. Let

$$y_\nu = \begin{pmatrix} y_{1,\nu} \\ y_{2,\nu} \end{pmatrix}, \quad \nu = 0, 1, \ldots, n-1,$$

be an arbitrary $2n$-dimensional vector satisfying the conditions (12.8) – (12.9). We define the operator L_h,

$$L_h(y_\nu) = t_\nu \equiv \begin{pmatrix} t_{1,\nu} \\ t_{2,\nu} \end{pmatrix},$$

by means of the relations

$$t_{1,\nu} = \frac{1}{h}(y_{2,\nu+1} - y_{2,\nu}) - p_\nu y_{1,\nu},$$

$$t_{2,\nu} = \frac{1}{h}(y_{1,\nu} - y_{1,\nu-1}) + r_\nu y_{2,\nu}.$$

Let z_ν be another $2n$-dimensional vector satisfying the conditions (12.8) – (12.9).

It follows from the previous computations that $(L_h y_\nu, z_\nu) = (y_\nu, L_h z_\nu)$, i.e. the operator L_h is selfadjoint.

Of course one can also see this by considering the matrix of L_h.

LEMMA 12.3. *The eigenvalues of the problem* (12.4) – (12.7) *are real.*

The proof is left to the reader.

We now turn to the proof of the expansion theorem. Here, as in the case of a Sturm-Liouville operator, we will first prove the Parseval equality, from which the expansion theorem can be obtained by the usual methods.

Let us denote the eigenvalues of the problem (12.4) – (12.7) in the order of increasing absolute values by $\lambda_{k,h}$ $(k = \pm 1, \pm 2, \cdots, \pm n)$, and corresponding eigenvectors by

$$y_{k,h}=\{y_{1,0}^{(k,h)},\ \ldots,\ y_{1,n-1}^{(k,h)};\ y_{2,0}^{(k,h)},\ \ldots,\ y_{2,n-1}^{(k,h)}\}.$$

We further denote the square of the norm of the vector $y_{k,h}$ by $\alpha_{k,h}^2$, i.e. we put

$$\alpha_{k,h}^2=\|y_{k,h}\|^2=(y_{k,h},\ y_{k,h})=\sum_{\nu=0}^{n-1} h\,\{[y_{1,\nu}^{(k,h)}]^2+[y_{2,\nu}^{(k,h)}]^2\}.$$

The vectors $y_{\pm 1,h}, y_{\pm 2,h}, \cdots, y_{\pm n,h}$ form an orthogonal basis in $2n$-dimensional vector space. Therefore we have, for any vector $f=\{f_{1,0}, f_{1,1}, \cdots, f_{1,n-1}; f_{2,0}, f_{2,1}, \cdots, f_{2,n-1}\}$, the Parseval equality

$$(12.21)\qquad (f,\ f)=\sum_{\nu=0}^{n-1} h\,(f_{1,\nu}^2+f_{2,\nu}^2)=\sum_{k=-n}^{n}{}' \frac{1}{\alpha_{k,h}^2}(f,\ y_{k,h})^2 .$$ [18]

Let us denote by

$$y_\nu(\lambda)=\begin{pmatrix} y_{1,\nu} \\ y_{2,\nu}\end{pmatrix}$$

the solution of the system (12.6) satisfying the initial conditions (12.4) – (12.5). If $h\rightarrow 0$ and $\nu h\rightarrow x$, then

$$y_\nu(\lambda)\rightarrow y(x,\ \lambda)=\begin{pmatrix} y_1(x,\ \lambda) \\ y_2(x,\ \lambda)\end{pmatrix}$$

uniformly in every bounded region of the λ-plane. It follows that the roots of equation (12.7) tend to the successive roots of the equation (10.8). The boundedness of the number of eigenvalues $\lambda_{k,h}$ in every fixed interval $(-N, N)$ can be proved in the same way as for the case of a Sturm-Liouville operator (cf. the Remark in §5), if for the vector-function

$$f(x)=\begin{pmatrix} f_1(x) \\ f_2(x)\end{pmatrix}$$

we take the function

$$f_{1,\delta}(x)\equiv 0,\qquad f_{2,\delta}(x)=\begin{cases}\dfrac{1}{\delta}, & 0\leqslant x<\delta,\\ 0, & \delta<x\leqslant\pi.\end{cases}$$

We will show that the Parseval equality for the problem (10.6) – (10.8) can be obtained from (12.21) by letting $h\rightarrow 0$.

---

[18] The $'$ indicates that we omit the value $k = 0$.

Suppose that the vector-function $f(x)$ $(0 \leq x \leq \pi)$ has a continuous first derivative and vanishes in neighborhoods of the points $x=0$ and $x=\pi$. We put $f_1(\nu h)=f_{1,\nu}$, $f_2(\nu h)=f_{2,\nu}$ $(\nu=0,1,\cdots,n-1)$ and write the Parseval equality (12.21) for the vector $f=\{f_{1,0},f_{1,1},\cdots,f_{1,n-1};$ $f_{2,0},f_{2,1},\cdots,f_{2,n-1}\}$:

$$
\begin{aligned}
(f,f) &= \sum_{\nu=0}^{n-1} h\{f_{1,\nu}^2+f_{2,\nu}^2\} = \sum_{k=-n}^{n} \frac{1}{\alpha_{k,h}^2}(f,y_{k,h})^2 \\
&= \sum_{|\lambda_{k,h}|\leqslant N} \frac{1}{\alpha_{k,h}^2}(f,y_{k,h})^2 + \sum_{|\lambda_{k,h}|>N} \frac{1}{\alpha_{k,h}^2}(f,y_{k,h})^2 = I_1+I_2,
\end{aligned} \tag{12.22}
$$

where N is an arbitrary positive number. Let us estimate I_1. To this end we transform the Fourier coefficients:[19]

$$
\begin{aligned}
(f,y_{k,h}) &= \sum_{\nu=0}^{n-1} h\{f_{1,\nu}y_{1,\nu}^{(k,h)}+f_{2,\nu}y_{2,\nu}^{(k,h)}\} \\
&= \frac{1}{\lambda_{k,h}}\sum_{\nu=0}^{n-1} hf_{1,\nu}\left\{\frac{y_{2,\nu+1}^{(k,h)}-y_{2,\nu}^{(k,h)}}{h}-p_\nu y_{1,\nu}^{(k,h)}\right\} \\
&\quad+\frac{1}{\lambda_{k,h}}\sum_{\nu=0}^{n-1} hf_{2,\nu}\left\{-\frac{y_{1,\nu}^{(k,h)}-y_{1,\nu-1}^{(k,h)}}{h}+r_\nu y_{2,\nu}^{(k,h)}\right\} \\
&= \frac{1}{\lambda_{k,h}}\left(-y_{1,n-1}^{(k,h)}f_{2,n}+y_{2,n}^{(k,h)}f_{1,n-1}+y_{1,-1}^{(k,h)}f_{2,0}-y_2^{(k,h)}f_{1,-1}\right) \\
&\quad+\frac{1}{\lambda_{k,h}}\sum_{\nu=0}^{n-1} hy_{1,\nu}^{(k,h)}\left\{\frac{f_{2,\nu+1}-f_{2,\nu}}{h}-p_\nu f_{1,\nu}\right\} \\
&\quad+\frac{1}{\lambda_{k,h}}\sum_{\nu=0}^{n-1} hy_{2,\nu}^{(k,h)}\left\{-\frac{f_{1,\nu}-f_{1,\nu-1}}{h}+r_\nu f_{2,\nu}\right\}.
\end{aligned} \tag{12.23}
$$

If h is sufficiently small, then $f_{2,0}=f_{2,n}=f_{1,n-1}=0$. We also put $f_{1,-1}=0$. Further, let the vector $g=\{g_{1,0},g_{1,1},\cdots,g_{1,n-1};\ g_{2,0},g_{2,1},\cdots,$ $g_{2,n-1}\}$ be defined by

$$
\begin{gathered}
g_{1,\nu}=\frac{1}{h}(f_{2,\nu+1}-f_{2,\nu})-p_\nu f_{1,\nu}, \\
g_{2,\nu}=-\frac{1}{h}(f_{1,\nu}-f_{1,\nu-1})+r_\nu f_{2,\nu} \\
(\nu=0,\ 1,\ \ldots,\ n-1).
\end{gathered}
$$

Then (12.23) can be written in the form

---

[19] The following transformations are analogous to the integration by parts of a definite integral.

$$(f,\ y_{k,h})=\frac{1}{\lambda_{k,h}}\sum_{\nu=0}^{n-1}h\,(y_{1,\nu}^{(k,h)}g_{1,\nu}+y_{2,\nu}^{(k,h)}g_{2,\nu})=\frac{1}{\lambda_{k,h}}(g,\ y_{k,h}).$$

Therefore by Bessel's inequality

$$I_2\leqslant\frac{1}{N^2}\sum_{|\lambda_{k,h}|>N}\frac{1}{\alpha_{k,h}^2}(g,\ y_{k,h})^2=\frac{1}{N^2}\sum_{k=-n}^{n}{}'\frac{1}{\alpha_{k,h}^2}(g,\ y_{k,h})^2$$

$$(12.24)\qquad =\frac{1}{N^2}\|g\|^2.$$

If $h\to 0$, then

$$\|g\|^2=\sum_{\nu=0}^{n-1}h\,(g_{1,\nu}^2+g_{2,\nu}^2)=\sum_{\nu=0}^{n-1}h\left\{\left[\frac{f_{2,\nu+1}-f_{2,\nu}}{h}-p_\nu f_{1,\nu}\right]^2\right.$$

$$\left.+\left[-\frac{f_{1,\nu}-f_{1,\nu-1}}{h}+r_\nu f_{2,\nu}\right]^2\right\}.$$

$$\to\int_0^\pi\left\{\left[\frac{df_2(x)}{dx}-p(x)f_1(x)\right]^2+\left[-\frac{df_1(x)}{dx}+r(x)f_2(x)\right]^2\right\}dx.$$

Therefore it follows from the inequality (12.24) that there exists a constant $A>0$ such that, uniformly as $h\to 0$, $I_2<A/N^2$. Consequently, we obtain from (12.22)

$$(12.25)\qquad \left|\sum_{\nu=0}^{n-1}h\,\{f_{1,\nu}^2+f_{2,\nu}^2\}-\sum_{|\lambda_{k,h}|\leqslant N}\frac{1}{\alpha_{k,h}^2}(f,\ y_{k,h})^2\right|<\frac{A}{N^2}.$$

Since as $h\to 0$

$$\sum_{\nu=0}^{n-1}h\,\{f_{1,\nu}^2+f_{2,\nu}^2\}\to\int_0^\pi\{f_1^2(x)+f_2^2(x)\}\,dx,$$

$$\alpha_{k,h}^2=\sum_{\nu=0}^{n-1}h\,\{[y_{1,\nu}^{(k,h)}]^2+[y_{2,\nu}^{(k,h)}]^2\}\to$$

$$\to\int_0^\pi\{y_1^2(x,\ \lambda_k)+y_2^2(x,\ \lambda_k)\}\,dx=\alpha_k^2,$$

$$(f,\ y_{k,h})=\sum_{\nu=0}^{n-1}h\,\{f_{1,\nu}y_{1,\nu}^{(k,h)}+f_{2,\nu}y_{2,\nu}^{(k,h)}\}$$

$$\to\int_0^\pi\{f_1(x)\,y_1(x,\ \lambda_k)+f_2(x)\,y_2(x,\ \lambda_k)\}\,dx,$$

where λ_k is the kth zero (in the order of increasing absolute value) of the equation (10.8), and since the number of roots $\lambda_{k,h}$ in a fixed interval $(-N, N)$ is bounded uniformly in h, passing to the limit $h \to 0$ in (12.25) we obtain

$$\left| \int_0^\pi \{f_1^2(x) + f_2^2(x)\}\, dx - \sum_{|\lambda_k| \leq N} \frac{1}{\alpha_k^2} \left\{ \int_0^\pi [f_1(x)\, y_1(x, \lambda_k) + f_2(x)\, y_2(x, \lambda_k)]\, dx \right\}^2 \right| < \frac{A}{N^2}.$$

Letting $N \to \infty$, it follows from this inequality that

$$\int_0^\pi \{f_1^2(x) + f_2^2(x)\}\, dx = \sum_{k=-\infty}^{\infty} \frac{1}{\alpha_k^2} \left\{ \int_0^\pi [f_1(x)\, y_1(x, \lambda_k) + f_2(x)\, y_2(x, \lambda_k)]\, dx \right\}^2, \tag{12.26}$$

which is the Parseval equality for functions which satisfy the conditions indicated earlier.[20] Extending the Parseval equality to arbitrary square-integrable vector-functions can be carried out in the same way as for the case of a Sturm-Liouville operator (cf. §5.3).

Let us denote by

$$\begin{pmatrix} u_k(x) \\ v_k(x) \end{pmatrix}$$

normalized eigen-vector-functions of the problem (10.6) + (10.7) + (10.8), i.e. we put

$$u_k(x) = \frac{1}{\alpha_k} y_1(x, \lambda_k), \qquad v_k(x) = \frac{1}{\alpha_k} y_2(x, \lambda_k).$$

Then the Parseval equality (12.26) assumes the form

$$\int_0^\pi \{f_1^2(x) + f_2^2(x)\}\, dx = \sum_{k=-\infty}^{\infty} a_k^2, \tag{12.26$'$}$$

where the a_k are the Fourier coefficients of the vector-function $f(x)$ with respect to the normalized eigen-vector-functions $\begin{pmatrix} u_k(x) \\ v_k(x) \end{pmatrix}$, i.e.

$$a_k = \int_0^\pi \{f_1(x)\, u_k(x) + f_2(x)\, v_k(x)\}\, dx.$$

---

[20] The $'$ on the summation can now be left out by suitably changing the enumeration of the eigenvalues.

THEOREM 12.1 (EXPANSION THEOREM). *If for a continuous vector-function*

$$f(x) = \begin{pmatrix} f_1(x) \\ f_2(x) \end{pmatrix}$$

the series

$$\sum_{k=-\infty}^{\infty} a_k u_k(x) \quad \text{and} \quad \sum_{k=-\infty}^{\infty} a_k v_k(x)$$

converge uniformly in the interval $[0, \pi]$, *then they represent the functions* $f_1(x)$ *and* $f_2(x)$ *respectively, i.e.*

$$f_1(x) = \sum_{k=-\infty}^{\infty} a_k u_k(x), \qquad f_2(x) = \sum_{k=-\infty}^{\infty} a_k v_k(x). \tag{12.27}$$

PROOF. Put

$$h_1(x) = f_1(x) - \sum_{k=-\infty}^{\infty} a_k u_k(x), \quad h_2(x) = f_2(x) - \sum_{k=-\infty}^{\infty} a_k v_k(x). \tag{12.28}$$

By hypothesis the vector-function

$$h(x) = \begin{pmatrix} h_1(x) \\ h_2(x) \end{pmatrix}$$

is continuous in $[0, \pi]$. Further, computing the Fourier coefficients of $h(x)$, it follows from (12.28) that

$$\begin{aligned} c_k &= \int_0^{\pi} \{h_1(x) u_k(x) + h_2(x) v_k(x)\}\, dx \\ &= \int_0^{\pi} \{f_1(x) u_k(x) - \sum_{n=-\infty}^{\infty} a_n u_n(x) u_k(x) + f_2(x) v_k(x) \\ &\quad - \sum_{n=-\infty}^{\infty} a_n v_n(x) v_k(x)\}\, dx = \int_0^{\pi} \{f_1(x) u_k(x) + f_2(x) v_k(x)\}\, dx \\ &\quad - \sum_{n=-\infty}^{\infty} a_n \int_0^{\pi} \{u_n(x) u_k(x) + v_n(x) v_k(x)\}\, dx = a_k - a_k = 0 \\ &\qquad (k = 0, \pm 1, \pm 2, \ldots). \end{aligned}$$

Then by the Parseval equality

$$\int_0^{\pi} \{h_1^2(x) + h_2^2(x)\}\, dx = 0,$$

i.e. $h(x) \equiv 0$. Therefore (12.27) follows from (12.28).

§13. Proof of the expansion theorem by the method of integral equations

1. In this section we will show that, as in the case of a Sturm-Liouville operator, one can prove the completeness of the eigenvector-functions of a Dirac system on a finite interval by the method of integral equations.

We will seek the solution of the system

$$y_2' - \{\lambda + p(x)\} y_1 = f_1(x), \tag{13.1}$$

$$y_1' + \{\lambda + r(x)\} y_2 = -f_2(x) \tag{13.2}$$

($f(x) = \begin{pmatrix} f_1(x) \\ f_2(x) \end{pmatrix} \not\equiv 0$ is a continuous vector-function) satisfying the boundary conditions

$$y_1(a) \sin\alpha + y_2(a)\cos\alpha = 0, \tag{13.3}$$

$$y_1(b) \sin\beta + y_2(b)\cos\beta = 0. \tag{13.4}$$

Let λ be a fixed complex number. We denote by

$$\varphi(x, \lambda) = \begin{pmatrix} \varphi_1(x, \lambda) \\ \varphi_2(x, \lambda) \end{pmatrix}$$

the solution of the system (10.6) satisfying the initial conditions

$$\varphi_1(a, \lambda) = \cos\alpha, \qquad \varphi_2(a, \lambda) = -\sin\alpha, \tag{13.5}$$

and by

$$\psi(x, \lambda) = \begin{pmatrix} \psi_1(x, \lambda) \\ \psi_2(x, \lambda) \end{pmatrix}$$

the solution of the same system, satisfying the initial conditions

$$\psi_1(b, \lambda) = \cos\beta, \qquad \psi_2(b, \lambda) = -\sin\beta. \tag{13.6}$$

If $\varphi(x,\lambda)$ and $\psi(x,\lambda)$ are linearly independent, i.e. if $\varphi(x,\lambda)$ is not an eigen-vector-function (if $\varphi(x,\lambda) = C\psi(x,\lambda)$, then $\varphi(x,\lambda)$ satisfies the boundary conditions (13.3) – (13.4), and is consequently an eigen-vector-function), then the Wronskian determinant is different from zero, i.e.

$$W\{\varphi, \psi\} = \begin{vmatrix} \varphi_1(x, \lambda) & \varphi_2(x, \lambda) \\ \psi_1(x, \lambda) & \psi_2(x, \lambda) \end{vmatrix} \neq 0.$$

Conversely, if for some λ the Wronskian determinant equals zero, then $\varphi(x,\lambda) = C\psi(x,\lambda)$, and so $\varphi(x,\lambda)$ is an eigen-vector-function. Thus the eigenvalues of the problem (10.6) + (13.3) + (13.4) coincide with the zeros of the Wronskian determinant, regarding the latter as a function of λ.

We will prove that the Wronskian determinant $W\{\varphi,\psi\}$ does not depend upon x. In fact, we have

$$(13.7) \qquad \begin{aligned} &\varphi_2' - \{\lambda + p(x)\}\varphi_1 = 0,\\ &\varphi_1' + \{\lambda + r(x)\}\varphi_2 = 0,\\ &\psi_2' - \{\lambda + p(x)\}\psi_1 = 0,\\ &\psi_1' + \{\lambda + r(x)\}\psi_2 = 0. \end{aligned}$$

Multiplying these equations by $\psi_1(x,\lambda)$, $-\psi_2(x,\lambda)$, $-\varphi_1(x,\lambda)$ and $\varphi_2(x,\lambda)$ respectively, and adding, we obtain

$$\varphi_2'(x,\lambda)\psi_1(x,\lambda) - \varphi_1'(x,\lambda)\psi_2(x,\lambda) - \\ - \psi_2'(x,\lambda)\varphi_1(x,\lambda) + \psi_1'(x,\lambda)\varphi_2(x,\lambda) = 0,$$

i.e.

$$\frac{d}{dx}\{\varphi_1(x,\lambda)\psi_2(x,\lambda) - \varphi_2(x,\lambda)\psi_1(x,\lambda)\} = \frac{d}{dx}W\{\varphi,\psi\} = 0,$$

which proves our assertion. Thus $W\{\varphi,\psi\}$ depends only upon λ; $W\{\varphi,\psi\} = \omega(\lambda)$.

We introduce the following notation: the transpose of a matrix $u = \begin{pmatrix} u_1 \\ u_2 \end{pmatrix}$ will be denoted by u^T, i.e. $u^T = (u_1, u_2)$. Therefore from the rules for matrix multiplication we have: if

$$u = \begin{pmatrix} u_1 \\ u_2 \end{pmatrix} \quad \text{and} \quad v = \begin{pmatrix} v_1 \\ v_2 \end{pmatrix},$$

then

$$u^T v = u_1 v_1 + u_2 v_2, \qquad u v^T = \begin{pmatrix} u_1 \\ u_2 \end{pmatrix}(v_1, v_2) = \begin{pmatrix} u_1 v_1 & u_1 v_2 \\ u_2 v_1 & u_2 v_2 \end{pmatrix}.$$

Let

$$(13.8) \qquad G(x, y; \lambda) = \begin{cases} \dfrac{1}{\omega(\lambda)}\psi(x,\lambda)\varphi^T(y,\lambda), & y < x, \\[2mm] \dfrac{1}{\omega(\lambda)}\varphi(x,\lambda)\psi^T(y,\lambda), & x < y. \end{cases}$$

We will show that the vector-function

$$(13.9) \qquad y(x;\lambda) = \begin{pmatrix} y_1(x,\lambda) \\ y_2(x,\lambda) \end{pmatrix} = \int_a^b G(x, y; \lambda) f(y)\,dy$$

is the solution of the system (13.1) – (13.2) satisfying the conditions (13.3) – (13.4).

Indeed, we have by definition

$$G(x, y; \lambda) = \begin{cases} \dfrac{1}{\omega(\lambda)} \begin{pmatrix} \varphi_1(x, \lambda)\psi_1(y, \lambda) & \varphi_1(x, \lambda)\psi_2(y, \lambda) \\ \varphi_2(x, \lambda)\psi_1(y, \lambda) & \varphi_2(x, \lambda)\psi_2(y, \lambda) \end{pmatrix}, & x < y, \\ \dfrac{1}{\omega(\lambda)} \begin{pmatrix} \varphi_1(y, \lambda)\psi_1(x, \lambda) & \varphi_2(y, \lambda)\psi_1(x, \lambda) \\ \varphi_1(y, \lambda)\psi_2(x, \lambda) & \varphi_2(y, \lambda)\psi_2(x, \lambda) \end{pmatrix}, & y < x. \end{cases}$$

Therefore

$$G(x, y; \lambda) f(y) =$$

$$= \begin{cases} \dfrac{1}{\omega(\lambda)} \begin{pmatrix} \varphi_1(x, \lambda)\psi_1(y, \lambda) f_1(y) + \varphi_1(x, \lambda)\psi_2(y, \lambda) f_2(y) \\ \varphi_2(x, \lambda)\psi_1(y, \lambda) f_1(y) + \varphi_2(x, \lambda)\psi_2(y, \lambda) f_2(y) \end{pmatrix}, & x < y, \\ \dfrac{1}{\omega(\lambda)} \begin{pmatrix} \varphi_1(y, \lambda)\psi_1(x, \lambda) f_1(y) + \varphi_2(y, \lambda)\psi_1(x, \lambda) f_2(y) \\ \varphi_1(y, \lambda)\psi_2(x, \lambda) f_1(y) + \varphi_2(y, \lambda)\psi_2(x, \lambda) f_2(y) \end{pmatrix}, & y < x. \end{cases}$$

Consequently, putting

$$[u^T v](y) = u_1(y) v_1(y) + u_2(y) v_2(y),$$

we obtain from (13.9)

$$(13.10)\quad y_1(x, \lambda) = \frac{1}{\omega(\lambda)} \left\{ \psi_1(x, \lambda) \int_a^x [\varphi^T f](y)\, dy + \varphi_1(x, \lambda) \int_x^b [\psi^T f](y)\, dy \right\},$$

$$(13.11)\quad y_2(x, \lambda) = \frac{1}{\omega(\lambda)} \left\{ \psi_2(x, \lambda) \int_a^x [\varphi^T f](y)\, dy + \varphi_2(x, \lambda) \int_x^b [\psi^T f](y)\, dy \right\}.$$

Now differentiating (13.10) with respect to x, we obtain

$$y_1'(x, \lambda) = \frac{1}{\omega(\lambda)} \left\{ \psi_1'(x, \lambda) \int_a^x [\varphi^T f](y)\, dy \right.$$

$$+ \varphi_1'(x, \lambda) \int_x^b [\psi^T f](y)\, dy - \varphi_1(x, \lambda)[\psi_1(x, \lambda) f_1(x)$$

$$\left. + \psi_2(x, \lambda) f_2(x)] + \psi_1(x, \lambda)[\varphi_1(x, \lambda) f_1(x) + \varphi_2(x, \lambda) f_2(x)] \right\}.$$

Inserting the values for $\varphi_1'(x, \lambda)$ and $\psi_1'(x, \lambda)$ from the system (13.7), we find that

$$y_1'(x,\lambda) = -\{\lambda + r(x)\}\frac{1}{\omega(\lambda)}\left\{\psi_2(x,\lambda)\int_a^x [\varphi^T f](y)\,dy + \varphi_1(x,\lambda)\int_x^b [\psi^T f](y)\,dy\right\}$$

$$(13.12) \qquad -\frac{1}{\omega(\lambda)}\{\varphi_1(x,\lambda)\psi_2(x,\lambda) - \varphi_2(x,\lambda)\psi_1(x,\lambda)\} f_2(x).$$

Taking into account (13.11) and the fact that

$$\varphi_1(x,\lambda)\psi_2(x,\lambda) - \varphi_2(x,\lambda)\psi_1(x,\lambda) = W(\varphi,\psi) = \omega(\lambda),$$

(13.12) can be rewritten in the form

$$y_1'(x,\lambda) = -\{\lambda + r(x)\}y_2(x,\lambda) - f_2(x),$$

i.e. equation (13.2) is satisfied. The proof that equation (13.1) is satisfied is analogous.

One can verify directly that the vector-function $y(x,\lambda)$ satisfies the boundary conditions (13.3) – (13.4).

We have thus proved:

THEOREM 13.1. *If λ is not an eigenvalue of the homogeneous boundary-value problem* (10.6) + (13.3) + (13.4), *then the system* (13.1) – (13.4) *is solvable for any function $f(x)$, and the solution is given by the formula* (13.9). *Conversely, if λ is an eigenvalue of the homogeneous boundary-value problem, then the system* (13.1) – (13.4), *generally speaking, is not solvable.*

If λ is not an eigenvalue of the homogeneous boundary-value problem, then the system (13.1) – (13.4) has a unique solution. In fact, the difference of two solutions of the inhomogeneous problem is obviously an eigen-vector-function of the homogeneous problem, which by virtue of our hypothesis must be identically zero.

We can assume that the number $\lambda = 0$ is not an eigenvalue, for in the contrary case we could choose a fixed number η and consider the boundary-value problem

$$\begin{aligned} y_2' - \{(\lambda - \eta) + p(x)\}\, y_1 &= 0, \\ y_1' + \{(\lambda - \eta) + r(x)\}\, y_2 &= 0, \\ y_1(a)\sin\alpha + y_2(a)\cos\alpha &= 0, \\ y_1(b)\sin\beta + y_2(b)\cos\beta &= 0. \end{aligned}$$

The eigenvectors of this problem are the same as those of the problem (10.6) + (13.3) + (13.4), while the eigenvalues are all shifted by an amount η. We can obviously choose η in such a way that the number zero will not be an eigenvalue of the new problem.

Put $G(x,y;0) = G(x,y)$. Then as has been proved, the vector-function

$$y(x) = \int_a^b G(x, y) f(y)\, dy \tag{13.9'}$$

is the solution of the system

$$y_2' - p(x) y_1 = f_1(x), \qquad y_1' + r(x) y_2 = -f_2(x),$$

satisfying the boundary conditions (13.3) – (13.4). We rewrite the system (13.1) in the form

$$y_2' - p(x) y_1 = f_1(x) + \lambda y_1, \qquad -y_1' - r(x) y_2 = f_2(x) + \lambda y_2,$$

or

$$\begin{pmatrix} -p(x) & \dfrac{d}{dx} \\ -\dfrac{d}{dx} & -r(x) \end{pmatrix} y = I(f(x) + \lambda y), \qquad I = \begin{pmatrix} 1 & 0 \\ 0 & 1 \end{pmatrix}. \tag{13.13}$$

If we put

$$B = \begin{pmatrix} -p(x) & \dfrac{d}{dx} \\ -\dfrac{d}{dx} & -r(x) \end{pmatrix},$$

then the system (13.13) assumes the form $By = I(f(x) + \lambda y)$.

On the basis of the foregoing we can assert that the system (13.1) – (13.4) is equivalent to the system of integral equations

$$y(x) = \int_a^b G(x, y)[f(t) + \lambda y(t)]\, dt$$

or

$$y(x) - \lambda \int_a^b G(x, t) y(t)\, dt = g(x),$$

where

$$g(x) = \int_a^b G(x, t) f(t)\, dt.$$

In particular, the homogeneous boundary-value problem (10.6) + (13.3) + (13.4) is equivalent to the integral equation

$$y(x) - \lambda \int_a^b G(x, t) y(t)\, dt = 0. \tag{13.14}$$

2. Let us denote by $v_0(x), v_{\pm 1}(x), \cdots$ a maximal orthonormal set of eigen-vector-functions of the boundary-value problem (10.6) + (13.3) + (13.4), and by $\lambda_0, \lambda_{\pm 1}, \cdots$ the corresponding eigenvalues.

We will consider the kernel

$$H(x, \xi) = \sum_{n=-\infty}^{\infty} \frac{v_n(x)\, v_n^T(\xi)}{\lambda_n}. \tag{13.15}$$

By virtue of the asymptotic formulas obtained in §11, the series for $H(x, \xi)$ converges uniformly outside every neighborhood of the diagonal $x = \xi$. Further, from the boundedness of the partial sums of the series $\sum_{n=1}^{\infty} \sin nx/n$ there follows the boundedness of the partial sums of the series (13.15), which can therefore be integrated term by term.

We now consider the kernel

$$Q(x, \xi) = H(x, \xi) - G(x, \xi) = -G(x, \xi) + \sum_{n=-\infty}^{\infty} \frac{v_n(x)\, v_n^T(\xi)}{\lambda_n}.$$

As is easily seen, this kernel is continuous[21] and satisfies the condition $Q(x, \xi) = Q^T(\xi, x)$. This condition guarantees the selfadjointness of the integral operator having $Q(x, \xi)$ as its kernel. Therefore we will call every matrix satisfying this condition a selfadjoint kernel.

By a well-known theorem from the theory of integral equations, every selfadjoint kernel $Q(x, \xi)$ which is not identically zero has at least one eigenfunction, i.e. there exists a number λ_0 and a function $u(x) \not\equiv 0$ which satisfy the equation

$$u(x) + \lambda_0 \int_a^b Q(x, \xi)\, u(\xi)\, d\xi = 0. \tag{13.16}$$

Thus if we show that the kernel $Q(x, \xi)$ has no eigenfunctions, we can conclude that $Q(x, \xi) \equiv 0$, i.e.

$$G(x, \xi) = \sum_{n=-\infty}^{\infty} \frac{v_n(x)\, v_n^T(\xi)}{\lambda_n}. \tag{13.15'}$$

It follows from (13.14) that

$$\int_a^b G(x, \xi)\, v_k(\xi)\, d\xi = \frac{1}{\lambda_k} v_k(x). \tag{13.17}$$

---

[21] $H(x, \xi)$ and $G(x, \xi)$ are continuous everywhere except on the diagonal $x = \xi$, where they both have the same jump:

$$G(x, x+0) - G(x, x-0) = H(x, x+0) - H(x, x-0) = \begin{pmatrix} 0 & 1 \\ -1 & 0 \end{pmatrix}.$$

Let us now calculate the integral

$$\int_a^b Q(x, \xi) v_k(\xi)\, d\xi \equiv q(x) = \begin{pmatrix} q_1(x) \\ q_2(x) \end{pmatrix}.$$

By the definition of the matrix $Q(x, \xi)$ we have

$$q(x) = \int_a^b H(x, \xi) v_k(\xi)\, d\xi - \int_a^b G(x, \xi) v_k(\xi)\, d\xi. \tag{13.18}$$

On the other hand, every term in the series (13.15) has the form

$$\frac{1}{\lambda_n} v_n(x) v_n^T(\xi) = \frac{1}{\lambda_n} \begin{pmatrix} v_{n1}(x) v_{n1}(\xi) & v_{n1}(x) v_{n2}(\xi) \\ v_{n2}(x) v_{n1}(\xi) & v_{n2}(x) v_{n2}(\xi) \end{pmatrix} \quad \left(v_n(x) = \begin{pmatrix} v_{n1}(x) \\ v_{n2}(x) \end{pmatrix} \right).$$

Therefore if we introduce the notation

$$\int_a^b H(x, \xi) v_k(\xi)\, d\xi \equiv h(x) = \begin{pmatrix} h_1(x) \\ h_2(x) \end{pmatrix},$$

then from the orthonormality of the eigen-vector-functions $v_n(x)$ we obtain

$$h_1(x) = \frac{1}{\lambda_k} \int_a^b \{ v_{k1}(x) v_{k1}^2(\xi) + v_{k1}(x) v_{k2}^2(\xi) \}\, d\xi = \frac{1}{\lambda_k} v_{k1}(x),$$

$$h_2(x) = \frac{1}{\lambda_k} \int_a^b \{ v_{k2}(x) v_{k1}^2(\xi) + v_{k2}(x) v_{k2}^2(\xi) \}\, d\xi = \frac{1}{\lambda_k} v_{k2}(x),$$

i.e. $h(x) = (1/\lambda_k) v_k(x)$. By virtue of (13.17) we then obtain from (13.18)

$$q(x) = \int_a^b Q(x, \xi) v_k(\xi)\, d\xi = \frac{1}{\lambda_k} v_k(x) - \frac{1}{\lambda_k} v_k(x) \equiv 0, \tag{13.18$'$}$$

i.e. the kernel $Q(x, \xi)$ is orthogonal to all the eigen-vector-functions of the boundary-value problem (10.6) + (10.3) + (13.4).

Let

$$u(x) = \begin{pmatrix} u_1(x) \\ u_2(x) \end{pmatrix}$$

be a solution of the integral equation (13.16). We will show that $u(x)$ is orthogonal to all of the $v_k(x)$. In fact, it follows from (13.16) that

$$u_1(x) + \lambda_0 \int_a^b \{q_{11}(x, \xi) u_1(\xi) + q_{12}(x, \xi) u_2(\xi)\} \, d\xi = 0,$$

$$u_2(x) + \lambda_0 \int_a^b \{q_{21}(x, \xi) u_1(\xi) + q_{22}(x, \xi) u_2(\xi)\} \, d\xi = 0.$$

$$\left(Q = \begin{pmatrix} q_{11} & q_{12} \\ q_{21} & q_{22} \end{pmatrix} \right)$$

Multiplying the first of these equations by $v_{n1}(x)$ and the second by $v_{n2}(x)$, integrating with respect to x from a to b, and then adding, we obtain

$$\int_a^b \{u_1(x) v_{n1}(x) + u_2(x) v_{n2}(x)\} \, dx$$
$$+ \lambda_0 \int_a^b \left\{ u_1(\xi) \int_a^b [q_{11}(x, \xi) v_{n1}(x) + q_{12}(x, \xi) v_{n2}(x)] \, dx \right.$$
$$\left. + u_2(\xi) \int_a^b [q_{21}(x, \xi) v_{n1}(x) + q_{22}(x, \xi) v_{n2}(x)] \, dx \right\} d\xi = 0$$

or, more concisely,

$$\text{(13.19)} \qquad \int_a^b u^T(x) v_n(x) \, dx + \lambda_0 \int_a^b u^T(\xi) \left\{ \int_a^b Q^T(x, \xi) v_n(x) \, dx \right\} d\xi = 0.$$

Since $Q(x, \xi) = Q^T(\xi, x)$, by virtue of (13.18′) we have

$$\int_a^b Q^T(x, \xi) v_n(x) \, dx = \int_a^b Q(\xi, x) v_n(x) \, dx = 0.$$

It therefore follows from (13.19) that

$$\int_a^b u^T(x) v_n(x) \, dx = 0.$$

From the definition of the matrix $Q(x, \xi)$ we then conclude that

$$u(x) + \lambda_0 \int_a^b Q(x, \xi) u(\xi) \, d\xi = u(x) - \lambda_0 \int_a^b G(x, \xi) u(\xi) \, d\xi = 0,$$

i.e. $u(x)$ is an eigen-vector-function of the boundary-value problem (10.6) + (13.3) + (13.4). But since $u(x)$ is orthogonal to all of the eigen-vector-functions $v_n(x)$, we have $u(x) \equiv 0$, and consequently $Q(x, \xi) \equiv 0$.

THEOREM 13.2 (EXPANSION THEOREM). *If*

$$f(x) = \begin{pmatrix} f_1(x) \\ f_2(x) \end{pmatrix}$$

has continuous first derivative and satisfies the boundary conditions (13.3) – (13.4), *then* $f(x)$ *can be expanded in an absolutely and uniformly convergent Fourier series with respect to the eigen-vector-functions of the boundary-value problem* (10.6) + (13.3) + (13.4):

$$f(x) = \sum_{n=-\infty}^{\infty} a_n v_n(x), \tag{13.20}$$

i.e.

$$\begin{aligned} f_1(x) &= \sum_{n=-\infty}^{\infty} a_n v_{n1}(x), \\ f_2(x) &= \sum_{n=-\infty}^{\infty} a_n v_{n2}(x), \end{aligned} \tag{13.20'}$$

where

$$a_n = \int_a^b f^T(x)\, v_n(x) \cdot dx. \tag{13.21}$$

PROOF. We put

$$f_2'(x) - p(x) f_1(x) = h_1(x), \quad f_1'(x) + r(x) f_2(x) = -h_2(x).$$

Then by formula (13.9′) we have

$$f(x) = \int_a^b G(x, \xi)\, h(\xi)\, d\xi, \quad h(x) = \begin{pmatrix} h_1(x) \\ h_2(x) \end{pmatrix},$$

or, using the expression for $G(x, \xi)$ from (13.15′),

$$f_1(x) = \sum_{n=-\infty}^{\infty} v_{n1}(x) \frac{1}{\lambda_n} \int_a^b h^T(\xi)\, v_n(\xi)\, d\xi = \sum_{n=-\infty}^{\infty} a_n v_{n1}(x),$$

$$f_2(x) = \sum_{n=-\infty}^{\infty} v_{n2}(x) \frac{1}{\lambda_n} \int_a^b h^T(\xi)\, v_n(\xi)\, d\xi = \sum_{n=-\infty}^{\infty} a_n v_{n2}(x).$$

Further, multiplying the first equality by $v_{n1}(x)$, the second by $v_{n2}(x)$, adding and then integrating with respect to x from a to b, by virtue of the orthonormality of the vector-functions $v_n(x)$ we obtain

$$a_n = \int_a^b f^T(x)\, v_n(x)\, dx,$$

which proves the theorem.

THEOREM 13.3. *For every function*

$$f(x) = \begin{pmatrix} f_1(x) \\ f_2(x) \end{pmatrix}$$

which is square integrable over the interval $[a, b]$ *one has the Parseval equality*

$$\int_a^b \{f_1^2(x) + f_2^2(x)\}\, dx = \sum_{n=-\infty}^{\infty} a_n^2. \tag{13.22}$$

PROOF. If $f(x)$ satisfies the hypothesis of the previous theorem, then (13.22) follows at once from the uniform convergence of the series (13.20′). In fact, multiplying the first series by $f_1(x)$, the second by $f_2(x)$, adding and then integrating, we obtain by virtue of (13.21)

$$\int_a^b \{f_1^2(x) + f_2^2(x)\}\, dx = \sum_{n=-\infty}^{\infty} a_n \int_a^b f^T(x)\, v_n(x)\, dx = \sum_{n=-\infty}^{\infty} a_n^2,$$

i.e. (13.22).

Extension of the Parseval equality to an arbitrary vector-function of the class $\mathscr{L}^2(a, b)$ can be carried out by the usual means (cf. §5).

3. Let us return to formula (13.9). The right side of this formula is called the resolvent. As was indicated, the resolvent exists for all λ which are not eigenvalues of the boundary-value problem (10.6) + (13.3) + (13.4).

We will now show how to obtain the Fourier series expansion of the resolvent if one knows the expansion of the vector-function $f(x)$.

The vector-function $y(x, \lambda)$, defined by (13.9), satisfies the boundary conditions (13.3) – (13.4). Therefore, integrating by parts, we obtain

$$\int_a^b \{[y_2'(x, \lambda) - p(x)\, y_1(x, \lambda)]\, v_{n1}(x) - [y_1'(x, \lambda) + r(x)\, y_2(x, \lambda)]\, v_{n2}(x)\}\, dx$$

$$= y_2(x, \lambda)\, v_{n1}(x)\big|_a^b - y_1(x, \lambda)\, v_{n2}(x)\big|_a^b - \int_a^b \{v_{n1}'(x) + r(x)\, v_{n2}(x)\}\, y_2(x, \lambda)\, dx$$

$$+ \int_a^b \{v_{n2}'(x) - p(x)\, v_{n1}(x)\}\, y_1(x, \lambda)\, dx = \lambda_n \int_a^b v_n^T(x)\, y(x, \lambda)\, dx \equiv \lambda_n d_n(\lambda).$$

(13.23)

Let

$$y(x, \lambda) = \sum_{n=-\infty}^{\infty} d_n(\lambda) v_n(x); \quad a_n = \int_a^b f^T(x) v_n(x)\, dx.$$

Then it follows from the system (13.1) – (13.2) and the equality (13.23) that

$$a_n = \int_a^b f^T(x) v_n(x)\, dx = \int_a^b \{[y_2'(x, \lambda) - p(x) y_1(x, \lambda)] v_{n1}(x) - [y_1'(x, \lambda) + r(x) y_2(x, \lambda)] v_{n2}(x)\}\, dx - \lambda \int_a^b y^T(x, \lambda) v_n(x) = (\lambda_n - \lambda) d_n(\lambda).$$

Hence $d_n(\lambda) = a_n/(\lambda_n - \lambda)$. Consequently the expansion of the resolvent has the form

$$y(x, \lambda) = \sum_{n=-\infty}^{\infty} \frac{a_n}{\lambda_n - \lambda} v_n(x), \tag{13.24}$$

i.e.

$$\begin{aligned} y_1(x, \lambda) &= \sum_{n=-\infty}^{\infty} \frac{a_n}{\lambda_n - \lambda} v_{n1}(x), \\ y_2(x, \lambda) &= \sum_{n=-\infty}^{\infty} \frac{a_n}{\lambda_n - \lambda} v_{n2}(x). \end{aligned} \tag{13.24'}$$

§14. Calculation of the regularized trace for a Sturm-Liouville operator

1. We will consider the boundary-value problem

$$y'' + \{\lambda + q(x)\} y = 0 \qquad (0 \leqslant x \leqslant \pi), \tag{14.1}$$

$$y'(0) - hy(0) = 0, \tag{14.2}$$

$$y'(\pi) + Hy(\pi) = 0, \tag{14.3}$$

where $q(x) \in C_2(0, \pi)$ and h, H are real finite numbers. We denote by $\lambda_0, \lambda_1, \cdots$ the eigenvalues of this problem. As was shown in §2, for sufficiently large n one has the asymptotic formula (cf. (2.13′))

$$\sqrt{\lambda_n} = n + \frac{c}{2n} + O\left(\frac{1}{n^3}\right), \tag{14.4}$$

where

$$c = \frac{2}{\pi}\left(h + H + \frac{1}{2}\int_0^\pi q(x)\, dx\right). \tag{14.5}$$

From (14.4) it follows that

$$\lambda_n = n^2 + c + O\left(\frac{1}{n^2}\right). \tag{14.4'}$$

Therefore

$$\sum_{n=0}^{\infty} (\lambda_n - n^2 - c) < \infty. \tag{14.6}$$

The series in (14.6) is called the regularized trace of the Sturm-Liouville operator being considered, and the purpose of this section is to find its sum.

2. Let us denote by $\varphi(x,\lambda)$ the solution of equation (14.1) which satisfies the initial conditions

$$\varphi(0,\lambda) = 1, \qquad \varphi'_x(0,\lambda) = h.$$

Then the eigenvalues λ_n are the roots of the entire analytic function $\varphi'_x(\pi,\lambda) + H\varphi(\pi,\lambda)$. Therefore

$$\varphi'_x(\pi,\lambda) + H\varphi(\pi,\lambda) = A\Phi(\lambda), \tag{14.7}$$

where

$$\Phi(\lambda) = \prod_{n=0}^{\infty}\left(1 - \frac{\lambda}{\lambda_n}\right)$$

and A is a constant which will be determined later.[22] If $\lambda_j = 0$, then the factor $1 - \lambda/\lambda_j$ is to be replaced by $-\lambda$.

We will study the asymptotic behaviour of both sides of (14.7) for large negative $\lambda = -\mu$. Then comparing the coefficients of like powers of μ on the right and left sides, we will in particular obtain the value of the sum (14.6).

We put[23]

$$\begin{aligned} \Phi(-\mu) &= \prod_{n=0}^{\infty}\left(1 + \frac{\mu}{\lambda_n}\right) = \left(1 + \frac{\mu}{\lambda_0}\right) \frac{\prod_{n=1}^{\infty}\left(1 + \frac{\mu}{\lambda_n}\right)}{\prod_{n=1}^{\infty}\left(1 + \frac{\mu}{n^2}\right)} \frac{\sinh \pi\sqrt{\mu}}{\pi\sqrt{\mu}} \\ &= C_1(\lambda_0 + \mu)\,\psi(\mu)\,\frac{\operatorname{sh}\pi\sqrt{\mu}}{\pi\sqrt{\mu}}, \end{aligned} \tag{14.8}$$

[22] The absence of exponential factors follows from Theorem 28, Chapter 1 of the monograph by Levin [1].

[23] By virtue of the equality

$$\sin z = z\prod_{n=1}^{\infty}\left(1 - \frac{z^2}{n^2\pi^2}\right).$$

where

$$C_1=\frac{1}{\lambda_0}\prod_{n=1}^{\infty}\frac{n^2}{\lambda_n},\qquad \psi(\mu)=\prod_{n=1}^{\infty}\left(1-\frac{n^2-\lambda_n}{\mu+n^2}\right). \tag{14.9}$$

We will study the asymptotic behaviour of the function $\psi(\mu)$ for large positive μ. To this end we consider $\log\psi(\mu)$:

$$\log\psi(\mu)=\sum_{n=1}^{\infty}\ln\left(1-\frac{n^2-\lambda_n}{\mu+n^2}\right)=-\sum_{n=1}^{\infty}\sum_{k=1}^{\infty}\frac{1}{k}\left(\frac{n^2-\lambda_n}{\mu+n^2}\right)^k$$
$$=-\sum_{k=1}^{\infty}\frac{1}{k}\sum_{n=1}^{\infty}\left(\frac{n^2-\lambda_n}{\mu+n^2}\right)^k.$$

Further estimates are based upon the following lemma.

LEMMA 14.1. *If* $|n^2-\lambda_n|\leq a$, *then*

$$\sum_{n=1}^{\infty}\frac{|n^2-\lambda_n|^k}{(\mu+n^2)^k}\leqslant\frac{\pi}{2}\cdot\frac{a^k}{\mu^{k-\frac{1}{2}}}. \tag{14.10}$$

In fact,

$$\sum_{n=1}^{\infty}\frac{|n^2-\lambda_n|^k}{(\mu+n^2)^k}\leqslant a^k\sum_{n=1}^{\infty}\frac{1}{(\mu+n^2)^k}\leqslant a^k\int_0^{\infty}\frac{dx}{(\mu+x^2)^k}$$
$$=\frac{a^k}{\mu^{k-\frac{1}{2}}}\int_0^{\infty}\frac{dt}{(1+t^2)^k}\leqslant\frac{\pi}{2}\cdot\frac{a^k}{\mu^{k-\frac{1}{2}}},$$

which proves the lemma.

By virtue of the bounds (14.10) we have

$$\sum_{k=2}^{\infty}\frac{1}{k}\sum_{n=1}^{\infty}\frac{|n^2-\lambda_n|^k}{(\mu+n^2)^k}\leqslant\frac{\pi}{2}\sum_{k=2}^{\infty}\frac{a^k}{\mu^{k-\frac{1}{2}}}=\frac{\pi}{2}\cdot\frac{a^2}{\mu^{3/2}}\sum_{k=0}^{\infty}\left(\frac{a}{\mu}\right)^k=O(\mu^{-3/2}). \tag{14.11}$$

Further,

$$-\sum_{n=1}^{\infty}\frac{n^2-\lambda_n}{\mu+n^2}=\sum_{n=1}^{\infty}(\lambda_n-n^2-c)\frac{1}{\mu+n^2}+c\sum_{n=1}^{\infty}\frac{1}{\mu+n^2}$$
$$=c\sum_{n=1}^{\infty}\frac{1}{\mu+n^2}+\frac{1}{\mu}\sum_{n=1}^{\infty}(\lambda_n-n^2-c) \tag{14.12}$$
$$-\frac{1}{\mu}\sum_{n=1}^{\infty}(\lambda_n-n^2-c)\frac{n^2}{\mu+n^2}.$$

Since $\sup_n|(\lambda_n-n^2-c)n^2|<\infty$, the bounds (14.10) imply the estimate

(14.13)
$$\frac{1}{\mu}\sum_{n=1}^{\infty}(\lambda_n - n^2 - c)\, n^2 \frac{1}{\mu + n^2} = O(\mu^{-3/2}).$$

Therefore by virtue of (14.11), (14.12) and (14.13) we obtain

(14.14)
$$\log\psi(\mu) = c\sum_{n=1}^{\infty}\frac{1}{\mu+n^2} + \frac{1}{\mu}\sum_{n=1}^{\infty}(\lambda_n - n^2 - c) + O(\mu^{-3/2}).$$

As is known,

$$\sum_{n=1}^{\infty}\frac{1}{\mu+n^2} = \frac{\pi\coth\pi\sqrt{\mu}}{2\sqrt{\mu}} - \frac{1}{2\mu} = \frac{\pi}{2\sqrt{\mu}} - \frac{1}{2\mu} + O(e^{-2\pi\sqrt{\mu}}).$$

It then follows from (14.14) that

$$\log\psi(\mu) = \frac{c\pi}{2\sqrt{\mu}} + \frac{1}{\mu}\left(s_\lambda - \lambda_0 + \frac{1}{2}c\right) + O(\mu^{-3/2}),$$

where

(14.15)
$$s_\lambda = \sum_{n=0}^{\infty}(\lambda_n - n^2 - c).$$

Consequently,

(14.16)
$$\begin{aligned}\psi(\mu) &= \exp\left\{\frac{c\pi}{2\sqrt{\mu}} + \frac{1}{\mu}\left(s_\lambda - \lambda_0 + \frac{1}{2}c\right) + O(\mu^{-3/2})\right\}\\ &= 1 + \frac{c\pi}{2\sqrt{\mu}} + \frac{1}{\mu}\left(s_\lambda - \lambda_0 + \frac{1}{2}c + \frac{c^2\pi^2}{8}\right) + O(\mu^{-3/2}).\end{aligned}$$

By virtue of (14.16) it follows from (14.8) that

(14.17)
$$\Phi(-\mu) = \frac{1}{2\pi}C_1 e^{\pi\sqrt{\mu}}\left\{\sqrt{\mu} + \frac{1}{2}c\pi + \frac{1}{\sqrt{\mu}}\left(s_\lambda + \frac{1}{2}c + \frac{c^2\pi^2}{8}\right) + O\left(\frac{1}{\mu}\right)\right\}.$$

3. We will now study the asymptotic behaviour of the function $\varphi'_x(\pi,\lambda) + H\varphi(\pi,\lambda)$ for large negative $\lambda = -\mu$, using Liouville's integral equation. In order to simplify the computations, we will assume that $h = 0$ and that the function $q(x)$ is such that $\int_0^\pi q(x)\,dx = 0$.[24]

By formula (2.4) we have

$$\varphi(x,-\mu) = \cosh\sqrt{\mu}\,x + \frac{1}{\sqrt{\mu}}\int_0^x q(t)\sinh\{\sqrt{\mu}\,(x-t)\}\,\varphi(t,-\mu)\,dt.$$

Using the estimate (2.7), we obtain

---

[24] For the general case see Levitan [**20**].

$$\varphi(\pi, -\mu) = \frac{1}{2} e^{\pi\sqrt{\mu}} \left\{ 1 + \frac{1}{\sqrt{\mu}} \int_0^\pi q(t) \sinh\{\sqrt{\mu}(\pi - t)\} \cosh \sqrt{\mu}\, t dt + O\left(\frac{1}{\mu}\right) \right\}$$

(14.18)

$$= \frac{1}{2} e^{\pi\sqrt{\mu}} \left\{ 1 + O\left(\frac{1}{\mu}\right) \right\}.$$

Further,

$$\varphi'_x(x, -\mu) = \sqrt{\mu} \sinh \sqrt{\mu}\, x + \int_0^x \cosh\{\sqrt{\mu}(x - t)\} q(t) \varphi(t, -\mu)\, dt.$$

Therefore

$$\varphi'_x(\pi, -\mu) = \sqrt{\mu} \sinh \sqrt{\mu}\, \pi + \int_0^\pi \cosh\{\sqrt{\mu}(\pi - t)\} q(t) \left[\cosh \sqrt{\mu}\, t \right.$$

(14.19) $$\left. + \frac{1}{\sqrt{\mu}} \int_0 \sinh\{\sqrt{\mu}(t - s)\} q(s) \cosh \sqrt{\mu}\, s\, ds \right] dt + O\left(\frac{e^{\pi\sqrt{\mu}}}{\mu}\right)$$

$$= \frac{1}{2} e^{\pi\sqrt{\mu}} \left\{ \sqrt{\mu} + \frac{1}{4\sqrt{\mu}} [q(0) + q(\pi)] + O\left(\frac{1}{\mu}\right) \right\}.$$

By (14.18) and (14.19) we have

$$\varphi'_x(\pi, -\mu) + H\varphi(\pi, -\mu)$$

(14.20)

$$= \frac{1}{2} e^{\pi\sqrt{\mu}} \left\{ \sqrt{\mu} + H + \frac{1}{4\sqrt{\mu}} [q(0) + q(\pi)] + O\left(\frac{1}{\mu}\right) \right\}.$$

It follows from (14.7), (14.17) and (14.20) that $AC_1 = \pi$, i.e. $A = \pi/C_1$. We therefore obtain from (14.17) and (14.20) that

$$H + \frac{1}{4\sqrt{\mu}} [q(0) + q(\pi)] + O\left(\frac{1}{\mu}\right) = \frac{c\pi}{2} + \frac{1}{\sqrt{\mu}} \left(s_\lambda + \frac{c}{2} + \frac{\pi^2 c^2}{8} \right) + O\left(\frac{1}{\mu}\right).$$

Equating the coefficients of $1/\sqrt{\mu}$ and taking account of (14.5), we obtain

$$s_\lambda = \frac{1}{4} [q(0) + q(\pi)] - \frac{H}{\pi} - \frac{H^2}{2},$$

where s_λ is defined by (14.15).

4. Calculation of the trace for a Dirac system. The asymptotic formulas for the eigenvalues of the boundary-value problem (10.6) – (10.8), obtained in §11, enable us to calculate the trace for a Dirac system. The trace of an operator is usually defined as the sum of its eigenvalues. If we assume that the functions $p(x)$ and $r(x)$ from the equation (10.6) are differentiable, then the formula (11.18) can be substantially sharpened, namely, one can write down the following terms of an expansion:

$$\lambda_n = \frac{\vartheta}{\pi} + n + \frac{C_1}{n} + \frac{C_2}{n^2} + O\left(\frac{1}{n^3}\right). \tag{14.21}$$

If we further assume that $\vartheta = 0$, i.e.

$$\beta - \alpha - \frac{1}{2}\int_0^\pi \{p(x) + r(x)\}\,dx = 0,$$

then (14.21) implies the convergence of the series

$$\lambda_0 + \sum_{n=1}^{\infty} (\lambda_n + \lambda_{-n}).$$

Proceeding in a manner analogous to that which led to the calculation of the sum (14.15), one can prove the following formula:

$$\lambda_0 + \sum_{n=1}^{\infty} (\lambda_n + \lambda_{-n}) = \frac{r(0) - p(0)}{4}\cos 2\alpha + \frac{r(\pi) - p(\pi)}{4}\cos 2\beta.$$

§15. Selfadjoint nth order differential operators

The theory developed in the preceding sections can be generalized to higher order differential operators. In this section we shall give a brief outline of the basic concepts and theorems of this subject, whose proofs the reader can find in the monographs of Naĭmark [1] and Coddington and Levinson [1].

1. Linear differential expressions and boundary conditions. *By a linear differential expression is meant an expression of the form*

$$l(y) = p_0(x)y^{(n)} + p_1(x)y^{(n-1)} + \cdots + p_n(x)y; \tag{15.1}$$

the functions $p_0(x), p_1(x), \cdots, p_n(x)$ are called the coefficients, and the number n is called the order of the differential expression.

We will assume that the functions $p_0^{-1}(x), p_1(x), \cdots, p_n(x)$ are continuous in some fixed finite interval $[a, b]$. Let $C^{(n)}$ denote the set of all functions having continuous derivatives up to order n inclusively in the interval $[a, b]$. For any function $y \in C^{(n)}$ the differential expression $l(y)$ is defined, and is a continuous function on the interval $[a, b]$.

We denote by

$$y_a, y'_a, \cdots, y_a^{(n-1)}; \qquad y_b, y'_b, \cdots, y_b^{(n-1)} \tag{15.2}$$

the values of the function $y(x)$ and its first $n-1$ derivatives at the points a and b. Let $U(y)$ denote a linear form in the variables (15.2), i.e.

$$\begin{aligned} U(y) = {} & \alpha_0 y_a + \alpha_1 y'_a + \ldots + \alpha_{n-1} y_a^{(n-1)} \\ & + \beta_0 y_b + \beta_1 y'_b + \ldots + \beta_{n-1} y_b^{(n-1)}. \end{aligned} \tag{15.3}$$

If we are given several such forms $U_\nu(y)$ $(\nu = 1, 2, \cdots, m)$, then the equalities

$$U_\nu(y) = 0 \qquad (\nu = 1, 2, \cdots, m) \tag{15.4}$$

are called *boundary conditions.*

Let us denote by D the set of all functions $y \in C^{(n)}$ which satisfy boundary conditions of the form (15.4). Obviously D is a subspace of $C^{(n)}$ which coincides with $C^{(n)}$ only in the absence of conditions (15.4).

Suppose that we are given some differential expression $l(y)$ and some manifold D defined by the conditions (15.4). With every function $y \in D$ we associate the function $u = l(y)$. This correspondence is a linear operator with domain D. We denote this operator by L. Thus, if $y \in D$ and $u = l(y)$, then by definition of the operator L we have $u = Ly$. *The operator L is called the differential operator generated by the differential expression $l(y)$ and the boundary conditions* (15.4).

Thus, and this is a very essential point, *one and the same differential expression can generate distinct differential operators, depending upon the choice of boundary conditions.*

2. Selfadjoint differential expressions. Suppose that the coefficients $p_k(x)$ $(k = 0, 1, \cdots, n)$ of the differential expression (15.1) have continuous derivatives in $[a, b]$ up to order $n - k$ inclusively.

Let y and z be two arbitrary functions from $C^{(n)}$. Integrating by parts k times, we obtain

$$\begin{aligned}\int_a^b p_{n-k}\bar{z}y^{(k)}dx &= [p_{n-k}\bar{z}y^{(k-1)} - (p_{n-k}\bar{z})' y^{(k-2)} + \ldots + (-1)^{k-1}(p_{n-k}\bar{z})^{(k-1)}y]_{x=a}^{x=b} \\ &\quad + (-1)^k \int_a^b y\,(p_{n-k}\bar{z})^{(k)}\,dx.\end{aligned} \tag{15.5}$$

Setting $k = n, n-1, \cdots, 0$ and adding the equalities thereby obtained, we arrive at the formula

$$\int_a^b l(y)\,\bar{z}\,dx = P(\eta, \zeta) + \int_a^b y\overline{l^*(z)}\,dx, \tag{15.6}$$

where

$$l^*(z) = (-1)^n(\bar{p}_0 z)^{(n)} + (-1)^{n-1}(\bar{p}_1 z)^{n-1} + \cdots + \bar{p}_n z, \tag{15.7}$$

and $P(\eta, \zeta)$ is a bilinear form in the variables

$$\eta = (y_a, y_a', \cdots, y_a^{(n-1)}; y_b, y_b', \cdots, y_b^{(n-1)})$$

and

$$\zeta = (z_a, z_a', \cdots, z_a^{(n-1)}; z_b, z_b', \cdots, z_b^{(n-1)}).$$

The differential expression $l^*(z)$ defined by the formula (15.7) *is called the adjoint of the differential expression $l(y)$; the formula* (15.6) *is called Lagrange's formula.*

Applying the preceding arguments to the integral $\int_a^b l^*(z) \cdot \bar{y} dx$, we arrive at a formula of the form

$$\int_a^b l^*(z) \cdot \bar{y}\, dx = Q(\eta, \zeta) + \int_a^b z \cdot \overline{l(y)}\, dx.$$

Consequently the differential expression $l(y)$ is the adjoint of the expression $l^*(z)$, i.e. $(l^*(y))^* = l(y)$. In other words, the differential expressions $l(y)$ and $l^*(y)$ are mutually adjoint.

The differential expression $l(y)$ is called selfadjoint if $l^*(y) = l(y)$.

THEOREM 15.1. *Every selfadjoint differential expression is a sum of differential expressions*

$$l_{2\nu}(y) = (py^{(\nu)})^{(\nu)};\quad l_{2\nu-1}(y) = \frac{1}{2}\{(ipy^{(\nu-1)})^{\nu} + (ipy^{(\nu)})^{(\nu-1)}\},$$

where the functions $p(x)$ appearing in these expressions assume only real values.

THEOREM 15.2. *Every selfadjoint differential expression with real coefficients is necessarily of even order and has the form*

$$l(y) = (p_0 y^{(\nu)})^{(\nu)} + (p_1 y^{(\nu-1)})^{(\nu-1)} + \cdots + (p_{\nu-1} y')' + p_\nu y,$$

where $p_0(x), p_1(x), \cdots, p_\nu(x)$ are real functions.

3. Selfadjoint differential operators. Let $U_1, U_2, \cdots, U_m$ be linearly independent forms in the variables $y_a, y_a', \cdots, y_a^{(n-1)}$; $y_b, y_b', \cdots, y_b^{(n-1)}$; if $m < 2n$ we supplement them by arbitrary forms $U_{m+1}, \cdots, U_{2n}$ so as to obtain a system of $2n$ linearly independent forms. By virtue of the linear independence of these forms the variables $y_a, y_a', \cdots, y_a^{(n-1)}$; $y_b, y_b', \cdots, y_b^{(n-1)}$ can be expressed as linear combinations of the forms $U_1, \cdots, U_{2n}$. Let us insert these expressions in the bilinear form $P(\eta, \zeta)$ in Lagrange's formula (15.6). Then $P(\eta, \zeta)$ becomes a linear homogeneous form in the variables $U_1, \cdots, U_{2n}$. The coefficients of the variables $U_1, \cdots, U_{2n}$ are linear homogeneous forms in the variables $z_a, z_a', \cdots, z_a^{(n-1)}$; $z_b, z_b', \cdots, z_b^{(n-1)}$. We denote these forms by $V_{2n}, V_{2n-1}, \cdots, V_1$ respectively. Then Lagrange's formula (15.6) can be rewritten in the form

$$\int_a^b l(y) \cdot \bar{z}\, dx = U_1 V_{2n} + U_2 V_{2n-1} + \ldots + U_{2n} V_1 + \int_a^b y \cdot \overline{l^*(z)}\, dx. \tag{15.8}$$

It is easy to show that the forms $V_1, V_2, \cdots, V_{2n}$ are linearly independent.

The boundary conditions

$$V_1 = 0, V_2 = 0, \cdots, V_{2n-m} = 0, \tag{15.9}$$

as well as every boundary condition which is equivalent to them, are said to be *adjoint to the boundary conditions*

$$U_1 = 0, U_2 = 0, \cdots, U_m = 0. \tag{15.10}$$

Boundary conditions are said to be selfadjoint if they are equivalent to their adjoint conditions.

The differential operator generated by the expression $l^*(y)$ *and the boundary conditions* (15.9) *is said to be the adjoint of the operator* L *generated by the expression* $l(y)$ *and the boundary conditions* (15.10). The adjoint of the operator L is denoted by L^*.

From the definition of the adjoint operator it follows that the operator L is the adjoint of L^*, i.e. $(L^*)^* = L$.

It follows from (15.8) and from the boundary conditions (15.9) and (15.10) that the operators L and L^* satisfy the relation

$$\int_a^b Ly \cdot \bar{z}\, dx = \int_a^b y \cdot \overline{L^* z}\, dx. \tag{15.11}$$

Let us introduce the notation

$$(y, z) = \int_a^b y(x) \overline{z(x)}\, dx.$$

Then (15.11) assumes the form

$$(Ly, z) = (y, L^* z). \tag{15.12}$$

The operator L *is said to be selfadjoint if* $L^* = L$.

From the definition of the operator L^* it follows that:

The operator L *is selfadjoint if and only if it is generated by a selfadjoint differential expression and selfadjoint boundary conditions.*

4. The eigenvalues and eigenfunctions of a selfadjoint operator. We can, without loss of generality in our further considerations, take the interval $[a, b]$ to be the interval $[0, 1]$. Suppose that the selfadjoint operator L is generated by the selfadjoint differential expression

(15.13) $$ly = (q_0(x)y^{(n)})^{(n)} + (q_1(x)y^{(n-1)})^{(n-1)} + \cdots + q_n(x)y,$$

where $q_j(x)$ is a real function from the class $C^{(n-j)}[0,1]$, and by the selfadjoint boundary conditions

$$U_j(y) = \sum_{m=1}^{2n} \{M_{jm}y^{(m-1)}(0) + N_{jm}y^{(m-1)}(1)\} = 0 \quad (j = 1, 2, \ldots, 2n),$$

which for brevity we will write as a single equation $U(y) = 0$.

Let us consider the eigenvalue problem

(15.14) $$ly = \lambda y, \qquad U(y) = 0.$$

For any λ this problem has the trivial solution $y(x) \equiv 0$.

If λ is such that the problem (15.14) has a nontrivial solution, then λ is called an eigenvalue of the operator L (or of the problem (15.14)), and the corresponding solutions are called eigenfunctions.

The following result holds.

Theorem 15.3. *The eigenvalues of the problem* (15.14) *are real and form an at most countable set which has no limit points on the finite line. Eigenfunctions corresponding to distinct eigenvalues are orthogonal.*

In fact, let $\varphi_j = \varphi_j(x, \lambda)$ $(j = 1, 2, \cdots, 2n)$ be the solutions of the equation

(15.15) $$ly = \lambda y$$

satisfying the initial conditions

$$\varphi_j^{(m-1)}(0, \lambda) = \delta_{jm} \qquad (j, m = 1, 2, \cdots, 2n).$$

The functions $\varphi_j(x, \lambda)$ are jointly continuous in the variables x, λ for $x \in [0,1]$ and for all λ, and for any fixed $x \in [0,1]$ are entire analytic functions of λ.

Since the functions $\varphi_j(x, \lambda)$ form a fundamental system of solutions of the equation (15.15), the problem (15.14) has λ as an eigenvalue when and only when there exist constants c_j, not all equal to zero, such that the function

$$y(x, \lambda) = \sum_{j=1}^{2n} c_j \varphi_j(x, \lambda)$$

satisfies the conditions $U_k(y) = 0$ $(k = 1, 2, \cdots, 2n)$, i.e. the system of equations

(15.16) $$\sum_{j=1}^{2n} c_j U_k(\varphi_j) = 0 \qquad (k = 1, 2, \ldots, 2n)$$

has a nontrivial solution. But this situation obtains if and only if the determinant of the system (15.16),

$$\Delta(\lambda) = \| U_k(\varphi_j) \|_{k,j=1}^{2n},$$

is equal to zero. Since $\varphi_j^{(m-1)}(0,\lambda)$ and $\varphi_j^{(m-1)}(1,\lambda)$ are entire functions of λ, the same is true of $\Delta(\lambda)$.

The function $\Delta(\lambda)$ can have only real roots (we are considering a selfadjoint problem) and therefore cannot vanish identically. The zeros of $\Delta(\lambda)$, which, as we have seen, are the eigenvalues of the problem (15.14), can cluster only at infinity.

The question arises as to whether there exists at least one system of selfadjoint boundary conditions. An exhaustive response to this turns out to be nonsimple. A complete discussion of this question is given in Chapter 11 of the book by Coddington and Levinson [**1**]. However various particular systems of selfadjoint boundary conditions can be pointed out rather easily. For example, it is easy to verify, using the identity (15.8), that the boundary conditions

$$y(0) = y'(0) = \cdots = y^{(n-1)}(0) = 0, \tag{15.17}$$

$$y(1) = y'(1) = \cdots = y^{(n-1)}(1) = 0 \tag{15.18}$$

are selfadjoint (for an l of the form (15.13)).

From now on we will consider only conditions of the form (15.17) – (15.18). It should be noted that the results obtained in the next section can be extended without essential changes to various other selfadjoint boundary conditions (but not to all of them).

Let us write in explicit form the equation which determines the eigenvalues for the case of the boundary conditions (15.17) – (15.18). It follows from (15.17) that an eigenfunction can be written in the form

$$y = \sum_{j=n+1}^{2n} c_j \varphi_j (x, \lambda).$$

Therefore (15.18) implies the system of equations

$$\sum_{j=n+1}^{2n} c_j \varphi_j^{(m)} (1, \lambda) = 0 \qquad (m = 0, 1, \ldots, n-1). \tag{15.19}$$

Since not all the c_j are to equal zero, (15.19) can be satisfied if and only if

$$\Delta(\lambda) = \begin{vmatrix} \varphi_{n+1}(1, \lambda), & \ldots, & \varphi_{2n}(1, \lambda) \\ \varphi'_{n+1}(1, \lambda), & \ldots, & \varphi'_{2n}(1, \lambda) \\ \cdot\ \cdot\ \cdot & \cdot\ \cdot\ \cdot & \cdot\ \cdot\ \cdot \\ \varphi_{n+1}^{(n-1)}(1, \lambda), & \ldots, & \varphi_{2n}^{(n-1)}(1, \lambda) \end{vmatrix} = 0. \tag{15.20}$$

§16. Proof of the completeness of the system of eigenfunctions of a selfadjoint differential operator of order $2n$ by the method of finite differences

1. Finite-difference approximation of a boundary-value problem. In this subsection we will consider the finite-difference analog of the boundary-value problem (15.14) and derive formulas which are analogous to those considered in the preceding section for differential operators. We will see that for selfadjoint finite-difference operators the completeness of the system of eigenvectors follows trivially from a calculation of the dimension of the appropriate vector space.

Basing ourselves on this, we will in the next subsection prove the completeness of the system of eigenfunctions of the operator (15.13) for the boundary conditions (15.17) – (15.18) by means of a passage to the limit.

We divide the interval $[0,1]$ into N equal parts by means of the points νh $(\nu = 0, 1, 2, \cdots, N-1)$, $h = 1/N$. Let $f(x)$ be a continuous real function defined on $[0,1]$, and put $f_\nu = f(\nu h)$ $(\nu = 0, 1, \cdots, N-1)$. The numbers $f_0, f_1, \cdots, f_{N-1}$ can be regarded as the coordinates of an N-dimensional vector $\bar{f}$.

Let us define the scalar product:

$$(\bar{f}, \bar{g})_h = \sum_{\nu=0}^{N-1} f_\nu g_\nu h. \tag{16.1}$$

We further introduce the difference operators

$$(\delta_+ \bar{f})_\nu = \frac{f_{\nu+1} - f_\nu}{h}, \qquad (\delta_- \bar{f})_\nu = \frac{f_\nu - f_{\nu-1}}{h} \qquad (\nu = 0, 1, \ldots, N-1).$$

In the definitions of $(\delta_- \bar{f})_0$ and $(\delta_+ \bar{f})_{N-1}$ there appear the quantities f_{-1} and f_N, whose indices "lie outside" the interval $[0,1]$, and it will therefore be necessary to define them in another way (we will see that these quantities, as well as certain others, can be defined from the boundary conditions).

In the sequel an important role is played by the following formulas for the "differentiation" of a product and "integration" by parts:

$$\begin{aligned} \delta_-(f_\nu \cdot g_\nu) &= f_{\nu-1}\delta_- g_\nu + g_\nu \delta_- f_\nu, \\ \delta_+(f_\nu \cdot g_\nu) &= f_{\nu+1}\delta_+ g_\nu + g_\nu \delta_+ f_\nu, \end{aligned} \tag{16.2}$$

$$\begin{aligned} (\delta_- \bar{f}, \bar{g})_h &= -(\bar{f}, \delta_+ \bar{g}) + f_{N-1} g_N - f_{-1} g_0, \\ (\delta_+ \bar{f}, \bar{g})_h &= -(\bar{f}, \delta_- \bar{g})_h + f_N g_{N-1} - f_0 g_{-1}. \end{aligned} \tag{16.3}$$

We now consider the finite-difference operator generated by the differential expression (15.13):

$$l_h\bar{y} = \delta_+^n(\bar{q}_0\delta_-^n\bar{y}) + \delta_+^{n-1}(\bar{q}_1\delta_-^{n-1}\bar{y}) + \cdots + \bar{q}_n\bar{y}. \tag{16.4}$$

$l_h\bar{y}$ is the vector with components $\{l_h y\}_\nu$ $(\nu = 0, 1, \cdots, N-1)$. These quantities obviously contain $y_{-1}, y_{-2}, \cdots, y_{-n}$; $y_N, y_{N+1}, \cdots, y_{N+n-1}$, which are not directly defined and, as we will see further on, can be defined (at least in part) from the boundary conditions.

It easily follows from (16.3) that the finite-difference expression (16.4) is formally selfadjoint (with respect to the scalar product (16.1)).

Applying the formulas (16.2) and (16.3), it is not difficult to deduce the formula

$$(l_h\bar{y}, \bar{z})_h = (\bar{y}, l_h\bar{z})_h + [\bar{y}, \bar{z}]_h(N) - [\bar{y}, \bar{z}]_h(0), \tag{16.5}$$

where $[\bar{y}, \bar{z}]_h(\nu)$ is a form in the quantities

$$\begin{gathered} y_{\nu-1}, \delta_- y_{\nu-1}, \cdots, \delta_-^{n-1}y_{\nu-1};\ z_{\nu-1}, \delta_- z_{\nu-1}, \cdots, \delta_-^{n-1}z_{\nu-1}, \\ \delta_-^n y_\nu, \delta_+\delta_-^n y_\nu, \cdots, \delta_+^{n-1}\delta_-^n y_\nu;\ \delta_-^n z_\nu, \delta_+\delta_-^n z_\nu, \cdots, \delta_+^{n-1}\delta_-^n z_\nu. \end{gathered} \tag{16.6}$$

It is not hard to see that the quantities $y_{\nu-1}, \delta_- y_{\nu-1}, \cdots, \delta_-^{n-1}y_{\nu-1}$; $z_{\nu-1}, \delta_- z_{\nu-1}, \cdots, \delta_-^{n-1}z_{\nu-1}$ appear as factors in the form $[\bar{y}, \bar{z}]_h(\nu)$, and the remaining difference ratios for y_ν and z_ν appear in more complicated expressions whose coefficients consist of difference ratios of the functions $q_0(x), q_1(x), \cdots, q_n(x)$. Therefore if we replace the derivatives in the boundary conditions in (15.14) by the difference ratios (16.6) without changing the numbers M_{jm} and N_{jm}, then in general, even if the differential problem is selfadjoint, we do not obtain selfadjoint boundary conditions for the difference problem. However for the case of the simplest boundary conditions (15.17) – (15.18) it is not hard to write down selfadjoint boundary conditions for the finite-difference problem. These conditions have the form

$$y_{-1} = \delta_- y_{-1} = \cdots = \delta_-^{n-1}y_{-1} = 0, \tag{16.7}$$

$$y_{N-1} = \delta_- y_{N-1} = \cdots = \delta_-^{n-1}y_{N-1} = 0. \tag{16.8}$$

In fact, as we already noted above, the quantities

$$\begin{gathered} y_{-1},\ \delta_- y_{-1},\ \ldots,\ \delta_-^{n-1}y_{-1}; \qquad z_{-1},\ \delta_- z_{-1},\ \ldots,\ \delta_-^{n-1}z_{-1}, \\ y_{N-1},\ \delta_- y_{N-1},\ \ldots,\ \delta_-^{n-1}y_{N-1}; \quad z_{N-1},\ \delta_- z_{N-1},\ \ldots,\ \delta_-^{n-1}z_{N-1} \end{gathered}$$

appear as factors in the right side of (16.5). Therefore if all of these quantities are equal to zero, then regardless of what the other quantities are equal to, one has $(l_h\bar{y}, \bar{z})_h = (\bar{y}, l_h\bar{z})_h$, i.e. the problem

$$l_h\bar{y} = \lambda\bar{y} \tag{16.9}$$

with the boundary conditions (16.7) – (16.8) is selfadjoint.

2. Completeness of the system of eigenfunctions. Giving ν the values $0, 1, 2, \cdots, N-1$, we obtain from (16.9) a system of homogeneous linear equations which connect the quantities $y_{\nu-n}, y_{\nu-n+1}, \cdots, y_\nu$, $y_{\nu+1}, \cdots, y_{\nu+n}$. It follows from the boundary conditions (16.7) that

$$y_{-1} = y_{-2} = \cdots = y_{-n} = 0. \tag{16.10}$$

Therefore from the equation (16.9) we can express $y_r, y_{r+1}, \cdots, y_{N+r-1}$ in terms of $y_0, y_1, \cdots, y_{n-1}$. The latter quantities can be regarded as arbitrary.

Let us denote by $\varphi_{j,h}(\nu, \lambda)$ $(j = 1, 2, \cdots, 2n;\ \nu = 0, 1, \cdots, N-1)$ the solution of equation (16.9) satisfying the initial conditions

$$\delta_-^{m-1} y_{-1} = \delta_{jm} \qquad (j, m = 1, 2, \cdots, 2n). \tag{16.11}$$

The conditions (16.11) contain the values of $\varphi_{m,h}(\nu, \lambda)$ for $\nu = -1, -2, \cdots, -2n$. Therefore in order that equation (16.9) have meaning, it is necessary to extend the coefficients of the differential operator (15.13) slightly to the left while preserving their differentiability properties.

It is known that if $h \to 0$ and $\nu h \to x$, then

$$\delta_-^{m-1} \varphi_{j,h}(\nu, \lambda) \to \varphi_j^{(m-1)}(x, \lambda). \tag{16.12}$$

Let us denote by $\lambda_{1,h} \leqq \lambda_{2,h} \leqq \cdots \leqq \lambda_{N,h}$ the eigenvalues of the problem (16.9) + (16.7) + (16.8). The eigenvalues $\lambda_{m,h}$ are the roots of the equation (cf. the derivation of equation (15.20))

$$\Delta_h(\lambda) \equiv \begin{vmatrix} \varphi_{n+1,h}(N-1, \lambda), & \ldots, & \varphi_{2n,h}(N-1, \lambda) \\ \delta_- \varphi_{n+1,h}(N-1, \lambda), & \ldots, & \delta_- \varphi_{2n,h}(N-1, \lambda) \\ \cdot\ \cdot\ \cdot & \cdot\ \cdot\ \cdot & \cdot\ \cdot\ \cdot \\ \delta_-^{n-1} \varphi_{n+1,h}(N-1, \lambda), & \ldots, & \delta_-^{n-1} \varphi_{2n,h}(N-1, \lambda) \end{vmatrix} = 0.$$

It follows from (16.12) that

$$\lim_{h \to 0} \Delta_h(\lambda) = \Delta(\lambda), \tag{16.13}$$

where $\Delta(\lambda)$ is defined by (15.20).

We now denote by $\psi_{m,h}(\nu)$ an eigenvector of the problem (16.9) + (16.7) + (16.8) corresponding to the eigenvalue $\lambda_{m,h}$. Since the vectors $\varphi_{j,h}(\nu, \lambda)$ form a fundamental system of solutions of equation (16.9) and the vectors $\psi_{m,h}(\nu)$ satisfy the initial conditions (16.7), we have

$$\psi_{m,h}(\nu) = \sum_{j=n+1}^{2n} r_{mj,h} \varphi_{j,h}(\nu, \lambda_m), \tag{16.14}$$

where the $r_{mj,h}$ are real constants.

Let $\bar{f}=(f_0,f_1,\cdots,f_{N-1})$ be an arbitrary N-dimensional vector. Since the vectors $\psi_{m,h}(\nu)$ $(\nu=0,1,\cdots,N-1;\ m=1,2,\cdots,N)$ form a complete orthonormal system of vectors in N-dimensional vector space, the Parseval equality holds:

$$(\bar{f},\bar{f})_h=\sum_{m=1}^{N}(\bar{f},\bar{\psi}_{m,h})_h^2. \tag{16.15}$$

Inserting (16.14) in (16.15), we can write (16.15) in the form

$$\|\bar{f}\|_h^2=\int_{-\infty}^{\infty}\sum_{i,j=n+1}^{2n}g_{ih}(\lambda)g_{jh}(\lambda)\,d\rho_{ij,h}(\lambda), \tag{16.16}$$

where $g_{ih}(\lambda)=(\bar{f},\bar{\varphi}_{i,h}(\lambda))_h$, and the matrix $\rho_h(\lambda)=\{\rho_{ij,h}(\lambda)\}_{i,j=n+1}^{2n}$ consists of jump functions whose jumps occur at the eigenvalues of the problem (16.9) + (16.7) + (16.8). The value of the jump at the point $\lambda_{p,h}$ is determined from the equality

$$\rho_{ij,h}(\lambda_{p,h}+0)-\rho_{ij,h}(\lambda_{p,h}-0)=\sum_k r_{ki,h}r_{kj,h},$$

where the sum is taken over all k such that $\lambda_{k,h}=\lambda_{p,h}$ (since the spectrum, generally speaking, contains nonsimple eigenvalues, different vectors $\psi_{k,h}(\nu)$ can correspond to one and the same eigenvalue $\lambda_{p,h}$).

Arguing now in precisely the same way as in the proof of the expansion theorem for the eigenfunctions of a singular differential operator (cf. Lemma 9.1, Chapter 2; cf. also §5 of Chapter 1 and §1 of Chapter 2), one can show that the increase of the function $\rho_{ii,h}(\lambda)$ in every finite interval (μ,ν) is bounded uniformly with respect to h. It follows that for bounded $\lambda_{p,h}$ the corresponding numbers $r_{mi,h}$ are also bounded uniformly with respect to h. Therefore we can find a sequence $h_s\to 0$ such that the limits

$$\lim_{s\to\infty}r_{ki,h_s}=r_{ki}\qquad(k,i=1,2,\cdots) \tag{16.17}$$

exist.

We will show that the limits $\lim_{h\to 0}\lambda_{p,h}$ $(p=1,2,\cdots)$ also exist. In fact, it follows from (16.13) and a well-known theorem of the theory of analytic functions that

$$\lim_{h\to 0}\lambda_{p,h}=\lambda_p, \tag{16.18}$$

where $\lambda_{p,h}$ and λ_p are the zeros of the functions $\Delta_h(\lambda)$ and $\Delta(\lambda)$ respectively, arranged in increasing order of their moduli. Replacing h

by h_s in (16.14) and letting $s \to \infty$, $\nu h_s \to x$, we obtain in the limit the functions

$$\psi_m(x) = \sum_{j=n+1}^{2n} r_{mj} \varphi_j(x, \lambda_m),$$

which, as is easy to verify, are orthonormal eigenfunctions of the problem (15.16) + (15.17) + (15.18). That the functions $\varphi_m(x)$ are solutions of equation (15.15) for $\lambda = \lambda_m$ follows from the fact that they are linear combinations of the functions $\varphi_j(x, \lambda_m)$. The boundary conditions (15.17) – (15.18), and also the orthonormality of the $\psi_m(x)$, are obtained from the corresponding relations for the $\psi_{m,h}(\nu)$ by a passage to the limit.

Arguing now as in the proof of Theorem 5.1, Chapter 1, we obtain from (16.15) the Parseval equality for any function $f(x) \in \mathcal{L}_2[0,1]$:

(16.19)
$$\int_0^1 f^2(x)\,dx = \sum_{m=1}^{\infty} \left(\int_0^1 f(x) \psi_m(x)\,dx \right)^2.$$

We will now show that the numbers λ_p coincide with all the roots of equation (15.20). Indeed, if the equation (15.20) had a root μ which differed from all the λ_p, then the problem (15.15) + (15.17) + (15.18) would have an eigenfunction $\psi(x)$ corresponding to the eigenvalue μ, and consequently orthogonal to all the $\psi_m(x)$. But it would then follow from (16.19) that $\psi(x) \equiv 0$.

§17. Calculation of the regularized trace for a fourth-order operator

Our aim in this section is to calculate the regularized trace of the operator

(17.1)
$$Ly = y^{\mathrm{IV}} + q(x)y = \lambda y \qquad (0 \leqq x \leqq \pi),$$

(17.2)
$$y(0) = y(\pi) = 0, \qquad y''(0) = y''(\pi) = 0.$$

Our arguments will be of a general nature, and we are considering an operator of the form (17.1) – (17.2) only to simplify the discussion. One can similarly work out formulas for the traces of operators defined by more general differential expressions and by arbitrary boundary conditions. Using the same methods, one can also determine trace formulas for certain singular problems.

To calculate the trace of the operator (17.1) – (17.2) it suffices to assume that the function $q(x)$ is once differentiable. But we will prove more than the formula for the so-called first trace. Therefore we will assume that $q(x)$ is a sufficiently smooth function. We will also assume, which is no loss of generality, that $\int_0^\pi q(x)\,dx = 0$.

1. Asymptotic behaviour of the solutions for large values of λ. It is well known[25] that if the complex s-plane ($s^4 = \lambda$) is divided into eight sectors S_l ($l = 0, 1, \cdots, 7$), defined by the inequalities

$$\frac{l\pi}{8} \leqslant \arg s \leqslant \frac{(l+1)\pi}{8} \qquad (l = 0, 1, \ldots, 7),$$

then in each of these sectors the equation (17.1) has four linearly independent solutions $y_k(x,s)$ ($k = 1, 2, 3, 4$), which are regular with respect to s in the sector S_l for $|s|$ sufficiently large and which satisfy the relation

$$y_k(x, s) = e^{\omega_k s x}\left[\sum_{\nu=0}^{N} \frac{u_{\nu,k}(x)}{s^\nu} + O\left(\frac{1}{s^{N+1}}\right)\right], \tag{17.3}$$

where the ω_k are the fourth roots of unity, while N is a positive integer which depends upon the smoothness of the function $q(x)$. If $q(x)$ has m continuous derivatives, then one can assert the existence of a representation (17.3) with $N = m + 3$. The functions $u_{\nu,k}(x)$ satisfy the following recursion relations:

$$\begin{aligned} &4u'_{\nu,k}(x) + 6\omega_k^3 u''_{\nu-1,k}(x) + 4\omega_k^2 u'''_{\nu-2,k}(x) \\ &+ \omega_k u^{\mathrm{IV}}_{\nu-3,k}(x) + \omega_k q(x)\, u_{\nu-3,k}(x) = 0, \qquad u_{\nu,k}(x) \equiv 0, \quad \nu < 0. \end{aligned} \tag{17.4}$$

Let us put, moreover, $u_{0,k}(x) \equiv 1$, $u_{\nu,h}(0) = 0$ for $\nu \geqq 1$. The functions $u_{\nu,k}(x)$ are now uniquely determined. Thus, for example, we find from (17.4) that

$$\begin{aligned} &u_{0,k}(x) = 1, \ u_{1,k}(x) = 0, \ u_{2,k}(x) = 0, \\ &u_{3,k} = -\frac{1}{4\omega_k^3}\int_0^x q(t)\,dt, \qquad u_{4,k} = \frac{3}{8} q(x), \ \ldots \end{aligned} \tag{17.5}$$

One can differentiate the formulas (17.3) with respect to x, whereby the following relations are found to be valid:

$$y_k^{(i)}(x, s) = e^{\omega_k s x}(\omega_k s)^i \left[\sum_{\nu=0}^{N} \frac{u^{(i)}_{\nu,k}(x)}{s^\nu} + O\left(\frac{1}{s^{N+1}}\right)\right],$$

where

$$\begin{aligned} &u^{(i)}_{0,k}(x) = 0, \quad u^{(i)}_{1,k}(x) = 0, \quad u^{(i)}_{2,k}(x) = 0, \quad u^{(i)}_{3,k}(x) = -\frac{1}{4\omega_k^3}\int_0^x q(t)\,dt, \\ &u'_{4k}(x) = \frac{1}{8} q(x), \ u''_{4,k}(x) = -\frac{1}{8} q(x), \\ &u'''_{4,k}(x) = -\frac{3}{8} q(x), \ \ldots \ (i = 0, 1, 2, 3). \end{aligned} \tag{17.6}$$

---

[25] Cf. J. Tamarkin [1], p. 64.

2. An entire function $f(s)$ whose zeros are the eigenvalues of the operator (17.1)—(17.2). In this section we will find an asymptotic representation of an entire function $f(s)$ whose zeros are the eigenvalues of the operator (17.1) – (17.2).

We introduce the following notation:

$$U_1(y_k) = U_{1k} = y_k(0), \quad U_2(y_k) = U_{2k} = y''(0), \tag{17.7}$$
$$U_3(y_k) = U_{3k} = y_k(\pi), \quad U_4(y_k) = U_{4k} = y''(\pi).$$

Now let $v_k(x,s)$ be the solution of equation (17.1), normalized at $x=0$ by the Cauchy conditions

$$v_k^{(p-1)}(0,s) = \delta_{k,p} \qquad (k,p = 1,2,3,4).$$

We denote by $\overline{\beta}_k(s)$ the vector $(y_k(0,s), y_k'(0,s), y_k''(0,s), y_k'''(0,s))$, where the primes denote differentiation with respect to x. Let

$$\overline{V}(x,s) = (v_1(x,s), v_2(x,s), v_3(x,s), v_4(x,s)).$$

We note that by virtue of the uniqueness theorem $y_k(x,s)$ is equal to the scalar product of the vectors $\overline{\beta}_k(s)$ and $\overline{V}(x,s)$, i.e.

$$y_k(x,s) = (\overline{V}(x,s), \overline{\beta}_k(s)).$$

Inserting the expressions (17.6) in $\Delta(s) = \det\|U_{ij}\|$ $(i,j = 1,2,3,4)$ (see (17.7)), we find without difficulty that $\Delta(s) = f(s)\delta(s)$, where $\delta(s)$ is the Wronskian determinant of the solutions $y_k(x,s)$ $(k = 1,2,3,4)$.

We have thus obtained a representation of an entire function $f(s)$, whose zeros are the eigenvalues of the operator (17.1) – (17.2), in the form

$$f(s) = \Delta(s)/\delta(s). \tag{17.8}$$

The representation (17.8) enables us to find the asymptotic behaviour of $f(s)$ for $s \to \infty$. On the basis of formulas (17.3), (17.5) and (17.6) we conclude that

$$\Delta(s) = \det\|U_i(y_j)\| = \det\|u_{i,j}\| = s^4 \sum_{k=1}^{4} e^{\alpha_k s} P_{k,N}(s), \tag{17.9}$$

where the α_k $(k = 1,2,3,4)$ are the fourth roots of minus one, numbered in increasing order of their argument, and $\alpha_5 = 0$. In each of the sectors S_l of the complex s-plane the functions $P_{k,N}(s)$ admit the representation

$$P_{k,N}(s) = \sum_{\nu=0}^{N} \frac{\beta_\nu^{(k)}}{s^\nu} + o\left(\frac{1}{s^{N+1}}\right). \tag{17.9'}$$

Formula (17.9′) is obtained by computing a fourth-order determinant.

We obtain in this connection that, for example,

(17.9″) $\beta_0^{(1)}=4,\ \beta_1^{(1)}=0,\ \beta_2^{(1)}=0,\ \beta_3^{(1)}=0,\ \beta_4^{(1)}=q(0)+q(\pi),\cdots.$

We can similarly compute the Wronskian determinant of the solutions $y_k(x,s)$:

$$\delta(s)=\sum_{\nu=0}^{N}\frac{\gamma_\nu}{s^\nu}+O\left(\frac{1}{s^{N+1}}\right), \tag{17.10}$$

where γ_0 is the Vandermonde determinant of the numbers ω_k.

We note that it follows from (17.6) that the first nonzero coefficient γ_ν ($\nu \geqq 1$) has index greater than four. We will use this remark in what follows.

3. The zeta-function associated with $f(s)$. Let P be the complex hull of the points $\bar{\alpha}_1,\bar{\alpha}_2,\bar{\alpha}_3,\bar{\alpha}_4,\bar{\alpha}_5$ in the complex plane. In our case P is a quadrangle. We will call the directions of the exterior normals of P critical directions. We delete from the plane four sectors T_m ($m=1,2,3,4$), with sufficiently small openings, whose bisectrices are parallel to the critical directions. The set of points remaining, which breaks up into four open sectors Ω_m ($m=1,2,3,4$), will be denoted by Ω.

The following assertion is easy to establish.

LEMMA 17.1. *For sufficiently large R, the intersection of the region $|s|>R$ with Ω contains no zeros of $f(s)$.*

PROOF. In fact, it is easy to verify that $\operatorname{Re}\alpha s=(\bar{\alpha},s)$, where the right side is the scalar product of the two-component vectors $\bar{\alpha}$ and s. Suppose now for definiteness that s belongs to Ω_1. Then it is geometrically clear that for some $\delta>0$ we will have the inequality $\operatorname{Re}\alpha_1 s-\operatorname{Re}\alpha_k s>\delta|s|$, $k\neq 1$. As a consequence we at once obtain from (17.9) $f(s)=Ce^{\alpha_1 s}(1+o(1))$. A similar estimate is valid in the other sectors Ω_m, which proves the lemma.

We now choose in one of the sectors (Ω_1 for definiteness) a ray l and construct a contour Γ_1 consisting of the ray l traversed twice and a circle γ with center at zero. In our case 0 is not an eigenvalue, hence $f(0)\neq 0$. It is now obvious that we can choose the ray l and the circle γ in such a way that all the zeros of $f(s)$ lie in the exterior of Γ_1.

Noting further that for $s\in\Gamma_1$ and $s\to\infty$

$$\frac{f'(s)}{f(s)}=\alpha_1+\frac{P'_{1,N}(s)}{P_{1,N(s)}}+O(e^{-\delta|s|})\sim\sum_{\nu=0}^{\infty}\frac{\omega_\nu^{(1)}}{s^\nu}, \tag{17.11}$$

we consider the integral

$$z_1(\sigma) = \frac{1}{2\pi i} \int_{\Gamma_1} s^{-\sigma} \frac{f'(s)}{f(s)} ds, \tag{17.12}$$

which by virtue of (17.11) converges in the halfplane $\operatorname{Re}\sigma > 1$. In (17.12) $s^{-\sigma} = e^{-\sigma \operatorname{Ln} s}$, where $\operatorname{Ln} s$ is a fixed branch of the logarithm in the exterior of Γ_1. We will call the function (17.12) the zeta-function associated with the function $f(s)$.

Lemma 17.2. *For* $\operatorname{Re}\sigma > 1$

$$z_1(\sigma) = \sum_{(l)} s_l^{-\sigma}, \tag{17.13}$$

where the s_l are the zeros of $f(s)$.

Proof. Since $f(s)$ is an entire function of order 1, $f'(s)/f(s)$ has the following expansion, which converges uniformly in every finite disc:

$$\frac{f'(s)}{f(s)} = \sum_{(l)} \left[\frac{1}{s - s_l} + \frac{1}{s_l} \right] + a. \tag{17.14}$$

Using the fact that for $|s_l| > R$ the zeros of $f(s)$ lie in the sectors T_m, it is not hard to obtain the estimate $|s - s_l| > \delta |s_l|$ $(\delta > 0)$ for all l and $s \in \Gamma_1$.

We split up the sum in the right side of (17.14) into two parts: $\sum_{(l)} = \sum_{(l')} + \sum_{(l'')}$, where the second sum contains all terms for which $|s_l| > R$. It is easily seen that

$$\sum_{(l'')} \leqslant \sum_{(l'')} \frac{|s|}{|s - s_l| |s_l|} < \frac{|s|}{\delta} \sum_{(l'')} \frac{1}{|s_l|^2} < \varepsilon |s| \tag{17.15}$$

for sufficiently large R. Multiplying (17.14) by $s^{-\sigma}$, $\operatorname{Re}\sigma > 2$, we integrate both sides over the contour Γ_1. The bound (17.15) permits us to interchange the order of summation and integration. Further, since

$$\frac{1}{2\pi i} \int_{\Gamma_1} s^{-\sigma} \left[\frac{1}{s - s_l} + \frac{1}{s_l} \right] ds = s_l^{-\sigma},$$

we arrive at (17.13) for $\operatorname{Re}\sigma > 2$. Noting, however, that both sides of (17.13) are defined and regular in the halfplane $\operatorname{Re}\sigma > 1$, we conclude that (17.13) holds for $\operatorname{Re}\sigma > 1$.

Lemma 17.3. *The zeta-function $z_1(\sigma)$ can be analytically extended to the halfplane $\operatorname{Re}\sigma > -N - 1$ as a regular function. If N in (17.3) can be chosen arbitrarily, then $z_1(\sigma)$ is an entire function.*

Proof. We break up the integral (17.12) into four integrals:

$$z_1(\sigma)=\frac{1}{2\pi i}\oint_{\gamma} s^{-\sigma}\frac{f'(s)}{f(s)}\,ds+\frac{1}{2\pi i}\int_{\Gamma_1'} s^{-\sigma}\left[\frac{f'(s)}{f(s)}-\sum_{\nu=0}^{\nu_0}\frac{\omega_\nu^{(1)}}{s^\nu}\right]ds$$

(17.16)
$$+\frac{1}{2\pi i}\int_{\Gamma_1} s^{-\sigma}\sum_{\nu=0}^{\nu_0}\frac{\omega_\nu^{(1)}}{s^\nu}\,ds-\frac{1}{2\pi i}\oint_{\gamma} s^{-\sigma}\sum_{\nu=0}^{\nu_0}\frac{\omega_\nu^{(1)}}{s^\nu}\,ds$$

$$=I_1(\sigma)+I_2(\sigma)+I_3(\sigma)+I_4(\sigma).$$

Here Γ_1' is that part of Γ_1 which traverses the ray l in opposite directions, and ν_0 is some positive integer less than $N+1$.

It is easy to see that $I_1(\sigma)$ and $I_4(\sigma)$ are entire functions of σ; $I_2(\sigma)$, by virtue of (17.11), can be analytically extended to the halfplane $\operatorname{Re}\sigma>-\nu_0$. Finally, $I_3(\sigma)$ equals zero for $\operatorname{Re}\sigma>1$, and consequently can be analytically extended to zero over the entire plane. Since ν_0 is arbitrary, the lemma is proved.

The representation (17.16) makes it possible to find the values of $z_1(\sigma)$ at the integers. We have the following result.

LEMMA 17.4. *For* $n=0,1,2,\cdots$

(17.17)
$$z_1(-n)=\omega_{n+1}^{(1)},$$

where the $\omega_\nu^{(1)}$ *are the coefficients in the expansion* (17.11).

For the proof we go back to formula (17.16). For σ a nonpositive integer we have $I_1(\sigma)=0$, since the integrand is a regular function of s inside γ; further, $I_2(\sigma)=0$ for integer σ as a consequence of the fact that it consists of the integral of a single-valued function along the ray l in opposite directions. Noting further that $I_3(\sigma)\equiv 0$, the question reduces to the calculation of $I_4(\sigma)$. This obviously leads us to formula (17.17), which proves the lemma.

We remark that the numbers $\omega_\nu^{(1)}$ can easily be found from the asymptotic behaviour of the function $f(s)$. Taking into account formulas (17.9), (17.9′), (17.9″) and (17.10), and also the remark following formula (17.10), we obtain, for example,

(17.18)
$$\omega_5^{(1)}=-[q(0)+q(\pi)].$$

4. The regularized trace of the operator (17.1)—(17.2). Let us assume, to simplify what follows, that $q(x)$ is an infinitely differentiable function. Then it is known that the roots of the function $f(s)$ have the following asymptotic representation:[26]

[26] Cf. Sadovnič̆iĭ [2], p. 296 = 266.

(17.19) $$s_{n,i} = n\alpha_i \left\{1 + \frac{C_{6,i}}{n^6} + \frac{C_{8,i}}{n^8} + \dots\right\}.$$

Here the index i $(i = 1, 2, 3, 4)$ denotes the index of the sector T_i in which the given series of roots lies. Raising both sides to the power $-\sigma$, we obtain

$$S_{n,i}^{-\sigma} \sim n^{-\sigma}\alpha_i^{-\sigma}\left\{1 + \sum_{k=1}^{\infty} \frac{C_{2k,i}}{n^{2k}}\right\}^{-\sigma}, \qquad C_{2,i} = C_{4,i} = 0.$$

Since the familiar Taylor expansion of the function $(1+x)^\alpha$ is valid for complex x and α, we can represent the third factor in the preceding formula by an asymptotic series and obtain

(17.20) $$S_{n,i}^{-\sigma} \sim \sum_{k=0}^{\infty} \frac{Q_{2k}^{(i)}(\sigma)}{n^{2k+\sigma}},$$

where

(17.20′) $$Q_0^{(i)}(\sigma) = \alpha_i^{-\sigma}, \quad Q_2^{(i)}(\sigma) = 0, \quad Q_4^{(i)}(\sigma) = 0, \cdots.$$

Let τ be a fixed, sufficiently large integer. It follows from (17.20) that

(17.21) $$\psi_\tau^{(1)}(\sigma) = \sum_{n=1}^{\infty}\sum_{i=1}^{4}\left[S_{n,i}^{-\sigma} - \sum_{k=0}^{\tau} \frac{Q_{2k}^{(i)}(\sigma)}{n^{2k+\sigma}}\right]$$

can be analytically extended to the halfplane $\operatorname{Re}\sigma > -2\tau - 1$, since the general term of the series (17.21) is $O(1/n^{2\tau+\sigma+2})$.

The superscript on $\psi_\tau^{(1)}(\sigma)$ indicates the dependence of this function upon the choice of the contour Γ_1.

Our aim is to determine the numbers $\frac{1}{4}\psi_\tau^{(1)}(-4m)$ $(m = 1, 2, \cdots;$ $2m < \tau + 1)$, which we will call regularized sums of the eigenvalues of the operator (17.1) – (17.2).

Let us consider the function

$$\Phi^{(1)}(\sigma) = \sum_{n=1}^{\infty}\sum_{i=1}^{4}\sum_{k=0}^{\tau} \frac{Q_{2k}^{(i)}(\sigma)}{n^{2k+\sigma}},$$

which is regular for $\operatorname{Re}\sigma > 1$. In accordance with (17.21) we have

$$\psi_\tau^{(1)}(\sigma) = z_1(\sigma) - \Phi_\tau^{(1)}(\sigma).$$

Since for an infinitely differentiable $q(x)$ the function $z_1(\sigma)$ is entire, then together with $\psi_\tau^{(1)}(\sigma)$ the function $\Phi_\tau^{(1)}(\sigma)$ can be analytically extended to the halfplane $\operatorname{Re}\sigma > -2\tau - 1$. We note that $\Phi_\tau^{(1)}(\sigma)$ can be expressed in terms of the Riemann zeta-function. In fact,

$$\Phi_\tau^{(1)}(\sigma) = \sum_{k=0}^{\tau} Q_{2k}(\sigma) \cdot \sum_{n=1}^{\infty} \frac{1}{n^{2k+\sigma}} = \sum_{k=0}^{\tau} Q_{2k}(\sigma)\,\zeta(2k+\sigma),$$

(17.22)

$$Q_{2k}(\sigma) = \sum_{i=1}^{4} Q_{2k}^{(i)}(\sigma).$$

Since the values of the zeta-function at the negative integers are well known, formula (17.22) enables us to find the values $\Phi_\tau^{(1)}(-4m)$, $m = 1, 2, \cdots$. Thus, taking account of (17.20′), we obtain from (17.22), for example ($\zeta(-4) = 0$),

(17.23) $\quad \Phi_2^{(1)}(-4) = Q_0(-4)\zeta(-4) + Q_2(-4)\zeta(-2) + Q_4(-4)\zeta(0) = 0.$

Recalling the formula (17.17) for $z_1(-n)$, we arrive at the following theorem.

THEOREM 17.1. *For any positive* m $(2m < \tau + 1)$ *we have the equality*

$$\sum_{n=1}^{\infty}\sum_{i=1}^{4}\left[S_{n,i}^{4m} - \sum_{k=0}^{\tau}\frac{Q_{2k}^{(i)}(-4m)}{n^{2k-4m}}\right] = \omega_{4m+1}^{(1)} - \Phi_\tau^{(1)}(-4m).$$

In particular, for $m = 1$, taking into account that the roots $S_{n,i}$ are symmetrically situated over all of the sectors T_i, and also formulas (17.20′), (17.18) and (17.23), we obtain, choosing τ equal to two, for example,

$$4\sum_{n=1}^{\infty}[\lambda_n - n^4] = \omega_5^{(1)} - \Phi_2^{(1)}(-4) = -[q(0) + q(\pi)].$$

Hence

$$\sum_{n=1}^{\infty}[\lambda_n - n^4] = -\frac{q(0) + q(\pi)}{4}.$$

Other regularized sums can be similarly calculated.

Bibliographical references

§§1–2. The transformation of the general second-order equation to canonical form and the asymptotic formulas for the eigenvalues and eigenfunctions are classical, and are basically due to Liouville.

§3. The results of the section are classical and were obtained by Sturm.

§4. The essential results of this theory were established by Hamel [**1**], Haupt [**1**] and Titchmarsh [**4**]. See also Hoheisel [**1**].

§5. The method of finite differences was first applied for the study of boundary-value problems for second-order equations by Plancherel [**1,2**]. He proved that the eigenvalues and eigenvectors of the finite-difference problem converge to the eigenvalues and eigenfunctions of the differential problem. He did not prove the completeness of the system of eigenfunctions by this route. This was done by Levitan [**18**].

§6. The results of this section are classical. The systematic application of the Green's function to the study of boundary-value problems was initiated by Hilbert [**1**]. We remark that the completeness and expansion theorems for the classical Sturm-Liouville problem were first proved by Steklov [**1**].

§7. See, for example, Hoheisel [**1**].

§8. The idea of applying contour integration for the proof of the expansion theorem goes back to Cauchy. This section was written with the help of §1.9 of the monograph by Titchmarsh [**1**].

§9. The results of this section are basically due to Hobson [**1**] and Haar [**1, 2**]. Another approach to equiconvergence theorems, which is also applicable to singular second-order differential operators, was developed by Levitan [**14, 15**].

§10. The canonical form (10.4) for a system of two first-order equations was given by Sargsjan [**7**], and (10.5), by Gasymov [**1**].

§11. The results of this section are basically due to Hurwitz [**1**]. The application of transformation operators is due to the authors.

§§12–13. The completeness of the system of eigenvector-functions of a system of two first-order equations was proved by Hurwitz [**1**]. Birkhoff and Langer [**1**] have considered a general first-order system. The finite-difference method was applied to the proof of completeness by Sargsjan [**7**]. The application of the method of integral equations is due to the authors and appears here for the first time.

§14. The regularized trace for the classical Sturm-Liouville problem was first calculated by Gel′fand and Levitan [**2**]. Cf. also Gel′fand [**1**], Levitan [**20**], Dikiĭ [**1**]. For Dirac systems the trace was calculated by Abdukadyrov [**1**].

§15. The results of this section are classical. In writing this section we made use of the monographs by Naĭmark [**1**] and by Coddington and Levinson [**2**].

§16. The application of the method of finite differences to the proof of the completeness of the eigenfunctions of an nth order differential operator is due to Levitan, and appears here for the first time.

§17. The formula for the first trace, for a differential operator defined by the two-term operator $l(y) = y^{(n)} + q(x)y$ with general regular boundary conditions, was first obtained by Ševčenko [**1**]. Then Sadovničiĭ [**1, 2**] proved trace formulas for more general cases. Finally, Lidskiĭ and Sadovničiĭ [**1–3**] indicated a procedure for calculating trace formulas for general problems involving ordinary differential equations on a finite interval.

CHAPTER 2

EIGENFUNCTION EXPANSIONS FOR A STURM-LIOUVILLE OPERATOR FOR THE CASE OF AN INFINITE INTERVAL

§1. Proof of the Parseval equality for the halfline

1. A Sturm-Liouville problem is said to be *regular* if the interval $[a,b]$ is finite and the function $q(x)$ is summable over this interval. In the contrary case, i.e. if the interval $[a,b]$ is infinite or the function $q(x)$ is not summable over $[a,b]$ (or both), then the Sturm-Liouville problem is said to be *singular*.

In §3 of Chapter 1 we obtained the expansion theorem for a regular Sturm-Liouville problem by means of a passage to the limit, replacing the differential equation by a finite-difference equation.

A similar idea makes it possible to obtain the expansion theorem for a singular problem, if we consider it as the limit of regular problems.

We will start by considering the case in which the interval is the halfline $[0, \infty)$, and the function $q(x)$ is continuous on this interval. Thus, we will consider the equation

$$y'' + \{\lambda - q(x)\}y = 0 \qquad (0 \leqq x < \infty). \tag{1.1}$$

We denote by α an arbitrary real number, and by $y(x,\lambda)$ the solution of (1.1) satisfying the initial conditions

$$y(0,\lambda) = \sin\alpha, \qquad y'(0,\lambda) = -\cos\alpha. \tag{1.2}$$

Obviously the function $y(x,\lambda)$ satisfies the boundary condition

$$y(0,\lambda)\cos\alpha + y'(0,\lambda)\sin\alpha = 0. \tag{1.3}$$

Let us denote by b an arbitrary positive number (subsequently b will increase unboundedly), by β an arbitrary real number, and adjoin to to the problem $(1.1)+(1.3)$ the boundary condition

$$y(b,\lambda)\cos\beta + y'(b,\lambda)\sin\beta = 0. \tag{1.4}$$

The problem $(1.1)+(1.3)+(1.4)$ is a regular Sturm-Liouville problem. Let us denote by $\lambda_{n,b}$ the eigenvalues of this problem and by $y_{n,b}(x)$

$=y(x,\lambda_{n,b})$ the corresponding eigenfunctions satisfying the initial conditions (1.2). If $f(x)\in\mathscr{L}^2(0,b)$ and

$$\alpha_{n,b}^2=\int_0^b y_{n,b}^2(x)\,dx,$$

then by the Parseval equality

$$\int_0^b f^2(x)\,dx=\sum_{n=1}^{\infty}\frac{1}{\alpha_{n,b}^2}\left(\int_0^b f(x)\,y_{n,b}(x)\,dx\right)^2. \tag{1.5}$$

If we introduce the monotonically nondecreasing jump function

$$\rho_b(\lambda)=-\sum_{\lambda<\lambda_{n,b}\leqslant 0}\frac{1}{\alpha_{n,b}^2}\qquad(\lambda\leqslant 0),$$

$$\rho_b(\lambda)=\sum_{0<\lambda_{n,b}\leqslant\lambda}\frac{1}{\alpha_{n,b}^2}\qquad(\lambda>0),$$

then (1.5) can be written in the form

$$\int_0^b f^2(x)\,dx=\int_{-\infty}^{\infty}F^2(\lambda)\,d\rho_b(\lambda), \tag{1.6}$$

where

$$F(\lambda)=\int_0^b f(x)\,y(x,\lambda)\,dx.$$

We will show that the Parseval equality for the problem (1.1) + (1.3) can be obtained from (1.6) by letting $b\to\infty$. To this end we prove a lemma.

LEMMA 1.1. *For any positive integer N there exists a positive constant $A=A(N)$, not depending upon b, such that*

$$\bigvee_{-N}^{N}\{\rho_b(\lambda)\}=\sum_{-N<\lambda_{n,b}\leqslant N}\frac{1}{\alpha_{n,b}^2}=\rho(N)-\rho(-N)<A. \tag{1.7}$$

PROOF. We will first consider the case $\sin\alpha\neq 0$. Since the function $y(x,\lambda)$ is continuous in the region $-N\leqq\lambda\leqq N$, $0\leqq x\leqq a$ (where a is an arbitrary fixed number), by virtue of the condition $y(0,\lambda)=\sin\alpha$ there exists a positive number h so small that for $|\lambda|\leqq N$

$$\left(\frac{1}{h}\int_0^h y(x,\lambda)\,dx\right)^2>\frac{1}{2}\sin^2\alpha. \tag{1.8}$$

If we now apply (1.6) to the function

$$f_h(x) = \begin{cases} \frac{1}{h}, & 0 \leqslant x < h, \\ 0, & x > h, \end{cases}$$

then, using the bound (1.8), we obtain

$$\int_0^h f_h^2(x)\,dx = \frac{1}{h} = \int_{-\infty}^{\infty} \left\{ \frac{1}{h} \int_0^h y(x, \lambda)\,dx \right\}^2 d\rho_b(\lambda)$$

$$\geqslant \int_{-N}^{N} \left\{ \frac{1}{h} \int_0^h y(x, \lambda)\,dx \right\}^2 d\rho_b(\lambda) > \frac{1}{2} \sin^2\alpha \int_{-N}^{N} d\rho_b(\lambda)$$

$$= \frac{1}{2} \sin^2\alpha \,\{\rho_b(N) - \rho_b(-N)\},$$

from which (1.7) follows.

For the case in which $\sin\alpha = 0$, we obtain the bound (1.7) by applying the Parseval equality to the function

$$f_h(x) = \begin{cases} \frac{1}{h^2}, & 0 \leqslant x < h, \\ 0, & x > h. \end{cases}$$

This proves the lemma.

We will now show that, using Lemma 1 1 and a well-known theorem on passing to the limit inside a Stieltjes integral, one can obtain the Parseval equality for the problem (1.1) + (1.3).

Suppose first that the function $f_n(x)$ vanishes outside the interval $0 \leqq x \leqq n$, $n < b$, has a continuous second derivative, and satisfies the boundary condition (1.3). Applying the Parseval equality (1.6) to this function, we obtain

$$\int_0^n f_n^2(x)\,dx = \int_{-\infty}^{\infty} F_n^2(\lambda)\,d\rho_b(\lambda), \tag{1.9}$$

where

$$F_n(\lambda) = \int_0^n f_n(x)\,y(x, \lambda)\,dx.$$

Since both $f_n(x)$ and $y(x,\lambda)$ satisfy the boundary condition (1.3) and $f_n(x)$ vanishes identically in a neighborhood of b, it follows from Green's identity that

$$F_n(\lambda) = -\frac{1}{\lambda} \int_0^b f_n(x)\,\{y''(x, \lambda) - q(x)\,y(x, \lambda)\}\,dx =$$

$$= -\frac{1}{\lambda}\int_0^b y(x, \lambda)\{f_n''(x) - q(x) f_n(x)\}\,dx.$$

Therefore for any finite $N > 0$ we have

$$\int_{|\lambda|>N} F_n^2(\lambda)\,d\rho_b(\lambda)$$
$$\leqslant \frac{1}{N^2}\int_{|\lambda|>N}\left\{\int_0^b y(x, \lambda)[f_n''(x) - q(x) f_n(x)]\,dx\right\}^2 d\rho_b(\lambda)$$
$$< \frac{1}{N^2}\int_{-\infty}^{\infty}\left\{\int_0^b y(x, \lambda)[f_n''(x) - q(x) f_n(x)]\,dx\right\}^2 d\rho_b(\lambda)$$
$$= \frac{1}{N^2}\int_0^n \{f_n''(x) - q(x) f_n(x)\}^2\,dx.$$

From this bound and from (1.9) we obtain

$$\left|\int_0^n f_n^2(x)\,dx - \int_{-N}^{N} F_n^2(\lambda)\,d\rho_b(\lambda)\right| < \frac{1}{N^2}\int_0^n \{f_n''(x) - q(x) f_n(x)\}^2\,dx. \tag{1.10}$$

By Lemma 1.1 the set of monotonic functions $\{\rho_b(\lambda)\}$ $(-N \leqq \lambda \leqq N)$ is bounded. We can therefore choose a sequence $\{b_k\}$ for which the functions $\rho_{b_k}(\lambda)$ converge to a monotonic function $\rho(\lambda)$.

Passing to the limit in (1.10) along the sequence $\{b_k\}$, we obtain the inequality

$$\left|\int_0^n f_n^2(x)\,dx - \int_{-N}^{N} F_n^2(\lambda)\,d\rho(\lambda)\right| \leqslant \frac{1}{N^2}\int_0^n \{f_n''(x) - q(x) f_n(x)\}^2\,dx.$$

Finally, letting $N \to \infty$ in this inequality, we obtain

$$\int_0^n f_n^2(x)\,dx = \int_{-\infty}^{\infty} F_n^2(\lambda)\,d\rho(\lambda),$$

i.e. the Parseval equality for functions which satisfy the conditions indicated above.

Now let $f(x)$ be an arbitrary square-integrable function: $f(x) \in \mathscr{L}^2(0, \infty)$. As is known, one can find a sequence $\{f_n(x)\}$ of functions which satisfy the preceding conditions (i.e., equal zero outside the interval $[0, n]$, satisfy the boundary condition (1.3) and have a continuous second derivative) as well as the condition

$$\lim_{n\to\infty}\int_0^\infty \{f(x)-f_n(x)\}^2\,dx=0.$$

Since

$$\int_0^\infty \{f_n(x)-f_m(x)\}^2\,dx\to 0 \quad (n,\ m\to\infty),$$

it follows that

$$\int_{-\infty}^\infty \{F_n(\lambda)-F_m(\lambda)\}^2\,d\rho(\lambda)=\int_0^\infty \{f_n(x)-f_m(x)\}^2\,dx\to 0.$$

Since the space of functions which are square-integrable (with respect to the monotonic function $\rho(\lambda)$) is complete,[1)] the preceding relation implies the existence of a limit function $F(\lambda)$ satisfying the Parseval equality

$$\int_0^\infty f^2(x)\,dx=\int_{-\infty}^\infty F^2(\lambda)\,d\rho(\lambda).$$

It remains to show that as $n\to\infty$ the function

$$F_n(\lambda)=\int_0^n f(x)\,y(x,\lambda)\,dx$$

converges in the mean (relative to $d\rho(\lambda)$) to $F(\lambda)$.

Let $g(x)$ be another square-integrable function, and let $G(\lambda)$ be constructed from $g(x)$ in the same way that $F(\lambda)$ was constructed from $f(x)$. It is obvious that

$$\int_0^\infty \{f(x)-g(x)\}^2\,dx=\int_{-\infty}^\infty \{F(\lambda)-G(\lambda)\}^2\,d\rho(\lambda).$$

Putting $g(x)=f(x)$ $(0\leqq x\leqq n)$, $g(x)=0$ $(x>n)$, we obtain

$$\int_{-\infty}^\infty \{F(\lambda)-F_n(\lambda)\}^2\,d\rho(\lambda)=\int_n^\infty f^2(x)\,dx\to 0 \quad (n\to\infty),$$

---

1) The usual proof of the completeness of the space $\mathscr{L}^2(-\infty,\infty)$ goes over word for word to the case of Lebesgue-Stieltjes integrals.

which proves the mean convergence of $F_n(\lambda)$ to $F(\lambda)$.

We have thus proved:

Theorem 1.1. *Let* $f(x) \in \mathscr{L}^2(0, \infty)$. *There exists a nondecreasing function* $\rho(\lambda)$,[2] *which does not depend upon* $f(x)$, *and a function* $F(\lambda)$ *(the generalized Fourier transform of* $f(x)$*) such that*

$$\int_0^\infty f^2(x)\,dx = \int_{-\infty}^{\infty} F^2(\lambda)\,d\rho(\lambda).$$

The function $F(\lambda)$ is the limit in quadratic mean (relative to $d\rho(\lambda)$) of the sequence of continuous functions

$$F_n(\lambda) = \int_0^n f(x)\,y(x, \lambda)\,dx,$$

i.e.

$$\lim_{n\to\infty} \int_{-\infty}^{\infty} \{F(\lambda) - F_n(\lambda)\}^2\,d\rho(\lambda) = 0.$$ [3]

2. Let $f(x)$, $g(x) \in \mathscr{L}^2(0, \infty)$ and let $F(\lambda)$ and $G(\lambda)$ be their (generalized) Fourier transforms. Obviously the functions $f(x) \pm g(x)$ have $F(\lambda) \pm G(\lambda)$ as their Fourier transforms. Therefore

$$\int_0^\infty \{f(x) + g(x)\}^2\,dx = \int_{-\infty}^{\infty} \{F(\lambda) + G(\lambda)\}^2\,d\rho(\lambda),$$

$$\int_0^\infty \{f(x) - g(x)\}^2\,dx = \int_{-\infty}^{\infty} \{F(\lambda) - G(\lambda)\}^2\,d\rho(\lambda).$$

Subtracting the lower relation from the upper, we obtain

$$\int_0^\infty f(x)\,g(x)\,dx = \int_{-\infty}^{\infty} F(\lambda)\,G(\lambda)\,d\rho(\lambda). \tag{1.11}$$

The relation (1.11) is called the generalized Parseval equality.

---

[2] We will see further on that the function $\rho(\lambda)$ may be nonunique.

[3] In our further considerations we will for brevity write this equality in the form

$$F(\lambda) = \underset{n\to\infty}{\text{l.i.m.}} \int_0^n f(x)\,y(x, \lambda)\,dx.$$

THEOREM 1.2 (EXPANSION THEOREM). *Let $f(x)$ be a continuous function $(0 \leqq x < \infty)$, and suppose that the integral*

$$\int_{-\infty}^{\infty} F(\lambda)\, y(x, \lambda)\, d\rho(\lambda)$$

converges absolutely and uniformly with respect to x in every finite interval. Then

$$f(x) = \int_{-\infty}^{\infty} F(\lambda)\, y(x, \lambda)\, d\rho(\lambda). \tag{1.12}$$

PROOF. Let $g(x)$ be a continuous function which vanishes outside the finite interval $[0, n]$. In this case (1.11) can be written as

$$\int_0^n f(x)\, g(x)\, dx = \int_{-\infty}^{\infty} F(\lambda) \left\{ \int_0^n g(x)\, y(x, \lambda)\, dx \right\} d\rho(\lambda).$$

By virtue of absolute convergence, the order of integration in the last integral can be reversed, and we obtain

$$\int_0^n f(x)\, g(x)\, dx = \int_0^n g(x) \left\{ \int_{-\infty}^{\infty} F(\lambda)\, y(x, \lambda)\, d\rho(\lambda) \right\} dx.$$

Since $g(x)$ is an arbitrary continuous function, and the functions $f(x)$ and $\int_{-\infty}^{\infty} F(\lambda) y(x, \lambda) d\rho(\lambda)$ are continuous (the latter's continuity follows from the assumption of uniform convergence of the integral), we obtain

$$f(x) = \int_{-\infty}^{\infty} F(\lambda)\, y(x, \lambda)\, d\rho(\lambda),$$

as was to be proved.

§2. The Weyl point and Weyl circle for a Sturm-Liouville operator

1. In this section we will as before consider the interval $[0, \infty)$ and a function $q(x)$ which is continuous on this interval.

Let $F(x)$ satisfy the differential equation

$$y'' + \{\lambda - q(x)\} y = 0, \tag{2.1}$$

and let $G(x)$ satisfy the same equation with λ' instead of λ. If b is a fixed number, then by the identity (1.4) of Chapter 1

$$(\lambda' - \lambda) \int_0^b F(x)\, G(x)\, dx$$

$$(2.2)\quad = \int_0^b \{F(x)[q(x)G(x) - G''(x)] - G(x)[q(x)F(x) - F''(x)]\}\,dx$$

$$= -\int_0^b \{F(x)G''(x) - G(x)F''(x)\}\,dx = W_0\{F, G\} - W_b\{F, G\},$$

where

$$W_x\{F, G\} = \begin{vmatrix} F(x) & G(x) \\ F'(x) & G'(x) \end{vmatrix}.$$

If, in particular, $\lambda = u + iv$, $\lambda' = \bar{\lambda} = u - iv$, then since $q(x)$ is real we can take $G(x) = \bar{F}(x)$, and it therefore follows from (2.2) that

$$(2.2')\quad 2v \int_0^b |F(x)|^2\,dx = iW_0\{F, \bar{F}\} - iW_b\{F, \bar{F}\}.$$

Let us denote by $\varphi(x) = \varphi(x, \lambda)$ and $\theta(x) = \theta(x, \lambda)$ the solutions of the equation (2.1) satisfying the initial conditions (with α a real number)

$$(2.3)\quad \begin{aligned} \varphi(0) &= \sin\alpha, & \varphi'(0) &= -\cos\alpha, \\ \theta(0) &= \cos\alpha, & \theta'(0) &= \sin\alpha. \end{aligned}$$

Since equation (2.1) does not contain the first derivative of y, by a well-known theorem of Liouville the Wronskian is a constant, and consequently

$$W_x\{\varphi, \theta\} = W_0\{\varphi, \theta\} = \sin^2\alpha + \cos^2\alpha = 1.$$

Therefore the general solution of equation (2.1) can be represented (up to a constant factor) in the form $\theta(x) + l\varphi(x)$.

Let us consider that solution which at the point b satisfies the condition

$$(2.3')\quad \{\theta(b) + l\varphi(b)\}\cos\beta + \{\theta'(b) + l\varphi'(b)\}\sin\beta = 0$$

with a real β. We define l from this condition:

$$(2.4)\quad l = -\frac{\theta(b)\operatorname{ctg}\beta + \theta'(b)}{\varphi(b)\operatorname{ctg}\beta + \varphi'(b)}.$$

If b is fixed and $\operatorname{ctg}\beta$ assumes all values from $-\infty$ to $+\infty$, then l will describe a circle C_b in the complex plane. We replace $\operatorname{ctg}\beta$ by the complex variable z and put

$$(2.4')\quad l = l(\lambda, z) = -\frac{\theta(b)z + \theta'(b)}{\varphi(b)z + \varphi'(b)}.$$

To the point $l=\infty$ there corresponds $z=-\varphi'(b)/\varphi(b)$. Therefore to the center of the circle C_b there corresponds the point $\bar z=-\bar\varphi'(b)/\bar\varphi(b)$. Inserting $\bar z$ in the formula for l, we find that the center of C_b is the point $-W_b\{\theta,\bar\varphi\}/W_b\{\varphi,\bar\varphi\}$. Further,

$$\operatorname{Im}\left\{-\frac{\varphi'(b)}{\varphi(b)}\right\}=\frac{1}{2}i\left\{\frac{\varphi'(b)}{\varphi(b)}-\frac{\bar\varphi'(b)}{\bar\varphi(b)}\right\}=-\frac{1}{2}i\,\frac{W_b\{\varphi,\bar\varphi\}}{|\varphi(b)|^2}.$$

Since $W_0\{\varphi,\bar\varphi\}=0$, by virtue of (2.2′) the last expression has the same sign as v. We have thus proved:

Lemma 2.1. *If $v>0$, then to the upper z-halfplane there corresponds the exterior of the circle C_b.*

Since the point $-\theta'(b)/\varphi'(b)$ lies on C_b (for $z=0$), we obtain for the radius r_b of C_b the expression

$$r_b=\left|\frac{\theta'(b)}{\varphi'(b)}-\frac{W_b\{\theta,\bar\varphi\}}{W_b\{\varphi,\bar\varphi\}}\right|=\left|\frac{W_b\{\theta,\varphi\}}{W_b\{\varphi,\bar\varphi\}}\right|=\left(2v\int_0^b|\varphi(x)|^2\,dx\right)^{-1}. \tag{2.5}$$

Further, by Lemma 2.1 l lies inside the circle C_b if $\operatorname{Im} z<0$, i.e. if $i(z-\bar z)>0$.

Solving (2.4′) with respect to z and inserting the value obtained in the inequality $i(z-\bar z)>0$, we obtain

$$i\left\{-\frac{l\varphi'(b)+\theta'(b)}{l\varphi(b)+\theta(b)}+\frac{\bar l\bar\varphi(b)+\bar\theta'(b)}{\bar l\bar\varphi(b)+\bar\theta(b)}\right\}>0,$$

from which we find that

$$\begin{aligned}i[|l|^2W_b\{\varphi,\bar\varphi\}+lW_b\{\varphi,\bar\theta\}+\bar lW_b\{\theta,\bar\varphi\}+W_b\{\theta,\bar\theta\}]\\=iW_b\{\theta+l\varphi,\bar\theta+\bar l\bar\varphi\}>0.\end{aligned}$$

Therefore, putting $F=\theta+l\varphi$ in (2.2′), we obtain

$$2v\int_0^b|\theta(x)+l\varphi(x)|^2\,dx<iW_0\{\theta+l\varphi,\bar\theta+\bar l\bar\varphi\}.$$

Since $W_0\{\varphi,\bar\theta\}=1$, $W_0\{\varphi,\bar\varphi\}=0$, $W_0\{\theta,\bar\theta\}=0$, and $W_0\{\theta,\bar\varphi\}=1$, it follows that $W_0\{\theta+l\varphi,\bar\theta+\bar l\bar\varphi\}=l-\bar l=2i\operatorname{Im} l$. Thus l lies inside C_b if $v>0$ and

$$\int_0^b|\theta(x)+l\varphi(x)|^2\,dx<-\frac{\operatorname{Im} l}{v}. \tag{2.6}$$

The same result can be obtained for the case $v<0$ (using the inequality

$\operatorname{Im} z > 0$). In both cases the sign of $\operatorname{Im} l$ is opposite to that of v. Consequently, if l lies inside C_b and $0 < b' < b$, then

$$\int_0^{b'} |\theta(x) + l\varphi(x)|^2 \, dx < \int_0^{b} |\theta(x) + l\varphi(x)|^2 \, dx < -\frac{\operatorname{Im} l}{v}.$$

Therefore l also lies inside $C_{b'}$. Hence if $b' < b$, then $C_{b'}$ contains C_b. It follows that as $b \to \infty$ the circle C_b either converges to a limit circle or a limit point. Let $m = m(\lambda)$ be the limit point or any point of the limit circle. Then for any b we have

$$\int_0^{b} |\theta(x) + m\varphi(x)|^2 \, dx \leqslant -\frac{\operatorname{Im}(m)}{v}. \tag{2.7}$$

Therefore

$$\int_0^{\infty} |\theta(x) + m\varphi(x)|^2 \, dx \leqslant \frac{\operatorname{Im}(m)}{v}.$$

We thus have the following result.

THEOREM 2.1. *For every nonreal value of* λ *equation* (2.1) *has a solution*

$$\psi(x, \lambda) = \theta(x, \lambda) + m(\lambda)\varphi(x, \lambda)$$

which is square integrable over the interval $[0, \infty)$.

For the case of a limit circle, r_b tends to a positive limit as $b \to \infty$. It therefore follows from (2.5) that $\varphi(x, \lambda) \in \mathscr{L}^2(0, \infty)$, i.e. in this case every solution of (2.1) belongs to $\mathscr{L}^2(0, \infty)$.

The question naturally arises: does the classification of limit circle or limit point depend upon the particular choice of λ, or does it depend only upon the operator $-d^2/dx^2 + q(x)$? The answer is given by the following result.

THEOREM 2.2. *If, for some complex* λ_0, *every solution of equation* (2.1) *belongs to* $\mathscr{L}^2(0, \infty)$, *then for any complex* λ *every solution of* (2.1) *belongs to* $\mathscr{L}^2(0, \infty)$. *In other words, if for some nonreal* λ_0 *the limit circle case holds, then this case holds for any nonreal* λ.

PROOF. Suppose that two linearly independent solutions φ and ψ of the equation $-d^2y/dx^2 + q(x)y = \lambda_0 y$, where $\operatorname{Im} \lambda_0 \neq 0$, belong to $\mathscr{L}^2(0, \infty)$. Let χ be any solution of the equation $-d^2y/dx^2 + q(x)y = \lambda y$. We write this equation in the form

$$-d^2y/dx^2+q(x)y=\lambda_0 y+(\lambda-\lambda_0)y.$$

Multiplying φ if necessary by a constant so as to have $W\{\varphi,\bar{\psi}\}=1$, and applying the formula of variation of constants, we obtain

$$\chi(x)=c_1\varphi(x)+c_2\psi(x)+(\lambda-\lambda_0)\int_c^x\{\varphi(x)\psi(t)-\varphi(t)\psi(x)\}\chi(t)\,dt, \tag{2.8}$$

where c, c_1 and c_2 are constants. If we use the notation

$$\|\chi(x)\|_c=\left\{\int_c^x|\chi(t)|^2dt\right\}^{1/2}$$

and if M is such that $\|\varphi(x)\|_c\leqq M$, $\|\psi(x)\|_c\leqq M$ for all $x\geqq c$, then the Cauchy-Bunjakovskiĭ inequality gives us

$$\int_c^x|\{\varphi(x)\psi(t)-\varphi(t)\psi(x)\}\chi(t)|\,dt\leqslant M\{|\varphi(x)|+|\psi(x)|\}\|\chi(x)\|_c.$$

Applying this inequality to the formula (2.8), and then applying Minkowski's inequality, we obtain

$$\|\chi(x)\|_c\leqq\{|c_1|+|c_2|\}M+2|\lambda-\lambda_0|M^2\|\chi(x)\|_c.$$

If c is so large that $|\lambda-\lambda_0|M^2<\frac{1}{4}$, then

$$\|\chi(x)\|_c\leqq 2\{|c_1|+|c_2|\}M.$$

Since the right side of this inequality does not depend upon x, it follows that $\chi(x)\in\mathscr{L}^2(0,\infty)$, which proves the theorem.

A useful sufficient condition for the operator $L\equiv-d^2/dx^2+q(x)$ to belong to the limit point case is given by the following theorem.

THEOREM 2.3. *If $q(x)\geqq-kx^2$, where k is some positive constant, then the limit point case holds for L.*

PROOF. We will show that the equation $-y''+q(x)y=0$ does not have two linearly independent solutions in the class $\mathscr{L}^2(0,\infty)$. Let us assume that $\varphi(x)$ is a real solution of the equation $Ly=0$, and let $\varphi(x)\in\mathscr{L}^2(0,\infty)$. From the equality $\varphi''=q\varphi$ we obtain $(c>0)$

$$\int_c^x\frac{\varphi''(t)\varphi(t)}{t^2}dt=\int_c^x\frac{q(t)}{t^2}\varphi^2(t)\,dt\geqslant-k\int_c^x\varphi^2(t)\,dt.$$

Integrating by parts and using the fact that $\varphi(x)\in\mathscr{L}^2(0,\infty)$, we prove the existence of a constant k_1 such that

$$-\frac{\varphi'(x)\varphi(x)}{x^2}+\int_c^x \frac{\{\varphi'(t)\}^2}{t^2}\,dt-\int_c^x \frac{2\varphi'(t)\varphi(t)}{t^3}\,dt<k_1. \tag{2.9}$$

Let

$$H(x)=\int_c^x \frac{\{\varphi'(t)\}^2}{t^2}\,dt.$$

Then using the Cauchy-Bunjakovskiĭ inequality, we obtain

$$\left|\int_c^x \frac{2\varphi'(t)\varphi(t)}{t^3}\,dt\right|^2 \leqslant k_2\left(\int_c^x \frac{|\varphi'(t)|\cdot|\varphi(t)|}{t}\,dt\right)^2 \leqslant k_2 H(x)\int_c^x \varphi^2(t)\,dt. \tag{2.10}$$

Thus by virtue of (2.10) it follows from (2.9) that there exists a constant k_3 such that

$$-\frac{\varphi'(x)\varphi(x)}{x^2}+H(x)-k_3H^{1/2}(x)<k_1. \tag{2.11}$$

If $H(x)\to\infty$ as $x\to\infty$, it follows from (2.11) that

$$\frac{\varphi'(x)\varphi(x)}{x^2}>\frac{1}{2}H(x)$$

for all sufficiently large x. This shows that the functions $\varphi(x)$ and $\varphi'(x)$ have the same sign for large x, and this contradicts the fact that $\varphi(x)\in\mathscr{L}^2(0,\infty)$. Thus $H(x)$ remains finite, so that

$$\int_c^\infty \frac{\{\varphi'(t)\}^2}{t^2}\,dt<\infty. \tag{2.12}$$

Let us assume that $\varphi(x)$ and $\psi(x)$ are two linearly independent solutions of the equation $Ly=0$ from $\mathscr{L}^2(0,\infty)$, i.e. we assume that the limit circle case holds for L. We can assume that these solutions are real and that

$$W\{\varphi,\psi\}=\varphi(x)\psi'(x)-\varphi'(x)\psi(x)=1.$$

It follows that

$$\varphi(x)\frac{\psi'(x)}{x}-\psi(x)\frac{\varphi'(x)}{x}=\frac{1}{x}.$$

By virtue of (2.12) and the Cauchy-Bunjakovskiĭ inequality, the left side of this equality is integrable over the interval (c,∞), and the right side is obviously nonintegrable. Thus the limit circle case cannot hold. This proves the theorem.

2. For given β the function $l=l(\lambda)$ is an analytic function of λ. It is even meromorphic with poles on the real axis. In fact, by virtue of (2.4) the poles of $l(\lambda)$ are the zeros of the function $\varphi(b,\lambda)\cos\beta+\varphi'(b,\lambda)\sin\beta$.

These zeros are the eigenvalues of equation (2.1) in the finite interval $[0,b]$, and are consequently simple.

It follows from the considerations of the previous subsection that for given λ the region in the l-plane covered by the disc C_b decreases as b increases. Therefore the family of analytic functions $l(\lambda)=l(\lambda,b,\beta)$ is uniformly bounded in every bounded region lying entirely in the upper (lower) λ-halfplane. By a well-known theorem of the theory of analytic functions this family is normal (i.e., one can choose a uniformly convergent subsequence from every infinite sequence of these functions) in every such region.

If for some nonreal λ one has the limit point case (by Theorem 2.2 this will be the case for all other nonreal λ), then as is easily seen the family of functions $l(\lambda)$ has a unique limit $m(\lambda)$ which is an analytic function by virtue of a familiar theorem of Weierstrass.

For the limit circle case the family $l(\lambda)$ has nonunique limits. From the normality of the family it follows that one can find an unboundedly increasing sequence of numbers b_k and a sequence of numbers β_k such that the limit

$$\lim_{k\to\infty} l(\lambda,b_k,\beta_k)=m(\lambda)$$

exists in each of the λ-halfspaces and is an analytic function. The number $m(\lambda)$ obviously lies on the limit of the circles.

3. We put $\psi_b(x,\lambda)=\theta(x)+l\varphi(x),\ \ l=l(\lambda,b)$.

LEMMA 2.2. *For every nonreal* λ

$$\psi_{b_k}(x,\lambda)\to\psi(x,\lambda),$$

$$\int_0^b|\psi_{b_k}(x,\lambda)|^2\,dx\to\int_0^\infty|\psi(x,\lambda)|^2\,dx\qquad(k\to\infty),$$

where $\{b_k\}$ *is as above.*

PROOF. It is obvious that

$$\psi_b(x,\lambda)=\psi(x,\lambda)+[l(\lambda)-m(\lambda)]\varphi(x,\lambda),$$

where $\psi(x,\lambda)\in\mathcal{L}^2(0,\infty)$.

For the case of a limit circle $l(\lambda)\to m(\lambda)$, and therefore $\psi_{b_k}(x,\lambda)\to\psi(x,\lambda)$; since $\varphi(x,\lambda)\in\mathcal{L}^2(0,\infty)$,

$$\int_0^b|\psi_{b_k}(x,\lambda)|^2\,dx\to\int_0^\infty|\psi(x,\lambda)|^2\,dx.$$

For the case of a limit point

$$|l(\lambda)-m(\lambda)|\leqslant r_{bk}=\left(2v\int_0^b|\varphi(x,\lambda)|^2\,dx\right)^{-1}\qquad(v\neq 0).$$

Since $r_{b_k}\to 0$, it follows that $\psi_{bk}(x,\lambda)\to\psi(x,\lambda)$,

$$\int_0^b|\{l(\lambda)-m(\lambda)\}\varphi(x,\lambda)|^2\,dx=|l(\lambda)-m(\lambda)|^2\int_0^b|\varphi(x,\lambda)|^2\,dx$$

$$\leqslant\left(4v^2\int_0^b|\varphi(x,\lambda)|^2\,dx\right)^{-1}$$

Therefore in this case also

$$\int_0^b|\psi_{bk}(x,\lambda)|^2\,dx\to\int_0^\infty|\psi(x,\lambda)|^2\,dx.$$

§3. Integral representation of the resolvent

Let $f(x)\in\mathscr{L}^2(0,\infty)$. We put ($\lambda$ is a nonreal number)

$$\Phi(x,\lambda)=\psi(x,\lambda)\int_0^x\varphi(y,\lambda)f(y)\,dy+\varphi(x,\lambda)\int_x^\infty\psi(y,\lambda)f(y)\,dy,$$

where $\varphi(x,\lambda)$ and $\psi(x,\lambda)$ are the functions introduced in the previous section. If $f(x)$ is a continuous function, then

$$\Phi'_x(x,\lambda)=\psi'_x(x,\lambda)\int_0^x\varphi(y,\lambda)f(y)\,dy+\varphi'_x(x,\lambda)\int_x^\infty\psi(y,\lambda)f(y)\,dy,$$

$$\Phi''(x,\lambda)=\psi''(x,\lambda)\int_0^x\varphi(y,\lambda)f(y)\,dy+\varphi''(x,\lambda)\int_x^\infty\psi(y,\lambda)f(y)\,dy$$

$$+\{\psi'(x,\lambda)\varphi(x,\lambda)-\varphi'(x,\lambda)\psi(x,\lambda)\}f(x)$$

$$=\{q(x)-\lambda\}\Phi(x,\lambda)+f(x),$$

since

$$\varphi(x,\lambda)\psi'(x,\lambda)-\varphi'(x,\lambda)\psi(x,\lambda)=W_x\{\varphi,\psi\}=W_0\{\varphi,\psi\}$$

$$=\begin{vmatrix}\sin\alpha & \cos\alpha+m(\lambda)\sin\alpha\\ -\cos\alpha & \sin\alpha-m(\lambda)\cos\alpha\end{vmatrix}=1.$$

Thus

$$\Phi''(x,\lambda) + \{\lambda - q(x)\}\Phi(x,\lambda) = f(x).$$

Moreover,

$$\Phi(0, \lambda) = \varphi(0, \lambda) \int_0^\infty \psi(y, \lambda) f(y)\, dy,$$

$$\Phi'(0, \lambda) = \varphi'(0, \lambda) \int_0^\infty \psi(y, \lambda) f(y)\, dy,$$

so that $\Phi(x, \lambda)$ satisfies the boundary condition

$$\Phi(0,\lambda)\cos\alpha + \Phi'(0,\lambda)\sin\alpha = 0.$$

Let $\lambda_{n,b}$ and $\varphi_{n,b}(x)$ be the eigenvalues and eigenfunctions of the boundary-value problem (1.1) + (1.3) + (1.4). Let $l(\lambda)$ and $\psi_b(x,\lambda)$ be defined as in the previous section. We put

$$G_b(x, y; z) = \begin{cases} \psi_b(x, z)\varphi(y, z), & y \leqslant x, \\ \varphi(x, z)\psi_b(y, z), & y > x, \end{cases}$$

$$R_{z,b}f = \int_0^b G_b(x, y; z) f(y)\, dy \qquad (z = u + iv).$$

Setting

$$\alpha_n^2 = \int_0^b \varphi_{n,b}^2(x)\, dx, \tag{3.1}$$

we obtain (cf. Chapter 1, §6)

$$\begin{aligned} R_{z,b}f &= \sum_{n=1}^\infty \frac{\varphi_{n,b}(x)\int_0^b f(y)\varphi_{n,b}(y)\, dy}{\alpha_n^2(z - \lambda_{n,b})} \\ &= \int_{-\infty}^\infty \frac{\varphi(x,\lambda)\int_0^b f(y)\varphi(y,\lambda)\, dy}{z - \lambda}\, d\rho_b(\lambda), \end{aligned} \tag{3.2}$$

where the function $\rho_b(\lambda)$ is defined in §1 of the present chapter.

LEMMA 3.1. *For every nonreal z and fixed x we have*

$$\int_{-\infty}^\infty \left|\frac{\varphi(x,\lambda)}{z-\lambda}\right|^2 d\rho_b(\lambda) < K. \tag{3.3}$$

Proof. Putting $f(y) = \varphi_{n,b}(y)$ in the first of the equalities (3.2) and taking account of (3.1) by virtue of the orthogonality of the eigenfunctions $\varphi_{n,b}(x)$ we obtain

$$\frac{1}{\alpha_n}\int_0^b G_b(x, y; z)\,\varphi_{n,b}(y)\,dy = \frac{\varphi_{n,b}(z)}{\alpha_n(z - \lambda_{n,b})}. \tag{3.4}$$

Now applying the Parseval equality to the function $G_b(x, y; z)$ (regarding it as a function of y) and taking (3.4) into account, we find that

$$\int_0^b |G_b(x, y; z)|^2\,dy = \sum_{n=1}^{\infty}\frac{\varphi_{n,b}^2(x)}{\alpha_n^2\,|z - \lambda_{n,b}|^2} = \int_{-\infty}^{\infty}\left|\frac{\varphi(x, \lambda)}{z - \lambda}\right|^2 d\rho_b(\lambda).$$

The assertion of the lemma follows from this equality, since by Lemma 2.2 the integral on the left converges.

Corollary I. *Suppose that the function $\rho(\lambda)$ is the same as in §1. Then*

$$\int_{-\infty}^{\infty}\left|\frac{\varphi(x, \lambda)}{z - \lambda}\right|^2 d\rho(\lambda) \leqslant K \tag{3.5}$$

with the same constant K as in the inequality (3.3).

In fact, for arbitrary $a > 0$ it follows from (3.3) that

$$\int_{-a}^{a}\left|\frac{\varphi(x, \lambda)}{z - \lambda}\right|^2 d\rho_b(\lambda) < K.$$

Letting first $b \to \infty$ and then $a \to \infty$, we obtain (3.5).

Corollary II. If $\sin\alpha \neq 0$, then putting $x = 0$ in (3.5), we obtain $\int_{-\infty}^{\infty}|z - \lambda|^{-2}d\rho(\lambda) < \infty$. It follows that for any $a > 0$

$$\int_{-\infty}^{-a}\frac{d\rho(\lambda)}{\lambda^2} < \infty, \qquad \int_a^{\infty}\frac{d\rho(\lambda)}{\lambda^2} < \infty. \tag{3.6}$$

If $\sin\alpha = 0$, then differentiating (3.4) we obtain

$$\frac{1}{\alpha_n}\int_0^b \frac{\partial}{\partial x} G_b(x, y; z)\,\varphi_{n,b}(y)\,dy = \frac{\varphi'_{n,b}(x)}{\alpha_n(z - \lambda_{n,b})}.$$

Hence by the Parseval equality we find that

$$\int_0^b \left|\frac{\partial}{\partial x} G_b(x, y; z)\right|^2 dy = \int_{-\infty}^{\infty}\left|\frac{\varphi'(x, \lambda)}{z - \lambda}\right|^2 d\rho_b(\lambda).$$

Proceeding now in a manner similar to before, it is not hard to obtain (3.6) for this case as well.

LEMMA 3.2. *Let* $f(x) \in \mathscr{L}^2(0, \infty)$ *and*

$$R_z f = \int_0^\infty G(x, y; z) f(y)\, dy,$$

where

$$G(x, y; z) = \begin{cases} \psi(x, z)\varphi(y, z), & y \leqslant x, \\ \varphi(x, z)\psi(y, z), & y > x. \end{cases}$$

Then

$$\int_0^\infty |R_z f|^2 dx \leqslant \frac{1}{v^2}\int_0^\infty f^2(x)\, dx.$$

PROOF. For every $b > 0$ it follows from (3.2), by the Parseval equality, that $(z = u + iv)$

$$\int_0^b |R_{z,b} f|^2 dx = \sum_{n=1}^\infty \frac{1}{\alpha_n^2 |z - \lambda_{n,b}|^2}\left\{\int_0^b f(y)\varphi_{n,b}(y)\, dy\right\}^2$$

$$\leqslant \frac{1}{v^2}\sum_{n=0}^\infty \frac{1}{\alpha_n^2}\left\{\int_0^b f(y)\varphi_{n,b}(y)\, dy\right\}^2 = \frac{1}{v^2}\int_0^b f^2(y)\, dy.$$

Let a be a fixed positive number. If $a < b$, then

$$\int_0^a |R_{z,b} f|^2 dx \leqslant \int_0^b |R_{z,b} f|^2 dx \leqslant \frac{1}{v^2}\int_0^b f^2(y)\, dy.$$

Letting b tend to ∞, we obtain

$$\int_0^a |R_z f|^2 dx \leqslant \frac{1}{v^2}\int_0^\infty f^2(y)\, dy,$$

and since a is arbitrary, the lemma is proved.

THEOREM 3.1 (INTEGRAL REPRESENTATION OF THE RESOLVENT). *For every function* $f(x) \in \mathscr{L}^2(0, \infty)$ *and every nonreal* z *one has the equality*

$$R_z f = \int_{-\infty}^\infty \frac{\varphi(x, \lambda) F(\lambda)}{z - \lambda}\, d\rho(\lambda), \tag{3.7}$$

where

$$F(\lambda) = \underset{n\to\infty}{\mathrm{l.i.m.}} \int_0^n f(x)\varphi(x, \lambda)\, dx.$$

Proof. Let us first assume that the function $f(x) = f_n(x)$ vanishes outside the finite interval $[0, n]$, satisfies the boundary condition (1.3) and has a continuous second derivative.

Let $b > n$ and let ν be an arbitrary positive number. We put

$$F_n(\lambda) = \int_0^n f_n(x)\varphi(x, \lambda)\, dx.$$

Then (3.2) can be rewritten in the form

$$\begin{aligned} R_{z,b}f_n &= \int_{-\infty}^{\infty} \frac{\varphi(x, \lambda) F_n(\lambda)}{z-\lambda}\, d\rho_b(\lambda) \\ &= \left\{\int_{-\infty}^{-\nu} + \int_{-\nu}^{\nu} + \int_{\nu}^{\infty}\right\} \frac{\varphi(x, \lambda) F_n(\lambda)}{z-\lambda}\, d\rho_b(\lambda) = I_1 + I_2 + I_3. \end{aligned} \tag{3.8}$$

Let us now estimate I_1. By virtue of (3.2) we have

$$\begin{aligned} I_1 &= \int_{-\infty}^{-\nu} \frac{\varphi(x, \lambda) F_n(\lambda)}{z-\lambda}\, d\rho_b(\lambda) \\ &= \sum_{\lambda_{k,b} < -\nu} \frac{\varphi_{k,b}(x)}{\alpha_k^2 (z-\lambda_{k,b})} \int_0^n f_n(y)\varphi_{k,b}(y)\, dy \\ &\leqslant \left(\sum_{\lambda_{k,b}<-\nu} \frac{\varphi_{k,b}^2(x)}{\alpha_k^2 |z-\lambda_{k,b}|^2}\right)^{1/2} \left(\sum_{\lambda_{k,b}<-\nu} \frac{1}{\alpha_k^2}\left[\int_0^n f_n(x)\varphi_{k,b}(x)\, dx\right]^2\right)^{1/2}. \end{aligned} \tag{3.9}$$

Further, integrating by parts twice, we obtain

$$\begin{aligned} \int_0^n f_n(x)\varphi_{k,b}(x)\, dx &= -\frac{1}{\lambda_{k,b}} \int_0^n f_n(x)\{\varphi''_{k,b}(x) - q(x)\varphi_{k,b}(x)\}\, dx \\ &= -\frac{1}{\lambda_{k,b}} \int_0^n \{f''_n(x) - q(x) f_n(x)\}\varphi_{k,b}(x)\, dx. \end{aligned} \tag{3.10}$$

Since by Lemma 3.1 the first sum in the right side of (3.9) converges, taking account of (3.10) we find from (3.9) that

$$I_1 \leqslant K^{1/2} \frac{1}{\nu}\left(\sum_{\lambda_{k,b}<-\nu} \frac{1}{\alpha_k^2}\left\{\int_0^n [f''_n(x) - q(x) f_n(x)]\varphi_{k,b}(x)\, dx\right\}^2\right)^{1/2}.$$

Therefore by Bessel's inequality

$$I_1 \leqslant \frac{K^{1/2}}{\nu}\left(\int_0^n \{f_n''(x) - q(x) f_n(x)\}^2\, dx\right)^{1/2} = \frac{C}{\nu}.$$

In the same way one can prove that $I_3 \leqq C/\nu$. It follows from these bounds that I_1 and I_3 tend to zero as $\nu \to \infty$, uniformly in b. We can therefore apply a generalization of Helly's theorem and obtain from (3.8) (see, by way of comparison, the proof of Theorem 1.1)

$$R_z f_n = \int_{-\infty}^{\infty} \frac{\varphi(x, \lambda) F_n(\lambda)}{z - \lambda}\, d\rho(\lambda). \tag{3.11}$$ [4)]

If $f(x)$ is an arbitrary square-integrable function on $[0, \infty)$, then, as is known, one can find a sequence $\{f_n(x)\}$ $(n = 1, 2, \cdots)$ of functions which satisfy the previous conditions and tend to $f(x)$ in quadratic mean as $n \to \infty$. By the Parseval equality the sequence of Fourier transforms $E_n(\lambda)$ of the functions $f_n(x)$ also converges in quadratic mean (with respect to the measure $d\rho(\lambda)$) to the Fourier transform of $f(x)$, which we denote by $F(\lambda)$. Therefore by Corollary I of Lemma 3.1 and Lemma 3.2 one can pass to the limit $n \to \infty$ in (3.11), and we obtain the assertion of the theorem.

Remark. By means of similar considerations one can obtain the formula

$$\int_0^{\infty} R_z(f)\, g(x)\, dx = \int_{-\infty}^{\infty} \frac{F(\lambda)\, G(\lambda)}{z - \lambda}\, d\rho(\lambda), \tag{3.7$'$}$$

where

$$F(\lambda) = \underset{n \to \infty}{\mathrm{l \cdot i \cdot m}} \int_0^n f(x)\, \varphi(x, \lambda)\, dx,$$

$$G(\lambda) = \underset{n \to \infty}{\mathrm{l \cdot i \cdot m}} \int_0^n g(x)\, \varphi(x, \lambda)\, dx.$$

Theorem 3.2. *Suppose that the function $f(x)$ satisfies the following conditions*:

1°. $f(x) \in \mathscr{L}^2(0, \infty)$,
2°. $\{f''(x) - q(x) f(x)\} \in \mathscr{L}^2(0, \infty)$,

---

4) The convergence of $R_{z,b} f_n$ to $R_z f_n$ follows from the first part of Lemma 2.2. We remark that from this point on we are considering $\rho(\lambda)$ corresponding to sequences $\{b_k\}$ and $\{\beta_k\}$ for which $\lim_{k \to \infty} l(\lambda, b_k, \beta_k) = m(\lambda)$ (cf. §2.2).

3°. $f(0)\cos\alpha + f'(0)\sin\alpha = 0$,

4°. $\lim_{x\to\infty} W\{f(x), E_\lambda(x)\} = 0$, *where* $E_\lambda(x) = \int_{+0}^{\lambda}\varphi(x,\lambda)\,d\rho(\lambda)$ $(\lambda \neq 0)$.

Put

$$g(\lambda) = \int_0^\infty f(y)\,E_\lambda(y)\,dy.$$

Then

$$f(x) = \int_{-\infty}^{\infty} \varphi(x, \lambda)\,dg(\lambda), \tag{3.12}$$

where the integral converges absolutely.

PROOF. If $f(x)$ vanishes outside some finite interval, then interchanging the order of integration we obtain

$$g(\lambda) = \int_0^\lambda F(\lambda)\,d\rho(\lambda), \qquad F(\lambda) = \int_0^\infty f(x)\,\varphi(x, \lambda)\,dx.$$

This formula for $g(\lambda)$ remains valid in the general case, in view of the convergence in the mean of the Fourier transform. Therefore formula (3.7) can be written in the form

$$R_z f = \int_{-\infty}^{\infty} \frac{\varphi(x, \lambda)\,dg(\lambda)}{z - \lambda}$$

By Lemma 3.1 the integral converges absolutely. Let $h(\lambda)$ be defined from the function $Lf = f''(x) - q(x)f(x)$ in the same way as $g(\lambda)$ was defined from $f(x)$. By condition 4° and Green's identity we have

$$h(\lambda) = \int_0^\infty \{f''(x) - q(x)f(x)\}\,E_\lambda(x)\,dx = -\int_0^\infty f(x)\left\{\int_{+0}^{\lambda} \mu\varphi(x, \mu)\,d\rho(\mu)\right\}dx.$$

If $f(x)$ vanishes outside some finite interval, then

$$h(\lambda) = -\int_{+0}^{\lambda} \mu F(\mu)\,d\rho(\mu) = -\int_{+0}^{\lambda} \mu\,dg(\mu). \tag{3.13}$$

This formula remains valid in the general case, since

$$\int_{+0}^{\lambda} \mu\varphi(x, \mu)\,d\rho(\mu) \in \mathscr{L}^2(0, \infty).$$

In fact,

$$\int_0^b \left\{ \int_{+0}^{\lambda} \mu \varphi(x, \mu)\, d\rho_b(\mu) \right\}^2 dx$$

$$= \int_0^b \left\{ \sum_{0 \leqslant \lambda_{n,b} \leqslant \lambda} \frac{\lambda_{n,b}}{\alpha_n^2} \varphi_{n,b}(x) \right\} \cdot \left\{ \sum_{0 \leqslant \lambda_{n,b} \leqslant \lambda} \frac{\lambda_{n,b}}{\alpha_n^2} \varphi_{n,b}(x) \right\} dx$$

$$= \sum_{0 \leqslant \lambda_{n,b} \leqslant \lambda} \frac{\lambda_{n,b}^2}{\alpha_n^2} = \int_{+0}^{\lambda} \mu^2\, d\rho_b(\mu).$$

Hence $(a < b)$

$$\int_0^a \left\{ \int_{+0}^{\lambda} \mu \varphi(x, \mu)\, d\rho_b(\mu) \right\}^2 dx \leqslant \int_{-0}^{\lambda} \mu^2 d\rho_b(\mu).$$

Letting first $b \to \infty$ and then $a \to \infty$, we obtain our assertion. Now, differentiating (3.13), we obtain

(3.14) $$dh(\lambda) = -\lambda dg(\lambda), \qquad dg(\lambda) = -dh(\lambda)/\lambda.$$

Since by condition 2° the integral

$$\int_{-\infty}^{\infty} \frac{\varphi(x, \lambda)\, dh(\lambda)}{z - \lambda}$$

converges absolutely, thanks to (3.14) the integral

$$\int_{-\infty}^{\infty} \varphi(x, \lambda)\, dg(\lambda)$$

converges absolutely. Therefore the theorem follows at once from Theorem 1.2.

§4. Orthogonality of the expansion

Theorem 1.1 establishes a mapping of the space of all $f(x) \in \mathcal{L}^2(0, \infty)$ into the space of functions $F(\lambda) \in \mathcal{L}^2_{\{\rho(\lambda)\}}(-\infty, \infty)$. An important additional fact is that the mapping in question is a mapping of $\mathcal{L}^2(0, \infty)$ onto $\mathcal{L}^2_{\{\rho(\lambda)\}}(-\infty, \infty)$. The spaces $\mathcal{L}^2(0, \infty)$ and $\mathcal{L}^2_{\{\rho(\lambda)\}}(-\infty, \infty)$ are Hilbert spaces, and therefore the aforementioned facts can be summed up by the statement that the correspondence $f \leftrightarrow F$ is a one-to-one mapping of the space $\mathcal{L}^2(0, \infty)$ onto the space $\mathcal{L}^2_{\{\rho(\lambda)\}}(-\infty, \infty)$ which is norm-preserving, i.e., it is a unitary mapping.

The purpose of this section is to prove this assertion.

We will first prove several lemmas.

Let $\varphi(x,\lambda)$ and $\rho(\lambda)$ be the same as in §3. We put

$$E_\Delta(x, y) = \int_\lambda^{\lambda+\Delta} \varphi(x, \mu)\varphi(y, \mu)\, d\rho(\mu).$$

LEMMA 4.1. *Let $\Delta = (\lambda, \lambda+\Delta)$ be a finite interval whose endpoints are points of continuity of the function $\rho(\lambda)$. For every fixed x the function $E_\Delta(x,y)$, as a function of y, belongs to $\mathscr{L}^2(0,\infty)$.*

PROOF. In the notation of the previous section we have

$$\int_0^b \left\{ \sum_{\lambda\leqslant\lambda_{n,b}\leqslant\lambda+\Delta} \frac{1}{a_n^2}\varphi_{n,b}(x)\varphi_{n,b}(y)\right\}\left\{ \sum_{\lambda\leqslant\lambda_{n,b}\leqslant\lambda+\Delta} \frac{1}{a_n^2}\varphi_{n,b}(y)\varphi_{n,b}(z)\right\} dy$$
$$= \sum_{\lambda\leqslant\lambda_{n,b}\leqslant\lambda+\Delta} \frac{1}{a_n^2}\varphi_{n,b}(x)\varphi_{n,b}(z).$$

If $a < b$ and $x = z$, then the preceding equality implies the inequality

$$\int_0^a \left\{ \int_\lambda^{\lambda+\Delta} \varphi(x, \mu)\varphi(y, \mu)\, d\rho_b(\mu)\right\}^2 dy \leqslant \int_\lambda^{\lambda+\Delta} \varphi^2(x, \mu)\, d\rho_b(\mu).$$

Letting $b\to\infty$ and passing to the limit under the integral sign, we obtain for fixed a

$$\int_0^a E_\Delta^2(x, y)\, dy \leqslant \int_\lambda^{\lambda+\Delta} \varphi^2(x, \lambda)\, d\rho(\lambda),$$

and since a is arbitrary, the lemma is proved.

LEMMA 4.2. *Let $f(x) \in \mathscr{L}^2(0,\infty)$ and*

$$E_\Delta f = \int_0^\infty E_\Delta(x, y) f(y)\, dy.$$

Then

$$\int_0^\infty \{E_\Delta f\}^2\, dx \leqslant \int_0^\infty f^2(x)\, dx.$$

PROOF. Suppose first that the function $f(x) = f_n(x)$ vanishes outside the finite interval $[0, n]$. For $b > n$ we have

$$\int_0^b \left\{ \int_0^n f_n(y) E_\Delta(x, y)\, dy \right\}^2 dx$$

$$= \int_0^b \left\{ \int_0^n f_n(y) \left[\sum_{\lambda \leqslant \lambda_{k,b} \leqslant \lambda+\Delta} \frac{1}{\alpha_k^2} \varphi_{k,b}(x) \varphi_{k,b}(y) \right] dy \right\}^2 dx$$

$$= \sum_{\lambda \leqslant \lambda_{k,b} \leqslant \lambda+\Delta} \frac{1}{\alpha_k^2} \left\{ \int_0^n f_n(y) \varphi_{k,b}(y)\, dy \right\}^2 \leqslant \int_0^n f_n^2(x)\, dx.$$

If $a < b$, then *a fortiori*

$$\int_0^a \left\{ \int_0^n f_n(y) \left[\int_\lambda^{\lambda+\Delta} \varphi(x, \mu) \varphi(y, \mu)\, d\rho_b(\mu) \right] dy \right\}^2 dx \leqslant \int_0^n f_n^2(x)\, dx.$$

Letting $b \to \infty$ for fixed a, we obtain

$$\int_0^a \left\{ \int_0^n E_\Delta(x, y) f_n(y)\, dy \right\}^2 dx \leqslant \int_0^n f_n^2(x)\, dx.$$

Passing now to the limit $a \to \infty$, we find that

$$\int_0^\infty \{E_\Delta f_n\}^2\, dx \leqslant \int_0^n f_n^2(x)\, dx,$$

which is the assertion of the lemma for the function $f_n(x)$. Now let $f(x)$ be an arbitrary square-integrable function. We put

$$f_n(x) = \begin{cases} f(x), & 0 \leqslant x \leqslant n, \\ 0, & x > n. \end{cases}$$

Then for fixed $a > 0$ it follows from the preceding inequality that

$$\int_0^a \left\{ \int_0^n E_\Delta(x, y) f_n(y)\, dy \right\}^2 dx \leqslant \int_0^n f_n^2(x)\, dx.$$

Letting $n \to \infty$, we obtain

$$\int_0^a \left\{ \int_0^\infty E_\Delta(x, y) f(y)\, dy \right\}^2 dx \leqslant \int_0^\infty f^2(x)\, dx,$$

from which the assertion of the lemma follows by letting $a \to \infty$.

LEMMA 4.3. *Let* $f(x) \in \mathcal{L}^2(0, \infty)$ *and*

$$F(\lambda) = \underset{n\to\infty}{\mathrm{l.i.m.}} \int_0^n f(x)\varphi(x, \lambda)\,dx.$$

Then

$$E_\Delta f = \int_\Delta F(\lambda)\varphi(x, \lambda)\,d\rho(\lambda). \tag{4.1}$$

PROOF. Suppose that $f(x) = f_n(x)$ vanishes outside the finite interval $[0, n]$. We put

$$F_n(\lambda) = \int_0^n f_n(y)\varphi(y, \lambda)\,dy.$$

Multiplying both sides of this equality by $\varphi(x, \lambda)$ and integrating with respect to λ over the interval $\Delta = (\lambda, \lambda + \Delta)$ (with respect to the measure defined by $\rho(\lambda)$), we obtain

$$\int_0^n f_n(y)E_\Delta(x, y)\,dy = \int_\Delta F_n(\lambda)\varphi(x, \lambda)\,d\rho(\lambda).$$

Let us now consider the general case. Let $f_n(y) = f(y)$ for $0 \leqq y \leqq n$, $f_n(y) = 0$ for $y > n$. If $n \to \infty$, then $F_n(\lambda)$ converges in quadratic mean to $F(\lambda)$. Therefore we can pass to the limit in the right side of the last equality. Passing to the limit in the left side is also possible on the basis of the preceding lemma, and we obtain (4.1).

LEMMA 4.4. *Let* $E_\Delta(x) = \int_\Delta \varphi(x, \lambda)\,d\rho(\lambda)$ *and let* $f(x) \in \mathcal{L}^2(0, \infty)$. *Then for almost all* λ *(with respect to the measure defined by* $\rho(\lambda)$*)*

$$F(\lambda) = \lim_{\Delta\to 0} \frac{1}{\rho(\Delta)} \int_0^\infty f(x)E_\Delta(x)\,dx.$$

PROOF. If $\sin\alpha \neq 0$, then the square integrability of the function $E_\Delta(x)$ follows from Lemma 4.1 (putting $y = 0$).

Suppose that $\sin\alpha = 0$. From the orthogonality of the functions $\varphi_{n,b}(x)$ it follows that

$$\int_0^b \left\{\frac{\partial E_\Delta(x, y)}{\partial x}\right\}^2 dy = \int_0^b \left\{\int_\Delta \varphi'_x(x, \lambda)\varphi(y, \lambda)\,d\rho_b(\lambda)\right\}^2 dy$$

$$= \int_0^b \left\{\sum_{\lambda \leqq \lambda_{n,b} \leqq \lambda+\Delta} \frac{1}{\alpha_n^2}\varphi_{n,b}(y)\frac{d\varphi_{n,b}(x)}{dx}\right\} dy = \int_\Delta \varphi_x'^2(x, \lambda)\,d\rho_b(\lambda).$$

Hence by means of the same arguments that were used in the proof of Lemma 4.1 it follows that for every fixed x we have

$$\frac{\partial E_\Delta(x, y)}{\partial x} \in \mathscr{L}^2(0, \infty).$$

In particular, for $x=0$ it follows that $E_\Delta(y) \in \mathscr{L}^2(0, \infty)$. Putting $x=0$ in (4.1),[5] we obtain the following equality:

$$\int_0^\infty f(x) E_\Delta(x)\, dx = \int_\Delta F(\lambda)\, d\rho(\lambda).$$

Dividing by $\rho(\Delta) = \rho(\lambda+\Delta) - \rho(\lambda)$, we obtain

$$\frac{1}{\rho(\Delta)} \int_0^\infty f(x) E_\Delta(x)\, dx = \frac{1}{\rho(\Delta)} \int_\Delta F(\lambda)\, d\rho(\lambda).$$

Finally, letting $\Delta \to 0$ and using Lebesgue's well-known theorem on the differentiability of a definite integral, we obtain the assertion of our lemma.

Theorem 4.1. *Let $f(x) \in \mathscr{L}^2(0, \infty)$ and let $\Delta = (\lambda, \lambda+\Delta)$ be a finite interval whose endpoints are points of continuity of $\rho(\lambda)$. Then for every nonreal z*

$$\begin{aligned} R_z E_\Delta f &= \int_0^\infty G(x, y; z) \left\{ \int_0^\infty E_\Delta(y, t) f(t)\, dt \right\} dy \\ &= \int_\Delta \frac{\varphi(x, \lambda) F(\lambda)}{z-\lambda}\, d\rho(\lambda), \end{aligned} \tag{4.2}$$

where $F(\lambda)$ is the Fourier transform of $f(x)$.

Proof. Let us first consider the case in which the function $f(x) = f_n(x)$ vanishes outside the finite interval $[0, n]$. Resorting again to the notation of §3, we obtain

$$\int_0^b G_b(x, y; z) \left\{ \int_0^n f_n(t) \left[\int_\Delta \varphi(y, \lambda) \varphi(t, \lambda)\, d\rho_b(\lambda) \right] dt \right\} dy$$
$$= \int_\Delta \frac{\varphi(x, \lambda)}{z-\lambda} \left\{ \int_0^n f_n(t) \varphi(t, \lambda)\, dt \right\} d\rho_b(\lambda).$$

---

[5] If $\sin\alpha = 0$, one has to first differentiate (4.1) with respect to x, which is possible since $\partial E_\Delta(x, y)/\partial x \in \mathscr{L}^2(0, \infty)$.

The convergence to the limit of the right side of this equality is obvious. Let us consider the left side. Let $a > 0$ be a fixed number. We have

$$\int_0^b G_b(x, y; z)\left\{\int_0^n f_n(t)\left[\int_\Delta \varphi(y, \lambda)\varphi(t, \lambda)\,d\rho_b(\lambda)\right]dt\right\}dy$$

$$(4.3)\qquad = \int_0^a G_b(x, y; z)\left\{\int_0^n f_n(t)\left[\int_\Delta \varphi(y, \lambda)\varphi(t, \lambda)\,d\rho_b(\lambda)\right]dt\right\}dy$$

$$+ \int_a^b G_b(x, y; z)\left\{\int_0^n f_n(t)\left[\int_\Delta \varphi(y, \lambda)\varphi(t, \lambda)\,d\rho_b(\lambda)\right]dt\right\}dy.$$

As $b \to \infty$, for fixed a the first term on the right tends to

$$\int_0^a G(x, y; z)\left\{\int_0^n f_n(t)\left[\int_\Delta \varphi(y, \lambda)\varphi(t, \lambda)\,d\rho(\lambda)\right]dt\right\}dy$$

(cf. Lemma 2.2).

We will show that the second term tends to zero as $a \to \infty$, $b > a$. For this we use the Cauchy-Bunjakovskiĭ inequality:

$$\left|\int_a^b G_b(x, y; z)\left\{\int_0^n f_n(t)\left[\sum_{\lambda\leqslant\lambda_{k,b}\leqslant\lambda+\Delta} \frac{1}{\alpha_k^2}\varphi_{k,b}(y)\varphi_{k,b}(t)\right]dt\right\}dy\right|$$

$$\leqslant \left\{\int_a^b |G_b(x, y; z)|^2\,dy\right\}^{1/2}$$

$$\times\left\{\int_a^b\left[\sum_{\lambda\leqslant\lambda_{k,b}\leqslant\lambda+\Delta} \frac{1}{\alpha_k}\varphi_{k,b}(y)\frac{1}{\alpha_k}\int_0^n f_n(t)\varphi_{k,b}(t)\,dt\right]^2 dy\right\}^{1/2}$$

$$= \left\{\int_a^b |G_b(x, y; z)|^2\,dy\right\}^{1/2}\left\{\sum_{\lambda\leqslant\lambda_{k,b}\leqslant\lambda+\Delta}\left[\frac{1}{\alpha_k^2}\int_0^n f(t)\varphi_{k,b}(t)\,dt\right]^2\right\}^{1/2}$$

$$\leqslant \left\{\int_a^b |G_b(x, y; z)|^2\,dy\right\}^{1/2}\cdot\left\{\int_0^n f^2(t)\,dt\right\}^{1/2}.$$

We will show that by virtue of Lemma 2.2 the integral

$$\int_a^b |G_b(x, y; z)|^2\,dy$$

can be made as small as desired.

In fact, let ϵ denote an arbitrary positive number and first choose a so large that

$$\int_a^\infty |\psi(y, z)|^2 dy < \varepsilon.$$

Now choose b so large that

$$\left| \int_0^a |\psi(y, z)|^2 dy - \int_0^a |\psi_b(y, z)|^2 dy \right| < \varepsilon,$$

$$\left| \int_0^b |\psi_b(y, z)|^2 dy - \int_0^a |\psi(y, z)|^2 dy \right| < \varepsilon,$$

which is possible by virtue of Lemma 2.2. We then obtain

$$\int_a^b |\psi_b(y, z)|^2 dy = \int_a^b |\psi_b(y, z)|^2 dy - \int_a^b |\psi(y, z)|^2 dy$$
$$+ \int_a^b |\psi(y, z)|^2 dy \leqslant \left| \int_0^b |\psi_b(y, z)|^2 dy - \int_0^b |\psi(y, z)|^2 dy \right|$$
$$+ \left| \int_0^a |\psi(y, z)|^2 dy - \int_0^a |\psi_b(y, z)|^2 dy \right| + \int_a^b |\psi(y, z)|^2 dy < 3\varepsilon.$$

We can therefore pass to the limit $b \to \infty$ in (4.3), and we obtain (4.2).

Let us now consider the general case. If $f_n(x) = f(x)$ for $0 \leq x \leq n$ and $f_n(x) = 0$ for $x > n$, then as was proved

$$R_z E_\Delta f_n = \int_\Delta \frac{\varphi(x, \lambda) F_n(\lambda)}{z - \lambda} d\rho(\lambda), \tag{4.4}$$

where

$$F_n(\lambda) = \int_0^n f(x) \varphi(x, \lambda) dx.$$

As $n \to \infty$ the sequence $\{F_n(\lambda)\}$ converges in the mean to $F(\lambda)$. Therefore the passage to the limit in the right side of (4.4) is obvious. The possibility of passing to the limit on the left side follows from Lemmas 4.2 and 3.2. This proves the theorem.

Remark. Let $g(x) \in \mathscr{L}^2(0, \infty)$. Multiplying both sides of (4.2) by $g(x)$ and integrating with respect to x from 0 to n, we obtain

$$\int_0^n R_z E_\Delta f \cdot g(x) dx = \int_\Delta \frac{G_n(\lambda) F(\lambda)}{z - \lambda} d\rho(\lambda),$$

where we have put

$$G_n(\lambda)=\int_0^n g(x)\varphi(x,\ \lambda)\,dx.$$

Letting $n\to\infty$, we obtain

$$\int_0^\infty R_z E_\Delta f\cdot g(x)\,dx=\int_\Delta \frac{G(\lambda)F(\lambda)}{z-\lambda}\,d\rho(\lambda), \tag{4.5}$$

where

$$G(\lambda)=\underset{n\to\infty}{\mathrm{l\cdot i\cdot m}}\int_0^n g(x)\varphi(x,\ \lambda)\,dx.$$

Let $f(x), g(x)\in\mathscr{L}^2(0,\infty)$. Multiplying both sides of (4.1) by $g(x)$ and integrating with respect to x from 0 to n, we obtain

$$\int_0^n g(x)\left\{\int_0^\infty E_\Delta(x,\ y)f(y)\,dy\right\}dx=\int_\Delta G_n(\lambda)F(\lambda)\,d\rho(\lambda).$$

Letting $n\to\infty$, we find that

$$\begin{aligned}\int_0^\infty g(x)\left\{\int_0^\infty E_\Delta(x,\ y)f(y)\,dy\right\}dx\\ =(E_\Delta f,\ g)=\int_\Delta F(\lambda)G(\lambda)\,d\rho(\lambda).\end{aligned} \tag{4.6}$$

The family of operators E_Δ has the following properties:

1. Selfadjointness, i.e. $(E_\Delta f,g)=(f,E_\Delta g)$.
2. Monotonicity, i.e. if $\Delta'\subset\Delta$, then $(E_{\Delta'}f,f)\leqq(E_\Delta f,f)$.
3. Completeness, i.e. $(E_{(-\infty,\infty)}f,g)=(f,g)=\int_0^\infty f(x)g(x)\,dx$.

Property 1 follows from (4.6) and the symmetry of the function $E_\Delta(x,y)$. Property 2 also follows at once from (4.6). Property 3 follows from (4.6) and from the Parseval equality (1.11).

Let us prove yet another property of the operators E_Δ, namely:

4. Orthogonality, i.e. $E_{\Delta'}E_\Delta=E_{\Delta'\cdot\Delta}$, where $\Delta'\cdot\Delta$ is the intersection of the intervals Δ' and Δ.

Let $E_\lambda=E_{(-\infty,\lambda)}$. By virtue of (4.6), the formula (4.5) can be written in the form

$$\int_0^\infty R_z E_\Delta f\cdot g(x)\,dx=\int_\Delta \frac{d_\lambda(E_\lambda f,\ g)}{z-\lambda}. \tag{4.7}$$

On the other hand, putting $E_\Delta f$ for f in formula (3.7′) and using (4.7), we obtain

$$\int_0^\infty R_z E_\Delta f \cdot g(x)\,dx = \int_{-\infty}^{\infty} \frac{d_\lambda (E_\lambda E_\Delta f, g)}{z-\lambda}. \tag{4.8}$$

It follows from (4.7) and (4.8), by virtue of the uniqueness theorem for the Stieltjes transform (cf. Chapter 14, Theorem 3.2), that

$$E_\lambda \cdot E_\Delta = E_{(-\infty,\lambda)} E_\Delta = E_{(-\infty,\lambda)\Delta},$$

which was to be proved.[6]

A family of operators E_λ having the properties 1–4 is called a *resolution of the identity.*

Lemma 4.5. *Let* $E_\Delta(x,y) = \int_\Delta \varphi(x,\lambda)\varphi(y,\lambda)\,d\rho(\lambda)$. *Then*

$$\begin{aligned} &\int_0^\infty E_\Delta(x,z)\,E_{\Delta'}(z,y)\,dz \\ &\quad = \int_{\Delta\cdot\Delta'} \varphi(x,\lambda)\,\varphi(y,\lambda)\,d\rho(\lambda) = E_{\Delta\cdot\Delta'}(x,y). \end{aligned} \tag{4.9}$$

The proof follows at once from orthogonality.

Lemma 4.6. *Let* $E_\Delta(x) = \int_\Delta \varphi(x,\lambda)\,d\rho(\lambda)$ *and let* $F_\Delta(\lambda)$ *be the Fourier transform of the function*[7] $E_\Delta(x)$. *For almost every* λ *(with respect to the measure defined by* $\rho(\lambda)$*)* $F_\Delta(\lambda) = 1$ *for* $\lambda \in \Delta$ *and equals* 0 *for* $\lambda \notin \Delta$.

Proof. If $\sin\alpha \neq 0$, then putting $x = y = 0$ in (4.9), we obtain

$$\int_0^\infty E_\Delta(z)\,E_{\Delta'}(z)\,dz = \int_{\Delta\cdot\Delta'} d\rho(\lambda). \tag{4.10}$$

If $\sin\alpha = 0$, then putting $x = y = 0$ in (4.9), we obtain zero on both sides. In this case we first have to differentiate (4.9) with respect to x and y, which is possible since (see the proof of Lemma 4.4)

$$\int_0^\infty \left\{\frac{\partial E_\Delta(x,z)}{\partial x}\right\}^2 dz < \infty,$$

---

[6] If $\Delta' = (\lambda', \lambda' + \Delta')$, then $E_{\Delta'} \cdot E_\Delta = \{E_{(-\infty,\lambda'+\Delta')} - E_{(-\infty,\lambda')}\} E_\Delta = E_{(-\infty,\lambda'+\Delta')} E_\Delta - E_{(-\infty,\lambda')} E_\Delta = E_{(-\infty,\lambda'+\Delta')\Delta} - E_{(-\infty,\lambda')\Delta} = E_{[(-\infty,\lambda'+\Delta')-(-\infty,\lambda')]\cdot\Delta} = E_{\Delta'\cdot\Delta}$.

[7] The existence of $F_\Delta(\lambda)$ follows from the proof of Lemma 4.4.

and then put $x=y=0$. We again obtain (4.10).

Dividing (4.10) by $\rho(\Delta')$, letting $\Delta'\to 0$ and using Lemma 4.4, we obtain the assertion of our lemma.

LEMMA 4.7. *Let $G(x,y;z)$ be as in §3. For each fixed x and z (z a nonreal number) the Fourier transform of $G(x,y;z)$ is $\varphi(x,\lambda)/(z-\lambda)$.*

PROOF. Putting $f(y)=E_\Delta(y)$ in (3.7), by virtue of Lemma 4.6 we obtain

$$\int_0^\infty G(x,\ y;\ z)\,E_\Delta(y)\,dy=\int_\Delta \frac{\varphi(x,\ \lambda)}{z-\lambda}\,d\rho(\lambda). \tag{4.11}$$

Dividing by $\rho(\Delta)$, we obtain

$$\frac{1}{\rho(\Delta)}\int_0^\infty G(x,\ y;\ z)\,E_\Delta(y)\,dy=\frac{1}{\rho(\Delta)}\int_\Delta \frac{\varphi(x,\ \lambda)}{z-\lambda}\,d\rho(\lambda).$$

Letting $\Delta\to 0$ and using Lemma 4.4, we obtain the assertion of the lemma.

We now turn to the proof of the basic theorem of this section.

THEOREM 4.2. *Let $\rho(\lambda)$ be as in Theorem 1.1.*[8] *For any square-integrable (with respect to the measure $d\rho(\lambda)$) function $F(\lambda)$ on $(-\infty,\infty)$ there exists a function $f(x)\in\mathscr{L}^2(0,\infty)$ such that*

$$f(x)=\underset{N\to\infty}{\mathrm{l\cdot i\cdot m}}\int_{-N}^{N} F(\lambda)\,\varphi(x,\ \lambda)\,d\rho(\lambda)$$

and such that the Parseval equality

$$\int_{-\infty}^{\infty} F^2(\lambda)\,d\rho(\lambda)=\int_0^\infty f^2(x)\,dx$$

holds.

PROOF. Let us first assume that $F_N(\lambda)$ is a continuous function which vanishes outside the finite interval $(-N,N)$. We put

$$g_N(x)=\int_{-N}^{N} F_N(\lambda)\,\varphi(x,\ \lambda)\,d\rho(\lambda).$$

We will show that

---

[8] See the footnote on p. 107.

$$\int_0^\infty g_N^2(x)\,dx < \infty. \tag{4.12}$$

Let $\lambda_1, \lambda_2, \cdots, \lambda_n$ be an arbitrary partition of the interval $(-N, N)$ and let $\xi_i \in [\lambda_{i-1}, \lambda_i]$. We put

$$g_N^{(n)}(x) = \sum_{i=1}^n F_N(\xi_i) \int_{\lambda_{i-1}}^{\lambda_i} \varphi(x, \lambda)\,d\rho(\lambda), \quad \lambda_0 = -N.$$

If $\max(\lambda_i - \lambda_{i-1}) \to 0$, then $g_N^{(n)}(x) \to g_N(x)$ $(n \to \infty)$ uniformly in every finite interval. By Lemma 4.5 and the Parseval equality

$$\int_0^\infty \{g_N^{(n)}(x)\}^2\,dx = \sum_{i=1}^n F_N^2(\xi_i) \int_{\lambda_{i-1}}^{\lambda_i} d\rho(\lambda) \to \int_{-N}^N F_N^2(\lambda)\,d\rho(\lambda). \tag{4.13}$$

Let $a > 0$ be a fixed number. There follows from (4.13) the existence of a fixed positive number K such that

$$\int_0^a \{g_N^{(n)}(x)\}^2\,dx < K. \tag{4.14}$$

Letting $n \to \infty$, we obtain $\int_0^a g_N^2(x)\,dx < K$, and since a is arbitrary, (4.12) follows.

We will now show that the Fourier transform of the function $g_N(x)$ coincides almost everywhere (with respect to the measure defined by $\rho(\lambda)$) with $F_N(\lambda)$. By (4.10)

$$\int_0^\infty g_N^{(n)}(x)\,E_\Delta(x)\,dx = \sum_i F_N(\xi_i) \int_{\Delta\cdot\Delta_i} d\rho(\lambda). \tag{4.15}$$

Let $n \to \infty$. From (4.14) and also from the fact that $E_\Delta(x) \in \mathcal{L}^2(0, \infty)$ it follows that one can pass to the limit in (4.15), to obtain

$$\int_0^\infty g_N(x)\,E_\Delta(x)\,dx = \int_\Delta F_N(\lambda)\,d\rho(\lambda).$$

Dividing by $\rho(\Delta)$ and letting $\Delta \to 0$, we conclude, using Lemma 4.4, that for almost all λ the Fourier transform of $g_N(x)$ coincides with $F_N(\lambda)$. Therefore by the Parseval equality (Theorem 1.1)

$$\int_0^\infty g_N^2(x)\,dx = \int_{-N}^N F_N^2(\lambda)\,d\rho(\lambda). \tag{4.16}$$

Now let $F(\lambda)$ be an arbitrary function from $\mathscr{L}^2_{\{\rho(\lambda)\}}(-\infty, \infty)$. We can find a sequence of continuous functions $F_N(\lambda)$, each of which vanishes outside the respective interval $(-N, N)$, which converges in the mean to $F(\lambda)$. Let

$$g_N(x) = \int_{-N}^{N} F_N(\lambda)\, \varphi(x, \lambda)\, d\rho(\lambda).$$

It follows from (4.16) $(N' > N)$ that

$$\int_0^\infty \{g_{N'}(x) - g_N(x)\}^2\, dx = \int_{-\infty}^{\infty} \{F_{N'}(\lambda) - F_N(\lambda)\}^2\, d\rho(\lambda).$$

Therefore the function $g_N(x)$ converges in quadratic mean to some function $f(x)$, and the Parseval equality

$$\int_0^\infty f^2(x)\, dx = \int_{-\infty}^{\infty} F^2(\lambda)\, d\rho(\lambda)$$

holds. It remains to show that as $N \to \infty$ the function

$$f_N(x) = \int_{-N}^{N} F(\lambda)\, \varphi(x, \lambda)\, d\rho(\lambda)$$

converges in the mean to $f(x)$.

Let $H(\lambda)$ be another function from $\mathscr{L}^2_{\{\rho(\lambda)\}}(-\infty, \infty)$ and let $h(x)$ be constructed from $H(\lambda)$ in the same way that $f(x)$ was constructed from $F(\lambda)$. It is obvious that

$$\int_0^\infty \{f(x) - h(x)\}^2\, dx = \int_{-\infty}^{\infty} \{F(\lambda) - H(\lambda)\}^2\, d\rho(\lambda).$$

Letting $H(\lambda) = F(\lambda)$ for $-N \leqq \lambda \leqq N$ and $H(\lambda) = 0$ for $|\lambda| > N$, we find that

$$\int_0^\infty \{f(x) - f_N(x)\}^2\, dx = \int_{-\infty}^{-N} F^2(\lambda)\, d\rho(\lambda) + \int_N^\infty F^2(\lambda)\, d\rho(\lambda).$$

Now letting $N \to \infty$, we establish that $f_N(x)$ converges in the mean to $f(x)$, which completes the proof of the theorem.

§5. Formulas for $\rho(\lambda)$ and $m(z)$

With the help of the integral representation for the resolvent, obtained in §3 (cf. Theorem 3.1), we can obtain useful formulas for the functions $\rho(\lambda)$ and $m(z)$. The formula for $\rho(\lambda)$ can be used as the basis of the definition of the spectrum of a Sturm-Liouville operator (cf. Chapter 4). We will first prove several lemmas.

Let $\psi(x,\lambda)$ be the function considered in §2 of the present chapter.

LEMMA 5.1. *For any fixed nonreal* λ *and* λ'

$$\lim_{x\to\infty} W\{\psi(x,\lambda),\psi(x,\lambda')\}=0.$$

PROOF. The function $\theta(x,\lambda)+l(\lambda)\varphi(x,\lambda)$ satisfies the boundary condition (2.3′). Therefore

$$W_b\{\theta(x,\lambda)+l(\lambda)\varphi(x,\lambda),\ \theta(x,\lambda')+l(\lambda')\varphi(x,\lambda')\}=0.$$

It follows that

$$W_b\{\psi(x,\lambda)+[l(\lambda)-m(\lambda)]\varphi(x,\lambda),\ \psi(x,\lambda')+[l(\lambda')-m(\lambda')]\varphi(x,\lambda')\}$$
$$=0,$$

i.e.

$$\begin{aligned} &W_b\{\psi(x,\lambda),\psi(x,\lambda')\}+[l(\lambda)-m(\lambda)]W_b\{\varphi(x,\lambda),\psi(x,\lambda')\}\\ &+[l(\lambda')-m(\lambda')]W_b\{\psi(x,\lambda),\varphi(x,\lambda')\}\\ &+[l(\lambda)-m(\lambda)][l(\lambda')-m(\lambda')]W_b\{\varphi(x,\lambda),\varphi(x,\lambda')\}=0. \end{aligned} \tag{5.1}$$

By virtue of the identity (2.2) we have

$$W_b\{\varphi(x,\lambda),\psi(x,\lambda')\}$$
$$=(\lambda'-\lambda)\int_0^b\varphi(x,\lambda)\psi(x,\lambda')\,dx+W_0\{\varphi(x,\lambda),\psi(x,\lambda')\}.$$

Therefore, on the basis of Theorem 2.1, as $b\to\infty$

$$W_b\{\varphi(x,\lambda),\psi(x,\lambda')\}=O\left\{\int_0^b|\varphi(x,\lambda)|^2\,dx\right\}+O(1).$$

For the limit-point case

$$|l(\lambda)-m(\lambda)|\leqslant r_b=\left\{2v\int_0^b|\varphi(x,\lambda)|^2\,dx\right\}^{-1},$$

so that

$$\lim_{b\to\infty}[l(\lambda)-m(\lambda)]W_b\{\varphi(x,\lambda),\psi(x,\lambda')\}=0.$$

This equality also remains valid for the limit circle case if $l(\lambda)\to m(\lambda)$, since in this case the integral $\int_0^b|\varphi(x,\lambda)|^2dx$ remains bounded. One can similarly bound the remaining terms in (5.1), and the lemma is proved.

LEMMA 5.2. *In the notation of the preceding lemma we have*

$$(5.2)\qquad \int_0^\infty \psi(x,\lambda)\psi(x,\lambda')\,dx = \frac{m(\lambda)-m(\lambda')}{\lambda'-\lambda}.$$

PROOF. By formula (2.2)

$$(5.3)\qquad \begin{aligned}(\lambda'-\lambda)\int_0^b \psi(x,\lambda)\psi(x,\lambda')\,dx \\ = W_0\{\psi(x,\lambda),\ \psi(x,\lambda')\} - W_b\{\psi(x,\lambda),\ \psi(x,\lambda')\}.\end{aligned}$$

Using the condition (2.3), we find that the first term on the right is equal to

$$\begin{aligned}&[\cos\alpha + m(\lambda)\sin\alpha][\sin\alpha - m(\lambda')\cos\alpha] \\ &\quad - [\cos\alpha + m(\lambda')\sin\alpha][\sin\alpha - m(\lambda)\cos\alpha] = m(\lambda) - m(\lambda').\end{aligned}$$

On the other hand, by the preceding lemma the second term on the right side of (5.3) tends to zero as $b\to\infty$. Therefore (5.2) follows from (5.3) if we pass to the limit $b\to\infty$ in the latter.

In particular, for $\lambda' = \bar\lambda$ $(\lambda = u + iv)$ it follows from (5.2) that

$$(5.4)\qquad \int_0^\infty |\psi(x,\lambda)|^2\,dx = -\frac{\operatorname{Im}\{m(\lambda)\}}{v},$$

so that the equal sign actually holds in (2.7).

LEMMA 5.3. *For fixed* u_1 *and* u_2 *and* $\delta\to 0$

$$(5.5)\qquad \int_{u_1}^{u_2} -\operatorname{Im}\{m(u+i\delta)\}\,du = O(1).$$

PROOF. Suppose that $\sin\alpha \neq 0$. By (4.7) (for $x = 0$) and the Parseval equality

$$(5.6)\qquad \int_0^\infty |\psi(y,z)|^2\,dy = \int_{-\infty}^\infty \frac{d\rho(\lambda)}{(u-\lambda)^2+v^2}\qquad (z = u+iv).$$

If $\sin\alpha = 0$, then we first have to prove that the Fourier transform of the function $\partial G(x,y;z)/\partial x$ is the function $\varphi_x'(x,\lambda)/(z-\lambda)$. This fact is easily established by differentiating both sides of (4.11) with respect to x, which is possible since for fixed x and z one has $\partial G(x,y;z)/\partial x \in \mathscr{L}^2(0,\infty)$. Then putting $x = 0$, we again obtain (5.6).

It follows from (5.4) and (5.6) that

$$-\operatorname{Im}\{m(u+i\delta)\}=\delta\int_{-\infty}^{\infty}\frac{d\rho(\lambda)}{(u-\lambda)^2+\delta^2}.$$

Integrating this equality with respect to u from u_1 to u_2, we find that

$$\int_{u_1}^{u_2}-\operatorname{Im}\{m(u+i\delta)\}\,du=\delta\int_{u_1}^{u_2}du\int_{-\infty}^{\infty}\frac{d\rho(\lambda)}{(u-\lambda)^2+\delta^2}.$$

Let (a,b) be a finite interval, with $a<u_1$, $b>u_2$. Then by virtue of (3.6)

$$\delta\int_{u_1}^{u_2}du\int_{-\infty}^{a}\frac{d\rho(\lambda)}{(u-\lambda)^2+\delta^2}=O(1),$$

$$\delta\int_{u_1}^{u_2}du\int_{b}^{\infty}\frac{d\rho(\lambda)}{(u-\lambda)^2+\delta^2}=O(1).$$

We further have

$$\delta\int_{u_1}^{u_2}du\int_{a}^{b}\frac{d\rho(\lambda)}{(u-\lambda)^2+\delta^2}=\int_{a}^{b}d\rho(\lambda)\int_{u_1}^{u_2}\frac{\delta\,du}{(u-\lambda)^2+\delta^2}=\int_{a}^{b}d\rho(\lambda)\int_{\frac{u_1-\lambda}{\delta}}^{\frac{u_2-\lambda}{\delta}}\frac{dv}{1+v^2}=O(1),$$

and the lemma is proved.

THEOREM 5.1. *Suppose that the endpoints of the interval $\Delta=(\lambda,\lambda+\Delta)$ are points of continuity of $\rho(\lambda)$. Then*

$$\rho(\Delta)=\rho(\lambda+\Delta)-\rho(\lambda)=\frac{1}{\pi}\lim_{\delta\to\infty}\int_{\Delta}\{-\operatorname{Im}[m(u+i\delta)]\}\,du. \tag{5.7}$$

PROOF. Suppose that the functions $f(x)$ and $g(x)$ belong to $\mathcal{L}^2(0,\infty)$ and vanish outside some finite interval. By (3.7′)

$$\Phi(\lambda)=\int_0^{\infty}R_zf\cdot g(x)\,dx=\int_{-\infty}^{\infty}\frac{F(\lambda)G(\lambda)}{z-\lambda}\,d\rho(\lambda)=\int_{-\infty}^{\infty}\frac{d\sigma(\lambda)}{z-\lambda},$$

where

$$\sigma(\Delta)=\int_{\Delta}F(\lambda)G(\lambda)\,d\rho(\lambda),$$

$$F(\lambda)=\int_0^{\infty}f(x)\varphi(x,\lambda)\,dx,$$

$$G(\lambda)=\int_0^{\infty}g(x)\varphi(x,\lambda)\,dx.$$

From the Stieltjes inversion formula (cf. Chapter 14, Theorem 3.1)

$$\sigma(\Delta)=\frac{1}{\pi}\lim_{\delta\to 0}\int_{\Delta}\{-\operatorname{Im}[\Phi(u+i\delta)]\}\,du. \tag{5.8}$$

We further have

$$\operatorname{Im}[\Phi(u+i\delta)]=\int_0^\infty g(x)\,dx\cdot\operatorname{Im}\Big\{\int_0^x[\theta(x,\ u+i\delta)+m(u+i\delta)\varphi(x,\ u+i\delta)]\varphi(y,\ u+i\delta)f(y)\,dy$$
$$+\int_x^\infty\varphi(x,u+i\delta)[\theta(y,u+i\delta)+m(u+i\delta)\varphi(y,u+i\delta)]f(y)\,dy\Big\}.$$

The functions $\varphi(x,u)$, $\theta(x,u)$, $f(x)$ and $g(x)$ are real. It therefore follows from the preceding equality, relation (5.8) and Lemma 5.3 that

$$\sigma(\Delta)=\frac{1}{\pi}\lim_{\delta\to 0}\int_{\Delta}\{-\operatorname{Im}[m(u+i\delta)]\}F(u)G(u)\,du. \tag{5.9}$$

We can choose $f(x)$ and $g(x)$ so that $F(u)$ and $G(u)$ differ arbitrarily little from unity in the fixed interval Δ (see the proof of Lemma 1.1). Therefore (5.7) follows from (5.9) and Lemma 5.3, which proves the theorem.

Theorem 5.2. *For every nonreal z we have*

$$m(z)=-\operatorname{ctg}\alpha+\int_{-\infty}^{\infty}\frac{d\rho(\lambda)}{z-\lambda}. \tag{5.10}$$

Proof. It follows from (3.7), by virtue of the arbitrariness of the function $f(x)$, that

$$G(x,\ y;\ z)=\int_{-\infty}^{\infty}\frac{\varphi(x,\ \lambda)\varphi(y,\ \lambda)}{z-\lambda}\,d\rho(\lambda). \tag{5.11}$$

On the other hand, by definition

$$G(x,\ y;\ z)=\begin{cases}[\theta(x,\ z)+m(z)\varphi(x,\ z)]\varphi(y,\ z), & y\leqslant x,\\ [\theta(y,\ z)+m(z)\varphi(y,\ z)]\varphi(x,\ z), & y>x.\end{cases}$$

Therefore by virtue of the condition (2.3) it follows from (5.11) that

$$G(0,\ 0;\ z)=[\cos\alpha+m(z)\sin\alpha]\sin\alpha=\int_{-\infty}^{\infty}\frac{\sin^2\alpha}{z-\lambda}\,d\rho(\lambda),$$

i.e.

$$m(z)=-\operatorname{ctg}\alpha+\int_{-\infty}^{\infty}\frac{d\rho(\lambda)}{z-\lambda},$$

as was to be proved.

We remark that formula (5.7) follows from (5.10) by virtue of the Stieltjes inversion formula (cf. Chapter 14, Theorem 3.1).

§6. Proof of the Parseval equality for the case of the entire line

We will consider the equation

$$y'' + [\lambda - q(x)]y = 0, \tag{6.1}$$

assuming now, however, that x varies in the interval $(-\infty, \infty)$, and $q(x)$ is a continuous function on this interval.

Let us denote by $\varphi(x, \lambda)$ the solution of (6.1) which satisfies the initial conditions

$$\varphi(0, \lambda) = 0, \qquad \varphi_x'(0, \lambda) = -1, \tag{6.2}$$

and by $\theta(x, \lambda)$ the solution of (6.1) for the initial conditions

$$\theta(0, \lambda) = 1, \qquad \theta_x'(0, \lambda) = 0. \tag{6.3}$$

Let $[a, b]$ be an arbitrary finite interval. We will consider the boundary-value problem defined by equation (6.1) and the boundary conditions

$$y(a)\sin\alpha + y'(a)\cos\alpha = 0, \qquad y(b)\sin\beta + y'(b)\cos\beta = 0, \tag{6.4}$$

where α and β are arbitrary real numbers.

Let $\lambda_1, \lambda_2, \cdots$ be the eigenvalues, and $y_1(x), y_2(x), \cdots$ corresponding orthonormalized eigenfunctions of this problem. Since $\varphi(x, \lambda)$ and $\theta(x, \lambda)$ are linearly independent solutions of (6.1), it follows that

$$y_n(x) = \beta_n\varphi(x, \lambda_n) + \gamma_n\theta(x, \lambda_n).$$

Suppose that $f(x)$ is square integrable over the interval $[a, b]$. Applying the Parseval equality to $f(x)$, we obtain

$$\begin{aligned} \int_a^b f^2(x)\,dx &= \sum_{n=1}^{\infty}\left\{\int_a^b f(x)\,[\beta_n\varphi(x, \lambda_n) + \gamma_n\theta(x, \lambda_n)]\,dx\right\}^2 \\ &= \sum_{n=1}^{\infty}\beta_n^2\left\{\int_a^b f(x)\,\varphi(x, \lambda_n)\,dx\right\}^2 \\ &\quad + 2\sum_{n=1}^{\infty}\beta_n\gamma_n\int_a^b f(x)\,\varphi(x, \lambda_n)\,dx \cdot \int_a^b f(x)\,\theta(x, \lambda_n)\,dx \\ &\quad + \sum_{n=1}^{\infty}\gamma_n^2\left\{\int_a^b f(x)\,\theta(x, \lambda_n)\,dx\right\}^2. \end{aligned} \tag{6.5}$$

If we now introduce jump functions whose jumps occur at the eigenvalues of the boundary-value problem (6.1) + (6.4):

$$\xi_{a,b}(\lambda) = \sum_{0\leqslant\lambda_n<\lambda} \beta_n^2 \quad \text{for} \quad \lambda > 0, \quad -\sum_{0\geqslant\lambda_n>\lambda} \beta_n^2 \quad \text{for } \lambda \leqslant 0,$$
$$\eta_{a,b}(\lambda) = \sum_{0\leqslant\lambda_n<\lambda} \beta_n\gamma_n \quad \text{for} \quad \lambda > 0, \quad -\sum_{0\geqslant\lambda_n>\lambda} \beta_n\gamma_n \quad \text{for } \lambda \leqslant 0,$$
$$\zeta_{a,b}(\lambda) = \sum_{0\leqslant\lambda_n<\lambda} \gamma_n^2 \quad \text{for} \quad \lambda > 0, \quad -\sum_{0\geqslant\lambda_n>\lambda} \gamma_n^2 \quad \text{for } \lambda \leqslant 0,$$

then (6.5) can be written in the form

$$\int_a^b f^2(x)\,dx = \int_{-\infty}^{\infty} \left\{\int_a^b f(x)\,\varphi(x,\lambda)\,dx\right\}^2 d\xi_{a,b}(\lambda)$$

$$(6.5') \qquad + 2\int_{-\infty}^{\infty} \left\{\int_a^b f(x)\,\varphi(x,\lambda)\,dx\right\}\left\{\int_a^b f(x)\,\theta(x,\lambda)\,dx\right\} d\eta_{a,b}(\lambda)$$

$$+ \int_{-\infty}^{\infty} \left\{\int_a^b f(x)\,\theta(x,\lambda)\,dx\right\}^2 d\zeta_{a,b}(\lambda).$$

Lemma 6.1. *For any positive number N there exists a positive number $A = A(N)$, not depending upon a and b, such that*

$$(6.6) \qquad \bigvee_{-N}^{N} \{\rho_{ij}(\lambda)\} < A \qquad (i, j = 1, 2),$$

where

$$(6.7) \qquad \rho_{11}(\lambda) = \xi_{a,b}(\lambda), \quad \rho_{12}(\lambda) = \rho_{21}(\lambda) = \eta_{a,b}(\lambda), \quad \rho_{22}(\lambda) = \zeta_{a,b}(\lambda).$$

Proof. To prove the inequality (6.6) it is sufficient to take $i = j$, since

$$(6.8) \qquad \bigvee_{-N}^{N} \{\rho_{12}(\lambda)\} \leqslant \frac{1}{2}\{\rho_{11}(N) - \rho_{11}(-N) + \rho_{22}(N) - \rho_{22}(-N)\}.$$

We put $\varphi(x,\lambda) = \psi_1(x,\lambda)$, $\theta(x,\lambda) = \psi_2(x,\lambda)$. Then by virtue of the notation (6.7) the equality (6.5) assumes the form

$$(6.9) \qquad \int_a^b f^2(x)\,dx = \int_{-\infty}^{\infty} \sum_{i,j=1}^{2} F_i(\lambda)\,F_j(\lambda)\,d\rho_{ij}(\lambda),$$

where

$$F_i(\lambda) = \int_a^b f(x)\,\psi_i(x,\lambda)\,dx.$$

Furthermore, since the functions $\psi_i^{(k-1)}(x,\lambda)$ $(i, k = 1, 2)$ are jointly continuous in the variables x, λ, and by the initial conditions (6.2) and (6.3) $\psi_i^{(k-1)}(0,\lambda) = \delta_{ik}$ ($\delta_{ik} = 1$ when $i = k$, $\delta_{ik} = 0$ when $i \neq k$), for any positive $\epsilon < 1$ and any given N there exists an $h > 0$ such that

$$|\psi_i^{(k-1)}(x,\lambda) - \delta_{ik}| < \epsilon \tag{6.10}$$

for $0 \leqq x \leqq h$ and $|\lambda| \leqq N$.

Let $g_h(x)$ be a twice continuously differentiable nonnegative function on the interval $[a,b]$ which vanishes outside the interval $[0,h]$ and is normalized so that

$$\int_0^h g_h(x)\,dx = 1. \tag{6.11}$$

Applying the Parseval equality (6.9) to the functions $g_h^{(k-1)}(x)$ ($k = 1,2$), we obtain

$$\int_0^h |g_h^{(k-1)}(x)|^2\,dx = \int_{-N}^{N} \sum_{i,j=1}^{2} G_{ik}(\lambda)\,G_{jk}(\lambda)\,d\rho_{ij}(\lambda), \tag{6.12}$$

where

$$G_{ik}(\lambda) = \int_0^h g_h^{(k-1)}(x)\,\psi_i(x,\lambda)\,dx = \pm\int_0^h g_h(x)\,\psi_i^{(k-1)}(x,\lambda)\,dx.$$

It follows by virtue of the bounds (6.10) and the relation (6.11) that

$$|G_{ik}(\lambda) - \delta_{ik}| < \epsilon \qquad (i,k = 1,2),$$

which together with (6.12) yields

$$\int_0^h |g_h^{(k-1)}(x)|^2\,dx \geqslant \int_{-N}^{N} \sum_{i,j=1}^{2} (\delta_{ik} - \varepsilon)(\delta_{jk} - \varepsilon)\,|d\rho_{ij}(\lambda)| \qquad (k = 1,\ 2). \tag{6.13}$$

Putting $k = 1$ in (6.13), we obtain

$$\int_0^h g_h^2(x)\,dx \geqslant (\varepsilon - 1)^2\{\rho_{11}(N) - \rho_{11}(-N)\} + 2\varepsilon(\varepsilon - 1)\bigvee_{-N}^{N}\{\rho_{12}(\lambda)\} + \varepsilon^2\{\rho_{22}(N) - \rho_{22}(-N)\},$$

from which it follows, by virtue of (6.8) and the fact that $0 < \epsilon < 1$, that

$$\int_0^h g_h^2(x)\,dx \geqslant (2\varepsilon^2 - 3\varepsilon + 1)\{\rho_{11}(N) - \rho_{11}(-N)\} + \epsilon(2\epsilon - 1)\{\rho_{22}(N) - \rho_{22}(-N)\}. \tag{6.14}$$

Similarly, putting $k=2$ in (6.13) and then using the inequality (6.8), we find that

$$\int_0^h |g_h'(x)|^2\,dx \geqslant (2\varepsilon^2-3\varepsilon+1)\{\rho_{22}(N)-\rho_{22}(-N)\} \tag{6.15}$$
$$+\epsilon(2\epsilon-1)\{\rho_{11}(N)-\rho_{11}(-N)\}.$$

Now adding inequalities (6.14) and (6.15), we obtain

$$\int_0^h g_h^2(x)\,dx+\int_0^h |g_h'(x)|^2\,dx$$
$$\geqslant (2\varepsilon-1)^2\{\rho_{11}(N)-\rho_{11}(-N)+\rho_{22}(N)-\rho_{22}(-N)\}.$$

which proves the bound (6.6) for $\rho_{11}(\lambda)$ and $\rho_{22}(\lambda)$, and thereby the lemma.

We will now show that, using Lemma 6.1 and a well-known theorem on passage to the limit under the integral for a Stieltjes integral, one can obtain the Parseval equality for equation (6.1) from (6.5), i.e. we have:

THEOREM 6.1. *Suppose that $f(x)\in\mathcal{L}^2(-\infty,\infty)$. There exist monotone functions $\xi(\lambda)$, $\zeta(\lambda)$ which are bounded in every finite interval, and a function $\eta(\lambda)$ which is of bounded variation over every finite interval, which functions do not depend upon $f(x)$, such that the Parseval equality*

$$\int_{-\infty}^{\infty} f^2(x)\,dx=\int_{-\infty}^{\infty} F^2(\lambda)\,d\xi(\lambda) \tag{6.16}$$
$$+2\int_{-\infty}^{\infty} F(\lambda)G(\lambda)\,d\eta(\lambda)+\int_{-\infty}^{\infty} G^2(\lambda)\,d\zeta(\lambda)$$

holds, where

$$F(\lambda)=\underset{n\to\infty}{\mathrm{l\cdot i\cdot m}}\int_{-n}^{n} f(x)\varphi(x,\lambda)\,dx,$$

$$G(\lambda)=\underset{n\to\infty}{\mathrm{l\cdot i\cdot m}}\int_{-n}^{n} f(x)\theta(x,\lambda)\,dx.$$

PROOF. Suppose first that the function $f(x)=f_n(x)$ vanishes outside the interval $-n\leqq x\leqq n$, $n<\min\{|a|,|b|\}$,[9] has a continuous second

[9] We may suppose without loss of generality that $a<0$, $b>0$. In fact, the change of variables

derivative and satisfies the boundary conditions (6.4). Applying the Parseval equality (6.5) to this function, we obtain

$$\int_{-n}^{n} f_n^2(x)\,dx = \sum_{k=1}^{\infty}\left\{\int_a^b f_n(x)\,y_k(x)\,dx\right\}^2. \tag{6.17}$$

Since both $f_n(x)$ and $y_k(x)$ satisfy the boundary conditions (6.4) and $f_n(x)$ vanishes identically in some neighborhoods of the points a and b, integrating by parts twice we obtain

$$\int_a^b f_n(x)\,y_k(x)\,dx = -\frac{1}{\lambda_k}\int_a^b f_n(x)\{y_k''(x) - q(x)\,y_k(x)\}\,dx$$

$$= -\frac{1}{\lambda_k}\int_a^b \{f_n''(x) - q(x)\,f_n(x)\}\,y_k(x)\,dx.$$

It follows that

$$\sum_{|\lambda_k|\geqslant\mu}\left\{\int_a^b f_n(x)\,y_k(x)\,dx\right\}^2$$

$$\leqslant \frac{1}{\mu^2}\sum_{|\lambda_k|\geqslant\mu}\left\{\int_a^b [f_n''(x) - q(x)\,f_n(x)]\,y_k(x)\,dx\right\}^2$$

$$\leqslant \frac{1}{\mu^2}\sum_{k=1}^{\infty}\left\{\int_a^b [f_n''(x) - q(x)\,f_n(x)]\,y_k(x)\,dx\right\}^2$$

$$= \frac{1}{\mu^2}\int_{-n}^{n}\{f_n''(x) - q(x)\,f_n(x)\}^2\,dx.$$

From this bound and from (6.17) we obtain

$$\left|\int_{-n}^{n} f_n^2(x)\,dx - \sum_{-\mu\leqslant\lambda_k\leqslant\mu}\left\{\int_a^b f_n(x)\,y_k(x)\,dx\right\}^2\right| < \frac{1}{\mu^2}\int_{-n}^{n}\{f_n''(x) - q(x)\,f_n(x)\}^2\,dx. \tag{6.18}$$

$$t = a_1 + \frac{x-a}{b-a}(b_1 - a_1),$$

where $a_1 < 0$, $b_1 > 0$, transforms the interval $[a, b]$ into the interval $[a_1, b_1]$ without altering the form of the boundary-value problem (6.1) + (6.4).

Further, we have

$$\sum_{-\mu\leqslant\lambda_k\leqslant\mu}\left\{\int_a^b f_n(x)\,y_k(x)\,dx\right\}^2$$

$$=\sum_{-\mu\leqslant\lambda_k\leqslant\mu}\left\{\int_a^b f_n(x)\,[\beta_k\varphi(x,\lambda_k)+\gamma_k\theta(x,\lambda_k)]\,dx\right\}^2$$

$$=\int_{-\mu}^{\mu}F_n^2(\lambda)\,d\xi_{a,b}(\lambda)+2\int_{-\mu}^{\mu}F_n(\lambda)\,G_n(\lambda)\,d\eta_{a,b}(\lambda)+\int_{-\mu}^{\mu}G_n^2(\lambda)\,d\zeta_{a,b}(\lambda),$$

where

$$F_n(\lambda)=\int_a^b f_n(x)\,\varphi(x,\lambda)\,dx,\quad G_n(\lambda)=\int_a^b f_n(x)\,\theta(x,\lambda)\,dx.$$

Then the inequality (6.18) can be rewritten in the form

$$\left|\int_{-n}^{n}f_n^2(x)\,dx-\int_{-\mu}^{\mu}[F_n^2(\lambda)\,d\xi_{a,b}(\lambda)+2F_n(\lambda)\,G_n(\lambda)\,d\eta_{a,b}(\lambda)+G_n^2(\lambda)\,d\zeta_{a,b}(\lambda)]\right|$$

$$(6.18') \qquad <\frac{1}{\mu^2}\int_{-n}^{n}\{f_n''(x)-q(x)\,f_n(x)\}^2\,dx.$$

By Lemma 6.1 the set of functions $\{\xi_{a,b}(\lambda)\}$ and $\{\zeta_{a,b}(\lambda)\}$, $-\mu\leqq\lambda\leqq\mu$, is bounded, and the set $\{\eta_{a,b}(\lambda)\}$, $-\mu\leqq\lambda\leqq\mu$, is bounded and has uniformly bounded variation. We can therefore find sequences $\{a_k\}$ and $\{b_k\}$, with $a_k\to-\infty$, $b_k\to+\infty$, such that the functions $\xi_{a_k,b_k}(\lambda)$, $\zeta_{a_k,b_k}(\lambda)$ and $\eta_{a_k,b_k}(\lambda)$ converge to $\xi(\lambda)$, $\zeta(\lambda)$ and $\eta(\lambda)$ respectively. Passing to the limit in (6.18′) along the sequences $\{a_k\}$ and $\{b_k\}$, we obtain

$$\left|\int_{-n}^{n}f_n^2(x)\,dx-\int_{-\mu}^{\mu}\{F_n^2(\lambda)\,d\xi(\lambda)+2F_n(\lambda)\,G_n(\lambda)\,d\eta(\lambda)+G_n^2(\lambda)\,d\zeta(\lambda)\}\right|$$

$$\leqslant\frac{1}{\mu^2}\int_{-n}^{n}\{f_n''(x)-q(x)\,f_n(x)\}^2\,dx.$$

Finally, letting $\mu\to\infty$ in this inequality. we obtain

$$\int_{-n}^{n}f_n^2(x)\,dx=\int_{-\infty}^{\infty}F_n^2(\lambda)\,d\xi(\lambda)+2\int_{-\infty}^{\infty}F_n(\lambda)\,G_n(\lambda)\,d\eta(\lambda)+\int_{-\infty}^{\infty}G_n^2(\lambda)\,d\zeta(\lambda),$$

which is (6.16) for functions satisfying the conditions specified above

The extension of the Parseval equality to an arbitrary function from $\mathscr{L}^2(-\infty, \infty)$ can be carried out by the usual methods.[10)]

Suppose that $f(x), g(x) \in \mathscr{L}^2(-\infty, \infty)$ and that $F_1(\lambda)$ and $G_1(\lambda)$ are constructed from $g(x)$ in the same way that $F(\lambda)$ and $G(\lambda)$ were constructed from $f(x)$. Obviously the transforms of the functions $f(x) \pm g(x)$ are the functions $F(\lambda) \pm F_1(\lambda)$ and $G(\lambda) \pm G_1(\lambda)$. Therefore

$$\int_{-\infty}^{\infty} \{f(x) + g(x)\}^2 \, dx = \int_{-\infty}^{\infty} \{F(\lambda) + F_1(\lambda)\}^2 \, d\xi(\lambda) + 2 \int_{-\infty}^{\infty} \{F(\lambda) + F_1(\lambda)\}\{G(\lambda) + G_1(\lambda)\} \, d\eta(\lambda) + \int_{-\infty}^{\infty} \{G(\lambda) + G_1(\lambda)\}^2 \, d\zeta(\lambda),$$

$$\int_{-\infty}^{\infty} \{f(x) - g(x)\}^2 \, dx = \int_{-\infty}^{\infty} \{F(\lambda) - F_1(\lambda)\}^2 \, d\xi(\lambda) + 2 \int_{-\infty}^{\infty} \{F(\lambda) - F_1(\lambda)\}\{G(\lambda) - G_1(\lambda)\} \, d\eta(\lambda) + \int_{-\infty}^{\infty} \{G(\lambda) - G_1(\lambda)\}^2 \, d\zeta(\lambda).$$

Subtracting the second equality from the first, we obtain

$$\int_{-\infty}^{\infty} f(x) g(x) \, dx = \int_{-\infty}^{\infty} F(\lambda) F_1(\lambda) \, d\xi(\lambda) + 2 \int_{-\infty}^{\infty} \{F(\lambda) G_1(\lambda) + F_1(\lambda) G(\lambda)\} \, d\eta(\lambda) + \int_{-\infty}^{\infty} G(\lambda) G_1(\lambda) \, d\zeta(\lambda). \tag{6.19}$$

---

10) One first has to prove the completeness of the space of vector-functions $\begin{pmatrix} F(\lambda) \\ G(\lambda) \end{pmatrix}$ with respect to the matrix measure

$$\rho(\lambda) = \begin{pmatrix} \xi(\lambda) & \eta(\lambda) \\ \eta(\lambda) & \zeta(\lambda) \end{pmatrix}.$$

Cf. Kac [1].

The equality (6.19) is called the generalized Parseval equality.

THEOREM 6.2 (EXPANSION THEOREM). *Let $f(x)$ be a continuous function on the line $-\infty < x < +\infty$, and suppose that the integrals*

$$\tag{6.20} \begin{gathered} \int_{-\infty}^{\infty} F(\lambda)\,\varphi(x,\ \varphi)\,d\xi(\lambda), \quad \int_{-\infty}^{\infty} F(\lambda)\,\theta(x,\ \lambda)\,d\eta(\lambda), \\ \int_{-\infty}^{\infty} G(\lambda)\,\varphi(x,\ \lambda)\,d\eta(\lambda), \quad \int_{-\infty}^{\infty} G(\lambda)\,\theta(x,\ \lambda)\,d\zeta(\lambda) \end{gathered}$$

converge absolutely and uniformly with respect to x in every finite interval. Then

$$\begin{aligned} f(x) = {} & \int_{-\infty}^{\infty} F(\lambda)\,\varphi(x,\ \lambda)\,d\xi(\lambda) + \int_{-\infty}^{\infty} F(\lambda)\,\theta(x,\ \lambda)\,d\eta(\lambda) \\ & + \int_{-\infty}^{\infty} G(\lambda)\,\varphi(x,\ \lambda)\,d\eta(\lambda) + \int_{-\infty}^{\infty} G(\lambda)\,\theta(x,\ \lambda)\,d\zeta(\lambda). \end{aligned}$$

PROOF. Let $g(x)$ be a continuous function which vanishes outside the finite interval $[-n, n]$. In this case the generalized Parseval equality (6.19) can be written as

$$\begin{aligned} \int_{-n}^{n} f(x)\,g(x)\,dx = {} & \int_{-\infty}^{\infty} F(\lambda)\left\{\int_{-n}^{n} g(x)\,\varphi(x,\ \lambda)\,dx\right\} d\xi(\lambda) \\ & + \int_{-\infty}^{\infty} F(\lambda)\left\{\int_{-n}^{n} g(x)\,\theta(x,\ \lambda)\,dx\right\} d\eta(\lambda) \\ & + \int_{-\infty}^{\infty} G(\lambda)\left\{\int_{-n}^{n} g(x)\,\varphi(x,\ \lambda)\,dx\right\} d\eta(\lambda) \\ & + \int_{-\infty}^{\infty} G(\lambda)\left\{\int_{-n}^{n} g(x)\,\theta(x,\ \lambda)\,dx\right\} d\zeta(\lambda). \end{aligned}$$

By virtue of the absolute convergence of the integrals (6.20) we can interchange the order of integration in the integrals on the right side, to obtain

$$\begin{aligned} \int_{-n}^{n} f(x)\,g(x)dx = \int_{-n}^{n} g(x)\Bigg\{ & \int_{-\infty}^{\infty} F(\lambda)\,\varphi(x,\ \lambda)\,d\xi(\lambda) + \int_{-\infty}^{\infty} F(\lambda)\,\theta(x,\ \lambda)\,d\eta(\lambda) \\ & + \int_{-\infty}^{\infty} G(\lambda)\,\varphi(x,\ \lambda)\,d\eta(\lambda) + \int_{-\infty}^{\infty} G(\lambda)\,\theta(x,\ \lambda)\,d\zeta(\lambda)\Bigg\}\,dx. \end{aligned}$$

Since the continuous function $g(x)$ is arbitrary, while $f(x)$ as well as the integrals (6.20) are continuous (the latter are continuous as functions of x thanks to their assumed uniform convergence), the assertion of the theorem follows from the last equality.

§7. Integral representation of the resolvent for the case of the entire line

1. Let the functions $\varphi(x,\lambda)$ and $\theta(x,\lambda)$ be as in the preceding section. Obviously $W\{\varphi,\theta\}=1$. Suppose that the endpoints of the interval $(\lambda,\lambda+\Delta)$ are points of continuity of the functions $\xi(\lambda)$, $\zeta(\lambda)$ and $\eta(\lambda)$, which are also defined as in the preceding section. We put

$$E_\Delta(x, y) = \int_\lambda^{\lambda+\Delta} \{\varphi(x, \lambda)\varphi(y, \lambda)\,d\xi(\lambda) + \varphi(x, \lambda)\theta(y,\lambda)\,d\eta(\lambda) + \varphi(y, \lambda)\theta(x, \lambda)\,d\eta(\lambda) + \theta(x, \lambda)\theta(y, \lambda)\,d\zeta(\lambda)\}.$$

Lemma 7.1. *For every fixed x the function $E_\Delta(x,y)$ belongs to $\mathscr{L}^2(-\infty,\infty)$, i.e.*

$$\int_{-\infty}^{\infty} E_\Delta^2(x, y)\,dy < +\infty.$$

Proof. Using the notation of the preceding section, by virtue of orthogonality we have

$$\int_a^b \Big[\sum_{\lambda<\lambda_n\leqslant\lambda+\Delta} \{\beta_n\varphi(x, \lambda_n) + \gamma_n\theta(x, \lambda_n)\}\{\beta_n\varphi(y, \lambda_n) + \gamma_n\theta(y, \lambda_n)\}\Big]^2 dy = \sum_{\lambda<\lambda_n\leqslant\lambda+\Delta} \{\beta_n\varphi(x, \lambda_n) + \gamma_n\theta(x, \lambda_n)\}^2.$$

If $a < a_1 < b_1 < b$, the preceding equality implies the inequality

$$\int_{a_1}^{b_1}\Big\{\int_\lambda^{\lambda+\Delta} [\varphi(x, \lambda)\varphi(y, \lambda)\,d\xi_{a,b}(\lambda) + \varphi(x, \lambda)\theta(y, \lambda)\,d\eta_{a,b}(\lambda) + \varphi(y, \lambda)\theta(x, \lambda)\,d\eta_{a,b}(\lambda) + \theta(x, \lambda)\theta(y, \lambda)\,d\zeta_{a,b}(\lambda)]\Big\}^2 dy$$
$$\leqslant \int_\lambda^{\lambda+\Delta} \{\varphi^2(x, \lambda)\,d\xi_{a,b}(\lambda) + 2\varphi(x, \lambda)\theta(x, \lambda)\,d\eta_{a,b}(\lambda) + \theta^2(x, \lambda)\,d\zeta_{a,b}(\lambda)\}.$$

For fixed a_1, b_1, letting $a\to-\infty$, $b\to+\infty$ and passing to the limit under the integral sign, we obtain

$$\int_{a_1}^{b_1} E_\Delta^2(x, y)\,dy \leqslant \int_\lambda^{\lambda+\Delta} \{\varphi^2(x, \lambda)\,d\xi(\lambda) + 2\varphi(x, \lambda)\theta(x, \lambda)\,d\eta(\lambda) + \theta^2(x, \lambda)\,d\zeta(\lambda)\},$$

and since a_1 and b_1 are arbitrary, the lemma is proved.

LEMMA 7.2. *Let $f(x) \in \mathscr{L}^2(-\infty, \infty)$. We put*

$$E_\Delta f = \int_{-\infty}^{\infty} E_\Delta(x, y) f(y)\, dy.$$

Then

$$\int_{-\infty}^{\infty} \{E_\Delta f\}^2\, dx \leqslant \int_{-\infty}^{\infty} f^2(y)\, dy.$$

PROOF. Suppose first that the function $f(x) = f_n(x)$ vanishes outside the finite interval $[-n, n]$. Resorting again to the notation of the preceding section and assuming that $n < \min\{|a|, |b|\}$, $a < 0$, $b > 0$, we obtain

$$\int_a^b \left\{ \int_{-n}^{n} \left[\sum_{\lambda < \lambda_n \leqslant \lambda + \Delta} y_k(x)\, y_k(t) \right] f_n(t)\, dt \right\}^2 dx$$

$$= \sum_{\lambda < \lambda_k \leqslant \lambda + \Delta} \left\{ \int_{-n}^{n} f_n(t)\, y_k(t)\, dt \right\}^2 \leqslant \int_{-n}^{n} f_n^2(t)\, dt.$$

If $a < a_1 < b_1 < b$, then the last inequality implies that

$$\int_{a_1}^{b_1} \left\{ \int_{-n}^{n} f_n(t) \left[\int_{\lambda}^{\lambda+\Delta} \{\varphi(x, \lambda)\, \varphi(t, \lambda)\, d\xi_{a,b}(\lambda) \right.\right.$$

$$+ \varphi(x, \lambda)\, \theta(t, \lambda)\, d\eta_{a,b}(\lambda) + \varphi(t, \lambda)\, \theta(x, \lambda)\, d\eta_{a,b}(\lambda)$$

$$\left.\left. + \theta(x, \lambda)\, \theta(t, \lambda)\, d\zeta_{a,b}(\lambda)\}\right] dt \right\}^2 dx \leqslant \int_{-n}^{n} f_n^2(t)\, dt.$$

Letting $a \to -\infty$, $b \to +\infty$ for fixed a_1, b_1, we obtain

$$\int_{a_1}^{b_1} \left\{ \int_{-n}^{n} E_\Delta(x, t)\, f_n(t)\, dt \right\}^2 dx \leqslant \int_{-n}^{n} f_n^2(t)\, dt, \tag{7.1}$$

from which the lemma follows for $f_n(x)$ by the arbitrariness of a_1 and b_1.

Suppose now that $f(x)$ is an arbitrary function from $\mathscr{L}^2(-\infty, \infty)$. We put

$$f_n(x) = \begin{cases} f(x), & -n \leqslant x \leqslant n, \\ 0, & |x| > n. \end{cases}$$

For fixed a_1 and b_1 it follows from (7.1) that

$$\int_{a_1}^{b_1}\left\{\int_{-n}^{n} E_\Delta(x,t) f_n(t)\,dt\right\}^2 dx \leqslant \int_{-n}^{n} f_n^2(t)\,dt.$$

Letting $n \to \infty$, we obtain

$$\int_{a_1}^{b_1}\left\{\int_{-\infty}^{\infty} E_\Delta(x,t) f(t)\,dt\right\}^2 dx \leqslant \int_{-\infty}^{\infty} f^2(t)\,dt,$$

which proves the lemma, since a_1 and b_1 are arbitrary.

2. In accordance with the arguments of §2 there exist functions $m_1(\lambda)$ and $m_2(\lambda)$ which are regular in the upper halfplane and are such that the function

$$\psi_1(x,\lambda) = \theta(x,\lambda) + m_1(\lambda)\varphi(x,\lambda)$$

belongs to $\mathscr{L}^2(-\infty, 0)$, and the function

$$\psi_2(x,\lambda) = \theta(x,\lambda) + m_2(\lambda)\varphi(x,\lambda)$$

belongs to $\mathscr{L}^2(0,\infty)$.

Since $W\{\varphi,\theta\} = 1$, it easily follows that

$$W\{\psi_1,\psi_2\} = m_1(\lambda) - m_2(\lambda). \tag{7.2}$$

Further, by virtue of (5.4) for $\lambda = u + iv$, we have

$$\int_{-\infty}^{0} |\psi_1(x,\lambda)|^2\,dx = \frac{\operatorname{Im}\{m_1(\lambda)\}}{v},$$

$$\int_{0}^{\infty} |\psi_2(x,\lambda)|^2\,dx = -\frac{\operatorname{Im}\{m_2(\lambda)\}}{v}.$$

Hence $\operatorname{Im} m_1(\lambda) > 0$ and $\operatorname{Im} m_2(\lambda) < 0$ for $v > 0$.

Put

$$G(x,y;\lambda) = \begin{cases} \dfrac{\psi_2(x,\lambda)\psi_1(y,\lambda)}{m_1(\lambda) - m_2(\lambda)}, & y \leqslant x, \\[2ex] \dfrac{\psi_1(x,\lambda)\psi_2(y,\lambda)}{m_1(\lambda) - m_2(\lambda)}, & y > x, \end{cases}$$

and let

$$\Phi(x,\lambda) = \int_{-\infty}^{\infty} G(x,y;\lambda) f(y)\,dy.$$

We will show that if $f(x)$ is a continuous function, then $\Phi(x,\lambda)$ is the solution of the equation

$$\Phi''(x,\lambda) + \{\lambda - q(x)\}\Phi(x,\lambda) = f(x).$$

In fact, from the definition of the Green's function $G(x,y;\lambda)$ we have

$$\Phi(x,\lambda)=\frac{\psi_2(x,\lambda)}{m_1(\lambda)-m_2(\lambda)}\int_{-\infty}^{x}\psi_1(y,\lambda)f(y)\,dy+\frac{\psi_1(x,\lambda)}{m_1(\lambda)-m_2(\lambda)}\int_{x}^{\infty}\psi_2(y,\lambda)f(y)\,dy.$$

Therefore

$$\Phi'_x(x,\lambda)=\frac{\psi'_2(x,\lambda)}{m_1(\lambda)-m_2(\lambda)}\int_{-\infty}^{x}\psi_1(y,\lambda)f(y)\,dy+\frac{\psi'_1(x,\lambda)}{m_1(\lambda)-m_2(\lambda)}\int_{x}^{\infty}\psi_2(y,\lambda)f(y)\,dy,$$

$$\Phi''_{xx}(x,\lambda)=\frac{\psi''_2(x,\lambda)}{m_1(\lambda)-m_2(\lambda)}\int_{-\infty}^{x}\psi_1(y,\lambda)f(y)\,dy+\frac{\psi''_1(x,\lambda)}{m_1(\lambda)-m_2(\lambda)}\int_{x}^{\infty}\psi_2(y,\lambda)f(y)\,dy$$

$$+\frac{1}{m_1(\lambda)-m_2(\lambda)}\{\psi_1(x,\lambda)\psi'_2(x,\lambda)-\psi'_1(x,\lambda)\psi_2(x,\lambda)\}f(x)$$
$$=\{q(x)-\lambda\}\Phi(x,\lambda)+f(x)$$

by virtue of (7.2).

Let $[a,b]$ be a finite interval, and α, β fixed real numbers. As in the preceding section, we denote by $\{\lambda_n\}$ and $\{y_n(x)\}$ the eigenvalues and eigenfunctions of the boundary-value problem (6.1) + (6.4). Further, let the functions $\psi_{1,a}(x,\lambda)$ and $\psi_{2,b}(x,\lambda)$ be defined in the same way as was the function $\psi_b(x,\lambda)$ for the case of the halfline $[0,\infty)$ (cf. §2.3).

LEMMA 7.3. *Let* $f(x)\in\mathscr{L}^2(-\infty,\infty)$, $z=u+iv$. *One has the bound*

$$\int_{-\infty}^{\infty}|R_zf|^2\,dx\leqslant\frac{1}{v^2}\int_{-\infty}^{\infty}f^2(x)\,dx,$$

where

$$R_zf=\Phi(x,z)=\int_{-\infty}^{\infty}G(x,y;z)f(y)\,dy.$$

The proof basically coincides with that of Lemma 3.2.

LEMMA 7.4. *For $a > 0$ we have*

$$\int_{-\infty}^{-a} \frac{d\xi(\lambda)}{\lambda^2} < \infty, \quad \int_{a}^{\infty} \frac{d\xi(\lambda)}{\lambda^2} < \infty, \quad \int_{-\infty}^{-a} \frac{d\zeta(\lambda)}{\lambda^2} < \infty,$$

$$\int_{a}^{\infty} \frac{d\zeta(\lambda)}{\lambda^2} < \infty, \quad \int_{-\infty}^{-a} \frac{|d\eta(\lambda)|}{\lambda^2} < \infty, \quad \int_{a}^{\infty} \frac{|d\eta(\lambda)|}{\lambda^2} < \infty.$$

PROOF. Let us denote by $G_{a,b}(x,t;z)$ the Green's function of the problem (6.1) + (6.4). Then similarly to formula (3.4) we have

$$\int_a^b G_{a,b}(x, t; z)\, y_n(t)\, dt = \frac{y_n(x)}{z - \lambda_n} \qquad (z = u + iv). \tag{7.3}$$

Differentiating this equality with respect to x. we obtain

$$\int_a^b \frac{\partial}{\partial x} G_{a,b}(x, t; z)\, y_n(t)\, dt = \frac{y_n'(x)}{z - \lambda_n}. \tag{7.3'}$$

Then by virtue of (7.3) and the Parseval equality we have (μ, ν fixed)

$$\sum_{\mu < \lambda_n \leqslant \nu} \frac{y_n^2(x)}{|z - \lambda_n|^2} \leqslant \int_a^b |G_{a,b}(x, t; z)|^2\, dt,$$

or in the notation of the preceding section

$$\int_{\mu}^{\nu} \frac{\varphi^2(x, \lambda)\, d\xi_{a,b}(\lambda) + 2\varphi(x, \lambda)\,\theta(x, \lambda)\, d\eta_{a,b}(\lambda) + \theta^2(x, \lambda)\, d\zeta_{a,b}(\lambda)}{(u - \lambda)^2 + v^2}$$
$$\leqslant \int_a^b |G_{a,b}(x, t; z)|^2\, dt.$$

Letting first $a \to -\infty$, $b \to +\infty$, and then $\mu \to -\infty$, $\nu \to +\infty$, we obtain

$$\int_{-\infty}^{\infty} \frac{\varphi^2(x, \lambda)\, d\xi(\lambda) + 2\varphi(x, \lambda)\,\theta(x, \lambda)\, d\eta(\lambda) + \theta^2(x, \lambda)\, d\zeta(\lambda)}{(u - \lambda)^2 + v^2} < \infty.$$

Putting now $x = 0$ in the last inequality, by virtue of the condition (6.3) we obtain the assertion of the lemma for the function $\zeta(\lambda)$. In the same way, applying the Parseval equality to the function $\partial G_{a,b}(x,t;z)/\partial x$, by virtue of (7.3′) and the condition (6.2) we obtain the assertion of the lemma for the function $\xi(\lambda)$. Finally, the assertion of the lemma for the function $\eta(\lambda)$ follows from the bound $|d\eta(\lambda)| \leq \{d\xi(\lambda) d\zeta(\lambda)\}^{1/2}$ and from the Cauchy-Bunjakovskiĭ inequality. This proves the lemma.

THEOREM 7.1 (INTEGRAL REPRESENTATION OF THE RESOLVENT). *Let $f(x) \in \mathcal{L}^2(-\infty, \infty)$. Then for any nonreal z we have*

$$R_z f = \int_{-\infty}^{\infty} \frac{d_\lambda E_\lambda f}{z - \lambda}, \tag{7.4}$$

where

$$E_\lambda f = \int_{-\infty}^{\infty} E_\lambda(x, y) f(y)\, dy,$$

$$E_\lambda(x, y) = \int_{+0}^{\lambda} \{\varphi(x, \lambda)\varphi(y, \lambda)\, d\xi(\lambda) + \varphi(x, \lambda)\theta(y, \lambda)\, d\eta(\lambda)$$
$$+ \varphi(y, \lambda)\theta(x, \lambda)\, d\eta(\lambda) + \theta(x, \lambda)\theta(y, \lambda)\, d\zeta(\lambda)\}.$$

The integral in the right side of (7.4) *converges absolutely.*

PROOF. If $f(x) = f_n(x)$ vanishes outside a finite interval and has a continuous second derivative, then the derivation of (7.4) coincides word for word with that of (3.7).

Let us now consider the general case. Let $f(x)$ be an arbitrary square-integrable function on the interval $(-\infty, \infty)$ and suppose that $f_n(x)$ converges in quadratic mean to $f(x)$ as $n \to \infty$ and also satisfies the conditions stated above. As has been proved,

$$R_z f_n = \int_{-\infty}^{\infty} \frac{d_\lambda E_\lambda f_n}{z - \lambda} = \int_{-\infty}^{\infty} \left\{ \frac{\varphi(x, \lambda) F_n(\lambda)\, d\xi(\lambda) + \varphi(x, \lambda) G_n(\lambda)\, d\eta(\lambda)}{z - \lambda} \right. \tag{7.5}$$
$$\left. + \frac{\theta(x, \lambda) F_n(\lambda)\, d\eta(\lambda) + \theta(x, \lambda) G_n(\lambda)\, d\zeta(\lambda)}{z - \lambda} \right\},$$

where

$$F_n(\lambda) = \int_{-\infty}^{\infty} f_n(x)\varphi(x, \lambda)\, dx, \quad G_n(\lambda) = \int_{-\infty}^{\infty} f_n(x)\theta(x, \lambda)\, dx.$$

We have to show that one can pass to the limit $n \to \infty$ in this equality. By virtue of Lemma 7.3 we have $R_z f_n \to R_z f$ as $n \to \infty$. It remains to consider the right side of the equality.

Let us consider the Hilbert space of pairs of functions.[11] The elements

[11] For the definition of Hilbert space, cf. Chapter 13.

of this Hilbert space are ordered pairs $\bar{f}=\{f_1(\lambda),f_2(\lambda)\}$ of complex-valued functions. The scalar product is defined by

$$(\bar{f},\bar{g})=\int_{-\infty}^{\infty}\{f_1(\lambda)\bar{g}_1(\lambda)\,d\xi(\lambda)+f_1(\lambda)\bar{g}_2(\lambda)\,d\eta(\lambda)$$

$$+f_2(\lambda)\bar{g}_1(\lambda)\,d\eta(\lambda)+f_2(\lambda)\bar{g}_2(\lambda)\,d\zeta(\lambda)\}.$$

It easily follows from the definitions of the functions $\xi(\lambda)$, $\zeta(\lambda)$ and $\eta(\lambda)$ that $(\bar{f},\bar{f})\geqq 0$.

Hence we obtain the Cauchy-Bunjakovskiĭ inequality in the usual way:

$$|(\bar{f},\bar{g})|\leqslant(\bar{f},\bar{f})^{1/2}\cdot(\bar{g},\bar{g})^{1/2}. \tag{7.6}$$

Let A be an arbitrary positive number. We put

$$R_zf_n=\left(\int_{-\infty}^{-A}+\int_{-A}^{A}+\int_{A}^{\infty}\right)\frac{d_\lambda E_\lambda f_n}{z-\lambda}.$$

From the definition of the function $E_\lambda f_n$, and by virtue of the inequality (7.6), there follows the bound

$$\left|\int_A^\infty\frac{d_\lambda E_\lambda f_n}{z-\lambda}\right|$$

$$\leqslant\left(\int_A^\infty\frac{\varphi^2(x,\lambda)\,d\xi(\lambda)+2\varphi(x,\lambda)\,\theta(x,\lambda)\,d\eta(\lambda)+\theta^2(x,\lambda)\,d\zeta(\lambda)}{|z-\lambda|^2}\right)^{1/2}$$

$$\times\left(\int_A^\infty\{F_n^2(\lambda)\,d\xi(\lambda)+2F_n(\lambda)\,G_n(\lambda)\,d\eta(\lambda)+G_n^2(\lambda)\,d\zeta(\lambda)\}\right)^{1/2}.$$

By Lemma 7.4 and the Parseval equality the last expression converges to zero uniformly in n as $A\to\infty$.

The analogous statement holds for the integral from $-\infty$ to $-A$. It remains to show that for fixed A, as $n\to\infty$

$$\int_{-A}^{A}\frac{d_\lambda E_\lambda f_n}{z-\lambda}\to\int_{-A}^{A}\frac{d_\lambda E_\lambda f}{z-\lambda}.$$

Integrating by parts, we obtain

$$\int_{-A}^{A}\frac{d_\lambda E_\lambda f_n}{z-\lambda}=\frac{E_Af_n}{z-A}-\frac{E_{-A}f_n}{z+A}-\int_{-A}^{A}\frac{E_\lambda f_n}{(z-\lambda)^2}\,d\lambda.$$

Since by Lemma 7.1 the function $E_\lambda(x,y)$ belongs to $\mathcal{L}^2(-\infty,\infty)$

as a function of y for fixed x, we can pass to the limit in the preceding expressions to obtain

$$\lim_{n\to\infty}\int_{-A}^{A}\frac{d_\lambda E_\lambda f_n}{z-\lambda}=\frac{E_A f}{z-A}-\frac{E_{-A}f}{z+A}-\int_{-A}^{A}\frac{E_\lambda f}{(z-\lambda)^2}\,d\lambda. \tag{7.7}$$

We will now show that as a function of λ, $E_\lambda f$ is of bounded variation in every finite interval. Suppose for definiteness that $\lambda > 0$, and assume first that $f(x) = f_n(x)$ vanishes outside a finite interval. $E_\lambda f_n$ is the limit, as $a \to -\infty$ amd $b \to +\infty$, of the sum

$$\sum_{0<\lambda_k\leqslant\lambda} y_k(x)\int_a^b f_n(t)\,y_k(t)\,dt.$$

By virtue of (7.2)

$$y_k(x)=(z-\lambda_k)\int_a^b G_{a,b}(x,t;z)\,y_k(t)\,dt.$$

It therefore follows from the Cauchy-Bunjakovskiĭ inequality that

$$\sum_{0<\lambda_k\leqslant\lambda}\left|\int_a^b f_n(t)\,y_k(t)\,y_k(x)\,dt\right|$$

$$=\sum_{0<\lambda_k\leqslant\lambda}\left|\int_a^b f_n(t)\,y_k(t)\,dt\right|\cdot|z-\lambda_k|\cdot\left|\int_a^b G_{a,b}(x,t;z)\,y_k(t)\,dt\right|$$

$$\leqslant \max_{0\leqslant\mu\leqslant\lambda}|z-\mu|\left(\int_a^b f_n^2(t)\,dt\right)^{1/2}\left(\int_a^b |G_{a,b}(x,t;z)|^2\,dt\right)^{1/2}.$$

The last expression is obviously bounded, uniformly with respect to n.

Thus $E_\lambda f$ is the limit of functions whose variations are uniformly bounded. Therefore this limit function also is of bounded variation.

Integrating the right side of (7.7) by parts in the opposite order, we obtain

$$\lim_{n\to\infty}\int_{-A}^{A}\frac{d_\lambda E_\lambda f_n}{z-\lambda}=\int_{-A}^{A}\frac{d_\lambda E_\lambda f}{z-\lambda},$$

and since A is arbitrary, this proves (7.4).

§8. Formulas for the functions $\xi(\lambda)$, $\zeta(\lambda)$ and $\eta(\lambda)$

As we will see in Chapter 4, on the halfline the representation of the function $\rho(\lambda)$ as the limit of the functions $\rho_b(\lambda)$ as $b \to \infty$ can in many cases be used to determine the spectrum of a Sturm-Liouville operator. A similar calculation on the entire axis turns out to be extremely difficult. This is why formulas for the functions $\xi(\lambda)$, $\zeta(\lambda)$ and $\eta(\lambda)$ analogous to the formula (5.7) are of particular value.

LEMMA 8.1. *For fixed* u_1 *and* u_2 *we have for* $v \to 0$ *the estimates*

$$\int_{u_1}^{u_2} \left\{ -\operatorname{Im} \frac{1}{m_1(u+iv) - m_2(u+iv)} \right\} du = O(1), \tag{8.1}$$

$$\int_{u_1}^{u_2} \left\{ -\operatorname{Im} \frac{m_1(u+iv)\, m_2(u+iv)}{m_1(u+iv) - m_2(u+iv)} \right\} du = O(1), \tag{8.2}$$

$$\int_{u_1}^{u_2} \left| \operatorname{Im} \frac{m_1(u+iv)}{m_1(u+iv) - m_2(u+iv)} \right| du = O(1). \tag{8.3}$$

PROOF. By the Parseval equality it follows from (7.2) that

$$\int_a^b |G_{a,b}(x, t; z)|^2 dt = \sum_{n=0}^{\infty} \frac{y_n^2(x)}{(u-\lambda_n)^2 + v^2} \qquad (z = u + iv).$$

Integrating with respect to u, we obtain

$$\int_{u_1}^{u_2} du \int_a^b |G_{a,b}(x, t; z)|^2 dt = \sum_{n=0}^{\infty} y_n^2(x) \int_{u_1}^{u_2} \frac{du}{(u-\lambda_n)^2 + v^n}. \tag{8.4}$$

We will show that for fixed u_1 and u_2, uniformly with respect to n, a and b,

$$\int_{u_1}^{u_2} \frac{du}{(u-\lambda_n)^2 + v^2} = O\left\{ \frac{1}{v} \cdot \frac{1}{\lambda_n^2 + 1} \right\}. \tag{8.5}$$

Let $-N \leqq u_1 < u_2 \leqq N$, where $N \geqq 1$. If $|\lambda_n| \geqq 2N$, then $|u - \lambda_n| \geqq \lambda_n/2$. Therefore $(v \leqq 1)$

$$\int_{u_1}^{u_2} \frac{v\, du}{(u-\lambda_n)^2 + v^2} \leqslant \int_{u_1}^{u_2} \frac{4du}{\lambda_n^2} \leqslant \frac{8N}{\lambda_n^2} \leqslant \frac{16N}{\lambda_n^2 + 1}. \tag{8.6}$$

If $|\lambda_n| < 2N$, then

$$\int_{u_1}^{u_3} \frac{v\, du}{(u-\lambda_n)^2 + v^2} \leqslant \int_{-\infty}^{\infty} \frac{v\, du}{(u-\lambda_n)^2 + v^2} = \pi \leqslant \pi \frac{4N^2 + 1}{\lambda_n^2 + 1}. \tag{8.7}$$

The estimate (8.5) follows from (8.6) and (8.7). From (8.4) and (8.5) we find that

$$\int_{u_1}^{u_2} du \int_a^b |G_{a,b}(x,\ t;\ z)|^2\, dt = O\left(\frac{1}{v}\sum_{n=0}^{\infty}\frac{y_n^2(x)}{\lambda_n^2+1}\right)$$

$$= O\left(\frac{1}{v}\int_a^b |G_{a,b}(x,\ t;\ i)|^2\, dy\right) = O\left(\frac{1}{v}\right) \qquad (v \to 0)$$

uniformly with respect to a, b and x; u_1 and u_2 fixed.

Now letting $a \to -\infty$, $b \to +\infty$, we obtain

$$\int_{u_1}^{u_2} du \int_{-\infty}^{\infty} |G(x,\ t;\ z)|^2\, dt = O\left(\frac{1}{v}\right). \tag{8.8}$$

For $x = 0$

$$G(0,\ t;\ z) = \begin{cases} \dfrac{\psi_1(t,\ z)}{m_1(z) - m_2(z)}, & y \leqslant 0, \\[2ex] \dfrac{\psi_2(t,\ z)}{m_1(z) - m_2(z)}, & y > 0. \end{cases}$$

By virtue of (5.4)

$$\begin{aligned} &\int_{-\infty}^{0} |\psi_1(x,\ z)|^2\, dx = \frac{\operatorname{Im}\{m_1(z)\}}{v}, \\ &\int_{0}^{\infty} |\psi_2(x,\ z)|^2\, dx = -\frac{\operatorname{Im}\{m_2(z)\}}{v}. \end{aligned} \tag{8.9}$$

Therefore putting $x = 0$ in (8.8), we obtain

$$\int_{u_1}^{u_2} \frac{\operatorname{Im}\{m_1(z) - m_2(z)\}}{|m_1(z) - m_2(z)|^2}\, du = O(1),$$

whence (8.1) follows.

By a similar argument we obtain with the help of (8.3)

$$\int_{u_1}^{u_2} du \int_{-\infty}^{\infty} |G'_x(x,\ t;\ z)|^2\, dt = O\left(\frac{1}{v}\right).$$

But

$$G'_x(0,\ t;\ z) = \begin{cases} \dfrac{-m_2(z)\,\psi_1(t,\ z)}{m_1(z) - m_2(z)}, & y < 0, \\[2ex] \dfrac{-m_1(z)\,\psi_2(t,\ z)}{m_1(z) - m_2(z)}, & y > 0. \end{cases}$$

Therefore by (8.9)

$$\int_{u_1}^{u_2} \frac{|m_2(z)|^2 \operatorname{Im}\{m_1(z)\} - |m_1(z)|^2 \operatorname{Im}\{m_2(z)\}}{|m_1(z) - m_2(z)|} du = O(1),$$

from which follows

$$\int_{u_1}^{u_2} \operatorname{Im}\left\{\frac{m_1(z)\, m_2(z)}{m_1(z) - m_2(z)}\right\} du = O(1),$$

i.e. (8.2). It remains to prove (8.3). To this end let us verify the inequality

$$\left[\operatorname{Im} \frac{m_1(z)}{m_1(z) - m_2(z)}\right]^2 \leqslant \operatorname{Im}\left\{\frac{1}{m_1(z) - m_2(z)}\right\} \cdot \operatorname{Im}\left\{\frac{m_1(z)\, m_2(z)}{m_1(z) - m_2(z)}\right\}. \tag{8.10}$$

We put $m_1(z) = r(\cos\varphi + i\sin\varphi)$; $m_2(z) = \rho(\cos\psi + i\sin\psi)$. By virtue of (8.9) $\operatorname{Im} m_1(z) > 0$, consequently $\sin\varphi > 0$; $\operatorname{Im} m_2(z) < 0$, consequently $\sin\psi < 0$, $v > 0$.

Multiplying all the numerators and denominators in (8.10) by $\overline{m_1(z)} - \overline{m_2(z)}$ and then multiplying out the resulting real denominators, we arrive at the inequality

$$[\operatorname{Im}\{m_1(z)\overline{m_2(z)}\}]^2 \leqslant \operatorname{Im}\{\overline{m_1(z)} - \overline{m_2(z)}\} \cdot \operatorname{Im}\{|m_1(z)|^2 m_2(z) - \overline{m_1(z)}\,|m_2(z)|^2\}$$

or in trigonometric form

$$r\rho\sin^2(\varphi - \psi) \leqslant (\rho\sin\psi - r\sin\varphi)(r\sin\psi - \rho\sin\varphi) = r\rho\sin^2\psi + r\rho\sin^2\varphi - (r^2 + \rho^2)\sin\varphi\sin\psi.$$

It follows that

$$r\rho[-2\sin^2\varphi\sin^2\psi - 2\sin\varphi\cos\psi\sin\psi\cos\varphi] \leqslant -(r^2 + \rho^2)\sin\varphi\sin\psi.$$

Cancelling out the positive quantity $-\sin\varphi\cdot\sin\psi$, we obtain

$$2r\rho\cos(\varphi - \psi) \leqq r^2 + \rho^2,$$

and this inequality is always valid.

The inequality (8.10) implies the bound

$$\int_{u_1}^{u_2} \left|\operatorname{Im} \frac{m_1(z)}{m_1(z) - m_2(z)}\right| du \leqq$$

$$\leqslant \int_{u_1}^{u_2} \left| \operatorname{Im} \frac{1}{m_1(z) - m_2(z)} \right|^{1/2} \left| \operatorname{Im} \frac{m_1(z)\, m_2(z)}{m_1(z) - m_2(z)} \right|^{1/2} du.$$

$$\leqslant \left(\int_{u_1}^{u_2} \operatorname{Im} \left\{ \frac{1}{m_1(z) - m_2(z)} \right\} du \cdot \int_{u_1}^{u_2} \operatorname{Im} \left\{ \frac{m_1(z)\, m_2(z)}{m_1(z) - m_2(z)} \right\} du \right)^{1/2},$$

which proves (8.3).

THEOREM 8.1. *Let* $\Delta = (\lambda, \lambda + \Delta)$ *be a finite interval whose endpoints are points of continuity of the functions* $\xi(\lambda)$, $\zeta(\lambda)$ *and* $\eta(\lambda)$.

Then we have ($\xi(\Delta) = \xi(\lambda + \Delta) - \xi(\lambda)$, *etc.*)

$$\xi(\Delta) = \frac{1}{\pi} \lim_{\delta \to 0} \int_{\Delta} \left\{ -\operatorname{Im} \frac{1}{m_1(u + i\delta) - m_2(u + i\delta)} \right\} du, \tag{8.11}$$

$$\eta(\Delta) = \frac{1}{\pi} \lim_{\delta \to 0} \int_{\Delta} \left\{ -\operatorname{Im} \frac{m_1(u + i\delta)}{m_1(u + i\delta) - m_2(u + i\delta)} \right\} du, \tag{8.12}$$

$$\zeta(\Delta) = \frac{1}{\pi} \lim_{\delta \to 0} \int_{\Delta} \left\{ -\operatorname{Im} \frac{m_1(u + i\delta)\, m_2(u + i\delta)}{m_1(u + i\delta) - m_2(u + i\delta)} \right\} du. \tag{8.13}$$

PROOF. Suppose that $f_1(x)$ and $f_2(x)$ vanish outside some finite interval and have continuous second derivatives. We put

$$F_1(\lambda) = \int_{-\infty}^{\infty} f_1(x)\, \varphi(x, \lambda)\, dx, \quad G_1(\lambda) = \int_{-\infty}^{\infty} f_1(x)\, \theta(x, \lambda)\, dx,$$

$$F_2(\lambda) = \int_{-\infty}^{\infty} f_2(x)\, \varphi(x, \lambda)\, dx, \quad G_2(\lambda) = \int_{-\infty}^{\infty} f_2(x)\, \theta(x, \lambda)\, dx.$$

It follows from Theorem 7.1 that

$$\Phi(z) = \int_{-\infty}^{\infty} R_z f_1 \cdot f_2(x)\, dx$$

$$= \int_{-\infty}^{\infty} \left[\frac{F_1(\lambda) F_2(\lambda)\, d\xi(\lambda) + \{F_1(\lambda) G_2(\lambda) + F_2(\lambda) G_1(\lambda)\}\, d\eta(\lambda)}{z - \lambda} \right.$$

$$\left. + \frac{G_1(\lambda) G_2(\lambda)\, d\zeta(\lambda)}{z - \lambda} \right] = \int_{-\infty}^{\infty} \frac{d\sigma(\lambda)}{z - \lambda},$$

where

$$\sigma(\Delta) = \int_{\Delta} \{F_1(\lambda) F_2(\lambda)\, d\xi(\lambda)$$

$$+ [F_1(\lambda) G_2(\lambda) + F_2(\lambda) G_1(\lambda)]\, d\eta(\lambda) + G_1(\lambda) G_2(\lambda)\, d\zeta(\lambda)\}.$$

By the Stieltjes inversion formula

$$\sigma(\Delta)=\frac{1}{\pi}\lim_{\delta\to 0}\int_{\Delta}\{-\operatorname{Im}\Phi(u+i\delta)\}\,du.$$

On the other hand,

$$\begin{aligned}&\operatorname{Im}\Phi(u+i\delta)\\&=\int_{-\infty}^{\infty}f_2(x)\,dx\,\operatorname{Im}\Bigg\{\int_{-\infty}^{x}\frac{\theta(x,u+i\delta)+m_2(u+i\delta)\varphi(x,u+i\delta)}{m_1(u+i\delta)-m_2(u+i\delta)}\\&\qquad\times[\theta(t,u+i\delta)+m_1(u+i\delta)\varphi(t,u+i\delta)]f_1(t)\,dt\\&\qquad+\int_{x}^{\infty}\frac{\theta(x,u+i\delta)+m_1(u+i\delta)\varphi(x,u+i\delta)}{m_1(u+i\delta)-m_2(u+i\delta)}\\&\qquad\qquad\times[\theta(t,u+i\delta)+m_2(u+i\delta)\varphi(t,u+i\delta)]f_1(t)\,dt\Bigg\}.\end{aligned}$$

Keeping in mind:

1. that the functions $\varphi(x,u)$ and $\theta(x,u)$ are real,
2. the equality

$$\operatorname{Im}\frac{m_2(z)}{m_1(z)-m_2(z)}=\operatorname{Im}\left\{\frac{m_1(z)}{m_1(z)-m_2(z)}-1\right\}=\operatorname{Im}\frac{m_1(z)}{m_1(z)-m_2(z)},$$

3. Lemma 8.1,

we obtain

$$\begin{aligned}\sigma(\Delta)&=\frac{1}{\pi}\lim_{\delta\to 0}\int_{\Delta}F_1(u)F_2(u)\left\{-\operatorname{Im}\frac{1}{m_1(u+i\delta)-m_2(u+i\delta)}\right\}du\\&+\frac{1}{\pi}\lim_{\delta\to 0}\int_{\Delta}\{F_1(u)G_2(u)+F_2(u)G_1(u)\}\\&\times\left\{-\operatorname{Im}\frac{m_1(u+i\delta)}{m_1(u+i\delta)-m_2(u+i\delta)}\right\}du\\&+\frac{1}{\pi}\lim_{\delta\to 0}\int_{\Delta}G_1(u)G_2(u)\left\{-\operatorname{Im}\frac{m_1(u+i\delta)\cdot m_2(u+i\delta)}{m_1(u+i\delta)-m_2(u+i\delta)}\right\}du.\end{aligned}\tag{8.14}$$

Choosing the functions $f_1(x)$ and $f_2(x)$ in such a way that $F_1(u)$ and $F_2(u)$ are close to zero in the interval Δ, and $G_1(u)$ and $G_2(u)$ are close to unity in Δ, by Lemma 8.1 we obtain the formula (8.13) from (8.14).

If $f_1(x)$ and $f_2(x)$ are such that $F_1(u)$ and $F_2(u)$ are close to zero, and $G_1(u)$ and $G_2(u)$ are close to unity, then integrating by parts it is easily seen that the functions

$$\int_{-\infty}^{\infty} f_1'(x)\,\varphi(x,\,u)\,dx, \qquad \int_{-\infty}^{\infty} f_2'(x)\,\varphi(x,\,u)\,dx$$

will be close to -1, and the functions

$$\int_{-\infty}^{\infty} f_1'(x)\,\theta(x,\,u)\,dx, \qquad \int_{-\infty}^{\infty} f_2'(x)\,\theta(x,\,u)\,dx$$

will be close to zero.

Choosing $f_1(x)$ and $f_2(x)$ so that $F_1(u)$ and $F_2(u)$ are close to unity, and $G_1(u)$ and $G_2(u)$ are close to zero, we obtain the formula (8.11) from (8.14).

Let us rewrite (8.14) in the form

$$\int_{\lambda}^{\lambda+\Delta} \{F_1(u)\,F_2(u)\,d\xi(u)$$

$$+ [F_1(u)\,G_2(u) + F_2(u)\,G(u)]\,d\eta(u) + G_1(u)\,G_2(u)\,d\zeta(u)\}$$

$$= \lim_{\delta\to 0}\int_{\lambda}^{\lambda+\Delta} F_1(u)\cdot F_2(u)\,d\,\frac{1}{\pi}\int_{+0}^{u}\left\{-\operatorname{Im}\frac{1}{m_1(u+i\delta)-m_2(u+i\delta)}\right\}du$$

$$+ \lim_{\delta\to 0}\int_{\lambda}^{\lambda+\Delta} G_1(u)\cdot G_2(u)\,d\,\frac{1}{\pi}\int_{+0}^{u}\left\{-\operatorname{Im}\frac{m_1(u+i\delta)\,m_2(u+i\delta)}{m_1(u+i\delta)-m_2(u+i\delta)}\right\}du$$

$$+ \frac{1}{\pi}\lim_{\delta\to 0}\int_{\lambda}^{\lambda+\Delta}\{F_1(u)\,G_2(u) + F_2(u)\,G_1(u)\}$$

$$\times\left\{-\operatorname{Im}\frac{m_1(v+i\delta)}{m_1(u+i\delta)-m_2(u+i\delta)}\right\}du.$$

Passing to the limit under the integral sign in the two integrals on the right (which is permissible by Helly's theorem) and using formulas (8.11) and (8.13), we obtain

$$\int_{\Delta} [F_1(u)\,G_2(u) + F_2(u)\,G_1(u)]\,d\eta(u)$$

$$= \frac{1}{\pi}\lim_{\delta\to 0}\int_{\Delta}\{F_1(u)\,G_2(u) + F_2(u)\,G_1(u)\}$$

$$\times\left\{-\operatorname{Im}\frac{m_1(u+i\delta)}{m_1(u+i\delta)-m_2(u+i\delta)}\right\}du.$$

Let $h > 0$. We put

$$f_1(x)=0, \quad x<0, \quad -\frac{1}{h}, \quad 0\leqslant x\leqslant h, \quad 0, \quad x>h,$$

$$f_2(x)=0, \quad x<0, \quad \frac{1}{h^2}, \quad 0\leqslant x\leqslant h, \quad 0, \quad x>h.$$

Then

$$F_1(u)G_2(u)+F_1(u)G_1(u)=-\frac{2}{h^2}\int_0^h \varphi(x, u)\,dx\cdot\frac{1}{h}\int_0^h \theta(x, u)\,dx.$$

By virtue of the initial condition $\theta(0,u)=1$, for sufficiently small h the integral $(1/h)\int_0^h\theta(x,u)\,dx$ differs from unity by arbitrarily little. Further, from the initial condition $\varphi(0,u)=0$ we obtain

$$-\frac{2}{h^2}\int_0^h \varphi(x, u)\,dx=-\frac{2}{h^2}\int_0^h \{\varphi(x, u)-\varphi(0, u)\}\,dx$$

$$=-\frac{2}{h^2}\int_0^h x\varphi'_x(\xi_x, u)\,dx.$$

By virtue of the initial condition $\varphi'_x(0,u)=-1$, for sufficiently small h the last expression will differ arbitrarily little from unity.

Thus, for sufficiently small h the expression $F_1(u)\,G_2(u)+F_2(u)\,G_1(u)$ differs arbitrarily little from unity. Therefore (8.12) follows from (8.25), which completes the proof of the theorem.

§9. The Parseval equality for an nth order selfadjoint differential operator

1. In this section the results of the first and sixth sections of the present chapter will be extended to an nth order selfadjoint differential operator. We will, in this connection, again use the method of passage to the limit.

Let $-\infty\leqq a<b\leqq\infty$ and suppose that we are given a differential expression

$$ly=p_0(x)y^{(n)}+p_1(x)y^{(n-1)}+\cdots+p_n(x)y \tag{9.1}$$

in the interval (a,b). We will assume that $p_k(x)$ is a complex-valued function of the class $C^{(n-k)}(a,b)$, and that $p_0(x)\neq 0$. We will also assume that the differential expression (9.1) is formally selfadjoint, i.e. coincides with its Lagrangian adjoint l^+:

$$l^+y=(-1)^n(\bar{p}_0(x)y)^{(n)}+(-1)^{n-1}(\bar{p}_1(x)y)^{(n-1)}+\cdots+\bar{p}_n(x)y.$$

For the case of even n the general form of such expressions is given in Chapter 1, §15. For the case of odd n some of the coefficients must necessarily be complex-valued.

As in §6, we will consider a selfadjoint problem on a finite closed interval $\delta=[a',b']$, $a<a'<b'<b$. Suppose that we are given n linearly independent selfadjoint boundary conditions on the interval δ:

$$U_{\delta_j}(y)=\sum_{k=1}^{n}\{M_{\delta_{jk}}y^{(k-1)}(a')+N_{\delta_{jk}}y^{(k-1)}(b')\}=0 \tag{9.2}$$
$$(j=1,\ 2,\ \ldots,\ n).$$

We will for brevity write the conditions (9.2) in the form $U_\delta(y)=0$. Then the boundary-value problem on δ can be written in the form

$$ly=\lambda y,\qquad U_\delta(y)=0. \tag{9.3}$$

The problem (9.3) is selfadjoint, and therefore for every $\delta\subset(a,b)$ there exists a complete orthonormal system of eigenfunctions $\{\chi_{\delta k}(x)\}$ and corresponding eigenvalues $\{\lambda_{\delta k}\}$.

In what follows the scalar product and the norm for functions from the space $\mathscr{L}^2(\delta)$ will be defined by

$$(u,\ v)_\delta=\int_\delta u(x)\,\bar v(x)\,dx,\qquad \|u\|_\delta=(u,\ u)_\delta^{1/2}.$$

For functions from $\mathscr{L}^2(a,b)$ these quantities will be denoted by (u,v) and $\|u\|$, i.e.

$$(u,\ v)=\int_a^b u(x)\,\bar v(x)\,dx,\qquad \|u\|=(u,\ u)^{1/2}.$$

2. The Parseval equality for the selfadjoint problem (9.3) can be written in the form

$$\|f\|_\delta^2=\sum_{k=1}^{\infty}|(f,\ \chi_{\delta k})|^2. \tag{9.4}$$

We will obtain the Parseval equality on the interval (a,b) by passing to the limit in (9.4), i.e. letting $\delta\to(a,b)$. To this end we denote by $\varphi_j=\varphi_j(x,\lambda)$ $(j=1,2,\cdots,n)$ the solutions of the equation $ly=\lambda y$ which, for some fixed c $(a<c<b)$, satisfy the initial conditions

$$\varphi_j^{(k-1)}(c,\lambda)=\delta_{jk}\qquad (j,k=1,2,\cdots,n), \tag{9.5}$$

where δ_{jk} is the Kronecker symbol. For fixed x the derivatives $\varphi_j^{(k-1)}(x,\lambda)$ are entire functions of λ. Since the solutions $\varphi_j(x,\lambda)$ are linearly independent, we have

$$\chi_{\delta k}(x) = \sum_{j=1}^{n} r_{\delta kj} \varphi_j(x, \lambda_{\delta k}), \tag{9.6}$$

where the $r_{\delta kj}$ are complex constants. Inserting (9.6) in (9.4), we write the Parseval equality in the form

$$\|f\|_\delta^2 = \int_{-\infty}^{\infty} \sum_{j,k=1}^{n} \bar{F}_{\delta j}(\lambda) F_{\delta k}(\lambda)\, d\rho_{\delta jk}(\lambda), \tag{9.7}$$

where

$$F_{\delta j}(\lambda) = \int_\delta f(x) \bar{\varphi}_j(x, \lambda)\, dx, \tag{9.8}$$

and the elements of the matrix $\rho_\delta(\lambda) = \|\rho_{\delta jk}(x)\|_{j,k=1}^{n}$ are jump functions with jumps at the eigenvalues of the problem (9.3). The magnitudes of the jumps are given by

$$\rho_{\delta jk}(\lambda + 0) - \rho_{\delta jk}(\lambda - 0) = \sum_m r_{\delta mj} r_{\delta mk}, \tag{9.9}$$

where the sum is taken over all m for which $\lambda_{\delta m} = \lambda_{\delta p}$ (in contrast with the case in §1, different functions $\chi_{\delta m}(x)$ can correspond to one and the same eigenvalue $\lambda_{\delta p}$). The matrix $\rho_\delta(\lambda)$ obviously has the following properties:

1. $\rho_\delta(\lambda)$ is a Hermitian matrix.
2. If $\lambda > \mu$, then the matrix $\rho_\delta(\Delta) = \rho_\delta(\lambda) - \rho_\delta(\mu)$ is nonnegative ($\Delta = (\mu, \lambda]$).
3. The total variation of $\rho_{\delta jk}(\lambda)$ is finite on every finite λ-interval.

For definiteness we will assume that $\rho_\delta(0)$ is the zero matrix and that $\rho_\delta(\lambda + 0) = \rho_\delta(\lambda)$.

The matrix $\rho_\delta(\lambda)$ is called the *spectral matrix* of the problem (9.3).

The following lemma is the analogue of Lemma 1.1.

LEMMA 9.1. *Let $\{\delta\}$ be a set of closed intervals which converge to the interval (a, b), and let $\{U_\delta(y) = 0\}$ be a corresponding set of selfadjoint boundary conditions. Then we can find a sequence $\{\delta_j\}$ in $\{\delta\}$ such that*

$$\lim_{j\to\infty} \delta_j = (a, b), \qquad \lim_{j\to\infty} \rho_{\delta_j}(\lambda) = \rho(\lambda),$$

where the second limit exists on the interval $(-\infty, \infty)$ and the limit matrix $\rho(\lambda)$ satisfies the conditions 1, 2 *and* 3 *listed above.*

PROOF. We will prove that for every fixed $A > 0$ there exists an $M = M(A) > 0$, not depending upon δ and U_δ, such that for $|\lambda| \leq A$

we have

$$\bigvee_{-A}^{A} \{\rho_{\delta jk}(\lambda)\} = \int_{-A}^{A} |d\rho_{\delta jk}(\lambda)| \leqslant M. \tag{9.10}$$

The lemma will then follow at once from Helly's selection theorem (see Helly's first theorem, Chapter 14, §2).

It suffices to prove the bound (9.10) for $j = k$, since it follows from (9.9) that

$$2\int_{-A}^{A} |d\rho_{\delta jk}(\lambda)| \leqslant \int_{-A}^{A} d\rho_{\delta jj}(\lambda) + \int_{-A}^{A} d\rho_{\delta kk}(\lambda). \tag{9.11}$$

We remark that the function $\varphi_j^{(k-1)}(x, \lambda)$ is jointly continuous in the variables x, λ and is equal to δ_{jk} for $x = c$. Therefore for given A and $\epsilon > 0$ (the number ϵ will be chosen later) there exists $h > 0$ such that

$$|\varphi_j^{(k-1)}(x, \lambda) - \delta_{jk}| < \epsilon \tag{9.12}$$

for $c \leqq x \leqq x + h$ and $|\lambda| \leqq A$.

Let $\tilde{f}(x)$ be a nonnegative function of the class $C^{(n)}(a, b)$ which vanishes together with its derivatives up to order $n-1$ inclusively outside the interval $(c, c+h)$, and is normalized so that

$$\int_a^b \tilde{f}(x)\, dx = 1. \tag{9.13}$$

Applying Bessel's inequality to the functions $(-1)^{m-1}\tilde{f}^{(m-1)}(x)$ $(m = 1, 2, \cdots, n)$, we obtain

$$\int_c^{c+h} |\tilde{f}^{(m-1)}(x)|^2\, dx \geqslant \int_{-A}^{A} \sum_{j,k=1}^{n} F_j(\lambda) F_k(\lambda)\, d\rho_{\delta jk}(\lambda), \tag{9.14}$$

where

$$F_k(\lambda) = (-1)^{m-1} \int_c^{c+h} \tilde{f}^{(m-1)}(x)\, \bar{\varphi}_k(x, \lambda)\, dx = \int_c^{c+h} \tilde{f}(x)\, \bar{\varphi}_k^{(m-1)}(x)\, dx.$$

It follows from (9.12) and the condition (9.13) that $|F_k(\lambda) - \delta_{km}| \leqq \epsilon$, whence

$$|F_k(\lambda)| \leqq \delta_{km} + \epsilon, \qquad |F_m(\lambda)| \geqq 1 - \epsilon. \tag{9.15}$$

Therefore

$$I = \int_{-A}^{A} \sum_{j,k=1}^{n} \bar{F}_j(\lambda) F_k(\lambda)\, d\rho_{\delta jk}(\lambda)$$

$$= \int_{-A}^{A} \sum_{j=1}^{n} |F_j(\lambda)|^2 d\rho_{\delta jj}(\lambda) + \int_{-A}^{A} \sum_{j \neq k} \bar{F}_j(\lambda) F_k(\lambda) d\rho_{\delta jk}(\lambda)$$

$$\geqslant \int_{-A}^{A} \sum_{j=1}^{n} |F_j(\lambda)|^2 d\rho_{\delta jj}(\lambda) - \int_{-A}^{A} \sum_{j \neq k} |F_j(x)| |F_k(\lambda)| |d\rho_{\delta jk}(\lambda)|.$$

Using the inequality (9.11), we obtain

$$I \geqslant \int_{-A}^{A} \sum_{j=1}^{n} |F_j(\lambda)|^2 d\rho_{\delta jj}(\lambda)$$

$$- \frac{1}{2} \int_{-A}^{A} \sum_{j \neq k} |F_j(\lambda)| |F_k(\lambda)| \{d\rho_{\delta jj}(\lambda) + d\rho_{\delta kk}(\lambda)\}$$

$$= 2 \int_{-A}^{A} \sum_{j=1}^{n} |F_j(\lambda)|^2 d\rho_{\delta jj}(\lambda)$$

$$- \frac{1}{2} \int_{-A}^{A} \sum_{j,\, k=1}^{n} |F_j(\lambda)| |F_k(\lambda)| \{d\rho_{\delta jj}(\lambda) + d\rho_{\delta kk}(\lambda)\}$$

$$= 2 \int_{-A}^{A} \sum_{j=1}^{n} |F_j(\lambda)|^2 d\rho_{\delta jj}(\lambda) - \int_{-A}^{A} \sum_{j,\, k=1}^{n} |F_j(\lambda)| |F_k(\lambda)| d\rho_{\delta jj}(\lambda).$$

Hence by virtue of the first of inequalities (9.15) there follows the inequality

$$I \geqslant 2 \int_{-A}^{A} |F_m(\lambda)|^2 d\rho_{\delta mm}(\lambda) - \int_{-A}^{A} \sum_{j,\, k=1}^{n} (\delta_{jm} + \varepsilon)(\delta_{km} + \varepsilon) d\rho_{\delta jj}(\lambda).$$

Finally, using the second of the inequalities (9.15), we obtain

$$I \geqslant 2(1-\varepsilon)^2 \int_{-A}^{A} d\rho_{\delta mm}(\lambda) - (1+n\varepsilon) \int_{-A}^{A} d\rho_{\delta mm}(\lambda)$$

$$- (1+n\varepsilon)\varepsilon \int_{-A}^{A} \sum_{j=1}^{n} d\rho_{\delta jj}(\lambda)$$

$$= \{1-(n+4)\varepsilon + 2\varepsilon^2\} \int_{-A}^{A} d\rho_{\delta mm}(\lambda) - (1+n\varepsilon)\varepsilon \int_{-A}^{A} \sum_{j=1}^{n} d\rho_{\delta jj}(\lambda).$$

It therefore follows from (9.14) that

$$\int_c^{c+h} |f^{(m-1)}(x)|^2\,dx$$

$$\geqslant \{1-(n+4)\varepsilon+2\varepsilon^2\}\int_{-A}^{A} d\rho_{\delta mm}(\lambda)-(1+n\varepsilon)\,\varepsilon\int_{-A}^{A}\sum_{j=1}^{n} d\rho_{\delta jj}(\lambda).$$

Summing with respect to m over the values $1,2,\cdots,n$, we obtain

$$\sum_{m=1}^{n}\int_c^{c+h} |f^{(m-1)}(x)|^2\,dx$$

$$\geqslant \{1-(n+4)\varepsilon+2\varepsilon^2\}\int_{-A}^{A}\sum_{m=1}^{n} d\rho_{\delta mm}(\lambda)-n(1+n\varepsilon)\,\varepsilon\int_{-A}^{A}\sum_{m=1}^{n} d\rho_{\delta mm}(\lambda)$$

$$=\{1-(2n+4)\varepsilon-(n^2-2)\varepsilon^2\}\int_{-A}^{A}\sum_{m=1}^{n} d\rho_{\delta mm}(\lambda). \tag{9.16}$$

If ϵ is chosen so that $1-2(n+2)\epsilon+(2-n^2)\epsilon^2>\frac{1}{2}$, then from (9.16) we obtain the bound (9.10), and consequently Lemma 9.1.

For every limit matrix $\rho(\lambda)$, whose existence is asserted by Lemma 9.1, we define $\mathcal{L}^2_{\{\rho(\lambda)\}}(-\infty,\infty)$ as the space of all vector-functions $g(\lambda)=\{g_j(\lambda)\}_{j=1}^n$ for which

$$\|g\|^2=\int_{-\infty}^{\infty}\sum_{j,k=1}^{n}\bar g_j(\lambda)\,g_k(\lambda)\,d\rho_{jk}(\lambda)<\infty.$$

One can show that the space $\mathcal{L}^2_{\{\rho(\lambda)\}}(-\infty,\infty)$ is complete (cf. Kac [1]).

Arguing now as in the proof of Theorem 6.1, one can obtain the following theorem.

THEOREM 9.1. *Let $\rho(\lambda)=\{\rho_{jk}(\lambda)\}_{j,k=1}^n$ be a limit matrix as defined in Lemma 9.1. If $f(x)\in\mathcal{L}^2(a,b)$, then there exists a vector-function $g(\lambda)\in\mathcal{L}^2_{\{\rho(\lambda)\}}(-\infty,\infty)$ such that if*

$$g_{\delta j}(\lambda)=\int_\delta f(x)\,\bar\varphi_j(x,\lambda)\,dx,\qquad \delta\subset(a,b),$$

then $\|g-g_\delta\|\to 0$ as $\delta\to(a,b)$. The Parseval equality $\|f\|=\|g\|$ and the expansion

$$f(x)=\int_{-\infty}^{\infty}\sum_{j,k=1}^{n} g_k(\lambda)\,\varphi_j(x,\lambda)\,d\rho_{jk}(\lambda)$$

hold, where the integral converges to the function $f(x)$ relative to the norm of the space $\mathscr{L}^2(a,b)$.

3. It was shown in §4 that for the spectral function of a second-order differential operator corresponding to a point of the limit circle one has the theorem on the inverse Fourier transform (cf. Theorem 4.2). In particular, for the case of a limit point the spectral function is unique and the inverse transform theorem holds for it.

For nth order operators the inverse transform theorem also holds under certain assumptions. In the following theorems, which we present without proofs (the reader can find the proofs of these theorems in the monograph by Coddington and Levinson [1], Chapter 10, §3), the assumptions correspond to the limit point case for a second-order differential operator.

THEOREM 9.2 (THE INVERSE TRANSFORM). *Let us assume that neither of the equations*

$$(l \pm i)y = 0 \tag{9.17}$$

has nontrivial solutions in $\mathscr{L}^2(a,b)$. For every vector-function $g(\lambda) \in \mathscr{L}^2_{\{\rho(\lambda)\}}(-\infty,\infty)$ there exists a function $f(x) \in \mathscr{L}^2(a,b)$ such that

$$f(x) = \int_{-\infty}^{\infty} \sum_{j,k=1}^{n} g_k(\lambda)\varphi_j(x,\lambda)\,d\rho_{jk}(\lambda),$$

where the integral converges to $f(x)$ in the norm of the space $\mathscr{L}^2(a,b)$ and

$$g_k(\lambda) = \int_a^b f(x)\bar{\varphi}_k(x,\lambda)\,dx.$$

THEOREM 9.3. *Let us assume that neither of the equations* (9.17) *has nontrivial solutions in $\mathscr{L}^2(a,b)$. Then the matrix $\rho(\lambda)$ is unique in the sense that if $\tilde{\rho}(\lambda)$ is any other matrix function for which Theorem* 9.1 *holds, then $\tilde{\rho}(\lambda) - \tilde{\rho}(\mu) = \rho(\lambda) - \rho(\mu)$ at all points of continuity of the matrices $\rho(\lambda)$ and $\tilde{\rho}(\lambda)$.*

Bibliographical references

§1. The results of this section are due to Weyl [**1—3**]. Another proof was given by Titchmarsh [**1**]. The proof of the Parseval equality presented here was given by Levitan [**1**] and independently by Levinson [**3, 4**] and Yosida [**1**].

§2. The results concerning the limit point and circle are due to Weyl [**2**]. Our discussion of Theorems 2.2 and 2.3 is based upon §2, Chapter

9 of the monograph by Coddington and Levinson [**1**]. We remark that Theorem 2.2 is a special case of a theorem on the deficiency indices of general symmetric operators in Hilbert space (cf., for example, Ahiezer and Glazman [**1**]). Theorem 2.3 is due to Titchmarsh [**5**]; a more general result was obtained by Levinson [**1**] and also by Sears and Titchmarsh [**1**].

§3. The integral representation of the resolvent was first proven by Weyl [**2**]. The proof presented here is due to Levitan [**1**]. Theorem 3.2 was obtained by Levitan [**1**]. Similar theorems have been obtained by Weyl and Titchmarsh.

§4. The results of this section show that to the points of Weyl's limit circle there correspond selfadjoint extensions of the Sturm-Liouville operator (cf. Chapter 13).

§5. Formula (5.7) is due to Titchmarsh [**1**].

§6. The basic result of this section is due to Weyl [**3**]. Another proof of the Parseval equality was given by Titchmarsh [**1**]. The proof presented here is due to Levitan [**1**].

§7. Weyl [**3**], Titchmarsh [**1**], Levitan [**1**].

§8. Formulas (8.11), (8.12) and (8.13) are due to Titchmarsh [**1**].

§9. Theorem 9.1 was first proven by M. G. Kreĭn [**1**] and Kodaira [**1**]. The proof discussed here, including the basic Lemma 9.1, is due to Levitan [**4**].

CHAPTER 3

EXPANSION IN THE SINGULAR CASE FOR A DIRAC SYSTEM

§1. Proof of the Parseval equality for a Dirac system

1. In §12 of Chapter 1 we obtained a theorem on expansion relative to the eigen-vector-functions of a regular problem for a Dirac system by means of a passage to the limit, replacing the differential equations by finite-difference equations.

As was the case for a Sturm-Liouville operator, considered in the preceding chapter, this idea enables us to obtain the expansion theorem for a singular problem for a Dirac system, if we regard it as the limit of regular problems.

We will consider the case in which the interval is the real halfline $[0, \infty)$.

Thus, we will consider the system $(0 \leq x < \infty)$

$$y_2' - \{\lambda + q_1(x)\}y_1 = 0, \tag{1.1}$$

$$y_1' + \{\lambda + q_2(x)\}y_2 = 0, \tag{1.2}$$

where the coefficients $q_1(x)$ and $q_2(x)$ are continuous in $[0, \infty)$.

Let us denote by α an arbitrary real number, and by

$$\varphi(x, \lambda) = \begin{pmatrix} \varphi_1(x, \lambda) \\ \varphi_2(x, \lambda) \end{pmatrix}$$

the solution of the system (1.1) – (1.2) which satisfies the initial conditions

$$\varphi_1(0, \lambda) = \cos\alpha, \qquad \varphi_2(0, \lambda) = -\sin\alpha. \tag{1.3}$$

Obviously $\varphi(x, \lambda)$ satisfies the boundary condition

$$y_1(0, \lambda)\sin\alpha + y_2(0, \lambda)\cos\alpha = 0. \tag{1.4}$$

We further denote by b an arbitrary positive number (further on b will be unboundedly increased) and by β an arbitrary real number, and adjoin to the problem (1.1) + (1.2) + (1.4) the boundary condition

$$y_2(b, \lambda)\cos\beta + y_1(b, \lambda)\sin\beta = 0. \tag{1.5}$$

The problem (1.1) + (1.2) + (1.4) + (1.5) is a regular problem for a Dirac system. Let us denote by $\lambda_{n,b}$ the eigenvalues of this problem, and by

$$\varphi_{n,b}(x) = \begin{pmatrix} \varphi^{(1)}_{n,b}(x) \\ \varphi^{(2)}_{n,b}(x) \end{pmatrix} = \varphi(x, \lambda_{n,b}) = \begin{pmatrix} \varphi_1(x, \lambda_{n,b}) \\ \varphi_2(x, \lambda_{n,b}) \end{pmatrix}$$

the corresponding eigen-vector-functions which satisfy the initial conditions (1.3). If

$$f(x) = \begin{pmatrix} f_1(x) \\ f_2(x) \end{pmatrix} \in \mathscr{L}^2(0, b),$$

i.e.

$$\int_0^b \{f_1^2(x) + f_2^2(x)\}\, dx < +\infty \quad \text{and} \quad \alpha^2_{n,b} = \int_0^b \{[\varphi^{(1)}_{n,b}(x)]^2 + [\varphi^{(2)}_{n,b}(x)]^2\}\, dx,$$

then by virtue of the Parseval equality (cf. (12.26), Chapter 1)

$$\int_0^b \{f_1^2(x) + f_2^2(x)\}\, dx = \sum_{n=-\infty}^{\infty} \frac{1}{\alpha^2_{n,b}} \left\{ \int_0^b [f_1(x)\, \varphi^{(1)}_{n,b}(x) + f_2(x)\, \varphi^{(2)}_{n,b}(x)]\, dx \right\}^2. \tag{1.6}$$

If we introduce the monotonically nondecreasing jump functions

$$\begin{aligned} \rho_b(\lambda) &= -\sum_{\lambda < \lambda_{n,b} < 0} \frac{1}{\alpha^2_{n,b}}, \qquad (\lambda \leqslant 0), \\ \rho_b(\lambda) &= \sum_{0 \leqslant \lambda_{n,b} < \lambda} \frac{1}{\alpha^2_{n,b}} \qquad (\lambda \geqslant 0), \end{aligned} \tag{1.7}$$

then the Parseval equality (1.6) can be written in the form

$$\int_0^b \{f_1^2(x) + f_2^2(x)\}\, dx = \int_{-\infty}^{\infty} F^2(\lambda)\, d\rho_b(\lambda), \tag{1.8}$$

where

$$F(\lambda) = \int_0^b \{f_1(x)\, \varphi_1(x, \lambda) + f_2(x)\, \varphi_2(x, \lambda)\}\, dx.$$

We will show that the Parseval equality for the problem (1.1) + (1.2) + (1.4) can be obtained from (1.8) by letting $b \to \infty$. We first prove a lemma.

Lemma 1.1. *For any positive number N there exists a positive constant $A = A(N)$, not depending upon b, such that*

$$(1.9) \qquad \bigvee_{-N}^{N} \{\rho_b(\lambda)\} = \sum_{-N\leqslant\lambda_{n,b}\leqslant N} \frac{1}{\alpha_{n,b}^2} = \rho_b(N) - \rho_b(-N) < A.$$

PROOF. Let $\sin\alpha \neq 0$. Since in the region $-N \leqq \lambda \leqq N$, $0 \leqq x \leqq a$ (a is an arbitrary positive number) the vector-function $\varphi(x,\lambda)$ is continuous, by virtue of the condition $\varphi_2(0,\lambda) = -\sin\alpha$ there exists a positive number h so small that for $|\lambda| \leqq N$ we have

$$(1.10) \qquad \left(\frac{1}{h}\int_0^h \varphi_2(x, \lambda)\,dx\right)^2 > \frac{1}{2}\sin^2\alpha.$$

Let the vector-function

$$f_h(x) = \begin{pmatrix} f_1(x) \\ f_2(x) \end{pmatrix}$$

be defined by

$$f_2(x) = \begin{cases} \frac{1}{h}, & 0 \leqslant x < h, \\ 0, & x \geqslant h, \end{cases} \qquad f_1(x) \equiv 0.$$

If we now apply the Parseval equality (1.8) to $f(x)$, then using the bound (10.10) we obtain

$$\int_0^h f_2^2(x)\,dx = \frac{1}{h} = \int_{-\infty}^{\infty} \left(\frac{1}{h}\int_0^h \varphi_2(x, \lambda)\,dx\right)^2 d\rho_b(\lambda)$$

$$\geqslant \int_{-N}^{N} \left(\frac{1}{h}\int_0^h \varphi_2(x, \lambda)\,dx\right)^2 d\rho_b(\lambda) > \frac{1}{2}\sin^2\alpha\,\{\rho_b(N) - \rho_b(-N)\},$$

from which the inequality (1.9) follows.

For the case in which $\sin\alpha = 0$ we obtain the inequality (1.9) by applying the Parseval equality to the vector-function

$$f_h(x) = \begin{pmatrix} f_1(x) \\ f_2(x) \end{pmatrix}$$

defined by

$$f_1(x) = \begin{cases} \frac{1}{h^2}, & 0 \leqslant x < h, \\ 0, & x \geqslant h, \end{cases} \qquad f_2(x) \equiv 0.$$

We will now show that using Lemma 1.1 and a well-known theorem on passage to the limit under the integral sign in a Stieltjes integral, one can obtain the Parseval equality for the problem (1.1) + (1.2) + (1.4).

Suppose first that the vector-function

$$f_n(x) = \begin{pmatrix} f_{1n}(x) \\ f_{2n}(x) \end{pmatrix}.$$

vanishes outside the interval $0 \leqq x \leqq n$, $n < b$, has a continuous first derivative and satisfies the boundary condition (1.4). Applying the Parseval equality (1.8) to this vector-function, we find that

$$\int_0^n \{f_{1n}^2(x) + f_{2n}^2(x)\}\, dx = \int_{-\infty}^{\infty} F_n^2(\lambda)\, d\rho_b(\lambda). \tag{1.11}$$

where

$$F_n(\lambda) = \int_0^b \{f_{1n}(x)\,\varphi_1(x, \lambda) + f_{2n}\varphi_2(x, \lambda)\}\, dx. \tag{1.12}$$

Since

$$\varphi(x,\lambda) = \begin{pmatrix} \varphi_1(x,\lambda) \\ \varphi_2(x,\lambda) \end{pmatrix}$$

satisfies the system (1.1) – (1.2), we have

$$\varphi_1(x, \lambda) = \frac{1}{\lambda}\left[\frac{d\varphi_2}{dx} - q_1(x)\,\varphi_1\right], \quad \varphi_2(x, \lambda) = \frac{1}{\lambda}\left[-\frac{d\varphi_1}{dx} - q_2(x)\varphi_2\right].$$

Therefore from (1.12) we obtain

$$F_n(\lambda) = \frac{1}{\lambda}\int_0^b f_n(x)\left[\frac{d\varphi_2}{dx} - q_1(x)\,\varphi_1(x, \lambda)\right] dx$$
$$+ \frac{1}{\lambda}\int_0^b f_{2n}(x)\left[-\frac{d\varphi_1}{dx} - q_2(x)\,\varphi_1(x, \lambda)\right] dx.$$

Further, since both the vector-functions $f_n(x)$ and $\varphi(x,\lambda)$ satisfy the boundary condition (1.4) and since $f_n(x)$ vanishes in a neighborhood of the point b, integrating the integrals on the right side of the last equality by parts, we obtain for $F_n(\lambda)$ the expression

$$F_n(\lambda) = \frac{1}{\lambda}\int_0^b \left\{\varphi_1(x, \lambda)\left[\frac{df_{2n}}{dx} - q_1(x)\, f_{1n}(x)\right]\right.$$
$$\left. + \varphi_2(x, \lambda)\left[-\frac{df_{1n}}{dx} - q_2(x)\, f_{2n}(x)\right]\right\} dx.$$

Therefore by the Parseval equality (1.8) we have for any finite $N > 0$

$$\int\limits_{|\lambda|>N} F_n^2(\lambda)\,d\rho_b(\lambda) \leqslant \frac{1}{N^2} \int\limits_{|\lambda|>N} \left\{ \int\limits_0^b \left[\varphi_1(x,\lambda)\left\{\frac{df_{2n}}{dx} - q_1(x) f_{1n}(x)\right\}\right.\right.$$

$$\left.\left. + \varphi_2(x,\lambda)\left\{-\frac{df_{1n}}{dx} - q_2(x) f_{2n}(x)\right\}\right] dx \right\}^2 d\rho_b(\lambda)$$

$$\leqslant \frac{1}{N^2} \int\limits_{-\infty}^{\infty} \left\{ \int\limits_0^b \left[\varphi_1(x,\lambda)\left\{\frac{df_{2n}}{dx} - q_1(x) f_{1n}(x)\right\}\right.\right.$$

$$\left.\left. + \varphi_2(x,\lambda)\left\{-\frac{df_{1n}}{dx} - q_2(x) f_{2n}(x)\right\}\right] dx \right\}^2 d\rho_b(\lambda)$$

$$= \frac{1}{N^2} \int\limits_0^n \left\{ \left[\frac{df_{2n}}{dx} - q_1(x) f_{1n}(x)\right]^2 + \left[-\frac{df_{1n}}{dx} - q_1(x) f_{2n}(x)\right]^2 \right\} dx.$$

From this bound and from (1.11) it follows that

$$\left| \int\limits_0^n \{f_{1n}^2(x) + f_{2n}^2(x)\}\,dx - \int\limits_{-N}^{N} F_n^2(\lambda)\,d\rho_b(\lambda) \right|$$

$$< \frac{1}{N^2} \int\limits_0^n \left\{ \left[\frac{df_{2n}}{dx} - q_1(x) f_{1n}(x)\right]^2 + \left[-\frac{df_{1n}}{dx} - q_2(x) f_{2n}(x)\right]^2 \right\} dx. \tag{1.13}$$

By Lemma 1.1 the set of functions $\{\rho_b(\lambda)\}$ $(-N \leqq \lambda \leqq N)$ is bounded. We can therefore find a sequence $\{b_k\}$ such that the functions $\rho_{b_k}(\lambda)$ converge to a monotone function $\rho(\lambda)$.

Passing to the limit in (1.13) along the sequence $\{b_k\}$, we obtain the inequality

$$\left| \int\limits_0^n \{f_{1n}^2(x) + f_{2n}^2(x)\}\,dx - \int\limits_{-N}^{N} F_n^2(\lambda)\,d\rho(\lambda) \right|$$

$$\leqslant \frac{1}{N^2} \int\limits_0^n \left\{ \left[\frac{df_{2n}}{dx} - q_1(x) f_{1n}(x)\right]^2 + \left[-\frac{df_{1n}}{dx} - q_2(x) f_{2n}(x)\right]^2 \right\} dx.$$

Finally, letting $N \to \infty$ in the last inequality, we obtain

$$\int\limits_0^n \{f_{1n}^2(x) + f_{2n}^2(x)\}\,dx = \int\limits_{-\infty}^{\infty} F_n^2(\lambda)\,d\rho(\lambda),$$

i.e. the Parseval equality for vector-functions

$$f_n(x) = \begin{pmatrix} f_{1n}(x) \\ f_{2n}(x) \end{pmatrix}$$

which vanish outside some finite interval, have continuous first derivatives and satisfy the boundary condition (1.4). The extension of the Parseval equality to any vector-function from $\mathscr{L}^2(0, \infty)$ for the problem (1.1) + (1.2) + (1.4) can be carried out in the very same way as for a Sturm-Liouville operator.

We have thus proven:

THEOREM 1.1. *Suppose that the vector-function*

$$f(x) = \begin{pmatrix} f_1(x) \\ f_2(x) \end{pmatrix}$$

belongs to $\mathscr{L}^2(0, \infty)$, *i.e.*

$$\int_0^\infty \{f_1^2(x) + f_2^2(x)\}\, dx < \infty.$$

There exists an increasing function $\rho(\lambda)$ $(-\infty < \lambda < \infty)$, *not depending upon* $f(x)$, *and a function* $F(\lambda)$ *(the generalized Fourier transform of the vector-function* $f(x)$ *with respect to the eigen-vector-functions of the problem* (1.1) + (1.2) + (1.4)) *such that*

$$\int_0^\infty \{f_1^2(x) + f_2^2(x)\}\, dx = \int_{-\infty}^{\infty} F^2(\lambda)\, d\rho(\lambda). \tag{1.14}$$

The function $F(\lambda)$ *is the limit in quadratic mean of the sequence of continuous functions*

$$F_n(\lambda) = \int_0^n \{f_1(x)\varphi_1(x, \lambda) + f_2(x)\varphi_2(x, \lambda)\}\, dx,$$

i.e.

$$\lim_{n\to\infty} \int_{-\infty}^{\infty} \{F(\lambda) - F_n(\lambda)\}^2\, d\rho(\lambda) = 0.$$

2. Suppose that the vector-functions

$$f(x) = \begin{pmatrix} f_1(x) \\ f_2(x) \end{pmatrix} \quad \text{and} \quad g(x) = \begin{pmatrix} g_1(x) \\ g_2(x) \end{pmatrix}$$

belong to $\mathscr{L}^2(0, \infty)$, and $F(\lambda)$ and $G(\lambda)$ are their Fourier transforms.

Obviously the Fourier transforms of the functions

$$f(x) \pm g(x) = \begin{pmatrix} f_1(x) \pm g_1(x) \\ f_2(x) \pm g_2(x) \end{pmatrix}$$

are the functions $F(\lambda) \pm G(\lambda)$. Therefore by (1.14)

$$\int_0^\infty \{[f_1(x) + g_1(x)]^2 + [f_2(x) + g_2(x)]^2\}\, dx = \int_{-\infty}^\infty \{F(\lambda) + G(\lambda)\}^2\, d\rho(\lambda),$$

$$\int_0^\infty \{[f_1(x) - g_1(x)]^2 + [f_2(x) - g_2(x)]^2\}\, dx = \int_{-\infty}^\infty \{F(\lambda) - G(\lambda)\}^2\, d\rho(\lambda).$$

Subtracting the second relation from the first, we obtain

$$\int_0^\infty \{f_1(x)\, g_1(x) + f_2(x)\, g_2(x)\}\, dx = \int_{-\infty}^\infty F(\lambda)\, G(\lambda)\, d\rho(\lambda). \tag{1.15}$$

The equality (1.15) is called the generalized Parseval equality.

THEOREM 1.2 (THE EXPANSION THEOREM). *Suppose that*

$$f(x) = \begin{pmatrix} f_1(x) \\ f_2(x) \end{pmatrix}$$

is a continuous vector-function $(0 \leqq x < \infty)$, *while* $\varphi(x, \lambda) = \begin{pmatrix} \varphi_1(x, \lambda) \\ \varphi_2(x, \lambda) \end{pmatrix}$ *and* $F(\lambda)$ *are as before, and that the integrals*

$$\int_{-\infty}^\infty F(\lambda)\, \varphi_1(x, \lambda)\, d\rho(\lambda), \qquad \int_{-\infty}^\infty F(\lambda)\, \varphi_2(x, \lambda)\, d\rho(\lambda) \tag{1.16}$$

converge absolutely and uniformly with respect to x in every finite interval. Then

$$f_1(x) = \int_{-\infty}^\infty F(\lambda)\, \varphi_1(x, \lambda)\, d\rho(\lambda), \quad f_2(x) = \int_{-\infty}^\infty F(\lambda)\, \varphi_2(x, \lambda)\, d\rho(\lambda).$$

PROOF. Let

$$g(x) = \begin{pmatrix} g_1(x) \\ g_2(x) \end{pmatrix}$$

be a continuous vector-function which equals zero outside the finite interval $[0, n]$. In this case the equality (1.15) can be written as

$$\int_0^n \{f_1(x)\,g_1(x) + f_2(x)\,g_2(x)\}\,dx$$
$$= \int_{-\infty}^{\infty} F(\lambda)\left\{\int_0^n [g_1(x)\,\varphi_1(x,\lambda) + g_2(x)\,\varphi_2(x,\lambda)]\,dx\right\} d\rho(\lambda).$$

Since we can interchange the order of integration in the last integral by virtue of its absolute convergence, we can rewrite the entire equality in the form

$$\int_0^n \left\{\left[f_1(x) - \int_{-\infty}^{\infty} F(\lambda)\,\varphi_1(x,\lambda)\,d\rho(\lambda)\right] g_1(x)\right.$$
$$\left. + \left[f_2(x) - \int_{-\infty}^{\infty} F(\lambda)\,\varphi_2(x,\lambda)\,d\rho(\lambda)\right] g_2(x)\right\} dx = 0.$$

Since $g(x)$ is arbitrary, and $f(x)$ and the functions (1.16) are continuous (the continuity of the latter follows from the assumed uniform convergence of the integrals in (1.16)), the coefficients of $g_1(x)$ and $g_2(x)$ in the integrand of the last integral are equal to zero, which proves the theorem.

§2. The Weyl circle and Weyl point for a Dirac system

In this section we will as before consider the interval $[0, \infty)$, and assume that the functions $q_1(x)$ and $q_2(x)$ are continuous in this interval.

If for some complex value $\lambda = \lambda_0$ every solution

$$\varphi(x,\lambda) = \begin{pmatrix} \varphi_1(x,\lambda) \\ \varphi_2(x,\lambda) \end{pmatrix}$$

of the system $\mathcal{L}y = \lambda y$ satisfies the condition

$$\int_0^{\infty} |\varphi(x,\lambda)|^2\,dx = \int_0^{\infty} \{|\varphi_1(x,\lambda)|^2 + |\varphi_2(x,\lambda)|^2\}\,dx < \infty,$$

i.e. $\varphi(x,\lambda) \in \mathcal{L}^2(0,\infty)$, then we say that the operator $\mathcal{L}$ falls into the limit circle case at infinity; in the contrary case $\mathcal{L}$ falls into the limit point case at infinity. To justify this definition we must show that this classification depends only upon the operator $\mathcal{L}$ and not upon the particular choice of the number λ_0. In the case at hand this can be proved in the same way as for the case of a Sturm-Liouville equation (cf. Theorem 2.2, Chapter 2). However we will not do this, since it will be proven in Chapter 13, §7 that under the assumptions made concerning the functions $q_1(x)$ and $q_2(x)$ the Dirac system falls into the limit point case.

Let $\varphi(x,\lambda)$ be the solution of the system (1.1) – (1.2) which satisfies the initial conditions (1.3), and $\theta(x,\lambda)$ the solution of the same system satisfying the initial conditions

$$\theta_2(0,\lambda) = \cos\alpha, \qquad \theta_1(0,\lambda) = \sin\alpha. \tag{2.1}$$

Now if $F(x)$ and $G(x)$ are solutions of the system (1.1) – (1.2) corresponding to λ and λ', then

$$\begin{aligned}
(\lambda'-\lambda)\int_{x_1}^{x_2}&\{F_1(x)G_1(x)+F_2(x)G_2(x)\}\,dx\\
&=\int_{x_1}^{x_2}\{F_1(x)[-G_2'(x)-q_2(x)G_1(x)]+F_2(x)[G_1'(x)-q_1(x)G_2(x)]\\
&\quad+G_1(x)[F_2'(x)+q_2(x)F_1(x)]\\
&\quad+G_2(x)[-F_2'(x)+q_1(x)F_2(x)]\}\,dx\\
&=\int_{x_1}^{x_2}[F_2'(x)G_1(x)+F_2(x)G_1'(x)-F_1'(x)G_2(x)-F_1(x)G_2'(x)]\,dx\\
&=\{F_2(x)G_1(x)-F_1(x)G_2(x)\}_{x_1}^{x_2}.
\end{aligned}\tag{2.2}$$

In particular, if $\lambda' = \lambda$ it follows from the last equality that $F_1(x)G_2(x) - F_2(x)G_1(x) = \text{const}$. Thus

$$\varphi_1(x,\lambda)\theta_2(x,\lambda) - \varphi_2(x,\lambda)\theta_1(x,\lambda) \equiv 1. \tag{2.3}$$

Hence $\varphi(x,\lambda)$ and $\theta(x,\lambda)$ are linearly independent solutions of the system (1.1) – (1.2), and therefore the general solution of this system can be represented (up to a constant factor) in the form $\theta(x,\lambda)+l\varphi(x,\lambda)$.

Let us consider those solutions which satisfy, at the point $x=b$, the boundary condition (β is a real number)

$$\{\theta_1(b,\lambda)+l\varphi_1(b,\lambda)\}\cos\beta + \{\theta_2(b,\lambda)+l\varphi_2(b,\lambda)\}\sin\beta = 0. \tag{2.4}$$

We define l from this condition:

$$l = -\frac{\theta_1(b,\lambda)\operatorname{ctg}\beta+\theta_2(b,\lambda)}{\varphi_1(b,\lambda)\operatorname{ctg}\beta+\varphi_2(b,\lambda)}. \tag{2.4$'$}$$

For complex values of λ the denominator of the right side of (2.4′) does not equal zero, since by Lemma 10.2, Chapter 1, the eigenvalues of a regular problem for a Dirac system are real. Therefore if we replace ctg β by a complex variable z, then $l=l(\lambda,z)$ is a meromorphic function of λ which is regular in the upper and lower halfplanes.

We have

$$l=l(\lambda,\ z)=-\frac{\theta_1(b,\ \lambda)z+\theta_2(b,\ \lambda)}{\varphi_1(b,\ \lambda)z+\varphi_2(b,\ \lambda)}. \tag{2.4''}$$

The real axis of the z-plane corresponds to some circle C_b in the l-plane. To the point $l=\infty$ there corresponds the point $z=-\varphi_2(b,\lambda)/\varphi_1(b,\lambda)$, and to the center of the circle there corresponds the point $\bar{z}=-\bar{\varphi}_2(b,\lambda)/\bar{\varphi}_1(b,\lambda)$. Hence the center of the circle lies at the point (noting that $\overline{\varphi(b,\lambda)}=\varphi(b,\bar{\lambda})$)

$$-\frac{\theta_1(b,\ \lambda)\varphi_2(b,\ \bar{\lambda})-\theta_2(b,\ \lambda)\varphi_1(b,\ \bar{\lambda})}{\varphi_1(b,\ \lambda)\varphi_2(b,\ \bar{\lambda})-\varphi_2(b,\ \lambda)\varphi_1(b,\ \bar{\lambda})}. \tag{2.5}$$

Further,

$$\begin{aligned}\operatorname{Im}\left\{-\frac{\varphi_2(b,\ \lambda)}{\varphi_1(b,\ \lambda)}\right\}&=\frac{1}{2}i\left\{\frac{\varphi_2(b,\ \lambda)}{\varphi_1(b,\ \lambda)}-\frac{\varphi_2(b,\ \bar{\lambda})}{\varphi_1(b,\ \bar{\lambda})}\right\}\\&=\frac{1}{2}i\,\frac{\varphi_2(b,\ \lambda)\varphi_1(b,\ \bar{\lambda})-\varphi_2(b,\ \bar{\lambda})\varphi_1(b,\ \lambda)}{|\varphi_1(b,\ \lambda)|^2}.\end{aligned} \tag{2.6}$$

If $\lambda'=\bar{\lambda}$, $\lambda=u+iv$, then by virtue of (2.2) ($G(x)=\bar{F}(x)$) it follows from (2.6) that

$$\operatorname{Im}\left\{-\frac{\varphi_2(b,\ \lambda)}{\varphi_1(b,\ \lambda)}\right\}=v\,\frac{\int_0^b\{\varphi_1(x,\ \lambda)\varphi_1(x,\ \bar{\lambda})+\varphi_2(x,\ \lambda)\varphi_2(x,\ \bar{\lambda})\}\,dx}{|\varphi_1(b,\ \lambda)|^2}.$$

This equality shows that if $v>0$, then to the upper z-halfplane there corresponds the exterior of the circle C_b.

Let us now compute the radius of the circle C_b. On the one hand, the point $-\theta_2(b,\lambda)/\varphi_2(b,\lambda)$ (for $z=0$) lies on the circle C_b, and on the other hand, as has already been mentioned, the center of this circle lies at the point (2.5). Therefore the radius of the circle is equal to

$$r_b=\left|\frac{\theta_1(b,\ \lambda)\varphi_2(b,\ \bar{\lambda})-\theta_2(b,\ \lambda)\varphi_1(b,\ \bar{\lambda})}{\varphi_1(b,\ \lambda)\varphi_2(b,\ \bar{\lambda})-\varphi_2(b,\ \lambda)\varphi_1(b,\ \bar{\lambda})}-\frac{\theta_2(b,\ \lambda)}{\varphi_2(b,\ \lambda)}\right|.$$

Hence, using the equalities (2.2) and (2.3), we find that

$$r_b=\frac{1}{2|v|\int_0^b\{|\varphi_1(x,\ \lambda)|^2+|\varphi_2(x,\ \lambda)|^2\}\,dx}=\frac{1}{2|v|\int_0^b|\varphi(x,\ \lambda)|^2\,dx}. \tag{2.7}$$

Further, l lies inside C_b for $v>0$ if $\operatorname{Im} z<0$, i.e. if $i(z-\bar{z})>0$.

Therefore, solving (2.4″) with respect to z, we obtain

$$i\left\{-\frac{l\varphi_2(b,\lambda)+\theta_2(b,\lambda)}{l\varphi_1(b,\lambda)+\theta_1(b,\lambda)}+\frac{\bar{l}\bar{\varphi}_2(b,\lambda)+\bar{\theta}_2(b,\lambda)}{\bar{l}\bar{\varphi}_1(b,\lambda)+\bar{\theta}_1(b,\lambda)}\right\}>0.$$

Hence we find that

$$\begin{aligned}(2.8)\qquad & i\{[l\varphi_1(b,\lambda)+\theta_1(b,\lambda)][\bar{l}\bar{\varphi}_2(b,\lambda)+\bar{\theta}_2(b,\lambda)]\\ & -[\bar{l}\bar{\varphi}_1(b,\lambda)+\bar{\theta}_1(b,\lambda)][l\varphi_2(b,\lambda)+\theta_2(b,\lambda)]\}>0.\end{aligned}$$

Putting now $\lambda = u + iv$, $\lambda' = u - iv$, $G(x) = \bar{F}(x)$ in the identity (2.2), we obtain

$$\begin{aligned}(2.9)\qquad 2v\int_0^b |F(x)|^2\,dx &= i\{F_1(0)\bar{F}_2(0)-F_2(0)\bar{F}_1(0)\}\\ &\quad - i\{F_1(b)\bar{F}_2(b)-F_2(b)\bar{F}_1(b)\}.\end{aligned}$$

Therefore if $F(x) = \theta(x,\lambda) + l\varphi(x,\lambda)$, it follows from (2.9) that

$$\begin{aligned}(2.9')\qquad & 2v\int_0^b |\theta(x,\lambda)+l\varphi(x,\lambda)|^2\,dx\\ & = iW\{\theta+l\varphi,\ \bar{\theta}+\bar{l}\bar{\varphi}\}_{x=0} - iW\{\theta+l\varphi,\ \bar{\theta}+\bar{l}\bar{\varphi}\}_{x=b}.\end{aligned}$$

Hence by virtue of (2.8) we obtain

$$(2.10)\qquad 2v\int_0^b |\theta(x,\lambda)+l\varphi(x,\lambda)|^2\,dx < iW\{\theta+l\varphi,\ \bar{\theta}+\bar{l}\bar{\varphi}\}_{x=0},$$

where $W\{f,g\} = f_1(x)g_2(x) - f_2(x)g_1(x)$. Since by the initial conditions (1.2) and (2.1)

$$W\{\theta,\bar{\theta}\}_{x=0} = W\{\varphi,\bar{\varphi}\}_{x=0} = 0;\quad W\{\varphi,\bar{\theta}\}_{x=0} = W\{\theta,\bar{\varphi}\}_{x=0} = 1,$$

and consequently

$$(2.10')\qquad W\{\theta+l\varphi,\bar{\theta}+\bar{l}\bar{\varphi}\} = l-\bar{l} = 2i\,\mathrm{Im}\,l,$$

it follows from (2.10) that

$$(2.11)\qquad \int_0^b |\theta(x,\lambda)+l\varphi(x,\lambda)|^2\,dx < -\frac{\mathrm{Im}\,l}{v}.$$

The same result can be obtained for the case $v < 0$ (using the inequality $\mathrm{Im}\,z > 0$). In both cases the sign of $\mathrm{Im}\,l$ is opposite to that of v.

Arguing similarly, it is easily shown that the point l lies on the circle C_b if and only if

$$(2.12)\qquad \int_0^b |\theta(x,\lambda)+l\varphi(x,\lambda)|^2\,dx = -\frac{\mathrm{Im}\,l}{v}.$$

Further, if l lies inside C_b and $0 < b' < b$, then

$$\int_0^{b'} |\theta(x, \lambda) + l\varphi(x, \lambda)|^2 dx < \int_0^{b} |\theta(x, \lambda) + l\varphi(x, \lambda)|^2 dx < -\frac{\operatorname{Im} l}{v}.$$

Therefore l also lies inside the circle $C_{b'}$. Hence if $b' < b$, then $C_{b'}$ contains C_b. It follows that as $b \to \infty$ the circle C_b either converges to a limit circle C_∞ or to a limit point m_∞. If C_b converges to a circle, then its radius $r_\infty = \lim r_b$ is of course positive, and it follows from (2.7) that $\varphi(x,\lambda) \in \mathscr{L}^2(0, \infty)$. If $m = m(\lambda)$ is any point on C_∞, then m lies inside any circle C_b for $b > 0$. Consequently, by virtue of (2.11)

$$\int_0^{b} |\theta(x, \lambda) + m\varphi(x, \lambda)|^2 dx \leqslant -\frac{\operatorname{Im} m}{v},$$

and letting $b \to \infty$, we find that $\theta(x,\lambda) + m\varphi(x,\lambda) \in \mathscr{L}^2(0, \infty)$. The same arguments hold if C_b converges to a point m_∞. Therefore if $\operatorname{Im}\lambda \neq 0$, then there always exists a solution of the system (1.1) – (1.2) from $\mathscr{L}^2(0, \infty)$. For the case $C_b \to C_\infty$ all solutions belong to $\mathscr{L}^2(0, \infty)$ if $\operatorname{Im}\lambda = v \neq 0$, since the solutions $\varphi(x,\lambda)$ and $\theta(x,\lambda) + m\varphi(x,\lambda)$ do, and this identifies the limit circle case with the existence of the circle C_∞. Correspondingly the limit point case is identified with the existence of a point m_∞. For the case $C_b \to m_\infty$ we have $\lim r_b = 0$, and it follows from (2.7) that the function $\varphi(x,\lambda)$ does not belong to $\mathscr{L}^2(0, \infty)$. Therefore in this case there is only one linearly independent solution from $\mathscr{L}^2(0, \infty)$ for $\operatorname{Im}\lambda \neq 0$.

In the limit-circle case l lies on C_b if and only if (2.12) is satisfied. Consequently, l lies on C_∞ if and only if

$$v \int_0^{\infty} |\theta(x, \lambda) + l\varphi(x, \lambda)|^2 dx = -\operatorname{Im} l.$$

Therefore by virtue of (2.9′) and (2.10′) it follows from (2.11) that l lies on the limit circle if and only if

$$\lim_{x\to\infty} W\{\theta(x, \lambda) + l\varphi(x, \lambda),\ \bar{\theta}(x, \lambda) + \bar{l}\bar{\varphi}(x, \lambda)\} = 0.$$

We have proved the following:

THEOREM 2.1. *If* $\operatorname{Im}\lambda \neq 0$ *and*

$$\varphi(x,\lambda) = \begin{pmatrix} \varphi_1(x,\lambda) \\ \varphi_2(x,\lambda) \end{pmatrix}, \quad \theta(x,\lambda) = \begin{pmatrix} \theta_1(x,\lambda) \\ \theta_2(x,\lambda) \end{pmatrix}$$

are linearly independent solutions of the system

$$y_2' - \{\lambda + q_1(x)\} y_1 = 0,$$
$$y_1' + \{\lambda + q_2(x)\} y_2 = 0,$$

satisfying the initial conditions

$$\varphi_1(0,\lambda) = \cos\alpha, \quad \varphi_2(0,\lambda) = -\sin\alpha,$$
$$\theta_1(0,\lambda) = \sin\alpha, \quad \theta_2(0,\lambda) = \cos\alpha,$$

then the solution $\psi(x,\lambda) = \theta(x,\lambda) + l\varphi(x,\lambda)$ *satisfies the boundary condition*

$$[\theta_1(b,\lambda) + l\varphi_1(b,\lambda)]\cos\beta + [\theta_2(b,\lambda) + l\varphi_2(b,\lambda)]\sin\beta = 0$$

if and only if l lies on the circle C_b of the complex l-plane whose equation is $W\{\psi(x,\lambda), \overline{\psi}(x,\lambda)\}_{x=b} = 0$. *If $b \to \infty$, then either C_b tends to a limit circle C_∞ or to a limit point m_∞. In the first case all solutions of the system belong to $\mathscr{L}^2(0,\infty)$. In the second case there is exactly one linearly independent solution from $\mathscr{L}^2(0,\infty)$ if $\operatorname{Im}\lambda \neq 0$. Moreover, in the limit-circle case a point lies on the limit circle C_∞ if and only if*

$$\lim_{x\to\infty} W\{\psi(x,\lambda), \overline{\psi}(x,\lambda\} = 0.$$

REMARK. In the limit-point case, if l is any point on C_b, then l tends to the limit point m_∞, and this is independent of the choice of β in the boundary condition (2.4). In particular, this holds when $\beta = 0$, and thus the limit point is given by

$$m_\infty(\lambda) = -\lim_{b\to\infty} \frac{\theta_1(b,\lambda)}{\varphi_1(b,\lambda)}.$$

Let us put $\psi_b(x,\lambda) = \theta(x,\lambda) + l\varphi(x,\lambda)$.

LEMMA 2.1. *For every nonreal λ (for $b \to \infty$)*

$$\psi_b(x,\lambda) \to \psi(x,\lambda), \qquad \int_0^b |\psi_b(x,\lambda)|^2\,dx \to \int^\infty |\psi(x,\lambda)|^2\,dx.$$

PROOF. It is obvious that

$$\psi_b(x,\lambda) = \psi(x,\lambda) + \{l(\lambda,b) - m(\lambda)\}\varphi(x,\lambda),$$

where $\psi(x,\lambda) \in \mathscr{L}^2(0,\infty)$ and $m(\lambda)$ is a point of the limit circle.[1)]

[1)] As was mentioned above, we will prove later that for a Dirac system only the limit-point case occurs (cf. Chapter 13, §7).

By virtue of (2.7) we have

$$|l(\lambda, b) - m(\lambda)| \leqslant 2r_b = \left(|v| \int_0^b |\varphi(x, \lambda)|^2 dx \right)^{-1}.$$

Therefore since $r_b \to 0$ as $b \to \infty$, we have $\psi_b(x, \lambda) \to \psi(x, \lambda)$. Further,

$$\int_0^b |\{l(\lambda, b) - m(\lambda)\} \varphi(x, \lambda)|^2 dx$$

$$= |l(\lambda, b) - m(\lambda)|^2 \int_0^b |\varphi(x, \lambda)|^2 dx \leqslant \left(v^2 \int_0^b |\varphi(x, \lambda)|^2 dx \right)^{-1}.$$

Consequently

$$\int_0^b |\psi_b(x, \lambda)|^2 dx \to \int_0^\infty |\psi(x, \lambda)|^2 dx.$$

§3. Integral representation of the resolvent. Formulas for the functions $\rho(\lambda)$ and $m(z)$

Let the vector-functions $\psi(x, \lambda)$ and $\varphi(x, \lambda)$ be as before. We put

$$G(x, y; \lambda) = \begin{cases} \psi(x, \lambda) \varphi^T(y, \lambda), & y < x, \\ \varphi(x, \lambda) \psi^T(y, \lambda), & y > x, \end{cases} \tag{3.1}$$

or in greater detail

$$G(x, y; \lambda) = \begin{cases} \begin{pmatrix} \psi_1(x, \lambda) \varphi_1(y, \lambda), & \psi_1(x, \lambda) \varphi_2(y, \lambda) \\ \psi_2(x, \lambda) \varphi_1(y, \lambda), & \psi_2(x, \lambda) \varphi_2(y, \lambda) \end{pmatrix} & (y < x), \\ \begin{pmatrix} \varphi_1(x, \lambda) \psi_1(y, \lambda), & \varphi_1(x, \lambda) \psi_2(y, \lambda) \\ \varphi_2(x, \lambda) \psi_1(y, \lambda), & \varphi_2(x, \lambda) \psi_2(y, \lambda) \end{pmatrix} & (y > x). \end{cases} \tag{3.1'}$$

Let

$$f(x) = \begin{pmatrix} f_1(x) \\ f_2(x) \end{pmatrix}$$

be an arbitrary vector-function in $\mathscr{L}^2(0, \infty)$. We will show that the vector-function

$$\Phi(x, \lambda) = \begin{pmatrix} \Phi_1(x, \lambda) \\ \Phi_2(x, \lambda) \end{pmatrix},$$

defined by

$$\Phi(x, \lambda) = \int_0^\infty G(x, y; \lambda) f(y) dy, \tag{3.2}$$

satisfies the system

$$(3.3)\qquad \begin{aligned} y_2' - \{\lambda + q_1(x)\}y_1 &= +f_1(x), \\ y_1' + \{\lambda + q_2(x)\}y_2 &= -f_2(x) \end{aligned}$$

and the boundary condition (1.4).

The Wronskian determinant $W\{\varphi, \psi\}$ does not depend upon x, and $W\{\varphi, \psi\} = 1$ (cf. §12.1, Chapter 1).

Now by the definition of the matrix $G(x, y; \lambda)$ we have

$$(3.4)\qquad \Phi_1(x, \lambda) = \psi_1(x, \lambda) \int_0^x [\varphi^T(y, \lambda) f(y)]\, dy + \varphi_1(x, \lambda) \int_x^\infty [\psi^T(y, \lambda) f(y)]\, dy,$$

$$(3.5)\qquad \Phi_2(x, \lambda) = \psi_2(x, \lambda) \int_0^x [\varphi^T(y, \lambda) f(y)]\, dy + \varphi_2(x, \lambda) \int_x^\infty [\psi^T(y, \lambda) f(y)]\, dy,$$

where

$$(3.6)\qquad [\varphi^T(y, \lambda) f(y)] = \varphi_1(x, \lambda) f_1(x) + \varphi_2(x, \lambda) f_2(x).$$

Differentiating (3.4) with respect to x, we obtain

$$\Phi_1'(x, \lambda) = \psi_1'(x, \lambda) \int_0^x [\varphi^T(y, \lambda) f(y)]\, dy + \varphi_1'(x, \lambda) \int_x^\infty [\psi^T(y, \lambda) f(y)]\, dy + \psi_1(x, \lambda) [\varphi^T(x, \lambda) f(x)] - \varphi_1(x, \lambda) [\psi^T(x, \lambda) f(x)].$$

Since $\psi(x, \lambda)$ and $\varphi(x, \lambda)$ are solutions of the system (1.1)–(1.2), inserting the values of the functions $\psi_1'(x, \lambda)$ and $\varphi_1'(x, \lambda)$ from the system (1.1)–(1.2) into the last equality and taking (3.5)–(3.6) into account, we obtain

$$\Phi_1'(x, \lambda) = -[\lambda + q_1(x)] \left\{ \psi_2(x, \lambda) \int_0^x [\varphi^T(y, \lambda) f(y)]\, dy + \varphi_2(x, \lambda) \int_x^\infty [\psi^T(y, \lambda) f(y)]\, dy \right\} + \{\psi_1(x, \lambda) \varphi_2(x, \lambda) - \psi_2(x, \lambda) \varphi_1(x, \lambda)\} f_2(x) = -[\lambda + q_2(x)] \Phi_2(x, \lambda) - f_2(x),$$

i.e. the second equation of the system (3.3) is satisfied. One can similarly show that $\Phi(x,\lambda)$ satisfies the first equation.

We further have from (3.4) – (3.5) that

$$\Phi_1(0, \lambda) = \varphi_1(0, \lambda) \int_0^\infty [\psi^T(y, \lambda) f(y)]\, dy,$$

$$\Phi_2(0, \lambda) = \varphi_2(0, \lambda) \int_0^\infty [\psi^T(y, \lambda) f(y)]\, dy,$$

so that $\Phi(x,\lambda)$ satisfies the boundary condition (1.4).

Let us denote by $\lambda_0, \lambda_{\pm 1}, \lambda_{\pm 2}, \cdots$ the eigenvalues, and by $\varphi_0(x)$, $\varphi_{\pm 1}(x), \varphi_{\pm 2}(x), \cdots$ the corresponding eigen-vector-functions of the boundary-value problem (1.1) + (1.2) + (1.4) + (1.5). Let $l(\lambda, b)$ and $\psi_b(x,\lambda)$ be the same as in the preceding section. We put

$$G_b(x, y; z) = \begin{cases} \psi_b(x, z)\, \varphi^T(y, z), & y < x, \\ \varphi(x, z)\, \psi_b^T(y, z), & y > x, \end{cases} \tag{3.7}$$

$$R_{z,b} f = \int_0^b G_b(x, y; z) f(y)\, dy.$$

Then, putting further

$$\alpha_n^2 = \int_0^b \{\varphi_{n1}^2(x) + \varphi_{n2}^2(x)\}\, dx,$$

by virtue of the expansion (13.24) of Chapter 1 we obtain

$$R_{z,b} f = -\sum_{n=-\infty}^{\infty} \frac{\varphi_n(x) \int_0^b f^T(y) \varphi_n(y)\, dy}{\alpha_n^2 (z - \lambda_n)} \tag{3.8}$$

$$= -\int_{-\infty}^{\infty} \frac{\varphi(x, \lambda) \int_0^b f^T(y) \varphi(y, \lambda)\, dy}{z - \lambda}\, d\rho_b(\lambda),$$

where $\rho_b(\lambda)$ is the same as in §1.

LEMMA 3.1. *For every nonreal z and fixed x we have the bound*

$$\int_{-\infty}^{\infty} \left| \frac{\varphi(x, \lambda)}{z - \lambda} \right|^2 d\rho_b(\lambda) < C, \tag{3.9}$$

where the constant C can be chosen to be independent of b.

PROOF. Putting $f(y)=\varphi_n(y)$ in formula (13.24) of Chapter 1, we obtain

$$\frac{1}{\alpha_n}\int_0^b G_b(x, y; z)\varphi_n(y)\,dy = -\frac{\varphi_n(x)}{\alpha_n(z-\lambda_n)}.$$

Therefore by the Parseval equality

$$\int_0^b |G_b(x, y; z)|^2\,dy = \sum_{n=-\infty}^{\infty}\frac{|\varphi_n(x)|^2}{\alpha_n^2|z-\lambda_n|^2} = \int_{-\infty}^{\infty}\frac{|\varphi(x,\lambda)|^2}{|z-\lambda|^2}\,d\rho_b(\lambda),\text{[2)]}$$

which by the definition of the matrix $G_b(x,y;z)$ is equivalent to

$$|\psi_b(x, z)|^2\int_0^x|\varphi(y, z)|^2\,dy + |\varphi(x, z)|^2\int_x^b|\psi_b(y, z)|^2\,dy = \int_{-\infty}^{\infty}\frac{|\varphi(x,\lambda)|^2}{|z-\lambda|^2}\,d\rho_b(\lambda).$$

Hence the bound (3.9) follows by virtue of Lemma 2.1.

COROLLARY I. *Let* $\rho(\lambda)$ *be as in* §1. *Then we have the bound*

$$\int_{-\infty}^{\infty}\frac{|\varphi(x,\lambda)|^2}{|z-\lambda|^2}\,d\rho(\lambda)\leqslant C \tag{3.10}$$

with the same constant C as in (3.9).

In fact, let $a>0$ be arbitrary. It follows from (3.9) that

$$\int_{-a}^{a}\frac{|\varphi(x,\lambda)|^2}{|z-\lambda|^2}\,d\rho_b(\lambda)< C. \tag{3.11}$$

Letting first $b\to\infty$, and then $a\to\infty$, we obtain (3.10).

COROLLARY II. *For any* $a>0$

$$\int_{-\infty}^{-a}\frac{d\rho(\lambda)}{\lambda^2}<+\infty,\qquad \int_a^{\infty}\frac{d\rho(\lambda)}{\lambda^2}<\infty. \tag{3.12}$$

PROOF. Since $\varphi_{n1}(0)$ and $\varphi_{n2}(0)$ cannot equal zero simultaneously, i.e. $|\varphi(0,\lambda)|^2\neq 0$, putting $x=0$ in (3.10) we obtain

[2)] Here we are using the following notation:

$$|G_b(x, y; z)|^2 = |g_{11,b}|^2+|g_{12,b}|^2+|g_{21,b}|^2+|g_{22,b}|^2;$$

$$|\varphi_n(x)|^2 = |\varphi_{n1}(x)|^2+|\varphi_{n2}(x)|^2.$$

$$\int_{-\infty}^{\infty} \frac{d\rho(\lambda)}{|z-\lambda|^2} < \infty,$$

from which our assertion follows.

LEMMA 3.2. *Suppose that*

$$f(x) = \begin{pmatrix} f_1(x) \\ f_2(x) \end{pmatrix} \in \mathcal{L}^2(0, \infty)$$

and $z = \mu + i\nu$. *Then we have the bound*

$$\int_0^\infty |R_z f|^2 dx \leqslant \frac{1}{\nu^2} \int_0^\infty |f(y)|^2 dy,$$

where

$$R_z f = \int_0^\infty G(x, y; z) f(y) dy,$$

and the matrix $G(x, y; z)$ *is defined by* (3.1).

PROOF. Let $b > 0$ be arbitrary. Then it follows from (3.8) by the Parseval equality that

$$\int_0^b |R_{z,b} f|^2 dx = \sum_{n=-\infty}^{\infty} \frac{1}{\alpha_n^2 |z-\lambda_n|^2} \left\{ \int_0^b f^T(y) \varphi_n(y) dy \right\}^2$$
$$\leqslant \frac{1}{\nu^2} \sum_{n=-\infty}^{\infty} \frac{1}{\alpha_n^2} \left\{ \int_0^b f^T(y) \varphi_n(y) dy \right\}^2 = \frac{1}{\nu^2} \int_0^b |f(y)|^2 dy.$$

If a is a fixed positive number and $a < b$, then

$$\int_0^a |R_{z,b} f|^2 dx \leqslant \int_0^b |R_{z,b} f|^2 dx \leqslant \frac{1}{\nu^2} \int_0^b |f(y)|^2 dy.$$

Letting $b \to \infty$, we obtain

$$\int_0^a |R_z f|^2 dx \leqslant \frac{1}{\nu^2} \int_0^\infty |f(y)|^2 dy,$$

whence the assertion of the lemma, since a is arbitrary.

THEOREM 3.1 (INTEGRAL REPRESENTATION OF THE RESOLVENT). *For every vector-function* $f(x) \in \mathcal{L}^2(0, \infty)$ *and every nonreal* z *we have*

$$R_z f = -\int_{-\infty}^{\infty} \frac{\varphi(x, \lambda) E(\lambda)}{z - \lambda} d\rho(\lambda), \tag{3.13}$$

where

$$E(\lambda) = \underset{n\to\infty}{\mathrm{l.i.m}} \int_0^n f^T(x)\varphi(x,\lambda)\,dx.$$

PROOF. Suppose first that $f(x) = f_n(x)$ vanishes outside the finite interval $[0, n]$, satisfies the boundary condition (1.4) and has a continuous first derivative. Let $b > n$ and μ be arbitrary positive numbers. We put

$$E_n(\lambda) = \int_0^n f_n^T(x)\varphi(x,\lambda)\,dx.$$

Then (3.8) can be rewritten in the form

$$\begin{aligned} R_{z,b}f_n &= -\int_{-\infty}^{\infty} \frac{\varphi(x,\lambda)E_n(\lambda)}{z-\lambda}\,d\rho_b(\lambda) \\ &= -\left\{\int_{-\infty}^{-\mu} + \int_{-\mu}^{\mu} + \int_{\mu}^{\infty}\right\} \frac{\varphi(x,\lambda)E_n(\lambda)}{z-\lambda}\,d\rho_b(\lambda) = I_1 + I_2 + I_3. \end{aligned} \tag{3.14}$$

We now estimate I_1. By virtue of (3.8) we have

$$\begin{aligned} |I_1| &= \left| \int_{-\infty}^{-\mu} \frac{\varphi(x,\lambda)E_n(\lambda)}{z-\lambda}\,d\rho_b(\lambda) \right| \\ &= \left| \sum_{\lambda_k < -\mu} \frac{\varphi_k(x)}{\alpha_k^2(z-\lambda_k)} \int_0^n f_n^T(x)\varphi_k(x)\,dx \right| \\ &\leqslant \left(\sum_{\lambda_k<-\mu} \frac{|\varphi_k(x)|^2}{\alpha_k^2(z-\lambda_k)^2} \right)^{1/2} \left(\sum_{\lambda_k<-\mu} \frac{1}{\alpha_k^2} \left[\int_0^n f_n^T(x)\varphi_k(x)\,dx \right]^2 \right)^{1/2}. \end{aligned} \tag{3.15}$$

Since the vector-function

$$\varphi_k(x) = \begin{pmatrix} \varphi_{k1}(x) \\ \varphi_{k2}(x) \end{pmatrix}$$

is a solution of the system (1.1)–(1.2), we have

$$\begin{aligned} \int_0^n f_n^T(x)\varphi_k(x)\,dx = \frac{1}{\lambda_k}\int_0^n \{ & f_{n1}(x)[\varphi_{k2}'(x) - q_1(x)\varphi_{k1}(x)] \\ & + f_{n2}(x)[-\varphi_{k1}'(x) - q_2(x)\varphi_{k2}(x)]\}\,dx. \end{aligned}$$

Integrating the last integral by parts and taking into account that the vector-functions $f_n(x)$ and $\varphi_k(x)$ satisfy the boundary condition (1.4), as well as the fact that $f_n(x)$ vanishes outside the interval $[0, n]$, we obtain

$$\int_0^n f_n^T(x)\varphi_k(x)\,dx = \frac{1}{\lambda_k}\int_0^n \{\varphi_{k1}(x)[f'_{n2}(x) - q_1(x) f_{n1}(x)] + \varphi_{k2}(x)[-f'_{n1}(x) - q_2(x) f_{n2}(x)]\}\,dx.$$

Therefore since by Lemma 3.1 the first sum in the right side of (3.15) is bounded, taking into account the last equality we obtain from (3.15)

$$(3.16)\qquad |I_1| \leqslant \frac{C^{1/2}}{\mu}\left(\sum_{\lambda_k<-\mu}\frac{1}{a_k^2}\left\{\int_0^n h_n^T(x)\varphi_k(x)\,dx\right\}^2\right)^{1/2},$$

where

$$h_n(x) = \begin{pmatrix} f'_{n2}(x) - q_1(x) f_{n1}(x) \\ -f'_{n1}(x) - q_2(x) f_{n2}(x)\end{pmatrix}.$$

Then by Bessel's inequality (3.16) implies the bound

$$|I_1| \leqslant \frac{C^{1/2}}{\mu}\left(\int_0^n |h_n(x)|^2\,dx\right) = \frac{C_1}{\mu}.$$

One can prove in the same way that $I_3 \leqq C_1/\mu$. It follows from these bounds that I_1 and I_3 tend to zero, as $\mu \to \infty$, uniformly with respect to b. We can therefore apply the generalized Helly theorem, and obtain from (3.14)

$$(3.17)\qquad R_z f_n = -\int_{-\infty}^{\infty}\frac{\varphi(x,\lambda) E_n(\lambda)}{z-\lambda}\,d\rho(\lambda).$$

For an arbitrary vector-function $f(x) \in \mathscr{L}^2(0, \infty)$ one can obtain (3.13) from (3.17) in exactly the same way as in the proof of Theorem 3.1, Chapter 2. This proves the theorem.

The following is an easy consequence of formula (3.13).

THEOREM 3.2. *For every nonreal z one has* $(\operatorname{Im} z_0 \neq 0)$

$$(3.18)\qquad m(z) - m(z_0) = \int_{-\infty}^{\infty}\left(\frac{1}{\lambda - z} - \frac{1}{\lambda - z_0}\right)d\rho(\lambda),$$

and if λ *and* μ *are points of continuity of* $\rho(\lambda)$, *then* $(\tau > 0)$

$$(3.19)\qquad \rho(\lambda) - \rho(\mu) = \frac{1}{\pi}\lim_{\tau\to 0}\int_\mu^\lambda \{-\operatorname{Im}[m(\sigma + i\tau)]\}\,d\sigma.$$

PROOF. Since $f(x)$ is an arbitrary vector-function, it follows from (3.13) that[3)]

$$G(x,y;z)=\int_{-\infty}^{\infty}\frac{\varphi(x,\lambda)\varphi^T(y,\lambda)}{\lambda-z}\,d\rho(\lambda).$$

Therefore

$$G(x,\ y;\ z)-G(x,\ y;\ z_0)=\int_{-\infty}^{\infty}\varphi(x,\ \lambda)\,\varphi^T(y,\ \lambda)\left(\frac{1}{\lambda-z}-\frac{1}{\lambda-z_0}\right)d\rho(\lambda).$$

The right and left sides of this relation are matrices. Hence corresponding elements of these matrices are equal. Equating the elements which lie at the intersection of the first row and first column, using (3.1′) and the definition of the product $\varphi(x,\lambda)\varphi^T(y,\lambda)$, and then putting $x=y=0$ and taking into account the initial conditions (1.3) and (2.1), we obtain

$$(\cos\alpha+m(z)\cos\alpha)\cos\alpha-(\cos\alpha+m(z_0)\cos\alpha)\cos\alpha$$
$$=\int_{-\infty}^{\infty}\cos^2\alpha\left(\frac{1}{\lambda-z}-\frac{1}{\lambda-z_0}\right)d\rho(\lambda),$$

which proves (3.18).

We now prove (3.19). Let $z=\sigma+i\tau$. It follows from (3.18) that

$$\chi(\sigma,\ \tau)=\frac{\operatorname{sign}\tau}{\pi}\,\frac{m(z)-m(\bar z)}{2i}=-\frac{1}{\pi}\int_{-\infty}^{\infty}\frac{|\tau|\,d\rho(\lambda)}{(\lambda-\sigma)^2+\tau^2}.$$

Then by the Stieltjes inversion formula (cf. Chapter 14, Theorem 3.1) it follows that if λ and μ are points of continuity of $\rho(\lambda)$, then

$$\rho(\lambda)-\rho(\mu)=\lim_{\tau\to0}\int_{\mu}^{\lambda}-\chi(\sigma,\ \tau)\,d\sigma. \tag{3.20}$$

On the other hand, by virtue of (2.12) $m(\bar z)=\overline{m(z)}$. Therefore

$$\chi(\sigma,\ \tau)=\frac{\operatorname{sign}\tau}{\pi}\cdot\frac{m(z)-\overline{m(z)}}{2i}=\frac{\operatorname{sign}\tau}{\pi}\operatorname{Im}\{m(z)\}. \tag{3.21}$$

For $\tau>0$ formula (3.19) follows from (3.20), by virtue of (3.21). This proves the theorem.

[3)] It follows from an asymptotic formula that this integral converges in the sense of the Cauchy principal value.

§4. Proof of the Parseval equality for the case of the entire line

We will again consider the system (1.1) – (1.2), assuming now, however, that x varies in the interval $(-\infty, \infty)$ and $q_1(x)$ and $q_2(x)$ are continuous on this interval.

We denote by

$$\varphi_1(x,\lambda) = \begin{pmatrix} \varphi_{11}(x,\lambda) \\ \varphi_{12}(x,\lambda) \end{pmatrix}$$

the solution of the system (1.1) – (1.2) which satisfies the initial conditions

$$\varphi_{11}(0,\lambda) = 1, \qquad \varphi_{12}(0,\lambda) = 0, \tag{4.1}$$

and by

$$\varphi_2(x,\lambda) = \begin{pmatrix} \varphi_{21}(x,\lambda) \\ \varphi_{22}(x,\lambda) \end{pmatrix}$$

the solution of the same system which satisfies the initial conditions

$$\varphi_{21}(0,\lambda) = 0, \qquad \varphi_{22}(0,\lambda) = 1. \tag{4.2}$$

Let $[a,b]$ be an arbitrary finite interval. We consider the boundary-value problem defined by the system (1.1) – (1.2) and the boundary conditions

$$\begin{gathered} y_1(a)\sin\alpha + y_2(a)\cos\alpha = 0, \\ y_1(b)\sin\beta + y_2(b)\cos\beta = 0, \end{gathered} \tag{4.3}$$

where α and β are arbitrary real numbers.

Let $\lambda_0, \lambda_{\pm 1}, \lambda_{\pm 2}, \cdots$ be the eigenvalues, and

$$y_0(x), y_{\pm 1}(x), y_{\pm 2}(x), \cdots \left(y_n(x) = \begin{pmatrix} y_{n1}(x) \\ y_{n2}(x) \end{pmatrix} \right)$$

corresponding eigen-vector-functions of this problem. Since the solutions $\varphi_1(x,\lambda)$ and $\varphi_2(x,\lambda)$ of the system (1.1) – (1.2) are linearly independent, we have

$$y_n(x) = \beta_n \varphi_1(x,\lambda_n) + \gamma_n \varphi_2(x,\lambda).$$

Further, in view of the fact that the problem under consideration is homogeneous, we may assume without loss of generality that $|\beta_n| \leq 1$, $|\gamma_n| \leq 1$. We put $\alpha_n^2 = \int_a^b |y_n(x)|^2 dx$.

Let

$$f(x) = \begin{pmatrix} f_1(x) \\ f_2(x) \end{pmatrix}$$

be square-integrable over the interval $[a,b]$. Applying the Parseval equality to $f(x)$, we obtain

$$\int_a^b |f(x)|^2\,dx = \sum_{n=-\infty}^{\infty} \frac{1}{\alpha_n^2}\left\{\int_a^b f^T(x)\,y_n(x)\,dx\right\}^2$$
$$= \sum_{n=-\infty}^{\infty} \frac{1}{\alpha_n^2}\left\{\int_a^b f^T(x)\,[\beta_n\varphi_1(x,\lambda_n)+\gamma_n\varphi_2(x,\lambda)]\,dx\right\}^2$$
$$= \sum_{n=-\infty}^{\infty} \frac{\beta_n^2}{\alpha_n^2}\left\{\int_a^b f^T(x)\,\varphi_1(x,\lambda_n)\,dx\right\}^2 \tag{4.4}$$
$$+2\sum_{n=-\infty}^{\infty} \frac{\beta_n\gamma_n}{\alpha_n^2}\left\{\int_a^b f^T(x)\,\varphi_1(x,\lambda_n)\,dx\right\}\left\{\int_a^b f^T(x)\,\varphi_2(x,\lambda_n)\,dx\right\}$$
$$+\sum_{n=-\infty}^{\infty} \frac{\gamma_n^2}{\alpha_n^2}\left\{\int_a^b f^T(x)\,\varphi_2(x,\lambda_n)\,dx\right\}^2.$$

Put

$$\rho_{11,a,b}(\lambda)=\sum_{0<\lambda_n\leqslant\lambda}\frac{\beta_n^2}{\alpha_n^2}\quad(\lambda>0);\qquad -\sum_{0\geqslant\lambda_n>\lambda}\frac{\beta_n^2}{\alpha_n^2}\quad(\lambda\leqslant 0),$$

$$\rho_{12,a,b}(\lambda)=\sum_{0<\lambda_n\leqslant\lambda}\frac{\beta_n\gamma_n}{\alpha_n^2}\quad(\lambda>0);\qquad -\sum_{0\geqslant\lambda_n>\lambda}\frac{\beta_n\gamma_n}{\alpha_n^2}\quad(\lambda\leqslant 0),$$

$$\rho_{12,a,b}(\lambda)=\rho_{21,a,b}(\lambda),$$

$$\rho_{22,a,b}(\lambda)=\sum_{0<\lambda_n\leqslant\lambda}\frac{\gamma_n^2}{\alpha_n^2}\quad(\lambda>0);\qquad -\sum_{0\geqslant\lambda_n<\lambda}\frac{\gamma_n^2}{\alpha_n^2}\quad(\lambda\leqslant 0).$$

Then (4.4) can be written in the form

$$\int_a^b |f(x)|^2\,dx=\int_{-\infty}^{\infty}\sum_{i,j=1}^{2}F_i(\lambda)\,F_j(\lambda)\,d\rho_{ij,a,b}(\lambda), \tag{4.5}$$

where

$$F_i(\lambda)=\int_a^b f^T(x)\,\varphi_i(x,\lambda)\,dx\qquad(i=1,2). \tag{4.6}$$

LEMMA 4.1. *For any positive number N there exists a positive constant $A=A(N)$, not depending upon a and b, such that*

$$\bigvee_{-N}^{N}\{\rho_{ij,a,b}(\lambda)\}<A\qquad(i,j=1,2).$$

PROOF. By virtue of the conditions (4.1)–(4.2) we have

$$\varphi_{ij}(0,\lambda)=\delta_{ij} \qquad (\delta_{ij}=1,\ i=j;\ \delta_{ij}=0,\ i\neq j).$$

Therefore for a given N there exists $h>0$ such that $(0<\epsilon<1)$

$$(4.7) \qquad |\varphi_{ij}(x,\lambda)-\delta_{ij}|<\epsilon,$$

where $0\leqq x\leqq h$ and $|\lambda|\leqq N$.

Let

$$f_h(x)=\begin{pmatrix} f_1(x) \\ f_2(x) \end{pmatrix}$$

be a nonnegative continuous function, where $f_2(x)\equiv 0$, and $f_1(x)$ vanishes outside the interval $[0,h]$ and is normalized so that

$$(4.8) \qquad \int_0^h f_1(x)\,dx=1.$$

We put

$$F_{ih}(\lambda)=\int_0^h f_h^T(x)\,\varphi_i(x,\lambda)\,dx=\int_0^h f_1(x)\,\varphi_{i1}(x,\lambda)\,dx \quad (i=1,\ 2).$$

Then from (4.7) and (4.8) we find that

$$(4.9) \qquad 1-\epsilon<F_{1h}(\lambda)<1+\epsilon,\quad |F_{2h}(\lambda)|<\epsilon \qquad (|\lambda|<N).$$

Further, applying the Parseval equality (4.5) to $f(x)$, we obtain[4)]

$$\int_0^h f_1^2(x)\,dx$$

$$\geqq \int_{-N}^{N} F_{1h}^2(\lambda)\,d\rho_{11}(\lambda)+2\int_{-N}^{N} F_{1h}(\lambda)F_{2h}(\lambda)\,d\rho_{12}(\lambda)+\int_{-N}^{N} F_{2h}^2(\lambda)\,d\rho_{22}(\lambda)$$

$$\geqq \int_{-N}^{N} F_{1h}^2(\lambda)\,d\rho_{11}(\lambda)-2\int_{-N}^{N} |F_{1h}(\lambda)|\,|F_{2h}(\lambda)|\,|d\rho_{12}(\lambda)|.$$

It follows from (4.9) that

$$(4.10) \qquad \int_0^h f_1^2(x)\,dx > \int_{-N}^{N}(1-\epsilon)^2 d\rho_{11}(\lambda)-\int_{-N}^{N} 2\epsilon(1+\epsilon)\,|d\rho_{12}(\lambda)|=I_1+I_2.$$

---

4) Throughout the proof of this lemma the indices a and b on the functions $\rho_{ij,a,b}(\lambda)$ will be omitted.

Let us calculate I_1 and I_2. Obviously

$$I_1 = (1-\epsilon)^2\{\rho_{11}(N) - \rho_{11}(-N)\}. \tag{4.11}$$

Further,

$$I_2 = -2\epsilon(1+\epsilon)\bigvee_{-N}^{N}\{\rho_{12}(\lambda)\}. \tag{4.12}$$

Since

$$2\bigvee_{-N}^{N}\{\rho_{12}(\lambda)\} \leqq \rho_{11}(N) - \rho_{11}(-N) + \rho_{22}(N) - \rho_{22}(-N), \tag{4.13}$$

it follows from (4.10) – (4.13) that

$$\int_0^h f_1^2(x)\,dx > (1-3\epsilon)\{\rho_{11}(N) - \rho_{11}(-N)\} - \epsilon(1+\epsilon)\{\rho_{22}(N) - \rho_{22}(-N)\}. \tag{4.14}$$

Suppose now that

$$g_h(x) = \begin{pmatrix} g_1(x) \\ g_2(x) \end{pmatrix}$$

satisfies the same conditions as does $f_h(x)$, but with $g_1(x)$ replaced by $g_2(x)$, i.e. $g_1(x) \equiv 0$ and $\int_0^h g_2(x)\,dx = 1$. Then proceeding as above we obtain the inequality

$$\int_0^h g_2^2(x)\,dx > (1-3\epsilon)\{\rho_{22}(N) - \rho_{22}(-N)\} - \epsilon(1+\epsilon)\{\rho_{11}(N) - \rho_{11}(-N)\}. \tag{4.15}$$

Adding the inequalities (4.14) and (4.15), we obtain

$$\int_0^h \{f_1^2(x) + g_2^2(x)\}\,dx > (1-4\epsilon-\epsilon^2)\{\rho_{11}(N) - \rho_{11}(-N) + \rho_{22}(N) - \rho_{22}(-N)\}.$$

from which follows the assertion of the lemma for the functions $\rho_{11}(\lambda)$ and $\rho_{22}(\lambda)$ by virtue of the monotonicity of these functions. The assertion of the lemma for the function $\rho_{12}(\lambda)$ follows from (4.13). This proves the lemma.

We will now show that, using Lemma 4.1 and a well-known theorem on passage to the limit under the integral sign for a Stieltjes integral, one can obtain from (4.5) the Parseval equality for the system (1.1) – (1.2), i.e.:

THEOREM 4.1. *Let*

$$f(x) = \begin{pmatrix} f_1(x) \\ f_2(x) \end{pmatrix} \in \mathscr{L}^2(-\infty, \infty).$$

There exist monotonic functions $\rho_{11}(\lambda)$ *and* $\rho_{22}(\lambda)$ *which are bounded over every finite interval, and a function* $\rho_{12}(\lambda)$ *which is of bounded variation over every finite interval, which functions do not depend upon* $f(x)$, *such that the Parseval equality*

$$\int_{-\infty}^{\infty} |f(x)|^2 dx = \int_{-\infty}^{\infty} \sum_{i,j=1}^{2} F_i(\lambda) F_j(\lambda) d\rho_{ij}(\lambda) \tag{4.16}$$

holds, where

$$F_i(\lambda) = \operatorname*{l.i.m}_{n\to\infty} \int_{-n}^{n} f^T(x) \varphi_i(x, \lambda)\, dx.$$

PROOF. Suppose first that the vector-function

$$f_n(x) = \begin{pmatrix} f_1(x) \\ f_2(x) \end{pmatrix}$$

vanishes outside the finite interval $-n \leqq x \leqq n$, $n < \min\{|a|, |b|\}$,[5] has a continuous first derivative and satisfies the boundary conditions (4.3). Applying the Parseval equality (4.4) to $f_n(x)$, we obtain

$$\int_{-n}^{n} |f_n(x)|^2 dx = \sum_{n=-\infty}^{\infty} \frac{1}{\alpha_k^2} \left\{ \int_a^b f_n^T(x) y_k(x)\, dx \right\}^2. \tag{4.17}$$

Since $f_n(x)$ is identically zero in neighborhoods of the points a and b, integrating by parts we obtain

$$\int_a^b f_n^T(x) y_k(x)\, dx = \frac{1}{\lambda_k} \int_a^b f_{n1}(x) [y'_{k2}(x) - q_1(x) y_{k1}(x)]\, dx$$

$$- \frac{1}{\lambda_k} \int_a^b f_{n2}(x) [y'_{k1}(x) + q_2(x) y_{k2}(x)]\, dx$$

$$= \frac{1}{\lambda_k} \int_a^b \{[f'_{n2}(x) - q_1(x) f_{n1}(x)] y_{k1}(x)$$

$$+ [-f'_{n1}(x) - q_2(x) f_{n2}(x)] y_{k2}(x)\}\, dx.$$

[5] Cf. the footnote on p. 000.

Hence by the Parseval equality it follows that

$$\sum_{|\lambda_k|\geqslant\mu} \frac{1}{\alpha_k^2}\left\{\int_a^b f_n^T(x)\, y_k(x)\, dx\right\}^2$$
$$\leqslant \frac{1}{\mu^2} \sum_{|\lambda_k|\geqslant\mu} \frac{1}{\alpha_k^2}\left\{\int_a^b [\{f'_{n2}(x) - q_1(x) f_{n1}(x)\}\, y_{k1}(x) + \{-f'_{n1}(x) - q_2(x) f_{n2}(x)\}\, y_{k2}(x)]\, dx\right\}^2$$
$$\leqslant \frac{1}{\mu^2} \sum_{k=-\infty}^{\infty} \frac{1}{\alpha_k^2}\left\{\int_a^b [\{f'_{n2}(x) - q_1(x) f_{n1}(x)\}\, y_{k1}(x) + \{-f'_{n1}(x) - q_2(x) f_{n2}(x)\}\, y_{k2}(x)]\, dx\right\}^2$$
$$= \frac{1}{\mu^2} \int_{-n}^{n} \{[f'_{n2}(x) - q_1(x) f_{n1}(x)]^2 + [f'_{n1}(x) + q_2(x) f_{n2}(x)]^2\}\, dx.$$

From this bound and from (4.17) we find that

$$\left|\int_{-n}^{n} |f_n(x)|^2\, dx - \sum_{-\mu\leqslant\lambda_k\leqslant\mu} \frac{1}{\alpha_k^2}\left\{\int_a^b f_n^T(x)\, y_k(x)\, dx\right\}^2\right|$$
$$\leqslant \frac{1}{\mu^2} \int_{-n}^{n} \{[f'_{n2}(x) - q_1(x) f_{n1}(x)]^2 + [f'_{n1}(x) + q_2(x) f_{n2}(x)]^2\}\, dx. \tag{4.18}$$

We further have

$$\sum_{-\mu\leqslant\lambda_k\leqslant\mu} \frac{1}{\alpha_k^2}\left\{\int_a^b f_n^T(x)\, y_k(x)\, dx\right\}^2$$
$$= \sum_{-\mu\leqslant\lambda_k\leqslant\mu} \frac{1}{\alpha_k^2}\left\{\int_a^b f_n^T(x)\, [\beta_n \varphi_1(x, \lambda_k) + \gamma_n \varphi_2(x, \lambda_k)]\, dx\right\}^2$$
$$= \int_{-\mu}^{\mu} \sum_{i,j=1}^{2} F_{in}(\lambda)\, F_{jn}(\lambda)\, d\rho_{ij,a,b}(\lambda),$$

where

$$F_{ni}(\lambda) = \int_a^b f_n^T(x)\, \varphi_i(x, \lambda)\, dx \qquad (i = 1,\ 2).$$

Therefore (4.18) can be rewritten in the form

$$\left| \int_{-n}^{n} |f_n(x)|^2 \, dx - \int_{-\mu}^{\mu} \sum_{i,j=1}^{2} F_{in}(\lambda) F_{jn}(\lambda) \, d\rho_{ij,a,b}(\lambda) \right| \tag{4.19}$$
$$\leqslant \frac{1}{\mu^2} \int_{-n}^{n} \{[f'_{n2}(x) - q_1(x) f_{n1}(x)]^2 + [f'_{n1}(x) + q_2(x) f_{n2}(x)]^2\} \, dx.$$

By Lemma 4.1 the set of monotonic functions $\rho_{ii,a,b}(\lambda)$ $(-\mu \leqq \lambda \leqq \mu)$, $i = 1, 2$, is bounded, and the set of functions $\rho_{12,a,b}(\lambda)$ is bounded and has bounded variation, in every finite interval, uniformly in a and b. Therefore we can find sequences $a_k \to -\infty$ and $b_k \to \infty$ such that the functions $\rho_{ij,a_k,b_k}(\lambda)$ converge respectively to $\rho_{ij}(\lambda)$. Passing to the limit in (4.19) along a_k and b_k, we obtain

$$\left| \int_{-n}^{n} |f_n(x)|^2 \, dx - \int_{-\mu}^{\mu} \sum_{i,j=1}^{2} F_{in}(\lambda) F_{jn}(\lambda) \, d\rho_{ij}(\lambda) \right|$$
$$\leqslant \frac{1}{\mu^2} \int_{-n}^{n} \{[f'_{n2}(x) - q_1(x) f_{n1}(x)]^2 + [f'_{n1}(x) + q_2(x) f_{n2}(x)]^2\} \, dx.$$

Finally, letting $\mu \to \infty$, we obtain

$$\int_{-n}^{n} |f_n(x)|^2 \, dx = \int_{-\infty}^{\infty} \sum_{i,j=1}^{2} F_{in}(\lambda) F_{jn}(\lambda) \, d\rho_{ij}(\lambda),$$

which is (4.16) for functions which satisfy the conditions indicated.

Extension of the Parseval equality to an arbitrary function from $\mathscr{L}^2(-\infty, \infty)$ can be carried out by the usual methods. This proves the theorem.

Let $f(x), g(x) \in \mathscr{L}^2(-\infty, \infty)$ and let the functions $G_i(\lambda)$ be constructed from $g(x)$ in the same way the $F_i(\lambda)$ were constructed from $f(x)$. Obviously the Fourier transforms of the functions $f(x) \pm g(x)$ are the functions $F_i(\lambda) \pm G_i(\lambda)$ $(i = 1, 2)$. Therefore

$$\int_{-\infty}^{\infty} |f(x) + g(x)|^2 \, dx = \int_{-\infty}^{\infty} \sum_{i,j=1}^{2} [F_i(\lambda) + G_i(\lambda)][F_j(\lambda) + G_j(\lambda)] \, d\rho_{ij}(\lambda),$$

$$\int_{-\infty}^{\infty} |f(x) - g(x)|^2 \, dx = \int_{-\infty}^{\infty} \sum_{i,j=1}^{2} [F_i(\lambda) - G_i(\lambda)][F_j(\lambda) - G_j(\lambda)] \, d\rho_{ij}(\lambda).$$

Subtracting the second relation from the first, we obtain

$$
\begin{aligned}
\int_{-\infty}^{\infty} f^T(x)\, g(x)\, dx &= \frac{1}{2}\int_{-\infty}^{\infty} \sum_{i,j=1}^{2} [F_i(\lambda) G_j(\lambda) + F_j(\lambda) G_i(\lambda)]\, d\rho_{ij}(\lambda) \\
&= \int_{-\infty}^{\infty} \sum_{i,j=1}^{2} F_i(\lambda) G_j(\lambda)\, d\rho_{ij}(\lambda).
\end{aligned} \tag{4.20}
$$

The equality (4.20) is called the generalized Parseval equality.

THEOREM 4.2 (EXPANSION THEOREM). *Let*

$$f(x) = \begin{pmatrix} f_1(x) \\ f_2(x) \end{pmatrix}$$

be a continuous vector-function, let $\varphi_i(x,\lambda)$ *and* $F_i(\lambda)$ $(i = 1, 2)$ *be as before, and suppose that the integrals*

$$\int_{-\infty}^{\infty} F_i(\lambda)\, \varphi_j(x, \lambda)\, d\rho_{ij}(\lambda) \qquad (i, j = 1, 2) \tag{4.21}$$

converge absolutely and uniformly with respect to x in every finite interval. Then

$$f(x) = \int_{-\infty}^{\infty} \sum_{i,j=1}^{2} F_i(\lambda)\, \varphi_j(x, \lambda)\, d\rho_{ij}(\lambda).$$

PROOF. Let $g(x)$ be a continuous vector-function which equals zero outside a finite interval $[-n, n]$. Applying the generalized Parseval equality (4.20) to $f(x)$ and $g(x)$, we obtain

$$\int_{-n}^{n} f^T(x)\, g(x)\, dx = \int_{-\infty}^{\infty} \sum_{i,j=1}^{2} F_i(\lambda) \left\{ \int_{-n}^{n} g^T(x)\, \varphi_j(x, \lambda)\, dx \right\} d\rho_{ij}(\lambda).$$

By virtue of absolute convergence we can interchange the order of integration in the right side, obtaining

$$\int_{-n}^{n} g^T(x)\, f(x)\, dx = \int_{-n}^{n} g^T(x) \left\{ \int_{-\infty}^{\infty} \sum_{i,j=1}^{2} F_i(\lambda)\, \varphi_j(x, \lambda)\, d\rho(\lambda) \right\} dx.$$

The assertion of the theorem follows from this equality by virtue of the arbitrariness of the vector-function $g(x)$, as well as the continuity of $f(x)$ and of the integrals (4.21) (the continuity of the latter follows from their assumed uniform convergence). This proves the theorem.

Bibliographical references

§1. The basic result of this section, i.e. Theorem 1.1, was first obtained by Conte and Sangren [**1**]. The proof presented here is due to Sargsjan [**7**].

§2. The results of this section are due to Titchmarsh [**4**] and Macleod [**1**]. It should be noted, however, that if the coefficients $p(x)$ and $r(x)$ are continuous, then only the limit point case occurs (cf. Chapter 13, §7).

§3. The integral representation of the resolvent was obtained by the authors and is published here for the first time. The remaining results of this section are due to Titchmarsh [**4**] and Macleod [**1**].

§4. Theorem 4.1 appears in a paper by Titchmarsh [**4**]. Lemma 4.1 and the proof of Theorem 4.1 presented here are due to the authors and are published here for the first time.

CHAPTER 4

INVESTIGATION OF THE SPECTRUM

By the spectrum corresponding to a problem for a Sturm-Liouville operator or a Dirac system for the case of the halfline $[0, \infty)$ is meant the complement of the set of points, in a neighborhood of which the spectral function $\rho(\lambda)$ is constant. For the case of the entire line $(-\infty, \infty)$ the spectrum is the complement of the set of points in a neighborhood of which the functions $\xi(\lambda)$, $\eta(\lambda)$ and $\zeta(\lambda)$ are all constant.[1] Obviously the spectrum is a closed set. By the point or discrete spectrum is meant the set of all points of discontinuity of the spectral function $\rho(\lambda)$ (for the case of the halfline $[0, \infty)$) or of the functions $\xi(\lambda)$, $\eta(\lambda)$ and $\zeta(\lambda)$ (for the case of the entire line $(-\infty, \infty)$), and by the continuous spectrum, the set of all points of continuity of $\rho(\lambda)$ which belong to the spectrum. The definition is analogous in the case of the entire line $(-\infty, \infty)$. The points of the discrete spectrum are also called the eigenvalues, and the solutions of the problem corresponding to such points are called eigenfunctions.

§1. Discrete spectrum

We will consider the problem

$$y'' + \{\lambda - q(x)\} y = 0, \tag{1.1}$$

$$y(0)\cos\alpha + y'(0)\sin\alpha = 0 \tag{1.2}$$

on the halfline.

In this section we will indicate various sufficient conditions for the discreteness of the spectrum of the problem (1.1) – (1.2).

LEMMA 1.1. *For the discreteness of the spectrum of the problem* (1.1) – (1.2) *it is sufficient that for every fixed* λ *any solution of the equation* (1.1) *have a finite number of zeros on the halfline* $[0, \infty)$.

PROOF. Let us consider the Sturm-Liouville problem on a finite

[1] In these definitions it is assumed that the functions $\rho(\lambda)$, $\xi(\lambda)$, $\eta(\lambda)$ and $\zeta(\lambda)$ of the corresponding problems are uniquely defined.

interval $[0,b]$, i.e. we adjoin to the problem (1.1) – (1.2) the boundary condition $y(b)\cos\beta + y'(b)\sin\beta = 0$. From Sturm's oscillation theorem (i.e. from Theorem 3.3, Chapter 1) it follows under the hypothesis of the lemma that the number of eigenvalues, $N_b(\lambda)$, of the problem on the finite interval $[0,b]$ remains bounded in every finite interval, uniformly with respect to b. Since the problem (1.1) – (1.2), i.e. the problem on $[0,\infty)$, is the limit of problems on $[0,b]$, it is obvious that the spectrum of the limit problem is discrete.

Remark. From Sturm's comparison theorem (cf. Theorem 3.1, Chapter 1) it follows that the hypothesis of Lemma 1.1 is surely fulfilled if

$$q(x) \to +\infty \quad \text{as} \quad x \to \infty. \tag{1.3}$$

In fact, for fixed λ there exists $x = x_\lambda$ such that for $x > x_\lambda$ the inequality $\lambda - q(x) < 0$ holds. Then by virtue of Sturm's comparison theorem any solution of the equation (1.1) has at most one zero in the interval (x_λ, ∞). But the number of zeros in the interval $(0, x_\lambda)$ is finite.

It turns out that the hypothesis of Lemma 1.1 is satisfied under substantially less restrictive conditions than (1.3). Namely, we have the following:

Lemma 1.2. *Suppose that the coefficient* $q(x)$ *satisfies the following conditions*:

1. *It is bounded from below, i.e. there exists a constant* $c > 0$ *such that for all* x

$$q(x) > -c. \tag{1.4}$$

2. *For any* $\omega > 0$

$$\lim_{x\to\infty} \int_x^{x+\omega} q(x)\, dt = \infty. \tag{1.5}$$

Then for every fixed λ *any solution of the equation* (1.1) *has a finite number of zeros on the halfline* $[0,\infty)$, *and consequently by Lemma* 1.1 *the spectrum of the problem* (1.1) – (1.2) *is discrete.*

Remark. A. M. Molčanov has proved [1] that when the condition (1.4) is satisfied, the condition (1.5) is necessary and sufficient for the discreteness of the spectrum of the problem (1.1) – (1.2). The sufficiency of the condition (1.5) follows from Lemma 1.2. The necessity of this condition is proved with the help of methods which we have not touched upon in this book (cf. the indicated paper by Mol̆čanov and the book by Glazman [1]).

PROOF OF LEMMA 1.2. We note first that we can assume without loss of generality that

$$q(x) \geqq 0. \tag{1.6}$$

Let us assume that for some $\lambda = \lambda_0 > 0$ there exists a solution $y = y(x)$ of the equation (1.1) which has an infinite number of zeros $\alpha_1 < \alpha_2 < \alpha_3 < \cdots$.

Let ω be a positive number satisfying the inequality $\omega < 1/(\lambda_0 + 1)$. We choose N so large that for $x > N$

$$\int_x^{x+\omega} q(t)\,dt > \omega(\lambda_0 + 1), \tag{1.7}$$

which is possible by virtue of the condition (1.5). We further choose n so large that $\alpha_n > N$, and then $m > n$ so large that $\alpha_m - \alpha_n > \omega$. We can assume without loss of generality that $\alpha_m - \alpha_n \geqq P\omega$, where P is an integer.

We rewrite the equation (1.1) (for $\lambda = \lambda_0$) in the form $y'' = [q(x) - \lambda_0]y$. We then multiply both sides of this equation by y and integrate from α_n to α_m. Integrating the left side by parts, we obtain

$$-\int_{\alpha_n}^{\alpha_m} [y'(t)]^2\,dt = \int_{\alpha_n}^{\alpha_m} q(t)\,y^2(t)\,dt - \lambda_0 \int_{\alpha_n}^{\alpha_m} y^2(t)\,dt. \tag{1.8}$$

Let us estimate the integral

$$\int_{\alpha_n}^{\alpha_m} q(t)\,y^2(t)\,dt.$$

To do this, we represent it in the form

$$\int_{\alpha_n}^{\alpha_m} q(t)\,y^2(t)\,dt = \sum_{k=1}^{P} \int_{\alpha_n+(k-1)\omega}^{\alpha_n+k\omega} q(t)\,y^2(t)\,dt.$$

From the theorem of the mean for integrals and from the condition (1.7) it follows that

$$\int_{\alpha_n+(k-1)\omega}^{\alpha_n+k\omega} q(t)\,y^2(t)\,dt > (\lambda_0 + 1)\,y^2(\xi_k)\,\omega, \tag{1.9}$$

$$\alpha_n + (k-1)\omega < \xi_k < \alpha_n + k\omega.$$

Therefore

$$\int_{\alpha_n}^{\alpha_m} q(t)\,y^2(t)\,dt > (\lambda_0+1)\sum_{k=1}^{p} y^2(\xi_k)\,\omega$$

(1.10)

$$= (\lambda_0+1)\int_{\alpha_n}^{\alpha_m} y^2(t)\,dt - (\lambda_0+1)\sum_{k=1}^{p}\int_{\alpha_n+(k-1)\omega}^{\alpha_n+k\omega} [y^2(t)-y^2(\xi_k)]\,dt.$$

We further have

$$|y^2(t)-y^2(\xi_k)| = 2\left|\int_{\xi_k}^{t} y'(u)\,y(u)\,du\right| \leqslant \int_{\xi_k}^{t} y^2(u)\,du + \int_{\xi_k}^{t} [y'(u)]^2\,du$$

$$\leqslant \int_{\alpha_n+(k-1)\omega}^{\alpha_n+k\omega} y^2(u)\,du + \int_{\alpha_n+(k-1)\omega}^{\alpha_n+k\omega} [y'(u)]^2\,du.$$

It therefore follows from (1.10) that

$$\int_{\alpha_n}^{\alpha_m} q(t)\,y^2(t)\,dt > (\lambda_0+1)\int_{\alpha_n}^{\alpha_m} y^2(t)\,dt$$

$$-(\lambda_0+1)\sum_{k=1}^{p}\int_{\alpha_n+(k-1)\omega}^{\alpha_n+k\omega}\left\{\int_{\alpha_n+(k-1)\omega}^{\alpha_n+k\omega} y^2(u)\,du\right\}dt$$

(1.11)

$$-(\lambda_0+1)\sum_{k=1}^{p}\int_{\alpha_n+(k-1)\omega}^{\alpha_n+k\omega}\left\{\int_{\alpha_n+(k-1)\omega}^{\alpha_n+k\omega} [y'(u)]^2\,du\right\}dt$$

$$= (\lambda_0+1)\int_{\alpha_n}^{\alpha_m} y^2(t)\,dt - (\lambda_0+1)\,\omega\int_{\alpha_n}^{\alpha_m} y^2(t)\,dt$$

$$-(\lambda_0+1)\,\omega\int_{\alpha_n}^{\alpha_m} [y'(t)]^2\,dt.$$

From (1.8) and (1.11) follows

$$-\int_{\alpha_n}^{\alpha_m} [y'(t)]^2\,dt > \{1-(\lambda_0+1)\,\omega\}\int_{\alpha_n}^{\alpha_m} y^2(t)\,dt - (\lambda_0+1)\,\omega\int_{\alpha_n}^{\alpha_m} [y'(t)]^2\,dt,$$

i.e.

$$\{1-(\lambda_0+1)\,\omega\}\int_{\alpha_n}^{\alpha_m} \{y^2(t)+[y'(t)]^2\}\,dt < 0,$$

which is impossible since $(\lambda_0+1)\omega<1$. This proves the lemma.

THEOREM 1.1. *If λ_0 is a point of the discrete spectrum and $\varphi(x, \lambda_0)$ is a corresponding eigenfunction, then $\varphi(x, \lambda_0) \in \mathscr{L}^2(0, \infty)$.*

PROOF. In the notation of Lemma 4.1, Chapter 2, we have

$$\int_0^a \left\{ \int_\Delta \varphi(x, \lambda) \varphi(y, \lambda)\, d\rho_b(\lambda) \right\}^2 dy \leqslant \int_\Delta \varphi^2(x, \lambda)\, d\rho_b(\lambda) \leqslant M,$$

where $a < b$, and M is a fixed positive number. Letting first $b \to \infty$ and then $a \to \infty$, we obtain

$$\int_0^\infty \left\{ \int_\Delta \varphi(x, \lambda) \varphi(y, \lambda)\, d\rho(\lambda) \right\}^2 dy \leqslant M.$$

Suppose that the interval Δ contains only one jump of the function $\rho(\lambda)$, whose magnitude we will denote by $d(\lambda_0)$. From the last inequality we obtain

$$d(\lambda_0) \varphi^2(x, \lambda_0) \int_0^\infty \varphi^2(y, \lambda_0)\, dy \leqslant M,$$

and since $\varphi(x, \lambda_0)$ is not identically zero, the theorem is proved.

§2. An oscillation theorem for the interval $(0, \infty)$

The purpose of this section is the proof of the following theorem.

THEOREM 2.1 (OSCILLATION THEOREM FOR THE INTERVAL $(0, \infty)$). *Suppose that the coefficient $q(x)$ from equation* (1.1) *satisfies the conditions* (1.4) *and* (1.5). *We denote by $\psi_n(x) = \varphi(x, \lambda_n)$*[2] *$(n = 0, 1, 2, \cdots)$ the eigenfunctions of the problem* (1.1)–(1.2). *Each function $\psi_n(x)$ has exactly n zeros in the interval $(0, \infty)$, and the zeros of the different eigenfunctions interlace.*

For the proof of this theorem we consider the Sturm-Liouville problem on a finite interval $[0, b]$:

$$y'' + \{\lambda - q(x)\} y = 0 \quad (0 \leqq x \leqq b), \tag{2.1}$$

$$y(0) \cos\alpha + y'(0) \sin\alpha = 0, \tag{2.2}$$

$$y(b) \cos\beta + y'(b) \sin\beta = 0. \tag{2.3}$$

Let us denote by $\lambda_{0;b}, \lambda_{1;b}, \lambda_{2;b}, \cdots$ the eigenvalues of the problem

---

[2] λ_n are the eigenvalues of the problem (1.1) – (1.2), and $\varphi(x, \lambda)$ is the solution of equation (1.1) satisfying the initial conditions $\varphi(0, \lambda) = \sin\alpha$, $\varphi'(0, \lambda) = -\cos\alpha$.

(2.1) – (2.3), and by $\psi_{n,b}(x) = \varphi(x, \lambda_{n;b})$ the corresponding eigenfunctions. As we already know, as $b \to \infty$ the numbers $\lambda_{n;b}$ tend to the eigenvalues λ_n $(n = 0, 1, 2, \cdots)$, while $\psi_{n;b}(x)$ tends, and indeed uniformly in every finite interval, to the eigenfunction $\psi_n(x)$ of the problem (1.1) – (1.2), and under the condition (1.4) the limit λ_n does not depend upon the number β from the condition (2.3). It will be convenient for us to put $\beta = 0$, i.e. to consider the boundary condition $y(b) = 0$ at the point b.

We denote by $x_{1,b}^{(n)}, x_{2,b}^{(n)}, \cdots, x_{n,b}^{(n)}$ the sequence of zeros of the function $\psi_{n,b}(x)$ in the interval $(0, b)$.

The proof of the theorem stated above is based on the following three lemmas.

LEMMA 2.1. *For every fixed n, with $b \to \infty$,*

$$\sup_b x_{n,b}^{(n)} < +\infty.$$

In fact, assuming the contrary, for any A there exists an n_0 and an unbounded increasing sequence $b_1, b_2, \cdots$ such that $x_{n_0,b_k}^{(n_0)} > A$. Since $\psi_{n_0,b_k}(x) = \varphi(x, \lambda_{n_0,b_k})$ is a solution of the equation $y'' + [\lambda_{n_0,b_k} - q(x)]y = 0$ and

$$\psi_{n_0,b_k}(x_{n_0,b_k}^{(n_0)}) = 0, \qquad \psi_{n_0,b_k}(b_k) = 0,$$

it follows from Sturm's comparison theorem that any solution of the equation $y'' - q(x)y = 0$ vanishes in the interval (A, b_k), which contradicts Lemma 1.2.

LEMMA 2.2. *For every fixed k, n with $b \to \infty$*

$$\inf_b (x_{k+1,b}^{(n)} - x_{k,b}^{(n)}) > 0.$$

Indeed, let us assume that the assertion of the lemma is false. Then there exist integers n_0 and $l_0 < n_0$ and an unbounded increasing sequence $\{b_k\}$ such that

$$\varliminf_{k\to\infty} (x_{l_0+1,b_k}^{(n_0)} - x_{l_0,b_k}^{(n_0)}) = 0. \tag{2.4}$$

Choosing a subsequence of $\{b_k\}$, we can replace the lower limit in (2.4) by the ordinary limit, and it follows from Lemma 2.1 that one can also assume the existence of a finite limit

$$\lim_{b_k\to\infty} x_{l_0,b_k}^{(n_0)} = x_{l_0}^{(n_0)}. \tag{2.5}$$

From (2.4) and (2.5) follows

$$\lim_{k\to\infty} x_{l_0+1,b_k}^{(n_0)} = x_{l_0}^{(n_0)}. \tag{2.6}$$

Let $\lambda > \lambda_{n_0,b_k}$ $(k = 1, 2, \cdots)$ but otherwise arbitrary. It follows from the comparison theorem that $\varphi(x, \lambda)$ has at least one zero in the interval $(x_{l_0,b_k}^{(n_0)}, x_{l_0+1,b_k}^{(n_0)})$. Passing to the limit $k \to \infty$, we obtain $\varphi(x_{l_0}^{(n_0)}, \lambda) = 0$ for all $\lambda > \lambda_{n_0}$, which is impossible since for fixed x and $\lambda \to \infty$ the function $\varphi(x, \lambda)$ is asymptotically equal to $C \cos\sqrt{\lambda}x$ or $C \sin\sqrt{\lambda}x$, where the constant C is different from zero.

Lemma 2.3. *For every fixed n, with* $b \to \infty$

$$\inf_b (\lambda_{n+1, b} - \lambda_{n, b}) > 0.$$

Proof. Let us assume the opposite. Then there exists a positive integer n_0 and an unbounded increasing sequence $\{b_k\}$ such that

$$\lim_{k\to\infty} (\lambda_{n_0+1, b_k} - \lambda_{n_0, b_k}) = 0. \tag{2.7}$$

Let

$$\lim_{k\to\infty} \lambda_{n_0, b_k} = \lambda_{n_0}. \tag{2.8}$$

It follows from (2.7) and (2.8) that

$$\lim_{k\to\infty} \lambda_{n_0+1, b_k} = \lambda_{n_0}. \tag{2.8'}$$

Let us consider the functions $\varphi(x, \lambda_{n_0, b_k})$ and $\varphi(x, \lambda_{n_0+1,b_k})$. Their zeros are $x_{i,b_k}^{(n_0)}$ $(i = 1, 2, \cdots, n_0)$ and $x_{i,b_k}^{(n_0+1)}$ $(i = 1, 2, \cdots, n_0 + 1)$.

We denote the limits of these zeros by $x_i^{(n_0)}$ and $x_i^{(n_0+1)}$.

For $x > x_{n_0,b_k}^{(n_0)}$ the function $\varphi(x, \lambda_{n_0,b_k})$ preserves its sign; let us assume for definiteness that it is positive. Therefore its limit, the function $\varphi(x, \lambda_{n_0})$, satisfies the inequality $\varphi(x, \lambda_{n_0}) \geqq 0$ for $x \geqq x_{n_0}^{(n_0)}$. Since $\varphi(x, \lambda_{n_0})$ is a solution of a second-order differential equation, for $x > x_{n_0}^{(n_0)}$ we have[3)]

$$\varphi(x, \lambda_{n_0}) > 0. \tag{2.9}$$

Since $\varphi(x, \lambda_{n_0+1,b_k})$ changes its sign one more time than does $\varphi(x, \lambda_{n_0,b_k})$, for $x > x_{n_0+1,b_k}^{(n_0+1)}$ we have $\varphi(x, \lambda_{n_0+1,b_k}) < 0$. Hence, as before, we conclude that for $x > x_{n_0+1}^{(n_0+1)}$, by virtue of (2.8′), we have

$$\varphi(x, \lambda_{n_0}) < 0,$$

which contradicts the inequality (2.9). This proves the lemma.

Proof of Theorem 2.1. We can now study the zeros of the eigenfunctions $\psi_n(x)$ of the problem (1.1) – (1.2), regarding this problem as the limit

---

3) If $\varphi(x, \lambda_0) \geqq 0$ for $x \geqq x_0$ and $\varphi(\bar{x}, \lambda_0) = 0$ for $\bar{x} > x_0$, then $\varphi'(\bar{x}, \lambda_0) = 0$, and consequently $\varphi(x, \lambda_0) \equiv 0$.

as $b \to \infty$ of the problem (2.1) $-$(2.3). We know that the limits

$$\lim_{b\to\infty} \lambda_{n,b} = \lambda_n \qquad (n = 0, 1, 2, \cdots) \tag{2.10}$$

exist. It follows from Lemma 2.3 that the numbers λ_n are distinct:

$$\lambda_0 < \lambda_1 < \lambda_2 < \cdots. \tag{2.11}$$

Further, by virtue of (2.10) we have, uniformly in every finite interval,

$$\lim_{b\to\infty} \varphi(x, \lambda_{n,b}) \equiv \lim_{b\to\infty} \psi_{n,b}(x) = \psi_n(x).$$

Therefore the zeros $x_{k,b}^{(n)}$ of the function $\psi_{n,b}(x)$ converge to the zeros $x_k^{(n)}$ of the function $\psi_n(x)$. It follows from Lemma 2.2 that the numbers $x_k^{(n)}$ are also distinct. Consequently, the function $\psi_n(x)$ has at least n zeros. We will show that the number of zeros is not greater than n, i.e. exactly equals n. Let us assume that for some n the function $\psi_n(x)$ has $n+1$ zeros: $x_1^{(n)}, \cdots, x_{n+1}^{(n)}$. It follows from Lemma 2.3 that for fixed n and sufficiently large b

$$\lambda_{n+1,b} - \lambda_{n,b} > \eta > 0,$$

where η does not depend upon b. Therefore there exists a number μ such that for all sufficiently large b

$$\lambda_{n,b} < \mu < \lambda_{n+1,b}, \quad \mu > \lambda_n.$$

From the second comparison theorem (cf. Theorem 3.2, Chapter 1) it follows that $\varphi(x, \mu)$ has not less than $n+1$ zeros in the interval $(0, \infty)$. Therefore for sufficiently large b the function $\varphi(x, \mu)$ has not less than $n+1$ zeros in the interval $(0, b)$, which contradicts the oscillation theorem.

The interlacing of the zeros of the functions $\psi_n(x)$ $(n = 0, 1, 2, \cdots)$ follows from (2.11) and the second comparison theorem. This proves the theorem.

§3. Investigation of the spectrum for the case $q(x) \in \mathscr{L}(0, \infty)$

1. The following result will be needed in the sequel.

LEMMA 3.1. *Suppose that the functions $h_1(x)$, $h_2(x)$, $g_1(x)$ and $g_2(x)$ are nonnegative on the interval $0 \leq x \leq X$, and suppose that $h_1(x)$ and $h_2(x)$ are continuous, and $g_1(x)$ and $g_2(x)$ integrable, on this interval.*

If

$$h_1(x),\ h_2(x) \leqslant C + \int_0^x [h_1(t)\, g_1(t) + h_2(t)\, g_2(t)]\, dt, \qquad 0 \leq x \leq X, \tag{3.1}$$

where C is a constant, then one has the inequalities

$$h_1(x),\ h_2(x) \leqslant C \exp\left\{\int_0^x [g_1(t) + g_2(t)]\right\} dt,\quad 0 \leqslant x \leqslant X. \tag{3.2}$$

PROOF. We put

$$y(x) = \int_0^x [h_1(s)\, g_1(s) + h_2(s)\, g_2(s)]\, ds, \tag{3.3}$$

$$\frac{dy}{dx} = h_1(x)\, g_1(x) + h_2(x)\, g_2(x). \tag{3.3$'$}$$

Multiplying the first of the inequalities (3.1) by $g_1(x)$, the second by $g_2(x)$, and adding, we obtain by virtue of (3.3) and (3.3′)

$$\frac{dy}{dx} \leqslant C[g_1(x) + g_2(x)] + y(x)[g_1(x) + g_2(x)].$$

This inequality can be rewritten in the form

$$\frac{d}{dx}\left\{y(x) \exp\left[-\int_0^x \{g_1(s) + g_2(s)\}\, ds\right]\right\}$$

$$\leqslant C[g_1(x) + g_2(x)] \exp\left\{-\int_0^x [g_1(s) + g_2(s)]\, ds\right\}.$$

Integrating this inequality from 0 to x, we find that

$$y(x) \exp\left\{-\int_0^x [g_1(s) + g_2(s)] ds\right\} \leqslant C - C \exp\left\{-\int_0^x [g_1(s) + g_2(s)]\, ds\right\}.$$

Hence

$$y(x) \leqslant C \exp\left\{\int_0^x [g_1(s) + g_2(s)]\, ds\right\} - C. \tag{3.4}$$

By virtue of (3.3), the inequalities (3.2) follow at once from (3.1) and (3.4), which proves the lemma.

2. Let $\varphi(x, \lambda)$ be the solution of equation (1.1) satisfying the conditions $\varphi(0, \lambda) = \sin \alpha$, $\varphi'(0, \lambda) = -\cos \alpha$. Rewriting equation (1.1) in the form $y'' + \lambda y = q(x) y$, applying the method of variation of constants, and putting $\lambda = s^2$, we obtain

$$\begin{aligned}\varphi(x, \lambda) &= \cos sx \sin \alpha - \frac{\sin sx}{s} \cos \alpha \\ &\quad + \frac{1}{s} \int_0^x \sin\{s(x - t)\}\, q(t)\, \varphi(t, \lambda)\, dt.\end{aligned} \tag{3.5}$$

Let $s = \sigma + i\tau$, $\tau \geqq 0$. We introduce the notation $\varphi_1(x,\lambda) = \varphi(x,\lambda)e^{-\tau x}$. Then it follows from (3.5) that

$$\varphi_1(x, \lambda) = e^{-\tau x}\cos sx \cdot \sin\alpha - e^{-\tau x}\frac{\sin sx}{s}\cos\alpha + \frac{1}{s}\int_0^x e^{-\tau(x-t)}\sin\{s(x-t)\}\,q(t)\,\varphi_1(t,\lambda)\,dt.$$

Since $|\cos sx| \leqq e^{\tau x}$, $|\sin sx| \leqq e^{\tau x}$, the last inequality implies the inequality

$$|\varphi_1(x,\lambda)| \leqslant 1 + \frac{1}{|s|} + \frac{1}{|s|}\int_0^x |q(t)\,\varphi_1(t,\lambda)|\,dt.$$

Applying Lemma 3.1 (putting $h_2(x) = g_2(x) \equiv 0$), we obtain

$$|\varphi_1(x,\lambda)| \leqslant \left(1 + \frac{1}{|s|}\right)\exp\left(\frac{1}{|s|}\int_0^x |q(t)|\,dt\right). \tag{3.6}$$

Since by hypothesis $q(t) \in \mathscr{L}(0,\infty)$, it follows from the last inequality that the function $\varphi_1(x,\lambda)$ is bounded for $0 \leqq x < \infty$ for $|s| \geqq \rho > 0$, $\tau \geqq 0$.

Let us first consider real positive values of s. Then for $s \geqq \rho$ the function $\varphi_1(x,\lambda)$ is bounded. Therefore from formula (3.5) we obtain (for $x \to \infty$)

$$\begin{aligned}\varphi(x,\lambda) &= \cos sx \sin\alpha - \frac{\sin sx}{s}\cos\alpha \\ &\quad + \frac{1}{s}\int_0^\infty \sin\{s(x-t)\}\,q(t)\,\varphi(t,\lambda)\,dt \\ &\quad - \frac{1}{s}\int_x^\infty \sin\{s(x-t)\}\,q(t)\,\varphi(t,\lambda)\,dt \\ &= \mu(\lambda)\cos sx + \nu(\lambda)\sin sx + o(1),\end{aligned} \tag{3.7}$$

where

$$\mu(\lambda) = \sin\alpha - \frac{1}{s}\int_0^\infty \sin st \cdot q(t)\,\varphi(t,\lambda)\,dt, \tag{3.8}$$

$$\nu(\lambda) = -\frac{\cos\alpha}{s} + \frac{1}{s}\int_0^\infty \cos st \cdot q(t)\,\varphi(t,\lambda)\,dt. \tag{3.9}$$

Since the integrals in (3.8) and (3.9) converge uniformly for $s \geqq \rho > 0$, it follows that $\mu(\lambda)$ and $\nu(\lambda)$ are continuous functions of s.

Similarly if $\theta(x,\lambda)$ is the solution of equation (1.1) satisfying the conditions $\theta(0,\lambda) = \cos\alpha$, $\theta'(0,\lambda) = -\sin\alpha$, then for $x \to \infty$

$$\theta(x,\lambda) = \mu_1(\lambda)\cos sx + \nu_1(1)\sin sx + o(1), \tag{3.7'}$$

where

$$\mu_1(\lambda) = \cos\alpha - \frac{1}{s}\int_0^\infty \sin st \cdot q(t)\,\theta(t,\lambda)\,dt, \tag{3.8'}$$

$$\nu_1(\lambda) = -\frac{\sin\alpha}{s} + \frac{1}{s}\int_0^\infty \cos st \cdot q(t)\,\theta(t,\lambda)\,dt. \tag{3.9'}$$

Further, differentiating (3.5) with respect to x, we find that

$$\begin{aligned}\varphi'_x(x,\lambda) &= -s\sin sx \sin\alpha - \cos sx\cos\alpha \\ &\quad + \int_0^x \cos\{s(x-t)\}\,q(t)\,\varphi(t,\lambda)\,dt.\end{aligned} \tag{3.10}$$

Estimating this function in the same way as was done above, we obtain for $x \to \infty$

$$\varphi'_x(x,\lambda) = -s\mu(\lambda)\sin sx + s\nu(\lambda)\cos sx + o(1). \tag{3.10'}$$

Similarly,

$$\theta'_x(x,\lambda) = -s\mu_1(\lambda)\sin sx + s\nu_1(\lambda)\cos sx + o(1).$$

Let us consider the Wronskian determinant of $\varphi(x,\lambda)$ and $\theta(x,\lambda)$. Since equation (1.1) does not contain the first derivative, it follows from a well-known formula of Liouville that the Wronskian determinant is equal to a constant. Therefore

$$\begin{aligned}W\{\varphi,\theta\} &= W\{\varphi,\theta\}_{x=0} = 1 \\ &= W\{\mu(\lambda)\cos sx + \nu(\lambda)\sin sx,\ \mu_1(\lambda)\cos sx + \nu_1(\lambda)\sin sx\} \\ &\qquad + o(1) = s\{\mu(\lambda)\nu_1(\lambda) - \mu_1(\lambda)\nu(\lambda)\} + o(1).\end{aligned}$$

From the equality

$$s\{\mu(\lambda)\nu_1(\lambda) - \mu_1(\lambda)\nu(\lambda)\} + o(1) = 1,$$

taking into account that the left side does not depend upon x, we obtain

$$\mu(\lambda)\nu_1(\lambda) - \mu_1(\lambda)\nu(\lambda) = 1/\sqrt{\lambda}. \tag{3.11}$$

From this last formula it follows, in particular, that the functions $\mu(\lambda)$ and $\nu(\lambda)$ cannot both vanish for one value of λ.

3. Let us now consider complex values of s. For a fixed positive τ it follows from (3.5) (for $x \to \infty$) that

$$\varphi(x, \lambda) = \frac{1}{2} e^{-isx} \sin\alpha + \frac{e^{-isx}}{2is} \cos\alpha + O(e^{-\tau x})$$
$$- \frac{1}{2is} \int_0^x e^{-is(x-t)} q(t) \varphi(t, \lambda)\, dt + O\left(\int_0^x e^{-\tau(x-t)} | q(t) \varphi(t, \lambda) |\, dt \right).$$

Since $|\varphi(t,\lambda)| = O(e^{\tau t})$, we have (putting, for example, $\delta = x/2$)

$$O\left(\int_0^x e^{-\tau(x-t)} | q(t) \varphi(t, \lambda) |\, dt \right) = O\left(\int_0^x e^{\tau(2t-x)} | q(t) |\, dt \right)$$
$$= O\left(e^{\tau(x-2\delta)} \int_0^{x-\delta} | q(t) |\, dt \right) + O\left(e^{\tau x} \int_{x-\delta}^x | q(t) |\, dt \right) = o(e^{\tau x}).$$

On the other hand,

$$\int_x^\infty e^{-is(x-t)} q(t) \varphi(t, \lambda)\, dt = O\left(e^{\tau x} \int_x^\infty | q(t) |\, dt \right) = o(e^{\tau x}).$$

Therefore

$$\varphi(x,\lambda) = e^{-isx}\{ M(\lambda) + o(1) \}, \tag{3.12}$$

where

$$M(\lambda) = \frac{1}{2} \sin\alpha + \frac{1}{2is} \cos\alpha - \frac{1}{2is} \int_0^\infty e^{ist} q(t) \varphi(t, \lambda)\, dt. \tag{3.13}$$

If we use formula (3.10), we obtain

$$\varphi_x'(x,\lambda) = -ise^{-isx}\{ M(\lambda) + o(1) \}. \tag{3.14}$$

Similarly

$$\theta(x,\lambda) = e^{-isx}\{ M_1(\lambda) + o(1) \}, \tag{3.12$'$}$$

where

$$M_1(\lambda) = \frac{1}{2} \cos\alpha - \frac{1}{2is} \sin\alpha - \frac{1}{2is} \int_0^\infty e^{ist} q(t) \theta(t, \lambda)\, dt. \tag{3.13$'$}$$

4. We now have at our disposal all the auxiliary tools for the investigation of the spectrum of a Sturm-Liouville operator for the case $q(x) \in \mathcal{L}(0, \infty)$. Let $b > 0$. We will consider the boundary-value problem (2.1)–(2.3) in the finite interval $[0, b]$.

We begin by investigating the negative part of the spectrum. It follows from formulas (3.12) and (3.13) that for every fixed $\lambda_0 < 0$ the number of roots of the equation $\varphi(b,\lambda) = 0$ which lie in the interval $-\infty < \lambda \leqq \lambda_0$

is bounded, uniformly with respect to b. Therefore the number of points of increase of the function $\rho_b(\lambda)$ (for the definition of $\rho_b(\lambda)$ see Chapter 2, §1) in the interval $-\infty < \lambda \leqq \lambda_0$ is bounded, uniformly with respect to b, and so the limit function $\rho(\lambda)$ has the same property. For the case in which $q(x) \to 0$ as $x \to \infty$, this fact easily follows from Sturmian theory (see the remark following the proof of Lemma 1.1).

If the number of zeros of the function $\varphi(x, 0)$ on the halfline $[0, \infty)$ is finite, then the number of negative eigenvalues is also finite. In fact, it follows in this case from Sturm's oscillation theorem that for any $b > 0$ the problem (2.1) – (2.3) has a bounded number of negative eigenvalues.

We have thus proved:

THEOREM 3.1. *If $q(x) \in \mathscr{L}(0, \infty)$, then the negative part of the spectrum of the problem* (1.1) – (1.2) *is discrete and bounded from below.*

The following lemma gives a simple sufficient criterion for the function $\varphi(x, 0)$ to have a finite number of zeros.

LEMMA 3.2. *If $(1 + x^2)q(x) \in \mathscr{L}(0, \infty)$, then the function $\varphi(x, 0)$ has a finite number of zeros in the interval $[0, \infty)$, and consequently the number of negative eigenvalues is finite.*

Indeed, as is known,[4] if $(1 + x^2)q(x) \in \mathscr{L}(0, \infty)$, then there exists a fundamental system of solutions y_1 and y_2 of the equation $y'' - q(x)y = 0$ such that

$$\lim_{x \to \infty} y_1(x) = 1, \qquad \lim_{x \to \infty} [y_2(x) - x] = 0.$$

Since $\varphi(x, 0) = C_1 y_1(x) + C_2 y_2(x)$, it follows that

$$\lim_{x \to \infty} \varphi(x, 0) = \infty$$

if $C_2 \neq 0$, and

$$\lim_{x \to \infty} \varphi(x, 0) = 1$$

if $C_2 = 0$. In either case $\varphi(x, 0)$ has a finite number of zeros on the half-line $[0, \infty)$.

Let us now consider the positive part of the spectrum. Since by virtue of (3.11) the functions $\mu(\lambda)$ and $\nu(\lambda)$ cannot both equal zero for one value of λ, we have $\mu^2(\lambda) + \nu^2(\lambda) > 0$. Therefore, putting

$$\frac{\mu(\lambda)}{[\mu^2(\lambda) + \nu^2(\lambda)]^{1/2}} = \sin \delta(\lambda), \qquad \frac{\nu(\lambda)}{[\mu^2(\lambda) + \nu^2(\lambda)]^{1/2}} = \cos \delta(\lambda),$$

---

[4] Cf. V. V. Stepanov, *Course in differential equations*, Moscow, 1945, p. 246.

we can rewrite (3.7) in the form (for $x \to \infty$)

$$\varphi(x,\lambda) = [\mu^2(\lambda) + \nu^2(\lambda)]^{1/2} \sin[sx + \delta(\lambda)] + o(1). \tag{3.15}$$

If we assume that[5)]

$$(1+x)\,q(x) \in \mathscr{L}(0,\infty), \tag{3.16}$$

then differentiating (3.5) with respect to s, we obtain after some simple transformations[6)]

$$\varphi_s'(x,\lambda) = \frac{\partial}{\partial s}\left\{[\mu^2(\lambda) + \nu^2(\lambda)]^{1/2} \sin[sx + \delta(\lambda)]\right\} + o(1). \tag{3.17}$$

Let us denote by b a large positive number and assume, for simplicity, that $\beta = 0$ in the boundary condition (2.3) (the case of an arbitrary real β can be handled similarly). Then the positive eigenvalues of the problem (2.1) – (2.3) are determined from the equation

$$\sin[sb + \delta(\lambda)] = o(1). \tag{3.18}$$

Let s_1 be a positive root of the equation (3.18). We note that $\delta(\lambda) = O(1)$. Then

$$s_1 b + \delta(\lambda_1) = m\pi + o(1) \quad (s_1 = \sqrt{\lambda_1}). \tag{3.19}$$

Let s_2 be the next root, after s_1, of (3.18). Then either

$$s_2 b + \delta(\lambda_2) = m\pi + o(1) \quad (s_2 = \sqrt{\lambda_2}), \tag{3.20}$$

or

$$s_2 b + \delta(\lambda_2) = (m+1)\pi + o(1). \tag{3.21}$$

We will show that the case (3.20) is impossible. In fact, if (3.20) held, then it would follow from Rolle's theorem that the equation $\varphi_s'(b,\lambda) = 0$ has a root s_3 lying between s_1 and s_2, and consequently satisfying the condition $s_3 b + \delta(\lambda_3) = m\pi + o(1)$, which is impossible by virtue of the asymptotic formula (3.17).

Subtracting (3.19) from (3.21) and taking into account that $\delta(\lambda)$ is a continuous function, we obtain for two successive eigenvalues λ_1 and λ_2 of the problem (2.1) – (2.3) the asymptotic formula

$$s_2 - s_1 = \frac{\pi}{b} + o\left(\frac{1}{b}\right) \qquad (s_i = \sqrt{\lambda_i};\ i = 1, 2). \tag{3.22}$$

---

5) We will subsequently get rid of the condition (3.16).

6) The boundedness of $\varphi_s'(x,\lambda)$ for large x follows from Lemma 3.1.

THEOREM 3.2. *If* $q(x) \in \mathscr{L}(0, \infty)$, $\lambda > 0$ *and* $\Delta > 0$, *then*

$$\rho(\lambda + \Delta) - \rho(\lambda) = \frac{1}{\pi} \int_{\lambda}^{\lambda+\Delta} \frac{d\lambda}{\sqrt{\lambda}\,[\mu^2(\lambda) + \nu^2(\lambda)]}, \tag{3.23}$$

i.e. the spectrum of the problem (1.1) – (1.2) *is continuous in the interval* $(0, \infty)$.

PROOF. Let us denote by $\lambda_{1,b}, \lambda_{2,b}, \cdots$ the eigenvalues of the problem (2.1) – (2.3), and by $\varphi(x, \lambda_{1,b}), \varphi(x, \lambda_{2,b}), \cdots$ the corresponding eigenfunctions. Let us first assume that $q(x)$ satisfies the condition (3.16). Then by the definition of the function $\rho_b(\lambda)$ (cf. Chapter 2, §1) and the asymptotic formula (3.22) we have ($s_{n,b} = \sqrt{\lambda_{n,b}}$)

$$\rho_b(\lambda + \Delta) - \rho_b(\lambda) = \sum_{\lambda < \lambda_n \leqslant \lambda + \Delta} \frac{1}{\int_0^b \varphi^2(x, \lambda_{n,b})\, dx}$$

$$= \sum_{\lambda < \lambda_{n,b} \leqslant \lambda + \Delta} \frac{s_{n+1,b} - s_{n,b}}{(s_{n+1,b} - s_{n,b}) \int_0^b \varphi^2(x, \lambda_{n,b})\, dx}$$

$$= \frac{1}{\pi} \sum_{\lambda < \lambda_{n,b} \leqslant \lambda + \Delta} \frac{\lambda_{n+1,b} - \lambda_{n,b}}{(s_{n+1,b} + s_{n,b}) \frac{1}{b} \int_0^b \varphi^2(x, \lambda_{n,b})\, dx + o(1)}.$$

It follows from the asymptotic formula (3.7) that

$$\frac{1}{b} \int_0^b \varphi^2(x, \lambda_{n,b})\, dx = \frac{1}{2} [\mu^2(\lambda_{n,b}) + \nu^2(\lambda_{n,b})] + o(1).$$

Therefore

$$\rho_b(\lambda + \Delta) - \rho_b(\lambda) = \frac{1}{\pi} \sum_{\lambda < \lambda_{n,b} \leqslant \lambda + \Delta} \left\{ \frac{1}{s_{n,b}[\mu^2(\lambda_{n,b}) + \nu^2(\lambda_{n,b})]} + o(1) \right\} \times (\lambda_{n+1,b} - \lambda_{n,b}).$$

Passing to the limit $b \to \infty$, we obtain the assertion of the theorem for the case in which $q(x)$ satisfies the condition (3.16).

Suppose now that the condition (3.16) is not satisfied. We put

$$q_n(x) = \begin{cases} q(x), & 0 \leq x \leq n, \\ 0, & x > n. \end{cases}$$

The function $q_n(x)$ obviously satisfies the condition (3.16). Therefore, denoting by $\rho_n(\lambda)$, $\mu_n(\lambda)$ and $\nu_n(\lambda)$ the functions corresponding to $q_n(x)$, we obtain

$$\rho_n(\lambda+\Delta)-\rho_n(\lambda)=\frac{1}{\pi}\int_{\lambda}^{\lambda+\Delta}\frac{d\lambda}{\sqrt{\lambda}\,[\mu_n^2(\lambda)+\nu_n^2(\lambda)]}.$$

Passing to the limit $n\to\infty$, we obtain (3.23).

§4. Transformation of the basic equation

In certain cases equation (1.1) with an unboundedly increasing $q(x)$ can be transformed, by means of an appropriate change of variables, to another form which makes it possible to deduce asymptotic formulas.

Let us first assume that λ is a real number, and that $q'(x)$ and $q''(x)$ are continuous functions. We introduce the notation

$$\xi(x)=\int_0^x\{\lambda-q(t)\}^{1/2}dt,\qquad \eta(x)=\{\lambda-q(x)\}^{1/4}y.$$

Then

$$\frac{d\eta}{d\xi}=\frac{d\eta}{dx}\cdot\frac{dx}{d\xi}=\left[\{\lambda-q(x)\}^{1/4}\frac{dy}{dx}-\frac{1}{4}\frac{q'(x)}{\{\lambda-q(x)\}^{3/4}}y\right]\frac{1}{\{\lambda-q(x)\}^{1/2}}$$

$$=\{\lambda-q(x)\}^{-1/4}\frac{dy}{dx}-\frac{1}{4}\frac{q'(x)}{\{\lambda-q(x)\}^{5/4}}y,$$

$$\frac{d^2\eta}{d\xi^2}=\Big\{[\lambda-q(x)]^{-1/4}\frac{d^2y}{dx^2}$$

$$-\frac{1}{4}\left[\frac{q''(x)}{\{\lambda-q(x)\}^{5/4}}+\frac{5}{4}\frac{q'^2(x)}{\{\lambda-q(x)\}^{9/4}}\right]y\Big\}\frac{1}{\{\lambda-q(x)\}^{1/2}}.$$

Therefore equation (1.1) can be transformed to the form

$$\frac{d^2\eta}{d\xi^2}+\eta+\left[\frac{1}{4}\cdot\frac{q''(x)}{\{\lambda-q(x)\}^2}+\frac{5}{16}\frac{q'^2(x)}{\{\lambda-q(x)\}^3}\right]\eta=0. \tag{4.1}$$

This equation is of the same form as (1.1), but in (4.1) the coefficient of η in the last term turns out to be small for large λ.

It follows from (4.1) that each of its solutions $\eta=\varphi(\xi)$ satisfies the integral equation

$$\varphi(\xi)=\varphi(0)\cos\xi+\varphi'(0)\sin\xi$$

$$-\int_0^{\xi}\sin(\xi-\tau)\left\{\frac{1}{4}\frac{q''(t)}{[\lambda-q(t)]^2}+\frac{5}{16}\frac{q'^2(t)}{[\lambda-q(t)]^3}\right\}\varphi(\tau)\,d\tau, \tag{4.2}$$

where $\tau=\xi(t)$. This equation can be used to obtain asymptotic formulas for $\varphi(\xi)$.

For nonreal λ or for $q(x) > \lambda$ the function ξ assumes nonreal values, and therefore formula (4.2) contains an integral along a complex path. There is, however, no necessity to introduce such integrals, since one can obtain a similar equation with a real independent variable, which in fact turns out to be x. To obtain this equation we assume that $q(0) = 0$ (this can always be achieved by an appropriate choice of the origin in the λ-plane). We then put

$$P(x) = \{\lambda - q(x)\}^{1/4} \frac{d}{dx}\left\{[\lambda - q(x)]^{-1/2} \frac{d\eta}{dx}\right\} - \frac{d^2y}{dx^2}. \tag{4.3}$$

Then

$$\begin{aligned} P(x) &= \{\lambda - q(x)\}^{1/4} \frac{d}{dx}\left\{[\lambda - q(x)]^{-1/4} \frac{dy}{dx} - \frac{1}{4}\frac{q'(x)}{[\lambda - q(x)]^{5/4}}\, y\right\} \\ &\quad - \frac{d^2y}{dx^2} = -\left\{\frac{1}{4}\cdot\frac{q''(x)}{[\lambda - q(x)]} + \frac{5}{16}\frac{q'^2(x)}{[\lambda - q(x)]^2}\right\} y. \end{aligned} \tag{4.3'}$$

By virtue of (1.1) and (4.3)

$$\begin{aligned} I &= \int_0^x \sin\{\xi(x) - \xi(t)\} \frac{P(t)}{\{\lambda - q(t)\}^{1/4}}\, dt \\ &= \int_0^x \sin\{\xi(x) - \xi(t)\} \frac{d}{dt}\left\{[\lambda - q(t)]^{-1/2} \frac{d\eta}{dt}\right\} dt \\ &\quad + \int_0^x \sin\{\xi(x) - \xi(t)\} \cdot \{\lambda - q(t)\}^{3/4} y\, dt = I_1 + I_2. \end{aligned}$$

Integrating by parts twice, we obtain

$$\begin{aligned} I_1 &= \left\{\sin[\xi(x) - \xi(t)]\,[\lambda - q(t)]^{-1/2} \frac{d\eta}{dt}\right\}_0^x \\ &\quad + \int_0^x \cos[\xi(x) - \xi(t)] \frac{d\eta}{dt}\, dt \\ &= -\eta'(0)\lambda^{-1/2} \sin\xi(x) + \{\cos[\xi(x) - \xi(t)]\,\eta(t)\}_0^x \\ &\quad - \int_0^x \sin[\xi(x) - \xi(t)]\,[\lambda - q(t)]^{1/2} \eta(t)\, dt \\ &= -\eta'(0)\lambda^{-1/2} \sin\xi(x) + \eta(x) - \eta(0)\cos\xi(x) - I_2. \end{aligned}$$

Therefore

$$\eta(x) = \eta(0)\cos\xi(x) + \eta'(0)\lambda^{-1/2}\sin\xi(x) + I,$$

i.e. $\eta(x)$ satisfies the following integral equation:

$$\eta(x) = \eta(0)\cos\xi(x) + \eta'(0)\lambda^{-1/2}\sin\xi(x)$$

(4.4)

$$+ \int_0^x \sin\{\xi(x) - \xi(t)\} R(t)\eta(t)\,dt,$$

where

(4.5) $$R(t) = -\left[\frac{1}{4}\cdot\frac{q''(t)}{\{\lambda - q(t)\}^{3/2}} + \frac{5}{16}\frac{q'^2(t)}{\{\lambda - q(t)\}^{5/2}}\right].$$

In formulas (4.4) and (4.5) we assume, for $\operatorname{Im}\lambda > 0$, that $0 < \arg\lambda < \pi$, and for those cases in which $q(t)$ varies from 0 to ∞, we assume that $\arg[\lambda - q(t)]$ varies from $\arg\lambda$ to π.

LEMMA 4.1. *Suppose that $q(x)$ tends monotonically to infinity, and suppose that*

(4.6) $$q'(x) = O([q(x)]^c), \qquad x > \infty,$$

where c is some fixed number satisfying the condition $0 < c < 3/2$, and $q''(x)$ is of fixed sign for large x. Under these assumptions the integral

(4.7) $$\int_0^\infty |R(x)|\,dx$$

converges uniformly with respect to λ in every region in which $[\lambda - q(x)] \geqq \delta > 0$ for $0 \leqq x < \infty$.

PROOF. By the estimate (4.6)

$$\int_{X_0}^X \frac{q'^2(x)}{\{q(x)\}^{5/2}}\,dx = O\left\{\int_{X_0}^X \frac{q'(x)}{[q(x)]^{5/2-c}}\,dx\right\} = O\left\{[q(x)]^{c-3/2}\Big|_{X_0}^X\right\} = O(1).$$

Moreover, we have

$$\int_{X_0}^X \frac{q''(x)}{\{q(x)\}^{3/2}}\,dx = \frac{q'(x)}{[q(x)]^{3/2}}\Big|_{X_0}^X + \frac{3}{2}\int_{X_0}^X \frac{q'^2(x)}{\{q(x)\}^{5/2}}\,dx = O(1).$$

The lemma follows from these two estimates by virtue of (4.5).

§5. The case $q(x) \to -\infty$

THEOREM 5.1. *Suppose that $q(x) \leqq 0$, $q'(x) < 0$ for $x \geqq 0$, $q(x) \to -\infty$ and $q''(x)$ is of fixed sign for large x. Moreover, suppose that the condition (4.6) is satisfied. If the integral*

(5.1) $$\int^\infty |q(x)|^{-1/2}\,dx$$

diverges, then the spectrum is continuous and fills up the entire axis $(-\infty, +\infty)$.

PROOF. We remark that the conditions of the theorem imply the convergence of the integral (4.7). Therefore it follows from equation (4.4) that for positive λ

$$\eta(x) = \eta(0)\cos\xi(x) + \eta'(0)\lambda^{-1/2}\sin\xi(x) + \int_0^\infty \sin\{\xi(x) - \xi(t)\} R(t)\eta(t)\,dt + o(1).$$

Hence, putting $\eta(x) = [\lambda - q(x)]^{1/4}\varphi(x,\lambda)$, where $\varphi(x,\lambda)$ is the same as in §3, we obtain

$$\varphi(x,\lambda) = \{\lambda - q(x)\}^{-1/4}\{\mu(\lambda)\cos\xi(x) + \nu(\lambda)\sin\xi(x)\} + o(1), \tag{5.2}$$

where

$$\mu(\lambda) = \lambda^{1/4}\sin\alpha - \int_0^\infty \sin\xi(t) R(t)\{\lambda - q(t)\}^{1/4}\varphi(t,\lambda)\,dt, \tag{5.3}$$

$$\nu(\lambda) = -\frac{q'(0)\sin\alpha}{4\lambda^{5/4}} - \frac{\cos\alpha}{\lambda^{1/4}} + \int_0^\infty \cos\xi(t) R(t)\{\lambda - q(t)\}^{1/4}\varphi(t,\lambda)\,dt. \tag{5.4}$$

Similarly, if $\theta(x,\lambda)$ is that solution of equation (1.1) for which $\theta(0,\lambda) = \cos\alpha$, $\theta'(0,\lambda) = \sin\alpha$, then

$$\theta(x,\lambda) = \{\lambda - q(x)\}^{-1/4}\{\mu_1(\lambda)\cos\xi(x) + \nu_1(\lambda)\sin\xi(x) + o(1)\}, \tag{5.5}$$

where $\mu_1(\lambda)$ and $\nu_1(\lambda)$ are obtained from $\mu(\lambda)$ and $\nu(\lambda)$ by replacing $\sin\alpha$, $-\cos\alpha$ and $\varphi(t,\lambda)$ by $\cos\alpha$, $\sin\alpha$ and $\theta(t,\lambda)$ respectively.

From these arguments it also follows that the integrals in (5.3) and (5.4) converge uniformly with respect to λ, and so $\mu(\lambda)$ and $\nu(\lambda)$ are continuous functions of λ.

Further, differentiating (4.4), we find that

$$\eta'(x) = \{\lambda - q(x)\}^{1/2}\left\{\lambda^{-1/2}\eta'(0)\cos\xi(x) - \eta(0)\sin\xi(x) + \int_0^x \cos\{\xi(x) - \xi(t)\} R(t)\eta(t)\,dt\right\}.$$

Therefore, proceeding as above, we obtain

$$\frac{d}{dx}\{[\lambda - q(x)]^{1/4}\varphi(x,\lambda)\} = \{\lambda - q(x)\}^{1/2}[\nu(\lambda)\cos\xi(x) - \mu(\lambda)\sin\xi(x) + o(1)],$$

from which, by virtue of (5.2) and (4.6), follows

$$(5.6)\qquad \varphi'(x,\lambda)=\{\lambda-q(x)\}^{1/4}\{\nu(\lambda)\cos\xi(x)-\mu(\lambda)\sin\xi(x)+o(1)\}.$$

Similarly,

$$(5.7)\qquad \theta'(x,\lambda)=\{\lambda-q(x)\}^{1/4}\{\nu_1(\lambda)\cos\xi(x)-\mu_1(\lambda)\sin\xi(x)+o(1)\}.$$

Thus

$$\lim_{x\to\infty} W\{\varphi,\theta\}=\mu(\lambda)\nu_1(\lambda)-\mu_1(\lambda)\nu(\lambda)=W\{\varphi,\theta\}_{x=0}=1.$$

Consequently $\mu(\lambda)$ and $\nu(\lambda)$ cannot both equal zero for the same positive value of λ.

The above arguments require some modification for $\lambda<0$. However we can choose X so that for $x\geqq X$ the inequality $\lambda-q(x)>0$ holds, and repeat all of the preceding argument for the interval (X,∞). As a consequence we arrive at the very same conclusions concerning the behaviour of the functions $\mu(\lambda)$ and $\nu(\lambda)$ for $\lambda\leqq 0$.

Let us first calculate the spectrum for the case $\lambda\geqq 0$. Let $b>0$ be arbitrary. We consider the boundary condition

$$(5.8)\qquad \varphi(b,\lambda)\cos\beta+\varphi'(b,\lambda)\sin\beta=0.$$

If $\sin\beta\neq 0$, then as $b\to\infty$ the condition (5.8) can be rewritten, using (5.2) and (5.6), in the form

$$(5.9)\qquad \nu(\lambda)\cos\xi(b)-\mu(\lambda)\sin\xi(b)=o(1).$$

For $\sin\beta=0$ we obtain

$$(5.9')\qquad \mu(\lambda)\cos\xi(b)+\nu(\lambda)\sin\xi(b)=o(1).$$

The analysis of these last two equations is identical. Let us consider (5.9′). Putting

$$\frac{\mu(\lambda)}{\{\mu^2(\lambda)+\nu^2(\lambda)\}^{1/2}}=\sin\omega(\lambda),\qquad \frac{\nu(\lambda)}{\{\mu^2(\lambda)+\nu^2(\lambda)\}^{1/2}}=\cos\omega(\lambda).$$

we can rewrite it in the form

$$(5.10)\qquad \sin\{\xi(b,\lambda)+\omega(\lambda)\}=o(1).$$

Let λ_1 and λ_2 be two successive positive roots of the equation $\varphi(b,\lambda)=0$ (or what is the same, of the equation (5.10)).

Using the asymptotic formula (5.2) and arguing in the same way as in the analysis of the equation (3.18), one can show that

$$\xi(b,\lambda_2)-\xi(b,\lambda_1)=\pi+o(1)\quad\left(\xi(x,\lambda)=\int_0^x\{\lambda-q(t)\}^{1/2}dt\right).$$

Therefore

$$
(5.11)\qquad \int_0^b \{\sqrt{\lambda_2 - q(t)} - \sqrt{\lambda_1 - q(t)}\}\, dt = \int_0^b \frac{\lambda_2 - \lambda_1}{\sqrt{\lambda_2 - q(t)} + \sqrt{\lambda_1 - q(t)}}\, dt = \pi + o(1).
$$

Further

$$
\int_0^b \frac{dt}{2\sqrt{\lambda_1 - q(\lambda)}} - \int_0^b \frac{dt}{\sqrt{\lambda_2 - q(t)} + \sqrt{\lambda_1 - q(t)}}
$$

$$
= (\lambda_2 - \lambda_1) \int_0^b \frac{dt}{2\sqrt{\lambda_1 - q(t)}\,[\sqrt{\lambda_2 - q(t)} + \sqrt{\lambda_1 - q(t)}]^2}
$$

$$
\leqslant \frac{\lambda_2 - \lambda_1}{2\sqrt{\lambda_1}} \int_0^b \frac{dt}{-q(t)} = o\left(\int_0^b \frac{dt}{\sqrt{\lambda_1 - q(t)}} \right).
$$

Hence it follows from (5.11) that

$$
(5.12)\qquad \lambda_0 - \lambda_1 = \frac{\pi}{\displaystyle\int_0^b \frac{dt}{2\sqrt{\lambda_1 - q(t)}}} + o\left\{ \left(\int_0^b \frac{dt}{\sqrt{\lambda_1 - q(t)}} \right)^{-1} \right\}.
$$

We will now show that for $b \to \infty$

$$
(5.13)\qquad \int_0^b \frac{\begin{matrix}\sin\\ \cos\end{matrix}\{2\xi(t, \lambda)\}}{\sqrt{\lambda - q(t)}}\, dt = O(1).
$$

Indeed, let us consider the integral with the sine (the integral with the cosine can be similarly estimated). Since

$$
\xi(t, \lambda) = \int_0^t \sqrt{\lambda - q(u)}\, du,
$$

we have

$$
\int_0^b \frac{\sin 2\xi(t, \lambda)}{\sqrt{\lambda - q(t)}}\, dt = \int_0^b \frac{\sin 2\xi(t, \lambda)}{\lambda - q(t)}\, d\xi(t, \lambda) = \int_0^{\xi(b, \lambda)} \frac{\sin 2\eta}{\lambda - q(t)}\, d\eta
$$

$$
(5.14)\qquad = \left\{ \int_0^{\frac{\pi}{2}} + \int_{\frac{\pi}{2}}^{\pi} + \cdots \right\} \frac{\sin 2\eta}{\lambda - q(t)}\, d\eta \qquad (\eta = \xi(t, \lambda)).
$$

If η increases, then t increases, and consequently $[\lambda - q(t)]^{-1}$ decreases.

Therefore

$$\int_{k\pi/2}^{(k+1)\pi/2} \frac{\sin 2\eta}{\lambda - q(t)}\, d\eta = \frac{1}{\lambda - q\left(k\frac{\pi}{2} + \theta\frac{\pi}{2}\right)} \int_{k\pi/2}^{(k+1)\pi/2} |\sin 2\eta|\, d\eta$$

$$= \frac{1}{\lambda - q\left(k\frac{\pi}{2} + \theta\frac{\pi}{2}\right)} \qquad (0 < \theta < 1).$$

As the number k increases, the last expression decreases monotonically, and the estimate (5.13) follows from the expansion (5.14), which is a sum with alternating signs.

From the estimates (5.13), the asymptotic formula (5.2) and the divergence of the integral (5.1) it follows that

$$\left(\int_0^b \frac{dt}{\sqrt{\lambda - q(\lambda)}}\right)^{-1} \int_0^b \varphi^2(t, \lambda)\, dt = \mu^2(\lambda) + \nu^2(\lambda) + o(1).$$

Hence we conclude from (5.12) that for $\lambda > 0$, $\Delta > 0$

$$\rho_b(\lambda + \Delta) - \rho_b(\lambda) = \sum_{\lambda < \lambda_{n,b} \leqq \lambda + \Delta} \frac{1}{\int_0^b \varphi^2(t, \lambda_{n,b})\, dt}$$

$$= \sum_{\lambda < \lambda_{n,b} \leqq \lambda + \Delta} \frac{\lambda_{n+1,b} - \lambda_{n,b}}{(\lambda_{n+1,b} - \lambda_{n,b}) \int_0^b \varphi^2(t, \lambda_{n,b})\, dt}$$

$$= \frac{1}{\pi} \sum_{\lambda < \lambda_{n,b} \leqq \lambda + \Delta} \left\{ \frac{1}{\mu^2(\lambda_{n,b}) + \nu^2(\lambda_{n,b})} + o(1) \right\} (\lambda_{n+1,b} - \lambda_{n,b})$$

$$\to \frac{1}{\pi} \int_\lambda^{\lambda+\Delta} \frac{d\lambda}{\mu^2(\lambda) + \nu^2(\lambda)},$$

i.e. for $\lambda \geqq 0$ the spectrum is continuous.

If $\lambda < 0$, then choosing X according to the condition that $\lambda - q(t) > 0$ for $t > X$, and replacing $\int_0^b [\lambda - q(t)]^{1/2} dt$ by $\int_X^b [\lambda - q(t)]^{1/2} dt$ everywhere in the asymptotic formulas, we arrive at the previous result. This proves the theorem.

THEOREM 5.2. *If all the hypotheses of Theorem* 5.1 *are satisfied, but the integral* (5.1) *converges, then the spectrum is discrete and has a unique limit point at infinity.*

PROOF. It follows from the convergence of the integral (5.1) and the definition of the function $\xi(x, \lambda)$ that as $x \to \infty$

$$\xi(x,\lambda)-\xi(x,0)=\int_0^x[\{\lambda-q(t)\}^{1/2}-\{-q(t)\}^{1/2}]\,dt$$

$$=\int_0^x\frac{\lambda\,dt}{\{\lambda-q(t)\}^{1/2}+\{-q(t)\}^{1/2}}\to\int_0^\infty\frac{\lambda\,dt}{\{\lambda-q(t)\}^{1/2}+\{-q(t)\}^{1/2}}<\infty.$$

Therefore the imaginary part of the function $\xi(x,\lambda)$ is uniformly bounded in every bounded portion of the λ-plane, in particular in every finite interval of the real axis.

It follows from (5.2) that the integral

$$\int_0^\infty \varphi^2(t,\lambda)\,dt \tag{5.15}$$

is uniformly bounded in every finite interval of variation of the real variable λ.

Let us assume that in some finite interval $(\lambda,\lambda+\Delta)$ the spectrum consists of an infinite set of points. Obviously in this case there must exist an increasing unbounded sequence $\{b_k\}$ of positive numbers such that the number of eigenvalues λ_{n,b_k} in the interval $(\lambda,\lambda+\Delta)$ grows unboundedly with k. Therefore since the integral (5.15) is bounded, the sum

$$\sum_{\lambda<\lambda_{n,b_k}\leqslant\lambda+\Delta}\frac{1}{\int_0^{b_k}\varphi^2(t,\lambda_{n,b_k})\,dt}$$

would grow unboundedly as $k\to\infty$, which contradicts Lemma 1.1 of Chapter 2.

§6. Investigation of the spectrum of a Dirac system for the case of summable coefficients

We will consider the problem

$$y_1'(x)-\{\lambda+q_1(x)\}y_2=0, \tag{6.1}$$

$$y_2'(x)+\{\lambda+q_2(x)\}y_1=0, \tag{6.2}$$

$$y_1(0)\cos\alpha+y_2(0)\sin\alpha=0 \tag{6.3}$$

on the halfline $[0,\infty)$. We will assume that the coefficients $q_1(x)$ and $q_2(x)$ belong to $\mathscr{L}(0,\infty)$.

We will denote by

$$\varphi(x,\lambda)=\begin{pmatrix}\varphi_1(x,\lambda)\\ \varphi_2(x,\lambda)\end{pmatrix}$$

the solution of the system (6.1) – (6.2) satisfying the initial conditions

$$\varphi_1(0,\lambda) = -\sin\alpha, \qquad \varphi_2(0,\lambda) = \cos\alpha. \tag{6.4}$$

Let us rewrite the system (6.1) – (6.2) in the following form:

$$y_1'(x) - \lambda y_2 = q_1(x)\, y_2, \tag{6.1'}$$

$$y_2'(x) + \lambda y_1 = -q_2(x) y_1. \tag{6.2'}$$

Regarding the right sides of (6.1′) and (6.2′) as known and applying the method of variation of constants, we find that

$$\begin{aligned}\varphi_1(x,\lambda) = \sin(\lambda x - \alpha) &+ \int_0^x \varphi_2(s,\lambda)\, q_1(s) \cos\lambda(x-s)\, ds \\ &- \int_0^x \varphi_1(s,\lambda)\, q_2(s) \sin\lambda(x-s)\, ds,\end{aligned} \tag{6.5}$$

$$\begin{aligned}\varphi_2(x,\lambda) = \cos(\lambda x - \alpha) &- \int_0^x \varphi_2(s,\lambda)\, q_1(s) \sin\lambda(x-s)\, ds \\ &- \int_0^x \varphi_1(s,\lambda)\, q_2(s) \cos\lambda(x-s)\, ds.\end{aligned} \tag{6.6}$$

Let $\lambda = \sigma + i\tau$ $(\tau > 0)$ and put

$$\varphi_1(x,\lambda) = h_1(x) e^{\tau x}, \qquad \varphi_2(x,\lambda) = h_2(x) e^{\tau x}. \tag{6.7}$$

Then it follows from (6.5) and (6.6) that

$$\begin{aligned}h_1(x) = e^{-\tau x} \sin(\lambda x - \alpha) &+ \int_0^x h_2(s) q_1(s) e^{-\tau(x-s)} \cos\lambda(x-s)\, ds \\ &- \int_0^x h_1(s)\, q_2(s)\, e^{-\tau(x-s)} \sin\lambda(x-s)\, ds,\end{aligned} \tag{6.5'}$$

$$\begin{aligned}h_2(x) = e^{-\tau x} \cos(\lambda x - \alpha) &- \int_0^x h_2(s)\, q_1(s)\, e^{-\tau(x-s)} \sin\lambda(x-s)\, ds \\ &- \int_0^x h_1(s)\, q_2(s)\, e^{-\tau(x-s)} \cos\lambda(x-s)\, ds.\end{aligned} \tag{6.6'}$$

Since

$$|\sin(\lambda x - \alpha)| \leqq e^{\tau x}, \qquad |\cos(\lambda x - \alpha)| \leqq e^{\tau x},$$

taking the absolute values of both sides of (6.5′) and (6.6′) we obtain

$$|h_1(x)| \leqslant 1 + \int_0^x \{|h_1(x)|\,|q_2(x)| + |h_2(x)|\,|q_1(s)|\}\,ds,$$

$$|h_2(x)| \leqslant 1 + \int_0^x \{|h_1(s)| \cdot |q_2(s)| + |h_2(s)| \cdot |q_1(s)|\}\,ds.$$

Therefore Lemma 3.1 is applicable, and we obtain

$$|h_1(x)|,\ |h_2(x)| \leqslant \exp\left\{\int_0^x [\,|q_1(s)| + |q_2(s)|\,]\,ds\right\}.$$

But since by hypothesis $q_1(x), q_2(x) \in \mathcal{L}(0, \infty)$, it follows from the last bounds that the vector-function

$$h(x) = \begin{pmatrix} h_1(x) \\ h_2(x) \end{pmatrix}$$

is bounded for $0 \leqq x < \infty$, $|\lambda| \geqq \rho > 0$ and $\tau \geqq 0$. Consequently, by virtue of (6.7) we have for large x

$$\varphi_1(x, \lambda) = O(e^{\tau x}), \qquad \varphi_2(x, \lambda) = O(e^{\tau x}). \tag{6.8}$$

Let us first consider real values of λ. Then by (6.8) the functions $\varphi_1(x,\lambda)$ and $\varphi_2(x,\lambda)$ are bounded. Therefore we obtain from (6.5) and (6.6) that for $x \to \infty$

$$\begin{aligned} \varphi_1(x, \lambda) &= \sin(\lambda x - \alpha) + \int_0^\infty \{\varphi_2(s, \lambda)\, q_1(s) \cos\lambda(x - s) \\ &\qquad - \varphi_1(s, \lambda)\, q_2(s) \sin\lambda(x - s)\}\,ds \\ &\qquad + O\left\{\int_x^\infty [\,|q_1(s)| + |q_2(s)|\,]\,ds\right\} \\ &= \mu(\lambda)\cos\lambda x + \nu(\lambda)\sin\lambda x + o(1), \end{aligned} \tag{6.9}$$

$$\begin{aligned} \varphi_2(x, \lambda) &= \cos(\lambda x - \alpha) - \int_0^\infty \{\varphi_2(s, \lambda)\, q_1(s) \sin\lambda(x - s) \\ &\qquad + \varphi_1(s, \lambda)\, q_2(s) \cos\lambda(x - s)\}\,ds \\ &\qquad + O\left\{\int_x^\infty [\,|q_1(s)| + |q_2(s)|\,]\,ds\right\} \\ &= \nu(\lambda)\cos\lambda x - \mu(\lambda)\sin\lambda x + o(1), \end{aligned} \tag{6.10}$$

where

$$(6.11)\quad \mu(\lambda) = -\sin\alpha + \int_0^\infty \{\varphi_2(s,\lambda)\, q_1(s)\cos\lambda s + \varphi_1(s,\lambda)\, q_2(s)\sin\lambda s\}\, ds,$$

$$(6.12\quad \nu(\lambda) = \cos\alpha + \int_0^\infty \{\varphi_2(s,\lambda)\, q_1(s)\sin\lambda s - \varphi_1(s,\lambda)\, q_2(s)\cos\lambda s\}\, ds.$$

Since the integrals in (6.11) and (6.12) converge uniformly with respect to λ, the functions $\mu(\lambda)$ and $\nu(\lambda)$ are continuous functions of λ.

Similarly, if

$$\theta(x,\lambda) = \begin{pmatrix}\theta_1(x,\lambda)\\ \theta_2(x,\lambda)\end{pmatrix}$$

is the solution of the system (6.1) – (6.2) satisfying the conditions

$$(6.4')\qquad \theta_1(0,\lambda) = -\cos\alpha, \qquad \theta_2(0,\lambda) = -\sin\alpha,$$

then

$$(6.13)\qquad \theta_1(x,\lambda) = \xi(\lambda)\cos\lambda x + \eta(\lambda)\sin\lambda x + o(1),$$

$$(6.14)\qquad \theta_2(x,\lambda) = \eta(\lambda)\cos\lambda x - \xi(\lambda)\sin\lambda x + o(1),$$

where

$$(6.13')\quad \xi(\lambda) = -\cos\alpha + \int_0^\infty \{\theta_2(s,\lambda)\, q_1(s)\cos\lambda s + \theta_1(s,\lambda)\, q_2(s)\sin\lambda s\}\, ds,$$

$$(6.14')\quad \eta(\lambda) = -\sin\alpha + \int_0^\infty \{\theta_2(s,\lambda)\, q_1(s)\sin\lambda s - \theta_1(s,\lambda)\, q_2(s)\cos\lambda s\}\, ds.$$

Then from (6.9), (6.10), (6.13) and (6.14) we have

$$(6.15)\qquad \begin{aligned} W\{\varphi,\theta\} &= \varphi_1(x,\lambda)\,\theta_2(x,\lambda) - \varphi_2(x,\lambda)\,\theta_1(x,\lambda)\\ &= \mu(\lambda)\,\eta(\lambda) - \nu(\lambda)\,\xi(\lambda) + o(1).\end{aligned}$$

Since by the initial conditions (6.4) and (6.4′)

$$W\{\varphi,\theta\}_{x=0} = \varphi_1(0,\lambda)\theta_2(0,\lambda) - \varphi_2(0,\lambda)\theta_1(0,\lambda) = 1,$$

it follows from (6.15) that for real λ

$$(6.16)\qquad \mu(\lambda)\eta(\lambda) - \nu(\lambda)\xi(\lambda) = 1.$$

It follows from (6.16) that $\mu(\lambda)$ and $\nu(\lambda)$ cannot both vanish for the same value of λ.

Let us now consider complex values of λ. For fixed positive τ we obtain from (6.5), for $x \to \infty$,

$$\varphi_1(x, \lambda) = -\frac{e^{-i\lambda x}}{2i}\cos\alpha - \frac{e^{-i\lambda x}}{2}\sin\alpha + O(e^{-\tau x})$$

$$+\frac{1}{2}\int_0^x e^{-i\lambda(x-s)}\{\varphi_2(s, \lambda)q_1(s) - i\varphi_1(s, \lambda)q_2(s)\}\,ds \tag{6.17}$$

$$+O\left\{\int_0^x e^{-\tau(x-s)}|\varphi_2(s, \lambda)q_1(s) - i\varphi_1(s, \lambda)q_2(s)|\,ds\right\}.$$

Further, since by virtue of (6.8) we have, for $x \to \infty$,

$$O\left\{\int_0^x e^{-\tau(x-s)}|\varphi_2(s, \lambda)q_1(s) - i\varphi_1(s, \lambda)q_2(s)|\,ds\right\}$$

$$= O\left\{\int_0^x e^{-\tau(x-2s)}[|q_1(s)| + |q_2(s)|]\,ds\right\}$$

$$= O\left\{e^{\tau(x-2\delta)}\int_0^{x-\delta}[|q_1(s)| + |q_2(s)|]\,ds\right\}$$

$$+ O\left\{e^{\tau x}\int_{x-\delta}^x [|q_1(s)| + |q_2(s)|]\,ds\right\} = o(e^{\tau x}),$$

and on the other hand

$$\int_x^\infty e^{-i\lambda(x-s)}\{\varphi_2(s, \lambda)q_1(s) - i\varphi_1(s, \lambda)q_2(s)\}\,ds$$

$$= O\left\{e^{\tau x}\int_x^\infty [|q_1(s)| + |q_2(s)|]\,ds\right\} = o(e^{\tau x}),$$

we have from (6.17) (for $x \to \infty$)

$$\varphi_1(x,\lambda) = e^{-i\lambda x}\{M_1(\lambda) + o(1)\}, \tag{6.18}$$

where

$$M_1(\lambda) = -\frac{\sin\alpha}{2} - \frac{\cos\alpha}{2i}$$

$$+\frac{1}{2}\int_0^\infty e^{i\lambda s}\{\varphi_2(s, \lambda)q_1(s) - iq_2(s)\varphi_1(s, \lambda)\}\,ds. \tag{6.19}$$

Similarly, from (6.6) and also from the analogous formulas for the solution $\theta(x,\lambda)$ we obtain

$$\varphi_2(x,\lambda) = e^{-i\lambda x}\{M_2(\lambda) + o(1)\}, \tag{6.20}$$

$$\theta_1(x,\lambda) = e^{-i\lambda x}\{N_1(\lambda) + o(1)\}, \tag{6.21}$$

$$\theta_2(x,\lambda) = e^{-i\lambda x}\{N_2(\lambda) + o(1)\}, \tag{6.22}$$

where

$$\begin{aligned} M_2(\lambda) = &-\frac{\sin\alpha}{2i} + \frac{\cos\alpha}{2} \\ &-\frac{1}{2}\int_0^\infty e^{i\lambda s}\{i\varphi_2(s,\ \lambda)\,q_1(s) + \varphi_1(s,\ \lambda)\,q_2(s)\}\,ds, \end{aligned} \tag{6.23}$$

$$\begin{aligned} N_1(\lambda) = &-\frac{\cos\alpha}{2} + \frac{\sin\alpha}{2i} \\ &+\frac{1}{2}\int_0^\infty e^{i\lambda s}\{\theta_2(s,\ \lambda)\,q_1(s) - i\theta_1(s,\ \lambda)\,q_2(s)\}\,ds, \end{aligned} \tag{6.24}$$

$$\begin{aligned} N_2(\lambda) = &-\frac{\sin\alpha}{2} - \frac{\cos\alpha}{2i} \\ &-\frac{1}{2}\int_0^\infty e^{i\lambda s}\{i\theta_2(s,\ \lambda)\,q_1(s) + \theta_1(s,\ \lambda)\,q_2(s)\}\,ds. \end{aligned} \tag{6.25}$$

THEOREM 6.1. *If the coefficients $q_1(x)$ and $q_2(x)$ belong to $\mathscr{L}(0,\infty)$, then the spectrum of the problem* (6.1) + (6.2) + (6.3) *is continuous and fills up the entire axis* $(-\infty,\infty)$.

PROOF. Let $\psi(x,\lambda) = \theta(x,\lambda) + m(\lambda)\varphi(x,\lambda)$ be that solution of the system (6.1) – (6.2) which belongs to $\mathscr{L}^2(0,\infty)$. Then using formulas (6.18), (6.20) – (6.22), we have

$$\psi_1(x,\lambda) = \theta_1(x,\lambda) + m(\lambda)\varphi_1(x,\lambda) = e^{-i\lambda x}\{N_1(\lambda) + m(\lambda)M_1(\lambda) + o(1)\},$$

$$\psi_2(x,\lambda) = \theta_2(x,\lambda) + m(\lambda)\varphi_2(x,\lambda) = e^{-i\lambda x}\{N_2(\lambda) + m(\lambda)M_2(\lambda) + o(1)\}.$$

Since for every nonreal λ there exists only one solution of the system (6.1) – (6.2) which belongs to $\mathscr{L}^2(0,\infty)$, it is obvious that $\varphi(x,\lambda)$ and $\theta(x,\lambda)$ do not belong to $\mathscr{L}^2(0,\infty)$. But since, by hypothesis, $\psi(x,\lambda) \in \mathscr{L}^2(0,\infty)$, we must have

$$m(\lambda) = -\frac{N_1(\lambda)}{M_1(\lambda)} = -\frac{N_2(\lambda)}{M_2(\lambda)}.$$

We further note that when λ tends to a real limit (i.e. $\tau \to 0$), it follows from (6.19), (6.23) – (6.25) and (6.11), (6.12), (6.13′) and (6.14′) that

$$M_1(\lambda) \to \frac{1}{2}\{\mu(\lambda) + i\nu(\lambda)\}, \quad N_1(\lambda) \to \frac{1}{2}\{\xi(\lambda) + i\eta(\lambda)\},$$

$$M_2(\lambda) \to \frac{1}{2}\{\nu(\lambda) - i\mu(\lambda)\}, \quad N_2(\lambda) \to \frac{1}{2}\{\eta(\lambda) - i\xi(\lambda)\}.$$

Consequently

$$\lim_{\tau \to 0} m(\lambda) = -\frac{\xi(\lambda) + i\eta(\lambda)}{\mu(\lambda) + i\nu(\lambda)}.$$

Therefore for real λ

$$\operatorname{Im}\{m(\lambda)\} = -\frac{1}{\mu^2(\lambda) + \nu^2(\lambda)}.$$

Since the functions $\mu(\lambda)$ and $\nu(\lambda)$ do not both vanish for one value of λ, it follows that $\operatorname{Im} m(\lambda)$ is a continuous function of λ on the entire axis $(-\infty, \infty)$. Then by virtue of (3.19), Chapter 3, we have

$$\rho(\lambda + \Delta) - \rho(\lambda) = \frac{1}{\pi} \int_{\lambda}^{\lambda+\Delta} \frac{d\lambda}{\mu^2(\lambda) + \nu^2(\lambda)},$$

which proves the theorem.

§7. Transformation of the basic system

In certain cases the system

(7.1) $$y_1' - \{\lambda + q_1(x)\} y_2 = 0,$$

(7.2) $$y_2' + \{\lambda + q_2(x)\} y_1 = 0$$

with coefficients $q_1(x)$ and $q_2(x)$ which increase unboundedly, can by means of an appropriate change of variables be transformed to a form which enables us to derive asymptotic formulas.

Let us assume that λ is a real number, and that the coefficients $q_1(x)$ and $q_2(x)$ are twice continuously differentiable. We put

(7.3) $$\alpha(x) = \int_0^x \{[\lambda + q_1(x)][\lambda + q_2(x)]\}^{1/2} dx,$$

(7.4) $$\begin{aligned} u(x) &= F(x, \lambda) y_2(x), \\ v(x) &= G(x, \lambda) y_2(x) - F^{-1}(x, \lambda) y_1(x), \end{aligned}$$

where

(7.5) $$F(x, \lambda) = \left\{\frac{\lambda + q_1(x)}{\lambda + q_2(x)}\right\}^{1/4}, \qquad G(x, \lambda) = \frac{F_x'(x, \lambda)}{\alpha'(x)}.$$

Then

$$\frac{du}{d\alpha}=\frac{du}{dx}\cdot\frac{dx}{d\alpha}=\{F'_x y_2+Fy'_2\}\cdot\frac{dx}{d\alpha}=\frac{F'_x}{\alpha'}y_2+\frac{F}{\alpha'}y'_2.$$

Hence

$$y'_2=\left\{\frac{du}{d\alpha}-\frac{F'_x}{\alpha'}y_2\right\}\frac{\alpha'}{F}.$$

Now inserting this expression for y'_2 into the equation (7.2), we obtain

$$\frac{du}{d\alpha}\cdot\frac{\alpha'}{F}-\frac{F'_x}{F}y_2+\{\lambda+q_2(x)\}y_1=0$$

or by virtue of (7.3) – (7.5)

$$\begin{aligned}\frac{du}{d\alpha}&=\frac{F'_x}{\alpha'}y_2-\frac{F}{\alpha'}\{\lambda+q_2(x)\}y_1\\&=G(x,\lambda)y_2(x)-F^{-1}(x,\lambda)y_1(x)=v(x).\end{aligned}\tag{7.6}$$

We further have

$$\frac{dv}{d\alpha}=\frac{dv}{dx}\cdot\frac{dx}{d\alpha}=\{G'_x y_2+Gy'_2+F^{-2}F'_x y_1-F^{-1}y'_1\}\frac{1}{\alpha'}$$

whence by virtue of (7.2) we find that

$$y'_1=-\frac{dv}{d\alpha}F\alpha'+G'_xFy^2-\left\{GF[\lambda+q_2(x)]-\frac{F'_x}{F}\right\}y_1.$$

Inserting the value of y'_1 from this equality in equation (7.1) and taking account of the first of the equalities (7.4), we obtain

$$\frac{dv}{d\alpha}=-\frac{\lambda+q_1(x)}{\alpha'F'}u+\frac{G'_x}{\alpha'F}u-\left\{\frac{G[\lambda+q_2(x)]}{\alpha'}-\frac{F'_x}{\alpha'F^2}\right\}y_1.\tag{7.7}$$

Since by (7.3) and (7.5) the coefficient of u in the first term on the right side of (7.7) equals -1, and the coefficient of y_1 is zero, equation (7.7) assumes the form

$$\frac{dv}{d\alpha}=-u+\frac{G'_x(x,\lambda)}{\alpha'(x)F(x,\lambda)}u.\tag{7.8}$$

Thus the system (7.1) – (7.2) is transformed to the form (cf. (7.6) and (7.8))

$$du/d\alpha=v,\tag{7.9}$$

$$dv/d\alpha=-u+R(x,\lambda)u,\tag{7.10}$$

where

$$R(x,\lambda)=\frac{G'(x,\lambda)}{\alpha'(x)F(x,\lambda)}\tag{7.11}$$

or, using (7.3) and (7.4)

$$R(x, \lambda) = \frac{q_1''(x)}{4[\lambda + q_1(x)]^2[\lambda + q_2(x)]} - \frac{q_2''}{4[\lambda + q_1(x)][\lambda + q_2(x)]^2}$$

$$(7.11') \quad -\frac{5}{16}\cdot\frac{q_1'^2(x)}{[\lambda + q_1(x)]^3[\lambda + q_2(x)]} - \frac{1}{8}\cdot\frac{q_1'(x)\,q_2'(x)}{[\lambda + q_1(x)]^2[\lambda + q_2(x)]}$$

$$+\frac{7}{16}\cdot\frac{q_2'^2(x)}{[\lambda + q_1(x)][\lambda + q_2(x)]}.$$

LEMMA 7.1. *If the vector-function*

$$\eta(x) = \begin{pmatrix}\eta_1(x)\\ \eta_2(x)\end{pmatrix}$$

is a solution of the system (7.9)–(7.10), *i.e. if*

$$(7.9') \qquad d\eta_1/d\alpha = \eta_2,$$

$$(7.10') \qquad d\eta_2/d\alpha = -\eta_1 + R(x, \lambda)\eta_1.$$

then it satisfies the system of integral equations

$$(7.12) \quad \eta_1(x, \lambda) = \eta_1(0)\cos\alpha(x) + \eta_2(0)\sin\alpha(x) - \int_0^x \eta_1(s, \lambda)P(s, \lambda)\sin[\alpha(s) - \alpha(x)]\,ds,$$

$$(7.13) \quad \eta_2(x, \lambda) = -\eta_1(0)\sin\alpha(x) + \eta_2(0)\cos\alpha(x) + \int_0^x \eta_1(s, \lambda)P(s, \lambda)\cos[\alpha(s) - \alpha(x)]\,ds,$$

where

$$(7.14) \qquad P(x, \lambda) = R(x, \lambda)\cdot\alpha'(x) = \frac{G'(x)}{F(x, \lambda)}.$$

PROOF. In fact, by equations (7.10′) and (7.14) we have

$$\int_0^x \eta_1(s, \lambda)P(s, \lambda)\sin[\alpha(s) - \alpha(x)]\,ds$$

$$= \int_0^x \left\{\frac{d\eta_2}{d\alpha}\alpha'(s) + \eta_1(s, \lambda)\alpha'(s)\right\}\sin[\alpha(s) - \alpha(x)]\,ds$$

$$(7.15) \quad = \int_0^x \frac{d\eta_2}{d\alpha}\alpha'(s)\sin[\alpha(s) - \alpha(x)]\,ds + \int_0^x \eta_1(s, \lambda)\alpha'(s)\sin[\alpha(s) - \alpha(x)]\,ds.$$

Now integrating the first integral in the right side of (7.15) by parts, then using equation (7.9′), and then integrating by parts once more, we find that

$$\int_0^x \frac{d\eta_2}{d\alpha} \sin[\alpha(s) - \alpha(x)]\, d\alpha(s)$$

$$= \eta_2(0) \sin \alpha(x) - \int_0^x \frac{d\eta_1}{d\alpha} \cos[\alpha(s) - \alpha(x)]\, d\alpha(s)$$

$$= \eta_2(0) \sin \alpha(x) - \eta_1(x, \lambda) + \eta_1(0) \cos \alpha(x)$$

$$- \int_0^x \eta_1(x, \lambda) \sin[\alpha(s) - \alpha(x)]\, d\alpha(s).$$

The validity of (7.12) follows from (7.15) by virtue of the preceding equality. The proof of (7.13) is similar.

LEMMA 7.2. *Suppose that* $q_1(x)$ *and* $q_2(x)$ *are monotonic functions, and that one of the following conditions is satisfied*:

1. $q_1(x) \geqq 0,\ q_2(x) \leqq 0,\ q_1'(x) > 0,\ q_2'(x) < 0,\ q_1(x) \to \infty,\ q_2(x) \to -\infty$;
2. $q_1(x) \leqq 0,\ q_2(x) \geqq 0,\ q_1'(x) < 0,\ q_2'(x) > 0,\ q_1(x) \to -\infty,\ q_2(x) \to \infty$;
3. $q_1(x), q_2(x) \geqq 0,\ q_1'(x), q_2'(x) > 0,\ q_1(x),\ q_2(x) \to \infty$,
4. $q_1(x), q_2(x) \leqq 0,\ q_1'(x), q_2'(x) < 0,\ q_1(x), q_2(x) \to -\infty$.

Suppose also that the following conditions are satisfied:

5.

$$q_1'(x) = O\{|q_1(x)|^c\}, \qquad q_2'(x) = O\{|q_2(x)|^c\}, \tag{7.16}$$

where $0 < c < 3/2$.

6. $q_1''(x)$ *and* $q_2''(x)$ *preserve their signs for large (in absolute value) values of* x.

Then the integral

$$\int_0^\infty |P(x, \lambda)|\, dx \tag{7.17}$$

converges uniformly with respect to λ *in every region* $|\lambda + q_1(x)|,\ |\lambda + q_2(x)| \geqq \delta > 0$ *for* $0 \leqq x < \infty$.

PROOF. By virtue of (7.11′), (7.14) and (7.3)

$$\begin{aligned} P(x, \lambda) = & \frac{q_1''(x)}{4[\lambda + q_1(x)]^{3/2}[\lambda + q_2(x)]^{1/2}} \\ & - \frac{q_2''(x)}{4[\lambda + q_1(x)]^{1/2}[\lambda + q_2(x)]^{3/2}} - \frac{5}{16} \cdot \frac{[q_1'(x)]^2}{[\lambda + q_1(x)]^{5/2}[\lambda + q_2(x)]^{1/2}} \\ & - \frac{1}{8} \cdot \frac{q_1'(x)\, q_2'(x)}{[\lambda + q_1(x)]^{3/2}[\lambda + q_2(x)]^{3/2}} + \frac{7}{16} \cdot \frac{[q_2'(x)]^2}{[\lambda + q_1(x)]^{1/2}[\lambda + q_2(x)]^{5/2}}. \end{aligned} \tag{7.18}$$

Obviously to prove the lemma it is sufficient to show that each of the terms is integrable on the semiaxis $[0, \infty)$.

Suppose that the conditions 1, 5 and 6 are satisfied (the proof for the remaining cases can be carried out in a completely analogous way). Then by (7.16) we have

$$\int_{X_0}^{X} \frac{\{q_1'(x)\}^2}{\{q_1(x)\}^{5/2}\{q_2(x)\}^{1/2}}\,dx = O\left\{\int_{X_0}^{X} \frac{q_1'(x)}{\{q_1(x)\}^{5/2-c}\{q_2(x)\}^{1/2}}\,dx\right\}.$$

Since by hypothesis $q_2(x)$ is a monotonic function, applying the theorem of the mean to the last integral we obtain $(X_0 < \xi < X)$

$$\int_{X_0}^{X} \frac{\{q_1'(x)\}^2}{\{q_1(x)\}^{5/2}\{q_2(x)\}^{1/2}}\,dx = O\left[\{q_2(\xi)\}^{-1/2}\int_{X_0}^{X} \frac{dq_1(x)}{\{q_1(x)\}^{5/2-c}}\right] \tag{7.19}$$
$$= O\left[\{q_2(\xi)\}^{-1/2}\{q_1(x)\}^{c-3/2}\right]_{X_0}^{X} = O(1).$$

Similarly we obtain the estimate

$$\int_{X_0}^{X} \frac{\{q_2'(x)\}^2}{\{q_1(x)\}^{1/2}\{q_2(x)\}^{5/2}}\,dx = O(1). \tag{7.20}$$

Further, again using the theorem of the mean and then integrating by parts, we find that $(X_0 < \xi < X)$

$$\int_{X_0}^{X} \frac{q_1''(x)}{\{q_1(x)\}^{3/2}\{q_2(x)\}^{1/2}}\,dx = \{q_2(\xi)\}^{-1/2}\int_{X_0}^{X} \frac{q_1''(x)}{\{q_1(x)\}^{3/2}}\,dx.$$
$$= \{q_2(\xi)\}^{-1/2}\left[\frac{q_1'(x)}{\{q_1(x)\}^{3/2}}\right]_{X_0}^{X} + \frac{3}{2}\int_{X_0}^{X} \frac{\{q_1'(x)\}^2}{\{q_1(x)\}^{5/2}}\,dx.$$

Therefore from the estimates (7.16) and (7.19) we conclude that

$$\int_{X_0}^{X} \frac{q_1''(x)}{\{q_1(x)\}^{3/2}\{q_2(x)\}^{1/2}}\,dx = O(1). \tag{7.21}$$

We can similarly establish that

$$\int_{X_0}^{X} \frac{q_2''(x)}{\{q_1(x)\}^{1/2}\{q_2(x)\}^{5/2}}\,dx = O(1). \tag{7.22}$$

Finally, using the estimate (7.16) and the theorem of the mean, we obtain

$$\int_{X_0}^{X} \frac{q_1'(x)\,q_2'(x)}{\{q_1(x)\}^{3/2}\{q_2(x)\}^{3/2}}\,dx = O\left[\int_{X_0}^{X} \frac{q_1'(x)}{\{q_1(x)\}^{3/2}\{q_2(x)\}^{3/2-c}}\,dx\right] \tag{7.23}$$

$$= O\left[\{q_2(\xi)\}^{c-3/2}\{q_1(x)\}^{-1/2}\right]_{x=X_0}^{X} = O(1).$$

The convergence of the integral (7.17) follows from the estimates (7.19) – (7.23) by virtue of (7.18). This proves the lemma.

§8. The case of a pure point spectrum

Our goal in this section is to establish various sufficient conditions for the discreteness of the spectrum of the problem

$$y_1' - \{\lambda + q_1(x)\} y_2 = 0, \tag{8.1}$$

$$y_2' + \{\lambda + q_2(x)\} y_1 = 0, \tag{8.2}$$

$$y_1(0)\cos\alpha + y_2(0)\sin\alpha = 0. \tag{8.3}$$

Theorem 8.1. *If the coefficients* $q_1(x)$ *and* $q_2(x)$ *satisfy the conditions* $1+5+6$ *or* $2+5+6$ *of Lemma* 7.2, *then the problem* (8.1) + (8.2) + (8.3) *has a pure point spectrum.*

Proof. For $\operatorname{Im}\lambda > 0$ we will assume that $0 < \arg\lambda < \pi$, and for those cases in which $q_1(x)$ or $q_2(x) \to -\infty$ we will assume that $\arg[\lambda + q_1(x)]$ and $\arg[\lambda + q_2(x)]$ vary from $\arg\lambda$ to π. Therefore

$$\frac{1}{2}\arg\lambda \leqslant \arg\{\lambda + q_1(x)\}^{1/2}, \quad \arg\{\lambda + q_2(x)\}^{1/2} \leqslant \frac{\pi}{2}.$$

It follows that the imaginary part of the function

$$\alpha(x) = \int_0^x \{[\lambda + q_1(s)][\lambda + q_2(s)]\}^{1/2}\, ds$$

is positive. Further, if condition 1 or 2 of Lemma 7.2 is satisfied, if λ is bounded and $x \to \infty$, then

$$\alpha(x) \sim i\int_0^x \{q_1(x)\}^{1/2}\{q_2(s)\}^{1/2}\, ds,$$

so that $e^{-i\alpha(x)} \to \infty$ as $x \to \infty$. Let $\xi(x,\lambda) = \eta(x,\lambda)e^{i\alpha(x)}$. Then it follows from (7.12) and (7.13) that

$$\xi_1(x,\lambda) = \eta_1(0)\, e^{i\alpha(x)}\cos\alpha + \eta_2(0)\, e^{i\alpha(x)}\sin\alpha(x)$$
$$- \int_0^x \xi_1(s,\lambda)\, e^{i\{\alpha(x)-\alpha(s)\}} P(s,\lambda)\sin\{\alpha(s) - \alpha(x)\}\, ds,$$

$$\xi_2(x,\lambda) = -\eta_1(0)\, e^{i\alpha(x)}\sin\alpha(x) + \eta_2(0)\, e^{i\alpha(x)}\cos\alpha(x)$$
$$+ \int_0^x \xi_1(s,\lambda)\, e^{i\{\alpha(x)-\alpha(s)\}} P(s,\lambda)\cos\{\alpha(s) - \alpha(x)\}\, ds.$$

Hence, since

$$|e^{i\alpha(x)}\sin\alpha(x)| \leqq 1, \qquad |e^{i\alpha(x)}\cos\alpha(x)| \leqq 1,$$

it follows that

$$|\xi_1(x,\lambda)| \leqslant |\eta_1(0)| + |\eta_2(0)| + \int_0^x |\xi_1(s,\lambda)|\,|P(s,\lambda)|\,ds,$$

$$|\xi_2(x,\lambda)| \leqslant |\eta_1(0)| + |\eta_2(0)| + \int_0^x |\xi_1(s,\lambda)|\,|P(s,\lambda)|\,ds,$$

and so by Lemma 3.1, which is applicable by virtue of Lemma 7.2,

$$|\xi_1(x,\lambda)|,\ |\xi_j(x,\lambda)| \leqslant \{|\eta_1(0)| + |\eta_2(0)|\} \exp\left\{\int_0^\infty |P(s,\lambda)|\,ds\right\}.$$

Consequently

$$\begin{aligned} &|\eta_1(x,\lambda)|,\ |\eta_2(x,\lambda)| \\ &\leqslant \{|\eta_1(0)| + |\eta_2(0)|\} \exp\left\{\int_0^\infty |P(s,\lambda)|\,ds\right\} |e^{-i\alpha(x)}| \qquad (8.4) \\ &= O(e^{-i\alpha(x)}) = O\{e^{\operatorname{Im}\alpha(x)}\}. \end{aligned}$$

Further, for fixed λ, $\operatorname{Im}\lambda > 0$, and for $x \to \infty$ it follows from (7.12) that

$$\begin{aligned} \eta_1(x,\lambda) &= \frac{\eta_1(0)}{2} e^{-i\alpha(x)} - \frac{\eta_2(0)}{2i} e^{-i\alpha(x)} + O(e^{-\operatorname{Im}\alpha(x)}) \\ &\quad + \frac{1}{2i}\int_0^x \eta_1(s,\lambda) P(s,\lambda) e^{i\{\alpha(s)-\alpha(x)\}} ds \qquad (8.5) \\ &\quad + O\left\{\int_0^x |\eta_1(s,\lambda)|\,|P(s,\lambda)|\, e^{-\operatorname{Im}[\alpha(x)-\alpha(s)]} ds\right\}. \end{aligned}$$

By (8.4) we have

$$\begin{aligned} &O\left\{\int_0^x |\eta_1(s,\lambda)|\,|P(s,\lambda)|\, e^{-\operatorname{Im}[\alpha(x)-\alpha(s)]} ds\right\} \\ &= O\left\{\int_0^x |P(s,\lambda)|\, e^{-\operatorname{Im}[\alpha(x)-2\alpha(s)]} ds\right\} \qquad (8.6) \\ &= O\left\{e^{-\operatorname{Im}[\alpha(x)-2\alpha(x-\delta)]}\int_0^{x-\delta} |P(s,\lambda)|\,ds\right\} + O\left\{e^{\operatorname{Im}\alpha(x)}\int_{x-\delta}^x |P(s,\lambda)|\,ds\right\}. \end{aligned}$$

Since

$$\alpha(x-\delta)=\alpha(x)-\int_{x-\delta}^{x}\{[\lambda+q_1(s)][\lambda+q_2(s)]\}^{1/2}\,ds,$$

it follows that

$$-\operatorname{Im}[\alpha(x)-2\alpha(x-\delta)]$$
$$=\operatorname{Im}\alpha(x)-2\operatorname{Im}\int_{x-\delta}^{x}\{[\lambda+q_1(s)][\lambda+q_2(s)]\}^{1/2}\,ds.$$

Therefore

$$\exp\{-\operatorname{Im}[\alpha(x)-2\alpha(x-\delta)]\}=o\{e^{\operatorname{Im}\alpha(x)}\}\quad(x\to\infty).$$

Consequently, using Lemma 7.2 again, for $x\to\infty$ we obtain from (8.6) the estimate

$$O\left\{\int_0^x|\eta_1(s,\lambda)||P(s,\lambda)|e^{-\operatorname{Im}[\alpha(x)-\alpha(s)]}ds\right\}=o\{e^{\operatorname{Im}\alpha(x)}\}. \tag{8.6'}$$

On the other hand, by Lemma 7.2 we have for $x\to\infty$

$$\begin{aligned}&\int_0^\infty\eta_1(s,\lambda)P(s,\lambda)e^{-i[\alpha(x)-\alpha(s)]}ds\\ &\qquad=O\left\{e^{\operatorname{Im}\alpha(x)}\int_x^\infty|P(s,\lambda)|\,ds\right\}=o\{e^{\operatorname{Im}\alpha(x)}\}.\end{aligned} \tag{8.7}$$

Thanks to the estimates (8.6′) and (8.7), the equality (8.5) can be rewritten in the form

$$\eta_1(x,\lambda)=e^{-i\alpha(x)}\{M^+(\lambda)+o(1)\}, \tag{8.8}$$

where

$$M^+(\lambda)=\frac{\eta_1(0)}{2}-\frac{\eta_2(0)}{2i}+\frac{1}{2i}\int_0^\infty\eta_1(s,\lambda)P(s,\lambda)e^{i\alpha(s)}ds. \tag{8.9}$$

Similarly we obtain from (7.13)

$$\eta_2(x,\lambda)=e^{-i\alpha(x)}\{N^+(\lambda)+o(1)\}, \tag{8.10}$$

where

$$N^+(\lambda)=\frac{\eta_1(0)}{2i}+\frac{\eta_2(0)}{2}+\frac{1}{2}\int_0^\infty\eta_1(s,\lambda)P(s,\lambda)e^{i\alpha(s)}ds. \tag{8.11}$$

Now if $\operatorname{Im}\alpha(x) \to -\infty$ as $x \to \infty$, and consequently $e^{i\alpha(x)} \to \infty$, then proceeding in the same way as above, we can give (7.12) – (7.13) the form

$$\eta_1(x,\lambda) = e^{i\alpha(x)}\{M^-(\lambda) + o(1)\}, \tag{8.12}$$

$$\eta_2(x,\lambda) = e^{i\alpha(x)}\{N^-(\lambda) + o(1)\}, \tag{8.13}$$

where

$$M^-(\lambda) = \frac{\eta_1(0)}{2} + \frac{\eta_2(0)}{2i} - \frac{1}{2i}\int_0^\infty \eta_1(s,\lambda)\,P(s,\lambda)\,e^{-i\alpha(s)}ds, \tag{8.14}$$

$$N^-(\lambda) = \frac{\eta_2(0)}{2} - \frac{\eta_1(0)}{2i} + \frac{1}{2}\int_0^\infty \eta_1(s,\lambda)\,P(s,\lambda)\,e^{-i\alpha(s)}ds. \tag{8.15}$$

Let us denote by

$$\varphi(x,\lambda) = \begin{pmatrix}\varphi_1(x,\lambda)\\ \varphi_2(x,\lambda)\end{pmatrix}$$

the solution of the system (8.1) – (8.2) satisfying the conditions

$$\varphi_1(0,\lambda) = -\sin\alpha, \qquad \varphi_2(0,\lambda) = \cos\alpha, \tag{8.16}$$

and by

$$\theta(x,\lambda) = \begin{pmatrix}\theta_1(x,\lambda)\\ \theta_2(x,\lambda)\end{pmatrix}$$

the solution satisfying the conditions

$$\theta_1(0,\lambda) = -\cos\alpha, \qquad \theta_2(0,\lambda) = -\sin\alpha. \tag{8.17}$$

Further, since it follows from (7.4) that

$$\begin{aligned}\eta_1(0) &= F(0)y_2(0),\\ \eta_2(0) &= G(0)y_2(0) - F^{-1}(0)y_1(0),\end{aligned} \tag{8.18}$$

then by (8.16) we have

$$\begin{aligned}\eta_1(0) &= F(0)\cos\alpha,\\ \eta_2(0) &= F^{-1}(0)\sin\alpha + G(0)\cos\alpha.\end{aligned} \tag{8.18$'$}$$

On the other hand, thanks to (7.4)

$$\begin{aligned}\varphi_1(x,\lambda) &= G(x,\lambda)\eta_1(x,\lambda) - F(x,\lambda)\eta_2(x,\lambda),\\ \varphi_2(x,\lambda) &= F^{-1}(x,\lambda)\eta_1(x,\lambda).\end{aligned} \tag{8.19}$$

Therefore, using (8.8) and (8.10), we obtain for the solution $\varphi(x,\lambda)$ the following asymptotic formula for $x\to\infty$;[7)]

$$\varphi_1^+(x,\lambda)=e^{i\alpha(x)}\{G(x,\lambda)M_\varphi^+(\lambda)-F(x,\lambda)N_\varphi^+(\lambda)+o(1)\}, \tag{8.20}$$

$$\varphi_2^+(x,\lambda)=e^{-i\alpha(x)}\{F^{-1}(x,\lambda)M_\varphi^+(\lambda)+o(1)\}, \tag{8.21}$$

where by virtue of (8.18′) and (8.19), using (8.9) and (8.11),

$$M_\varphi^+(\lambda)=\frac{1}{2i}\left\{-F^{-1}(0)\sin\alpha+[iF(0)-G(0)]\cos\alpha+\int_0^\infty \varphi_2^+(s,\lambda)P(s,\lambda)F(s,\lambda)e^{i\alpha(s)}ds\right\}, \tag{8.22}$$

$$N_\varphi^+(\lambda)=\frac{1}{2}\left\{F^{-1}(0)\sin\alpha-[iF(0)-G(0)]\cos\alpha+\int_0^\infty \varphi_2^+(s,\lambda)P(s,\lambda)F(s,\lambda)e^{i\alpha(s)}ds\right\}. \tag{8.23}$$

We can similarly obtain the following asymptotic formula for the solution $\theta(x,\lambda)$ as $x\to\infty$:

$$\theta_1^+(x,\lambda)=e^{-i\alpha(x)}\{G(x,\lambda)M_\theta^+(\lambda)-F(x,\lambda)N_\theta^+(\lambda)+o(1)\}, \tag{8.24}$$

$$\theta_2^+(x,\lambda)=e^{-i\alpha(x)}\{F^{-1}(x,\lambda)M_\theta^+(\lambda)+o(1)\}, \tag{8.25}$$

where $M_\theta^+(\lambda)$ and $N_\theta^+(\lambda)$ are obtained respectively from $M_\varphi^+(\lambda)$ and $N_\varphi^+(\lambda)$ by replacing $\sin\alpha$, $\cos\alpha$ and $\varphi_2^+(x,\lambda)$ by $\cos\alpha$, $-\sin\alpha$ and $\theta_2^+(x,\lambda)$.

Let $\psi(x,\lambda)=\theta(x,\lambda)+m(\lambda)\varphi(x,\lambda)$ be that solution of the system (8.1) – (8.2) which belongs to $\mathcal{L}^2(0,\infty)$. Then using the asymptotic formulas (8.20) – (8.21) and (8.24) – (8.25), we obtain

$$\begin{aligned}\psi_1(x,\lambda)&=\theta_1^+(x,\lambda)+m^+(\lambda)\varphi_1^+(x,\lambda)\\&=e^{-i\alpha(x)}\{G(x,\lambda)M_\theta^+(\lambda)-F(x,\lambda)N_\theta^+(\lambda)\\&\qquad+m^+(\lambda)[G(x,\lambda)M_\varphi^+(\lambda)-F(x,\lambda)N_\varphi^+(\lambda)]+o(1)\},\\ \psi_2(x,\lambda)&=\theta_2^+(x,\lambda)+m^+(\lambda)\varphi_2^+(x,\lambda)\\&=e^{-i\alpha(x)}\{F^{-1}(x,\lambda)M_\theta^+(\lambda)+m^+(\lambda)F^{-1}(x,\lambda)M_\varphi^+(\lambda)+o(1)\}.\end{aligned}$$

Since $\psi(x,\lambda)\in\mathcal{L}^2(0,\infty)$, it follows from the last two formulas that

$$m^+(\lambda)=-\frac{M_\theta^+(\lambda)}{M_\varphi^+(\lambda)}=-\lim_{x\to\infty}\frac{G(x,\lambda)M_\theta^+(\lambda)-F(x,\lambda)N_\theta^+(\lambda)}{G(x,\lambda)M_\varphi^+(\lambda)-F(x,\lambda)N_\varphi^+(\lambda)}. \tag{8.26}$$

[7)] The sign $+$ signifies that we are considering the case in which $\operatorname{Im}\alpha(x)\to+\infty$ as $x\to\infty$.

Similar formulas also hold for the case in which $\operatorname{Im}\alpha(x) \to -\infty$, with $m^+(\lambda)$, $M^+_{\varphi,\theta}(\lambda)$ and $N^+_{\varphi,\theta}(\lambda)$ in formulas (8.26) replaced, respectively, by $m^-(\lambda)$, $M^-_{\varphi,\theta}(\lambda)$ and $N^-_{\varphi,\theta}(\lambda)$, where the last two functions are obtained from (8.14) and (8.15) in the same way that $M^+_{\varphi,\theta}(\lambda)$ and $N^+_{\varphi,\theta}(\lambda)$ were obtained from (8.9) and (8.11).

Let λ be a fixed real number. According to condition 1 of Lemma 7.2, $q_1(x)$ tends monotonically to $+\infty$, and $q_2(x)$ tends monotonically to $-\infty$. Let us first assume that for all x we have $\lambda+q_1(x)>0$ and $\lambda+q_2(x)<0$. Then the function $\alpha(x)$ is pure imaginary, so that $\alpha(x) = e^{i\pi/2}|\alpha(x)|$. Let us assume the same thing if condition 2 is satisfied. Then, as is easily seen,

$$F(x,\lambda)=e^{i\pi/4}|F(x,\lambda)| \quad\text{and}\quad G(x,\lambda)=e^{-i\pi/4}|G(x,\lambda)|.$$

Consequently it follows from (8.22) that the function $e^{-i\pi/4}M_\varphi^+(\lambda)$ is real. One similarly proves that the functions $M_\theta^+(\lambda)$, $N_\varphi^+(\lambda)$, $N_\theta^+(\lambda)$, $M_\varphi^-(\lambda)$, $M_\theta^-(\lambda)$, $N_\varphi^-(\lambda)$ and $N_\theta'(\lambda)$ become real when they are multiplied by $e^{-i\pi/4}$ or $e^{i\pi/4}$.

We have thus established that the function $e^{-i\pi/4}M_\varphi^+(\lambda)$ is in fact real and continuous on the entire real axis. Since, moreover, $M_\varphi^+(\lambda)$ is regular in the upper λ-halfplane, it follows from the symmetry principle that $M_\varphi^+(\lambda)$ is an entire function. We can similarly conclude that $M_\theta^+(\lambda)$ is an entire function. Consequently, by formula (8.26) $m^+(\lambda)$ is meromorphic. The eigenvalues coincide with its poles, and consequently the spectrum is a pure point spectrum. The remaining cases can be worked out in similar fashion.

The preceding arguments can all be carried through from the same point of view for the case in which, when $q_1(x)\to+\infty$, $q_2(x)\to-\infty$ or when $q_1(x)\to-\infty$, $q_2(x)\to+\infty$, the inequalities $\lambda+q_1(x)>0$, $\lambda+q_2(x)<0$ or $\lambda+q_1(x)<0$, $\lambda+q_2(x)>0$ do not hold for all x. It is obvious that one can then choose X such that for all $x>X$ and for fixed real λ we will have $\lambda+q_1(x)>0$, $\lambda+q_2(x)<0$ (for the case of condition 1) and $\lambda+q_1(x)<0$, $\lambda+q_2(x)>0$ (for the case of condition 2), following which all the preceding deductions can be repeated for the interval (X,∞). The proof of the theorem is complete.

§9. Other cases for the spectrum

In this section we will continue the study of the spectrum of the problem $(8.1)+(8.2)+(8.3)$. Let the functions

$$\varphi(x,\lambda)=\begin{pmatrix}\varphi_1(x,\lambda)\\ \varphi_2(x,\lambda)\end{pmatrix} \quad\text{and}\quad \theta(x,\lambda)=\begin{pmatrix}\theta_1(x,\lambda)\\ \theta_2(x,\lambda)\end{pmatrix}$$

be as before. We denote by $\eta_\varphi(x,\lambda)$ and $\eta_\theta(x,\lambda)$ the solutions of the system (7.9′) − (7.10′) which are obtained from (7.4) by replacing $y(x) = \begin{pmatrix} y_1(x) \\ y_2(x) \end{pmatrix}$ by $\varphi(x,\lambda)$ and $\theta(x,\lambda)$, respectively. Then by virtue of (8.16) and (8.17) it follows from (8.18) that

$$\begin{aligned} \eta_{1\varphi}(0) &= F(0)\cos\alpha, \\ \eta_{2\varphi}(0) &= G(0)\cos\alpha + F^{-1}(0)\sin\alpha, \end{aligned} \tag{9.1}$$

$$\begin{aligned} \eta_{1\theta}(0) &= -F(0)\sin\alpha, \\ \eta_{2\theta}(0) &= -G(0)\sin\alpha + F^{-1}(0)\cos\alpha. \end{aligned} \tag{9.2}$$

Further, from (8.19) we obtain the following expressions for $\varphi(0,\lambda)$ and $\theta(0,\lambda)$:

$$\begin{aligned} \varphi_1(0,\lambda) &= G(0)\eta_{1\varphi}(0) - F(0)\eta_{2\varphi}(0), \\ \varphi_2(0,\lambda) &= F^{-1}(0)\eta_{1\varphi}(0), \end{aligned} \tag{9.3}$$

$$\begin{aligned} \theta_1(0,\lambda) &= G(0)\eta_{1\theta}(0) - F(0)\eta_{2\theta}(0), \\ \theta_2(0,\lambda) &= F^{-1}(0)\eta_{1\theta}(0). \end{aligned} \tag{9.4}$$

By virtue of the conditions (8.16) and (8.17) we have

$$W\{\varphi,\theta\}_{x=0} = \varphi_1(0,\lambda)\theta_2(0,\lambda) - \varphi(0,\lambda)\theta_1(0,\lambda) = 1.$$

Therefore, using the expressions (9.3) and (9.4), we obtain

$$\begin{aligned} W\{\varphi,\theta\} &= [G(0)\eta_{1\varphi}(0) - F(0)\eta_{2\varphi}(0)]F^{-1}(0)\eta_{1\theta}(0) \\ &\quad - [G(0)\eta_{1\theta}(0) - F(0)\eta_{2\theta}(0)]F^{-1}(0)\eta_{1\varphi}(0) \\ &= \eta_{1\varphi}(0)\eta_{2\theta}(0) - \eta_{2\varphi}(0)\eta_{1\theta}(0) = 1. \end{aligned} \tag{9.5}$$

Theorem 9.1. *If the coefficients $q_1(x)$ and $q_2(x)$ satisfy conditions* 3, 5 *and* 6 *or* 4, 5 *and* 6 *of Lemma* 7.2, *then the spectrum of the problem* (8.1) + (8.2) + (8.3) *is continuous and fills up the entire axis* $(-\infty, \infty)$.

Proof. Let λ be a fixed real number. Since, by conditions 3 and 4 of Lemma 7.2, as $x \to \infty$ the functions $q_1(x)$ and $q_2(x)$ simultaneously tend either to $+\infty$ or to $-\infty$, we will assume first that for all x we have simultaneously either $\lambda + q_1(x) > 0$, $\lambda + q_2(x) > 0$ or $\lambda + q_1(x) < 0$, $\lambda + q_2(x) < 0$. In both cases we obtain that both $\alpha(x)$ and $F(x,\lambda)$ (cf. formulas (7.3) and (7.5)) are real. Thus $\operatorname{Im}\alpha(x) = 0$. Therefore it follows from the estimate (8.4) that $|\eta_1(x,\lambda)|$ and $|\eta_2(x,\lambda)|$ are bounded. On the other hand, the hypotheses of the theorem guarantee the convergence of the integral (7.18). Then formulas (7.12) and (7.13) can be rewritten in the following form for $x \to \infty$:

$$\eta_1(x,\lambda) = \mu(\lambda)\sin\alpha(x) + \nu(\lambda)\cos\alpha(x) + o(1), \tag{9.6}$$

$$\eta_2(x,\lambda) = -\nu(\lambda)\sin\alpha(x) + \mu(\lambda)\cos\alpha(x) + o(1), \tag{9.7}$$

where

$$\mu(\lambda) = \eta_2(0) + \int_0^\infty \eta_1(s,\lambda) P(s,\lambda)\cos\alpha(s)\,ds, \tag{9.8}$$

$$\nu(\lambda) = \eta_1(0) - \int_0^\infty \eta_1(s,\lambda) P(s,\lambda)\sin\alpha(s)\,ds. \tag{9.9}$$

Since the integrals in (9.8) and (9.9) converge uniformly with respect to λ, it follows that $\mu(\lambda)$ and $\nu(\lambda)$ are continuous and bounded functions of λ.

Further, by virtue of the asymptotic formulas (9.6) and (9.7), it follows from (9.5) that for $x \to \infty$

$$\begin{aligned} W\{\varphi,\theta\} = {} & [\mu_\varphi(\lambda)\sin\alpha(x) + \nu_\varphi(\lambda)\cos\alpha(x)] \\ & \times[-\nu_\theta(\lambda)\sin\alpha(x) + \mu_\theta(\lambda)\cos\alpha(x)] \\ & -[-\nu_\varphi(\lambda)\sin\alpha(x) + \mu_\varphi(\lambda)\cos\alpha(x)] \\ & \times[\mu_\theta(\lambda)\sin\alpha(x) + \nu_\theta(x)\cos\alpha(x)] + o(1) \\ & = \mu_\theta(\lambda)\nu_\varphi(\lambda) - \mu(\lambda)\nu_\theta(\lambda) + o(1), \end{aligned}$$

where $\mu_\varphi(\lambda)$, $\nu_\varphi(\lambda)$ and $\mu_\theta(\lambda)$, $\nu_\theta(\lambda)$ are obtained from formulas (9.8) and (9.9) by replacing $\eta_1(x)$, $\eta_2(x)$ respectively by $F(x,\lambda)\varphi_2(x,\lambda)$, $G(x,\lambda)\varphi_2(x,\lambda) - F^{-1}(x,\lambda)\varphi_1(x,\lambda)$ and $F(x,\lambda)\theta_2(x,\lambda)$, $G(x,\lambda)\theta_2(x,\lambda) - F^{-1}(x,\lambda)\theta_1(x,\lambda)$. Consequently, for $x \to \infty$

$$\mu_\theta(\lambda)\nu_\varphi(\lambda) - \mu_\varphi(\lambda)\nu_\theta(\lambda) = 1,$$

and therefore neither $\mu_\varphi(\lambda)$, $\nu_\varphi(\lambda)$ nor $\mu_\theta(\lambda)$, $\nu_\theta(\lambda)$ can vanish for one value of λ.

By (9.1) and (9.2) we have from (8.22)

$$M_\varphi^+(\lambda) = \frac{1}{2i}\left[i\eta_{1\varphi}(0) - \eta_{2\varphi}(0) + \int_0^\infty \eta_{1\varphi}(s,\lambda) P(s,\lambda) e^{i\alpha(s)} ds\right]$$

or

$$M_\varphi^+(\lambda) = -\frac{1}{2i}[\mu_\varphi(\lambda) - i\nu_\varphi(\lambda)].$$

We can similarly obtain the formulas

$$M_{\theta}^{+}(\lambda) = -\frac{1}{2i}[\mu_{\theta}(\lambda) - i\nu_{\theta}(\lambda)],$$

$$N_{\varphi}^{+}(\lambda) = \frac{1}{2}[\mu_{\varphi}(\lambda) - i\nu_{\varphi}(\lambda)],$$

$$N_{\theta}^{+}(\lambda) = \frac{1}{2}[\mu_{\theta}(\lambda) - i\nu_{\theta}(\lambda)].$$

Similar expressions hold for the functions $M^{-}(\lambda)$ and $N^{-}(\lambda)$.

Then from formula (8.26) we obtain for $m^{+}(\lambda)$ the expression

$$m^{+}(\lambda) = -\frac{\mu_{\theta}(\lambda)\mu_{\varphi}(\lambda) - \nu_{\theta}(\lambda)\nu_{\varphi}(\lambda)}{\mu_{\varphi}^{2}(\lambda) + \nu_{\varphi}^{2}(\lambda)} - \frac{i}{\mu_{\varphi}^{2}(\lambda) + \nu_{\varphi}^{2}(\lambda)}.$$

Consequently

$$\operatorname{Im} m^{+}(\lambda) = -\frac{1}{\mu_{\varphi}^{2}(\lambda) + \nu_{\varphi}^{2}(\lambda)}.$$

Similarly

$$\operatorname{Im} m^{-}(\lambda) = \frac{1}{\mu_{\varphi}^{2}(\lambda) + \nu_{\varphi}^{2}(\lambda)}.$$

In both cases $\operatorname{Im} m(\lambda)$ is a continuous and bounded function of λ. Therefore the assertion of the theorem, i.e. the continuity of the spectrum, follows from (3.19), Chapter 3, since by this formula

$$\rho(\lambda + \Delta\lambda) - \rho(\lambda) = \pm\frac{1}{\pi}\int_{\lambda}^{\lambda+\Delta}\frac{d\lambda}{\mu_{\varphi}^{2}(\lambda) + \nu_{\varphi}^{2}(\lambda)}.$$

For the case in which the inequalities $\lambda + q_1(x) > 0$, $\lambda + q_2(x) > 0$ or $\lambda + q_1(x) < 0$, $\lambda + q_2(x) < 0$ do not hold for all x, the proof can be carried out in exactly the same way as that of Theorem 8.1. This proves the theorem.

Bibliographical references

§1. Lemma 1.1 is due to Levitan [**1**], and Lemma 1.2 is due to the authors and is published here for the first time.

§2. Theorem 2.1 under the condition $q(x) \to +\infty$ was obtained by Weyl [**1**], and under the condition (1.5) of A. M. Molčanov the theorem was proved by the authors and is published here for the first time.

§3. The results of this section are essentially due to Weyl [**1**]; cf. also Titchmarsh [**1**]. The method discussed here for calculating the function $\rho(\lambda)$ is due to Levitan [**1**].

§4. The results of this section are due to Titchmarsh [**1**].

§5. Theorems 5.1 and 5.2 are due to Titchmarsh [**1**]. The proofs of these theorems given here are due to Levitan [**1**].

§6. The results of this section are due to Conte and Sangren [**2**]. This section was written with the help of their work.

§§7, 8, 9. These sections were written with the help of the paper by Roos and Sangren [**1**]. The discussion of their results presented here is due to the authors.

CHAPTER 5

EXAMPLES

§1. The classical Fourier integral

In this chapter we will consider a number of examples which illustrate the theory discussed above.

For the Sturm-Liouville problem the simplest example is the case of the classical Fourier integral, corresponding to $q(x) \equiv 0$. If the interval under consideration is the real semiaxis $[0, \infty)$, and the boundary conditions at zero are given by (2.3), Chapter 2, then in this case

$$\theta(x, \lambda) = \cos\alpha \cdot \cos\sqrt{\lambda}x + \lambda^{-1/2}\sin\alpha \cdot \sin\sqrt{\lambda}x,$$
$$\varphi(x, \lambda) = \sin\alpha \cdot \cos\sqrt{\lambda}x - \lambda^{-1/2}\cos\alpha \cdot \sin\sqrt{\lambda}x.$$

The function $\psi(x,\lambda) = \theta(x,\lambda) + m(\lambda)\varphi(x,\lambda)$ for $\operatorname{Im}\lambda > 0$ can only differ from $e^{i\sqrt{\lambda}x}$ by a constant factor, as $e^{-i\sqrt{\lambda}x}$ does not belong to $\mathscr{L}^2(0, \infty)$. Consequently

$$m(\lambda) = \frac{\sin\alpha - i\sqrt{\lambda}\cos\alpha}{\cos\alpha + i\sqrt{\lambda}\sin\alpha}.$$

If $\operatorname{ctg}\alpha \leqq 0$, then for $\lambda < 0$ the denominator of this function cannot vanish. Therefore in this case

$$-\operatorname{Im} m(\lambda) = \begin{cases} \dfrac{\sqrt{\lambda}}{\cos^2\alpha + \lambda\sin^2\alpha}, & \lambda > 0, \\ 0, & \lambda < 0 \end{cases} \qquad (\operatorname{ctg}\alpha < 0),$$

and since a negative spectrum is absent the inversion formulas have the form

$$F(\lambda) = \int_0^\infty f(x)\{\sin\alpha\cos\sqrt{\lambda}x - \lambda^{-1/2}\cos\alpha \cdot \sin\sqrt{\lambda}x\}\,dx, \tag{1.1}$$

$$f(x) = \frac{1}{\pi}\int_0^\infty F(\lambda)\{\sin\alpha\cos\sqrt{\lambda}x$$
$$- \lambda^{-1/2}\cos\alpha \cdot \sin\sqrt{\lambda}x\}\frac{\sqrt{\lambda}\,d\lambda}{\cos^2\alpha + \lambda\sin^2\alpha}.$$

If $\operatorname{ctg}\alpha = h > 0$, then for $\lambda = -h^2$ the denominator of the function $m(\lambda)$ vanishes. Consequently in this case there is one negative eigenvalue $\lambda = -h^2$, and the corresponding normalized eigenfunction is $e^{-hx}/\sqrt{2h}$. Therefore to the right side of the expansion (1.1) one has to add the term

$$\frac{1}{2h} e^{-hx} \int_0^\infty f(t)\, e^{-ht} dt,$$

and the expansion assumes the form

$$\begin{aligned} f(x) = {} & \frac{1}{2h} e^{-hx} \int_0^\infty f(t)\, e^{-ht} dt \\ & + \frac{1}{\pi} \int_0^\infty F(\lambda) \left\{ \sin\alpha \cdot \cos\sqrt{\lambda}x - \lambda^{-1/2} \cos\alpha \cdot \sin\sqrt{\lambda}x \right\} \frac{\sqrt{\lambda}\, d\lambda}{\cos^2\alpha + \lambda \sin^2\alpha}. \end{aligned} \tag{1.2}$$

The expansions (1.1) and (1.2) can, of course, be justified under the usual assumptions for Fourier integrals by a direct consideration of the integrand.

We note that formula (1.2) can also be obtained from formula (1.12), Chapter 2, for $q(x) \equiv 0$. In fact, in this case

$$\mu(\lambda) = \sin\alpha, \qquad \nu(\lambda) = -\cos\alpha/\sqrt{\lambda}$$

and consequently

$$\rho'(\lambda) = \frac{1}{\pi} \frac{1}{\sqrt{\lambda}\,(\mu^2(\lambda) + \nu^2(\lambda))} = \frac{\sqrt{\lambda}}{\pi(\cos^2\alpha + \lambda \sin^2\alpha)}.$$

§2. Expansion in a Fourier-Bessel series

1. We consider the Bessel equation

$$x \frac{d^2z}{dx^2} + \frac{dz}{dx} + \left(\lambda x - \frac{p^2}{x}\right) z = 0, \tag{2.1}$$

where λ, x and p are real numbers. By means of the substitution $y = \sqrt{x}\, z$ equation (2.1) reduces to the form

$$\frac{d^2y}{dx^2} + \left(\lambda - \frac{p^2 - \frac{1}{4}}{x^2} \right) y = 0. \tag{2.2}$$

This equation has the following linearly independent solutions ($\sqrt{\lambda} = s$):

$$y_1(x, \lambda) = \sqrt{x}\, J_p(sx), \quad y_2(x, \lambda) = \sqrt{z}\, Y_p(sx),$$

where $J_p(t)$ is the Bessel function of the first kind and

$$Y_p(t) = \frac{J_p(t) \cos p\pi - J_{-p}(t)}{\sin p\pi}$$

is the Bessel function of the second kind. From well-known power-series expansions of Bessel functions it follows that for $x \to 0$

$$y_1 = O\left(x^{p+\frac{1}{2}}\right), \quad y_2 = O\left(x^{-p+\frac{1}{2}}\right).$$

It follows from these estimates that for $p \geqq 1$ only y_1 is square integrable in a neighborhood of zero. On the other hand, if $0 \leqq p < 1$, then both solutions are square integrable.

Let $0 < a < b$ and α, β be arbitrary real numbers. We adjoin to (2.1) the boundary conditions

$$y(a)\cos\alpha + y'(a)\sin\alpha = 0, \tag{2.3}$$

$$y(b)\cos\beta + y'(b)\sin\beta = 0. \tag{2.4}$$

Since the equation (2.2) does not have singularities in the interval $[a, b]$, together with the boundary conditions (2.3) and (2.4) it determines a regular Sturm-Liouville problem, to which the classical theory discussed in Chapter 1 is applicable.

Let us denote by $\lambda_1, \lambda_2, \cdots$ the eigenvalues, and by $\varphi_1(x), \varphi_2(x), \cdots$ corresponding eigenfunctions of this problem. In order to obtain a transcendental equation whose roots are the numbers λ_n, we will find the solution of equation (2.2) which satisfies the boundary condition (2.4). It is easily seen that this solution is the function

$$y(x, \lambda) = Cy_1(x, \lambda) + y_2(x, \lambda).$$

where

$$C = -\frac{y_2(b, \lambda)\operatorname{ctg}\beta + y_2'(b, \lambda)}{y_1(b, \lambda)\operatorname{ctg}\beta + y_1'(b, \lambda)}. \tag{2.5}$$

Therefore the eigenvalues λ_n are the roots of the equation

$$y(a, \lambda)\cos\alpha + y'(a, \lambda)\sin\alpha = 0,$$

which can be written in the form

$$C\{y_1(a, \lambda)\operatorname{ctg}\alpha + y_1'(a, \lambda)\} + \{y_2(a, \lambda)\operatorname{ctg}\alpha + y_2'(a, \lambda)\} = 0, \tag{2.6}$$

where C is defined by (2.5). Since $y_2(x, \lambda) \to \infty$ as $x \to 0$, and $y_1(x, \lambda) \to 0$, for every fixed λ the number of zeros of the function $y(x, \lambda)$ in the interval $[a, b]$ remains bounded as $a \to 0$. It follows (cf. Chapter 4) that the equation (2.2) and the boundary condition (2.4) determine in the interval $(0, b]$ a discrete spectrum of eigenvalues with a unique limit point at infinity. The corresponding eigenfunctions must have integrable squares. For the case $p \geqq 1$ this condition uniquely replaces the boundary condition (2.3), since in this case only $y_1(x, \lambda) = \sqrt{x}\, J_p(\sqrt{\lambda} x)$ is square integrable. Thus, to determine the eigenvalues in the interval $[0, b]$

we have to insert for $y(x,\lambda)$, in the boundary condition (2.4), the function $\sqrt{x}\,J_p(\sqrt{\lambda}x)$, and we obtain the following transcendental equation:

$$J_p(sb)\left\{\sqrt{b}\cos\beta+\frac{1}{2\sqrt{b}}\sin\beta\right\}+\sqrt{b}\,J'_p(sb)\sin\beta=0. \tag{2.7}$$

If $\sin\beta=0$, then we obtain the simpler transcendental equation

$$J_p(sb)=0. \tag{2.8}$$

Let us denote the eigenvalues in the latter case (i.e. the roots of equation (2.8)) by s_n. By a well-known formula[1)]

$$\int_0^b xJ_p^2(sx)\,dx=\frac{b^2}{2}\left\{[J'_p(sb)]^2+\left(1-\frac{p^2}{s^2b^2}\right)[J_p(sb)]^2\right\}.$$

Putting here $s=s_n$, we obtain

$$\int_0^b xJ_p^2(s_nx)\,dx=\frac{b^2}{2}[J'_p(s_nb)]^2.$$

Therefore the normalized eigenfunctions have the form

$$\sqrt{2x}\,J_p(s_nx)/b\,J'_p(s_nb),$$

and a Fourier-Bessel expansion can be written in the form

$$f(x)=\frac{2}{b^2}\sum_{n=1}^{\infty}\frac{\sqrt{x}\,J_p(s_nx)}{J'_p(s_nb)}\int_0^b\sqrt{t}\,J_p(s_nt)f(t)\,dt. \tag{2.9}$$

2. For $p<1$ all the solutions of equation (2.2) are square integrable. Therefore in this case the requirement of square integrability does not replace the boundary condition (2.3). We will show that, suitably changing α in (2.3) together with a (which we will require to tend to zero), one can obtain the missing boundary condition.

Since the equation (2.2) does not contain a first derivative, the Wronskian determinant of $y_1(x,\lambda)$ and $y_2(x,\lambda)$ is constant, i.e.

$$\begin{aligned}W\{y_1,y_2\}&=y_1(x,\lambda)\,y_2'(x,\lambda)-y_2(x,\lambda)\,y_1'(x,\lambda)\\&=\sqrt{x}\,J_p(sx)\{\sqrt{x}\,Y_p(sx)\}'-\sqrt{x}\,Y_p(sx)\{\sqrt{x}\,J_p(sx)\}'\\&=sx\{J_p(sx)\,Y'_p(sx)-Y_p(sx)\,J'_p(sx)\}=\text{const.}\end{aligned}$$

To complete the calculation of $W\{y_1,y_2\}$ we will use the well-known

---

1) See, for example, Watson [1], §5.1.

asymptotic formulas for Bessel functions of a large argument (cf., for example, Watson [1], §7.21):

(2.10)
$$J_p(t) = \sqrt{\frac{2}{\pi t}}\left[\cos\left(t - \frac{p\pi}{2} - \frac{\pi}{4}\right) + O\left(\frac{1}{t}\right)\right],$$
$$Y_p(t) = \sqrt{\frac{2}{\pi t}}\left[\sin\left(t - \frac{p\pi}{2} - \frac{\pi}{4}\right) + O\left(\frac{1}{t}\right)\right],$$

(2.10′)
$$J'_p(t) = -\sqrt{\frac{2}{\pi t}}\left[\sin\left(t - \frac{p\pi}{2} - \frac{\pi}{4}\right) + O\left(\frac{1}{t}\right)\right],$$
$$Y'_p(t) = \sqrt{\frac{2}{\pi t}}\left[\cos\left(t - \frac{p\pi}{2} - \frac{\pi}{4}\right) + O\left(\frac{1}{t}\right)\right].$$

Inserting these expressions in the formula for $W\{y_1, y_2\}$, and then letting $x \to \infty$, we obtain $W\{y_1, y_2\} = 2/\pi$.

Suppose, for definiteness, that $\beta = \pi/2$, i.e. $y'(b) = 0$. We can also assume without loss of generality that $y(b) = 1$. It is easily shown that

(2.11)
$$y(x, \lambda) = \frac{1}{W\{y_1, y_2\}}\{y_1(x, \lambda)\, y'_2(b, \lambda) - y'_1(b, \lambda)\, y_2(x, \lambda)\}$$
$$= \frac{\pi}{2} s\sqrt{bx}\,\{J_p(sx)\, Y'_p(sb) - Y_p(sx)\, J'_p(sb)\}.$$

The eigenvalues λ_n are now determined from the equation

$$y(a, \lambda)\operatorname{ctg}\alpha + y'(a, \lambda) = 0.$$

Suppose first that $0 < p < 1$. Since

$$Y_p(t) = \{J_p(t)\cos p\pi - J_{-p}(t)\}\operatorname{cosec} p\pi,$$

it follows from (2.11) that

(2.11′)
$$y(x, \lambda) = -\frac{\pi s\sqrt{bx}}{2\sin p\pi}\{J_p(sx)\, J'_{-p}(bs) - J_{-p}(sx)\, J'_p(bs)\}.$$

For fixed s and $x \to 0$

$$J_p(sx) = \frac{(sx)^p}{2^p\Gamma(1+p)} + O(x^{p+2}), \qquad J'_p(sx) = \frac{px^{p-1}s^p}{2^p\Gamma(1+p)} + O(x^{p+1}).$$

Therefore

(2.12)
$$y(a, \lambda)\operatorname{ctg}\alpha + y'(a, \lambda) =$$
$$= -\frac{\pi\sqrt{b}\, s}{2\sin p\pi}\left\{\frac{s^p J'_{-p}(bs)}{2^p\Gamma(1+p)}\left[a^{\frac{1}{2}+p}\operatorname{ctg}\alpha + \left(\frac{1}{2}+p\right)a^{-\frac{1}{2}+p}\right]\right.$$
$$-\frac{s^{-p}J'_p(bs)}{2^{-p}\Gamma(1-p)}\left[a^{\frac{1}{2}-p}\operatorname{ctg}\alpha + \left(\frac{1}{2}-p\right)a^{-\frac{1}{2}+p}\right]$$
$$\left. + O\left(a^{\frac{5}{2}-p}\,|\operatorname{ctg}\alpha|\right) + O\left(a^{\frac{3}{2}-p}\right)\right\}.$$

Let us denote by c an arbitrary constant, and let a satisfy the equation

$$\frac{a^{\frac{1}{2}-p}\operatorname{ctg}\alpha+\left(\frac{1}{2}-p\right)a^{-\frac{1}{2}-p}}{2^{-p}\Gamma(1-p)}=c\,\frac{a^{\frac{1}{2}+p}\operatorname{ctg}\alpha+\left(\frac{1}{2}+p\right)a^{-\frac{1}{2}+p}}{2^{p}\Gamma(1+p)}. \tag{2.13}$$

Solving for $\operatorname{ctg}\alpha$, we obtain

$$\operatorname{ctg}\alpha=\frac{2^{p}\Gamma(1+p)\left(\frac{1}{2}-p\right)a^{-\frac{1}{2}-p}-c2^{-p}\Gamma(1-p)\left(\frac{1}{2}+p\right)a^{-\frac{1}{2}+p}}{c2^{-p}\Gamma(1-p)\,a^{\frac{1}{2}+p}-2^{p}\Gamma(1+p)\,a^{\frac{1}{2}-p}}=O\left(\frac{1}{a}\right).$$

Therefore

$$a^{\frac{1}{2}+p}\operatorname{ctg}\alpha+\left(\frac{1}{2}+p\right)a^{-\frac{1}{2}+p}$$

$$=\frac{-2^{p+1}p\Gamma(1+p)}{c2^{-p}\Gamma(1-p)\,a^{\frac{1}{2}+p}-2^{p}\Gamma(1+p)\,a^{\frac{1}{2}-p}}\sim a^{p-\frac{1}{2}},$$

and consequently we can neglect the O-terms in (2.12) if $3/2-p>p-\frac{1}{2}$, i.e. $p<1$. From (2.12) and (2.13) there follows a transcendental equation which replaces the boundary condition (2.3):

$$s^{p}J'_{-p}(bs)-cs^{-p}J'_{p}(bs)=0,$$

or

$$\lambda^{p}J'_{-p}(bs)-cJ'_{p}(bs)=0. \tag{2.14}$$

We denote the roots of the equation (2.14) by s_n. The numbers s_n are the eigenvalues of the equation (2.2) with the boundary conditions (2.4) (with $\beta=\pi/2$) and (2.14). We obtain the corresponding eigenfunctions $y_n(x)$ by replacing s in (2.11′) by s_n:

$$\begin{aligned} y_n(x)&=\frac{\pi}{2}\,\frac{\sqrt{bx}\,s_n}{2\sin p\pi}\{J_p(s_nx)\,J'_{-p}(bs_n)-J_{-p}(s_nx)\,J'_p(bs_n)\}\\ &=\frac{\pi}{2}\,\frac{\sqrt{bx}\,s_nJ'_p(bs_n)}{2\sin p\pi}\{c\lambda_n^{-p}J_p(s_nx)-J_{-p}(s_nx)\}. \end{aligned} \tag{2.15}$$

If $c=\infty$, then we obtain the usual Fourier-Bessel series expansion of order p, and if $c=0$, of order $-p$.

Now let $p=0$. In this case

$$Y_0(sx)=\frac{2}{\pi}\left\{\gamma+\ln\frac{sx}{2}\right\}\{1+O(x^2)\},$$

$$Y'_0(sx)=\frac{2}{\pi xs}+O(x\,|\ln x|),$$

where γ is the Euler constant. Therefore

$$
\begin{aligned}
&y(a, \lambda)\operatorname{ctg}\alpha + y'(a, \lambda)\\
&= \frac{\pi\sqrt{bs}}{2}\left\{\left[Y_0'(bs) - J_0'(bs)\frac{2}{\pi}\left(\gamma + \ln\frac{as}{2}\right)\right]\right.\\
&\quad\times\left(a^{\frac{1}{2}}\operatorname{ctg}\alpha + \frac{1}{2}a^{-\frac{1}{2}}\right) - \frac{2}{2a^{\frac{1}{2}}}J_0'(bs)\\
&\quad\left. + O\left(a^{\frac{5}{2}}|\ln a|\cdot|\operatorname{ctg}\alpha|\right) + O\left(a^{\frac{3}{2}}|\ln a|\right)\right\}\\
&= \frac{\pi\sqrt{bs}}{2}\left\{\left[J_0'(bs) - \frac{2}{\pi}J_0'(bs)\ln s\right]\left(a^{\frac{1}{2}}\operatorname{ctg}\alpha + \frac{1}{2}a^{-\frac{1}{2}}\right)\right.\\
&\quad - J_0'(bs)\left[\frac{2}{\pi}\left(\gamma + \ln\frac{a}{2}\right)\left(a^{\frac{1}{2}}\operatorname{ctg}\alpha + \frac{1}{2}a^{-\frac{1}{2}}\right) + \frac{2}{\pi a^{1/2}}\right]\Bigg\}\\
&\quad + O\left(a^{\frac{5}{2}}|\ln a|\cdot|\operatorname{ctg}\alpha|\right) + O\left(a^{\frac{3}{2}}|\ln a|\right).
\end{aligned}
$$

Putting

$$
\frac{2}{\pi}\left(\gamma + \ln\frac{a}{2}\right)\left(a^{\frac{1}{2}}\operatorname{ctg}\alpha + \frac{1}{2}a^{-\frac{1}{2}}\right) + \frac{2}{\pi a^{1/2}} = c\left(a^{\frac{1}{2}}\operatorname{ctg}\alpha + \frac{1}{2}a^{-\frac{1}{2}}\right)
$$

and carrying out computations similar to the preceding, we obtain the following equation for the determination of the eigenvalues:

$$
cJ_0'(bs) - \left\{Y_0'(bs) - \frac{2}{\pi}J_0'(bs)\ln s\right\} = 0.
$$

For $c = \infty$ we obtain the usual Fourier-Bessel series expansion.

One can similarly study the expansion for $\beta = 0$. We obtain the following equation for the determination of the eigenvalues:

$$
c\lambda^{-p}J_p(bs) - J_{-p}(bs) = 0 \quad (p > 0),
$$

$$
cJ_0(bs) - \left\{Y_0(bs) - \frac{2}{\pi}J_0(bs)\ln s\right\} = 0.
$$

3. Using ordinary methods, one can show that eigenfunctions corresponding to distinct eigenvalues are orthogonal. Further, for every function $f(x)$ which is square integrable over the interval $[0, b]$ the Parseval equality holds (cf. Theorem 5.1, Chapter 1). From the Parseval equality it easily follows (cf. Theorem 5.2, Chapter 1) that if the Fourier-Bessel series of a continuous function $f(x)$ converges uniformly, then for[2] $x > 0$ its sum equals $f(x)$.

[2] For $x = 0$ the terms of the series (2.9) equal zero. If $f(0) \neq 0$, then since the sum of the series (2.9) is zero at $x = 0$, it cannot equal $f(0)$.

We will give a simple criterion for the uniform (and absolute) convergence of a Fourier-Bessel series. For definiteness we will consider the ordinary expansion in a Fourier-Bessel series.

THEOREM 2.1. *If the function $f(x)$ satisfies the following conditions:*
1. *$f(x)$ has a continuous second derivative in the interval* $[0, b]$,
2. $f(b) = 0$,
3. $f(0) = f'(0) = 0$,

then the Fourier-Bessel series of $f(x)$ converges absolutely and uniformly.

PROOF. Let

$$f(x) = \sum_{n=1}^{\infty} \sqrt{x}\, J_p(s_n x) \frac{1}{\alpha_n^2} \int_0^b \sqrt{t}\, J_p(s_n t) f(t)\, dt, \tag{2.16}$$

where

$$\alpha_n^2 = \int_0^b t J_p^2(s_n t)\, dt.$$

By the asymptotic formula (2.10)

$$\alpha_n^2 = \frac{2}{\pi} \int_0^b \cos^2\left(s_n t - \frac{p\pi}{2} - \frac{\pi}{4}\right) dt + O\left(\frac{1}{\lambda_n}\right) = O(1).$$

Further, integrating by parts twice we obtain

$$a_n = \int_0^b \sqrt{t}\, J_p(s_n t) f(t)\, dt$$

$$= -\frac{1}{\lambda_n} \int_0^b \left\{ [\sqrt{t}\, J_p(s_n t)]'' - \frac{p^2 - \frac{1}{4}}{t^2} [\sqrt{t}\, J_p(s_n t)] \right\} f(t)\, dt \tag{2.17}$$

$$= \frac{1}{\lambda_n} \int_0^b \sqrt{t}\, J_p(s_n t) \left\{ f''(t) - \frac{p^2 - \frac{1}{4}}{t^2} f(t) \right\} dt,$$

where the integrated terms vanish by conditions 2 and 3.

It is also easy to see that condition 3 implies the boundedness of the function $f(t)/t^2$. It therefore follows from (2.17) that

$$a_n = O(1/\lambda_n). \tag{2.18}$$

From the asymptotic formula (2.10) it easily follows that $\lambda_n = O(n^2)$ as $n \to \infty$.

Therefore it follows from (2.18) that the Fourier-Bessel series (2.16) converges absolutely and uniformly, and the theorem is proved.

Using the asymptotic formulas (2.10), one can show, just as for the classical Sturm-Liouville problem (cf. Chapter 1, §9) that the Fourier-Bessel series converges under the same conditions as does the ordinary Fourier cosine series, for all[3] $x > 0$.

§3. Expansion in a Fourier-Hankel integral

1. Let us now consider an expansion with respect to the eigenfunctions of equation (2.2) in the infinite interval $(0, \infty)$. As was shown in §1, Chapter 2, an eigenfunction expansion in the interval $(0, \infty)$ can be obtained from an expansion in a finite interval $[0, b]$ as $b \to \infty$. We will first consider the case $p \geqq 1$, and suppose for definiteness that $\beta = \pi/2$ in the boundary condition (2.4).

Using the asymptotic formulas for the Bessel functions $J_p(x)$ and $J_p'(x)$, one can show that

$$s_{n+1} - s_n = \frac{\pi}{b} + o\left(\frac{1}{b}\right),$$

where s_n is the nth root of the equation $J_p(bs) = 0$.

Arguing further as in §1, Chapter 2 (see also subsection 2 of the present section), one can obtain the formulas

$$\text{(3.1)} \qquad \begin{aligned} f(x) &= \int_0^\infty sF(s)\sqrt{x}\,J_p(sx)\,ds, \\ F(s) &= \int_0^\infty f(x)\sqrt{x}\,J_p(sx)\,dx \end{aligned}$$

and the Parseval formula

$$\text{(3.2)} \qquad \int_0^\infty f^2(x)\,dx = \int_0^\infty sF^2(s)\,ds.$$

Formulas (3.1) are called the Fourier-Hankel inversion formulas.

2. Suppose now that $0 < p < 1$. As was shown in the previous section, in this case the eigenvalues for a finite interval $[0, b]$ are determined

[3] To strengthen the expansion theorem for Fourier-Bessel series we need asymptotic formulas for large n. Since the solutions of equation (2.2) depend upon the product $\sqrt{\lambda}x$ and since we are using formulas (2.10), which are valid for large values of the argument, these formulas are uniform with respect to $x \geqq \delta > 0$ for any $\delta > 0$.

from the equation

$$\lambda^p J'_{-p}(bs) - cJ'_p(bs) = 0, \tag{3.3}$$

where c is an arbitrary real number.

Let us denote by λ_1 and λ_2 successive positive roots of the equation (3.3). From the asymptotic formulas (2.10) and (2.10′) it easily follows that for $b \to \infty$

$$\sqrt{\lambda_2} - \sqrt{\lambda_1} = \frac{\pi}{b} + o\left(\frac{1}{b}\right). \tag{3.4}$$

By (2.15) and the fact that the boundary-value problem is homogeneous, the functions

$$\psi_n(x) = cJ_p(s_n x) - \lambda_n^p J_{-p}(s_n x)$$

also are eigenfunctions. To determine the spectrum of the boundary-value problem under study for the interval $[0, \infty)$, we need to calculate the limit

$$\lim_{b\to\infty} \frac{1}{b} \int_0^b \psi_n^2(x)\, dx.$$

It follows from the asymptotic formulas (2.10) that for fixed $\lambda > 0$ and large x

$$\begin{aligned}
\psi_n(x) &= \sqrt{\frac{2}{\pi}}\, s_n^{-\frac{1}{2}} \left\{c \cos\left(s_n x - \frac{p\pi}{2} - \frac{\pi}{4}\right) - \lambda_n^p \cos\left(s_n x + \frac{p\pi}{2} - \frac{\pi}{4}\right) + O\left(\frac{1}{x}\right)\right\} \\
&= \sqrt{\frac{2}{\pi}}\, s_n^{-\frac{1}{2}} \left\{c \cos\left(s_n x - \frac{p\pi}{2} - \frac{\pi}{4}\right)\right. \\
&\qquad \left. - \lambda_n^p \cos\left(s_n x - \frac{p\pi}{2} - \frac{\pi}{4} + p\pi\right) + O\left(\frac{1}{x}\right)\right\} \\
&= \sqrt{\frac{2}{\pi}}\, s_n^{-\frac{1}{2}} \left\{(c - \lambda_n^p \cos p\pi) \cos\left(s_n x - \frac{p\pi}{2} - \frac{\pi}{4}\right)\right. \\
&\qquad \left. - \lambda_n^p \sin\left(s_n x - \frac{p\pi}{2} - \frac{\pi}{4}\right) \sin p\pi + O\left(\frac{1}{x}\right)\right\}.
\end{aligned}$$

Therefore

$$\begin{aligned}
\lim_{b\to\infty} \frac{1}{b} \int_0^b \psi_n^2(x)\, dx &= \frac{1}{\pi}\, s_n^{-1} \{(c - \lambda_n^p \cos p\pi)^2 + \lambda_n^{2p} \sin^2 p\pi\} \\
&= \frac{1}{\pi s_n} (c^2 - 2c\lambda_n^p \cos p\pi + \lambda_n^{2p}).
\end{aligned}$$

Putting, as in §1, Chapter 2,

$$\Delta\rho_b(\lambda) = \sum_{s < s_n \leq s+\Delta} \frac{1}{\int_0^b \psi_n^2(x)\, dx},$$

we obtain, by virtue of (3.4),

$$\Delta\rho_b(\lambda) = \sum_{s < s_n \leq s+\Delta} \frac{s_{n+1} - s_n}{\left\{\frac{\pi}{b} + o\left(\frac{1}{b}\right)\right\} \int_0^b \psi_n^2(x)\, dx}$$

$$= \sum_{s < s_n \leq s+\Delta} \left\{\frac{s_n}{c^2 - 2c\lambda_n^p \cos p\pi + \lambda_n^{2p}} + o(1)\right\}(s_{n+1} - s_n) \to$$

$$\to \int_s^{s+\Delta} \frac{s\, ds}{c^2 - 2cs^{2p}\cos p\pi + s^{4p}}\cdot$$

Consequently, for $\lambda > 0$ the spectrum is continuous. If there is no negative spectrum, then the expansion formula has the form

$$f(x) = \int_0^\infty \frac{\sqrt{x}\,\{cJ_p(xs) - s^{2p}J_{-p}(xs)\}}{c^2 - 2cs^{2p}\cos p\pi + s^{4p}}\, s\, ds \int_0^\infty \sqrt{t}\,\{cJ_p(ts) - s^{2p}J_{-p}(ts)\}\, f(t)\, dt. \tag{3.5}$$

Let us now investigate, for which c there will be a negative spectrum. Let $\lambda < 0$, $\sqrt{\lambda} = i\sigma$ $(\sigma > 0)$. From equation (3.3) it follows that

$$(-1)^p\sigma^{2p}J'_{-p}(bi\sigma) - cJ'_p(bi\sigma) = 0. \tag{3.6}$$

As is known,

$$J_p(iz) = e^{\frac{1}{2}p\pi i} I_p(z), \tag{3.7}$$

where

$$I_p(z) = \sum_{k=0}^\infty \frac{1}{k!\,\Gamma(p+k+1)}\left(\frac{z}{2}\right)^{p+2k}.$$

Differentiating (3.7), we obtain

$$J'_p(iz) = e^{\frac{1}{2}(p\pi - 1)i} I'_p(z).$$

Therefore from equation (3.6) follows

$$e^{p\pi i}\sigma^{2p}e^{\frac{1}{2}(-p\pi-1)i} I'_{-p}(b\sigma) - ce^{\frac{1}{2}(p\pi-1)i} I'_p(b\sigma) = 0$$

or $\sigma^{2p}I'_{-p}(b\sigma) - cI'_p(b\sigma) = 0$, i.e.

$$I_p'(b\sigma)/I_{-p}'(b\sigma) = \sigma^{2p}/c. \tag{3.8}$$

From the asymptotic formula for a Bessel function with a pure imaginary argument it follows (cf., for example, Watson [1], §7.21) that as $b \to \infty$ the left side of (3.8) tends to unity. Therefore if $c < 0$, then for sufficiently large b equation (3.8) has no solutions, and consequently in this case a negative spectrum is absent, and the expansion formula can be written in the form (3.5).

For $c > 0$ we obtain one negative eigenvalue $\sigma = c^{1/2p}$.

Let us determine the corresponding eigenfunction. Since $\sqrt{\lambda} = ic^{1/2p}$, it follows from (2.11) that

$$\psi(x, \lambda) = \sqrt{x}\left\{cJ_p\left(xic^{\frac{1}{2p}}\right) - (i)^{2p}\, cJ_{-p}\left(xic^{\frac{1}{2p}}\right)\right\} = -\frac{2}{\pi}\sin p\pi\,\sqrt{x}\, cK_p\left(xc^{\frac{1}{2p}}\right).$$

The factor $-(2c/\pi)\sin p\pi$ can be discarded, and we obtain the eigenfunction $\sqrt{x}\,K_p(xc^{1/2p})$. To normalize this function we use a well-known formula from the theory of Bessel functions:[4)]

$$\int_0^\infty xK_p^2\left(xc^{\frac{1}{2p}}\right)dx = \frac{1}{c^p}\int_0^\infty tK_p^2(t)\,dt = \frac{p}{2c^p}\,\frac{\pi}{\sin p\pi}.$$

Thus, for the case $c > 0$ we have to add to the right side of (3.5) the term

$$\frac{2c^p \sin p\pi}{p\pi}\sqrt{x}\,K_p\left(xc^{\frac{1}{2p}}\right)\int_0^\infty \sqrt{t}\,K_p\left(tc^{\frac{1}{2p}}\right)f(t)\,dt.$$

Similar calculations can be carried out for the case $p = 0$. We give only the final result:

$$f(x) = \int_0^\infty \frac{\sqrt{x}\left\{cJ_0(xs) - Y_0(xs) + \frac{2}{\pi}J_0(xs)\ln s\right\}}{\left\{c + \frac{2}{\pi}\ln s\right\}^2 + 1}\, s\,ds$$

$$\times \int_0^\infty \sqrt{t}\left\{cJ_0(ts) - Y_0(ts) + \frac{2}{\pi}J_0(ts)\ln s\right\}f(t)\,dt$$

$$+ 2\sqrt{x}\,K_0\left(xe^{-\frac{1}{2}\pi c}\right)\int_0^\infty \sqrt{t}\,K_0\left(te^{-\frac{\pi c}{2}}\right)f(t)\,dt.$$

4) See Titchmarsh [3], §8.19.

Thus, in this case the spectrum contains one negative value for any $c < +\infty$. For $c = +\infty$ the additional term to the integral drops out, and we obtain the ordinary expansion in a Fourier-Hankel integral with respect to Bessel functions of zero order.

3. From (3.2) there easily follows the generalized Parseval equality

$$\int_0^\infty f(x)\, g(x)\, dx = \int_0^\infty F(s)\, G(s)\, s\, ds, \tag{3.9}$$

where $G(s)$ is defined by $g(x)$ in the same way that $F(s)$ was defined by $f(x)$.

Just as in §1, Chapter 2 (cf. Theorem 1.2) the following theorem can be obtained from the generalized Parseval equality (3.9).

THEOREM 3.1. *If $f(x)$ is continuous for all $x > 0$ and the integral*

$$\int_0^\infty sF(s)\sqrt{x}\, J_p(sx)\, ds \tag{3.10}$$

converges uniformly in every finite interval, then

$$f(x) = \int_0^\infty sF(s)\sqrt{x}\, J_p(sx)\, ds. \tag{3.11}$$

Similarly to the way in which Theorem 2.1 was proven, one can prove the following theorem.

THEOREM 3.2. *If for all $x \geqq 0$*

1. *$f(x)$ has a continuous second derivative,*
2. *$f(0) = f'(0) = 0$,*
3. *$f(+\infty) = 0$, and $f(x)$, $f'(x)$, $f''(x) \in \mathscr{L}(0, \infty)$, then the integral* (3.10) *converges absolutely and uniformly for all $x \geqq 0$, and therefore the expansion* (3.11) *is valid.*

4. As is known, many differential equations can be reduced to the Bessel equation by an appropriate change of variables. Let us consider, in particular, the equation

$$\frac{d^2y}{dx^2} + \frac{2p+1}{x}\frac{dy}{dx} + \lambda y = 0. \tag{3.12}$$

This equation was already considered by Euler. If we make the substitution $z = x^{p+1/2}y$, we obtain the equation

$$\frac{d^2z}{dx^2}+\left(\lambda-\frac{p^2-\frac{1}{4}}{x^2}\right)z=0.$$

As we have seen, the solutions of this equation are the functions $\sqrt{x}\,J_p(\sqrt{\lambda}x)$ and $\sqrt{x}\,Y_p(\sqrt{\lambda}x)$. Therefore equation (3.12) has the solutions

$$J_p(\sqrt{\lambda}x)/(\sqrt{\lambda}x)^p, \qquad Y_p(\sqrt{\lambda}x)/(\sqrt{\lambda}x)^p.$$

Let us put

$$j_p(\sqrt{\lambda}\,x)=\frac{2^p\Gamma(p+1)\,J_p(\sqrt{\lambda}\,x)}{(\sqrt{\lambda}\,x)^p}=\sum_{k=0}^{\infty}(-1)^k\frac{\Gamma(p+1)}{\Gamma(p+1+1)\,k!}\lambda^k\left(\frac{x}{2}\right)^{2k}.$$

It is easily seen that the function $j_p(\sqrt{\lambda}x)$ satisfies the initial conditions $j_p(0)=1$, $j_p'(0)=0$.

Let us write down the formula for expansion in a Fourier-Hankel integral with respect to the functions $j_p(\sqrt{\lambda}x)$. To this end we will first write down the ordinary Fourier-Hankel inversion formulas, i.e. the formulas (3.1),

$$f(x)=\int_0^{\infty}sF(s)\sqrt{x}\,J_p(xs)\,dx,\qquad F(s)=\int_0^{\infty}\sqrt{x}\,f(x)\,J_p(xs)\,dx.$$

Let $f(x)=x^{p+1/2}g(x)$. Then

$$F(s)=\int_0^{\infty}\sqrt{x}\,f(x)\,J_p(xs)\,dx$$

$$=\frac{s^p}{2^p\Gamma(p+1)}\int_0^{\infty}g(x)\,j_p(sx)\,x^{2p+1}dx=\frac{s^p}{2^p\Gamma(p+1)}\,\Phi(s).$$

Therefore

$$x^{p+\frac{1}{2}}g(x)=\frac{1}{2^p\Gamma(p+1)}\int_0^{\infty}sF(s)\sqrt{x}\,(sx)^p\,j_p(sx)\,ds$$

$$=\frac{1}{2^{2p}\Gamma^2(p+1)}\int_0^{\infty}\Phi(s)\,s^{2p+1}x^{p+\frac{1}{2}}\,j_p(sx)\,dx$$

$$=\frac{x^{p+\frac{1}{2}}}{2^{2p}\Gamma^2(p+1)}\int_0^{\infty}\Phi(s)\,j_p(sx)\,s^{2p+1}ds.$$

Thus the expansion in terms of the functions $j_p(sx)$ has the form

$$\Phi(s) = \int_0^\infty g(x)\, j_p(sx)\, x^{2p+1} dx,$$

$$g(x) = \frac{1}{2^{2p}\Gamma^2(p+1)} \int_0^\infty \Phi(s)\, j_p(sx)\, s^{2p+1} ds.$$

§4. Expansion in terms of the orthogonal Čebyšev-Hermite functions

1. To begin with, we construct the so-called Čebyšev-Hermite polynomials. To this end we consider the function

$$\psi(x, t) = e^{-t^2+2tx} = e^{x^2} e^{-(t-x)^2}.$$

Expanding $\psi(x,t)$ in powers of t, we obtain

$$\psi(x, t) = e^{x^2} e^{-(t-x)^2} = \sum_{n=0}^{\infty} \frac{H_n(x)}{n!} t^n,$$

from which it follows at once that

$$H_n(x) = \left.\frac{\partial^n \psi(x, t)}{\partial t^n}\right|_{t=0} = (-1)^n e^{x^2} \frac{d^n e^{-x^2}}{dx^n}. \tag{4.1}$$

The functions $H_n(x)$ are called the Čebyšev-Hermite polynomials. We will derive a differential equation for these polynomials. From the relation $\partial\psi(x,t)/\partial x = 2t\psi(x,t)$ follows

$$H_n'(x) = 2nH_{n-1}(x) \qquad (n \geqq 1), \tag{4.2}$$

and from the relation $\partial\psi(x,t)/\partial t + 2(t-x)\psi(x,t) = 0$ follows

$$H_{n+1}(x) - 2xH_n(x) + 2nH_{n-1}(x) = 0 \qquad (n \geqq 1).$$

Combining the last two formulas, we obtain the following linear homogeneous second-order differential equation with respect to $H_n(x)$:

$$H_n''(x) - 2xH_n'(x) + 2nH_n(x) = 0 \qquad (n \geqq 0). \tag{4.3}$$

The first few Čebyšev-Hermite polynomials have the form

$$H_0(x) = 1, \qquad H_1(x) = 2x, \qquad H_2(x) = 4x^2 - 2,$$

$$H_3(x) = 8x^3 - 12, \qquad H_4(x) = 16x^4 - 48x^2 + 12.$$

A general expression for the Čebyšev-Hermite polynomial $H_n(x)$ is

$$H_n(x) = (2x)^n - \frac{n(n-1)}{1!}(2x)^{n-2} + \frac{n(n-1)(n-2)(n-3)}{2!}(2x)^{n-4} + \dots$$

The last term is equal to

$$(-1)^{\frac{n}{2}} \frac{n!}{(n/2)!} \text{ for even } n,$$

$$(-1)^{\frac{n-1}{2}} \frac{n!}{\left(\frac{n-1}{2}\right)!} 2x \text{ for odd } n.$$

We will show that the Čebyšev-Hermite polynomials are orthogonal over the entire line $(-\infty, \infty)$ with respect to the weight $p(x) = \exp(-x^2)$. In fact, by formula (4.1) we have

$$\int_{-\infty}^{\infty} H_m(x) H_n(x) e^{-x^2} dx = (-1)^n \int_{-\infty}^{\infty} H_m(x) \frac{d^n e^{-x^2}}{dx^n} dx.$$

Integrating the last integral by parts for $n > m$ and taking account of (4.2) and the fact that all derivatives of $\exp(-x^2)$ vanish as $x \to \infty$. we find that

$$\int_{-\infty}^{\infty} H_n(x) H_n(x) e^{-x^2} dx = (-1)^{n-1} 2m \int_{-\infty}^{\infty} H_{m-1}(x) \frac{d^{n-1} e^{-x^2}}{dx^{n-1}} dx = \dots$$

$$\dots = (-1)^{n-m} 2^m m! \int_{-\infty}^{\infty} H_0(x) \frac{d^{n-m} e^{-x^2}}{dx^{n-m}} = 0.$$

For $n = m$ one can compute in the same way that

$$\int_{-\infty}^{\infty} H_n^2 e^{-x^2} dx = 2^n n! \sqrt{\pi}.$$

Therefore the functions

$$\varphi_n(x) = \frac{H_n(x) e^{-\frac{x^2}{2}}}{\sqrt{2^n n! \sqrt{\pi}}} \quad (n = 0, 1, 2, \dots) \tag{4.4}$$

constitute an orthonormal system on the real line $(-\infty, \infty)$, and are called the orthonormal Čebyšev-Hermite functions.

2. Let $q(x) = x^2$, $-\infty < x < \infty$. In this case we are led to solve the equation

$$y'' + (\lambda - x^2) y = 0. \tag{4.5}$$

Since $q(x) = x^2 \to +\infty$ as $x \to \pm\infty$, we know from the results of Chapter 4 that the spectrum of the operator $l(y) \equiv -y'' + x^2 y$ on the entire line is discrete. Let us denote by $\lambda_1, \lambda_2, \cdots$ the spectrum of the equation (4.5) on the real line.

We will show that the orthonormalized eigenfunctions, i.e. the solutions of (4.5) in $\mathscr{L}^2(-\infty, \infty)$ (i.e. the square integrable solutions of (4.5)) coincide with the functions $\varphi_n(x)$ defined by (4.4).

In fact, let $y = \exp(-x^2/2)z$. Then equation (4.5) assumes the form

$$z'' - 2xz' + (\lambda - 1)z = 0. \tag{4.6}$$

If

$$z = \sum_{k=0}^{\infty} a_k x^k \tag{4.7}$$

is a solution of equation (4.6), we obtain from (4.6) the following recursion relation for the nonvanishing coefficients:

$$\frac{a_{k+2}}{a_k} = \frac{2k + 1 - \lambda}{(k+1)(k+2)}. \tag{4.8}$$

The relation (4.8) shows that equation (4.6) has two solutions; an even one $z_1(x)$ and an odd one $z_2(x)$. So in the power series (4.7) representing these solutions there appear either solely even or solely odd powers of x. From (4.8) it follows, first of all, that the series (4.7) either terminates (namely, if $\lambda = 2k + 1$ is an odd nonnegative integer), and in this case equation (4.6) coincides with equation (4.3), with the solution thus being a Čebyšev-Hermite polynomial, or has an infinite number of nonvanishing coefficients and converges for all values of x.

In the second case all a_k from some k on have the same sign. Suppose they are positive; then they satisfy the inequality $a_{k+2} > a_k/(k+2)$ and, as is easily seen, for sufficiently large x

$$z = \sum_{k=0}^{\infty} a_k x^k > Ce^{\frac{x^2}{2}}, \qquad C \neq 0,$$

and consequently $y = z\exp(-x^2/2)$ is not square integrable over the real line. This establishes that the orthogonal Čebyšev-Hermite functions are the unique solutions in $\mathscr{L}^2(-\infty, \infty)$ of our problem.

The completeness of the orthogonal Čebyšev-Hermite functions in $\mathscr{L}^2(-\infty, \infty)$ follows from the general theory discussed in Chapter 2.

§5. Expansion in terms of the Legendre polynomials and the associated Legendre functions

1. The Legendre polynomials are defined by the following formula:

$$P_0(x) = 1, \quad P_n(x) = \frac{1}{2^n n!}\frac{d^n (x^2 - 1)^n}{dx^n} \qquad (n = 1, 2, \ldots).$$

Obviously the function $P_n(x)$ is a polynomial of degree n.

We will prove that the polynomials $P_n(x)$ form an orthogonal system on the interval $[-1,1]$. In fact, putting $u_n(x) = (x^2-1)^n$, we have for any nonnegative integer $m < n$

$$\int_{-1}^{1} P_n(x)\, x^m dx = \frac{1}{2^n n!} \int_{-1}^{1} u_n^{(n)}(x)\, x^m dx = 0,$$

which is easily verified by repeated integration by parts, bearing in mind that all the derivatives of $u_n(x)$ of order up to $n-1$ vanish at the endpoints of the interval of integration. It follows that

$$\int_{-1}^{1} P_n(x)\, P_m(x)\, dx = 0 \qquad (m < n),$$

i.e. two distinct polynomials are indeed orthogonal. To compute the normalization constants we obtain by repeated integration by parts

$$\int_{-1}^{1} u_n^{(n)}(x)\, u_n^{(n)}(x)\, dx = -\int_{-1}^{1} u_n^{(n-1)}(x)\, u_n^{(n+1)}(x)\, dx$$

$$= \int_{-1}^{1} u_n^{(n-2)}(x)\, u_n^{(n+2)} dx = \dots = (-1)^n \int_{-1}^{1} u_n^{(2n)}(x)\, dx$$

$$= (2n)! \int_{-1}^{1} (1-x)^n (1+x)^n\, dx.$$

Further, since

$$\int_{-1}^{1} (1-x)^n (1+x)^n\, dx = \frac{n}{n+1} \int_{-1}^{1} (1-x)^{n-1} (1+x)^{n+1}\, dx = \dots$$

$$\dots = \frac{n(n-1)\dots 2\cdot 1}{(n+1)(n+2)\dots 2n} \int_{-1}^{1} (1+x)^{2n}\, dx = \frac{(n!)^2}{(2n)!\,(2n+1)} \cdot 2^{2n+1},$$

it follows that

$$\int_{-1}^{1} P_n^2(x)\, dx = \frac{1}{2^{2n}(n!)^2} \int_{-1}^{1} \{u_n^{(n)}(x)\}^2\, dx = \frac{2}{2n+1}.$$

Therefore the normalized polynomials have the form

$$\varphi_n(x) = \sqrt{\frac{2n+1}{2}}\, P_n(x) = \sqrt{\frac{2n+1}{2}} \cdot \frac{1}{2^n n!} \cdot \frac{d^n (x^2-1)^n}{dx^n} \tag{5.1}$$

$$(n = 0,\ 1,\ 2,\ \dots).$$

The Legendre polynomials $P_n(x)$ have the property that for any n

(5.2) $$P_n(1) = 1,$$

which is easily seen by forming the nth derivative of the expression $(x-1)^n(x+1)^n$ by Leibnitz' rule and then putting $x = 1$.

2. The Legendre polynomials play an important role in potential theory, where they appear as the coefficients in an expansion of the so-called generating function. In fact, the generating function in potential theory is defined as the reciprocal of the distance between two points, one of which is at unit distance from the origin, the other at a distance $u < 1$ from the origin, and whose radius vectors form an angle arc cos x, i.e.

$$\psi(x, u) = \frac{1}{\sqrt{1-2xu+u^2}}.$$

Now, inserting in the expansion

$$\frac{1}{\sqrt{1+t}} = 1 - \frac{1}{2}t + \frac{3}{8}t^2 - \frac{5}{16}t^3 + \ldots + (-1)^n \frac{(2n-1)!!}{(2n)!!} t^n + \ldots$$

$t = -2xu + u^2$ and grouping like powers of u, we obtain an expansion of the generating function in the form

(5.3) $$\psi(x, u) = \frac{1}{\sqrt{1-2xu+u^2}} = \sum_{n=0}^{\infty} Q_n(x) u^n.$$

Let us show that the functions $Q_n(x)$ are identical with the Legendre polynomials defined earlier. To do this we will first prove that the $Q_n(x)$ form an orthogonal system over the interval $[-1, 1]$. It follows from (5.3) that

(5.4) $$\frac{1}{\sqrt{1-2xu+u^2}} \cdot \frac{1}{\sqrt{1-2xv+v^2}} = \sum_{n,m=0}^{\infty} Q_n(x) Q_m(x) u^n u^m.$$

Integrating the left side of (5.4) with respect to x from -1 to 1, we obtain

$$\int_{-1}^{1} \frac{dx}{\sqrt{(1-2xu+u^2)(1-2xv+v^2)}} = \frac{1}{\sqrt{uv}} \ln \frac{1+\sqrt{uv}}{1-\sqrt{uv}},$$

whence, by virtue of the expansion

$$\ln \frac{1+t}{1-t} = 2t\left(1 + \frac{1}{3}t^2 + \frac{1}{5}t^4 + \ldots + \frac{1}{2n+1}t^{2n} + \ldots\right),$$

follows

$$\int_{-1}^{1} \frac{dx}{\sqrt{(1-2xu+u^2)(1-2xv+v^2)}} = \sum_{n=0}^{\infty} \frac{2}{2n+1} u^n v^n. \tag{5.5}$$

Now, integrating the right side of (5.4) with respect to x from -1 to 1 and comparing the result with the expansion (5.5), we obtain

$$\int_{-1}^{1} Q_n(x)\, Q_m(x)\, dx = \begin{cases} 0, & \text{for} \quad m \neq n, \\ \dfrac{2}{2n+1}, & \text{for} \quad m = n. \end{cases}$$

i.e. the functions $Q_n(x)$ form an orthogonal system on the interval $[-1,1]$. It follows from the construction of the $Q_n(x)$ that they are polynomials. On the other hand, just like the Legendre polynomials $P_n(x)$, the polynomials $Q_n(x)$ can be regarded as a system obtained from the system $1, x, x^2, \cdots$ by orthogonalizing the latter. Since the process of orthogonalization shows that, aside from constant factors, there exists only one system of orthogonal polynomials which contains polynomials of all degrees,[5)] to show that $Q_n(x) \equiv P_n(x)$ for every n it suffices to note that, like the polynomials $P_n(x)$, the polynomials $Q_n(x)$ satisfy the condition (5.2), i.e. $Q_n(1) = 1$. In fact, putting $x = 1$ in (5.3), we obtain

$$\psi(1, u) = \frac{1}{1-u} = 1 + u + u^2 + \ldots + u^n + \ldots = \sum_{n=0}^{\infty} Q_n(1)\, u^n,$$

which proves our assertion.

3. We will obtain a recursive formula and a differential equation for the Legendre polynomials.

Differentiating (5.3) with respect to u and then replacing $Q_n(x)$ by $P_n(x)$, we obtain

$$\frac{\partial \psi(x, u)}{\partial u} = \frac{x-u}{\sqrt{1-2xu+u^2} \cdot (1-2xu+u^2)} = \sum_{n=0}^{\infty} nP_n(x) u^{n-}$$

i.e.

$$(x-u) \sum_{n=0}^{\infty} P_n(x) u^n = (1-2xu+u^2) \sum_{n=0}^{\infty} nP_n(x) u^{n-1},$$

from which, equating the coefficients of u^n, we obtain the following formula for three successive Legendre polynomials:

$$(n+1) P_{n+1}(x) - x(2n+1) P_n(x) + nP_{n-1}(x) = 0.$$

---

5) See, for example, Courant and Hilbert [1], Chapter II, §1.2.

The first few Legendre polynomials have the form

$$P_0(x)=1,\qquad P_1(x)=x,\qquad P_2(x)=\frac{3}{2}x^2-\frac{1}{2},$$
$$P_3(x)=\frac{5}{2}x^3-\frac{3}{2}x,\qquad P_4(x)=\frac{35}{8}x^4-\frac{15}{4}x^2+\frac{3}{8}.$$

The Legendre polynomials

$$y(x)=\frac{1}{2^n n!}\frac{d^n(x^2-1)^n}{dx^n}$$

satisfy a homogeneous second-order linear differential equation:

$$(x^2-1)y''+2xy'-n(n+1)y=0 \tag{5.6}$$

or

$$[(x^2-1)y']'-n(n+1)y=0. \tag{5.6'}$$

This can be proved by differentiating the obvious equality $(x^2-1)u' = 2nxu$, where $u=(x^2-1)^n$, $n+1$ times, and then replacing $u^{(n)}(x)$ by $2^n n!y$.

It follows from equation (5.6) that all roots of the polynomial $P_n(x)$ are distinct, since $P_n(x)$ is a solution of a second-order equation. It further follows from the definition of the polynomial $P_n(x)$, on the basis of Rolle's theorem, that all of its roots are real and lie in the interval $[-1,1]$.

4. Let us consider a Sturm-Liouville problem in standard form:

$$\frac{d^2u}{dt^2}+\left\{\lambda+\left(\frac{1}{4}\operatorname{tg}^2 t+\frac{1}{2}\right)\right\}u=0 \qquad \left(-\frac{\pi}{2}\leqslant t\leqslant\frac{\pi}{2}\right), \tag{5.7}$$
$$u\left(-\frac{\pi}{2}\right)=u\left(\frac{\pi}{2}\right)=0.$$

Since here $q(t)=-\frac{1}{4}(\operatorname{tg}^2 t+\frac{1}{2})\to-\infty$ as $t\to\pm\pi/2$, we have the singular case. Let us find the eigenvalues and eigenfunctions of the problem (5.7). By means of the change of variables

$$x=\sin t,\qquad y=u/\sqrt{\cos t} \tag{5.8}$$

the problem (5.7) transforms into the equation

$$(1-x^2)y''-2xy'+\lambda y=0, \tag{5.9}$$

the interval $[-\pi/2,\pi/2]$ transforms into the interval $[-1,1]$, and the boundary conditions become the condition that $y(x)$ remain finite at both singular points of equation (5.9), i.e. at the points ±1. We already know that the numbers $\lambda=n(n+1)$ are eigenvalues, and the Legendre polynomials $P_n(x)$ eigenfunctions, of this problem.

We will show that the Legendre polynomials are the only solutions of this eigenvalue problem which are bounded on the interval $[-1,1]$. Note that if $u=f(x)$ satisfies the differential equation (5.9), so does the function $f(-x)$, and consequently so do the functions $f(x)+f(-x)$ and $f(x)-f(-x)$, the first of which is even, and the second, odd, and at least one of these does not vanish identically, since by hypothesis the function $u=f(x)$ is not identically zero. Thus it is sufficient to show that every even and every odd solution u of the equation (5.9) which is continuous for $-1\leqq x\leqq 1$ is a Legendre polynomial and that λ must be a number of the form $n(n+1)$. Representing u as a power series $u=\sum_{n=0}^{\infty}a_n x^n$, we obtain from (5.9) the following recursion formula for the coefficients:

$$a_{n+2}=\frac{n(n+1)-\lambda}{(n+1)(n+2)}a_n \qquad (n=0,\ 1,\ 2,\ \ldots). \tag{5.10}$$

If $u(x)$ is an even function, then all the a_n with odd n equal zero; if $u(x)$ is an odd function, then the coefficients a_n with even n are zero. For the case $n-2h>0$ we have from (5.10)

$$\begin{aligned} a_n=\frac{1}{n}\left[1-\frac{\lambda}{(n-1)(n-2)}\right]\left[1-\frac{\lambda}{(n-3)(n-4)}\right]\cdots \\ \cdots\left[1-\frac{\lambda}{(n-2h+1)(n-2h)}\right]ka_k, \end{aligned} \tag{5.11}$$

where $k=n-2h$. The series for $u(x)$ terminates if and only if λ is a number of the form $n(n+1)$. Obviously in this case $u(x)$ is the nth Legendre polynomial. For all other values of λ we obtain an infinite series, which from elementary tests is seen to converge for $|x|<1$. Let us give k a fixed value so large that all the factors in the product (5.11) are positive (a_k can be assumed positive). Since as n increases the product of the square brackets in (5.11), according to well-known theorems, tends to a positive limit, it follows that for $n>k$ surely $a_n>c/n$, where c is a positive constant. Therefore $\sum_{\nu=k}^{\infty}a_\nu x^\nu$ cannot be bounded in the neighborhood of the point $x=1$. It follows that a number $\lambda\neq n(n+1)$ cannot be an eigenvalue.

From (5.1) and (5.8) we conclude that the eigenvalues of the problem (5.7) are the numbers $\lambda_n=n(n+1)$, and the normalized eigenfunctions are

$$u_n(t)=\sqrt{\frac{2n+1}{n}}\,P_n(\sin t)\sqrt{\cos t}=\sqrt{\frac{2n+1}{n}}\cdot\frac{\sqrt{\cos t}}{2^n\cdot n!}\,\frac{d^n(-\cos^2 t)^n}{d^n\sin t}. \tag{5.12}$$

The completeness in $\mathscr{L}^2(-1,1)$ of the system (5.1) follows from the Weierstrass approximation theorem.

5. From the differential equation (5.9) for the Legendre polynomials it is easy to derive other orthogonal systems. Differentiating (5.9) with respect to x, we obtain a differential equation for the function u', and just as before, it turns out that only for $\lambda = n(n+1)$ does there exist a solution which is regular at both endpoints of $[-1, 1]$, namely $P_n'(x)$. The equation thus obtained for $P_n'(x)$ is no longer selfadjoint, but it can be transformed to a selfadjoint one by introducing the function $P_n'(x)\sqrt{1-x^2} = z_n(x)$ as the unknown. Then the new equation assumes the form

$$\{(1-x^2)z'\}' - \frac{z}{1-x^2} + \lambda z = 0.$$

Its eigenvalues are the numbers $\lambda_n = n(n+1)$ $(n = 1, 2, \cdots)$, and its eigenfunctions, $z_n(x) = \sqrt{1-x^2}P_n'(x)$.

The functions $z_n(x) = P_{n,1}(x)$ are called the associated Legendre functions of order one. The functions $P_n(x) = P_{n,0}(x)$ will in this connection be called the Legendre functions of zero order. The Legendre functions $P_{n,1}(x)$ satisfy the orthogonality condition

$$\int_{-1}^{1} P_{n,1}(x)\, P_{m,1}(x)\, dx = 0 \quad \text{for} \quad m \neq n.$$

In the same way, differentiating equation (5.9) h times, we obtain for the functions

$$(\sqrt{1-x^2})^h \frac{d^h}{dx^h} P_n(x) = P_{n,h}(x)$$

the differential equation

$$\{(1-x^2)z'\}' - \frac{h^2 z}{1-x^2} + \lambda z = 0 \tag{5.12'}$$

having the eigenvalues $\lambda_n = n(n+1)$ $(n = h, h+1, \cdots)$ and the corresponding eigenfunctions $P_{n,h}(x)$, which are also pairwise orthogonal over the interval $[-1, 1]$ and are called the associated Legendre functions of order h. Their normalization can be carried out by means of the following easily proved equality:

$$\int_{-1}^{1} P_{n,h}^2(x)\, dx = \frac{2}{2n+1}\frac{(n+h)!}{(n-h)!}.$$

That all the eigenvalues and eigenfunctions of the differential equation (5.12′) have been obtained in this way can be proved in the same way as for the Legendre functions of zero order.

§6. Expansion in terms of Čebyšev-Laguerre polynomials

1. The Čebyšev-Laguerre polynomial of degree n is defined by

$$\mathscr{L}_n(x) = e^x d^n(x^n e^{-x})/dx^n.$$

It is easily seen that $\mathscr{L}_n(x)$ is in fact a polynomial of degree n. Indeed, by Leibnitz' formula we have

$$\begin{aligned}\mathscr{L}_n(x) &= e^x \frac{d^n (x^n e^{-x})}{dx^n} \\ &= \sum_{k=0}^{n} (-1)^k \binom{n}{k} n(n-1)\dots(k+1)x^k \\ &= \sum_{k=0}^{n} (-1)^{n-k} \binom{n}{k} n(n-1)\dots(n-k+1)x^{n-k} \\ &= \sum_{k=0}^{n} (-1)^{n-k} \frac{[n(n-1)\dots(n-k+1)]^2}{k!} x^{n-k} \\ &= (-1)^n \left(x^n - \frac{n^2}{1!} x^{n-1} + \frac{n^2(n-1)^2}{2!} x^{n-2} - \dots + (-1)^n n!\right).\end{aligned}$$

The first few Čebyšev-Laguerre polynomials have the form

$$\mathscr{L}_0(x) = 1, \quad \mathscr{L}_1(x) = -x+1, \quad \mathscr{L}_2(x) = x^2 - 4x + 2,$$
$$\mathscr{L}_3(x) = -x^3 + 9x^2 - 18x + 6,$$
$$\mathscr{L}_4(x) = x^4 - 16x^3 + 72x^2 - 96x + 24.$$

In order to obtain recursion formulas and a differential equation for the functions $\mathscr{L}_n(x)$, we consider the function

$$\psi(x, t) = \frac{1}{1-t} e^{-\frac{xt}{1-t}}.$$

We will prove that for any n

$$\mathscr{L}_n(x) = \partial^n \psi(x,t)/\partial t^n |_{t=0}.$$

In fact, by the definition of the function $\mathscr{L}_n(x)$ we have

$$\sum_{n=0}^{\infty} \frac{\mathscr{L}_n(x)}{n!} t^n = \sum_{n=0}^{\infty} \sum_{k=0}^{n} (-1)^k \binom{n}{k} \frac{1}{k!} x^k t^n$$

$$= \sum_{k=0}^{\infty} \frac{(-1)^k x^k}{k!} \sum_{n=k}^{\infty} \binom{n}{k} t^n = \sum_{k=0}^{\infty} \frac{(-1)^k x^k}{k!} \cdot \frac{t^k}{(1-t)^{k+1}} \tag{6.1}$$

$$= \frac{1}{1-t} e^{-\frac{xt}{1-t}} = \psi(x, t),$$

which proves our assertion by virtue of the uniqueness of the expansion of $\psi(x,t)$ in powers of t.

From the obvious relation

$$(1-t)^2\,\partial\psi(x,t)/\partial t = (1-t-x)\psi(x,t)$$

there follows, by virtue of (6.1), the identity

$$(1-2t+t^2)\sum_{n=0}^{\infty}\frac{\mathscr{L}_n(x)}{(n-1)!}t^{n-1} = (1-t-x)\sum_{n=0}^{\infty}\frac{\mathscr{L}_n(x)}{n!}t^n.$$

Equating coefficients of like powers of t, we arrive at the recursive formula

$$\mathscr{L}_{n+1}(x) - (2n+1-x)\,\mathscr{L}_n(x) + n^2\mathscr{L}_{n-1}(x) = 0 \quad (n \geqq 1).$$

This formula, and the formula $\mathscr{L}_n'(x) - n\,\mathscr{L}_{n-1}'(x) = -n\,\mathscr{L}_{n-1}(x)$ $(n \geqq 1)$, which follows in similar fashion from the relation

$$(1-t)\,\partial\psi(x,t)/\partial x = -t\psi(x,t)$$

leads to the formula $x\,\mathscr{L}_n'(x) = n\,\mathscr{L}_n(x) - n^2\mathscr{L}_{n-1}(x)$ $(n \geqq 1)$ and to the homogeneous second-order linear differential equation

$$xy'' + (1-x)y' + ny = 0 \tag{6.2}$$

for the Čebyšev-Laguerre polynomials.

The Čebyšev-Laguerre polynomials form an orthogonal system on the halfline $(0, \infty)$ with weight e^{-x}, i.e.

$$\int_0^\infty e^{-x}\mathscr{L}_n(x)\,\mathscr{L}_n(x)\,dx = 0 \qquad (m < n).$$

This follows from the equality

$$\int_0^\infty e^{-x}x^k\mathscr{L}_n(x)\,dx = \int_0^\infty x^k\frac{d^n}{dx^n}(x^ne^{-x})\,dx = -k\int_0^\infty x^{k-1}\frac{d^{n-1}}{dx^{n-1}}(x^ne^{-x})\,dx$$

$$= k(k-1)\int_0^\infty x^{k-2}\frac{d^{n-2}}{dx^{n-2}}(x^ne^{-x})\,dx = \dots$$

$$\dots = (-1)^k\,k!\int_0^\infty\frac{d^{n-k}}{dx^{n-k}}(x^ne^{-x})\,dx = 0 \quad \text{for} \quad n > k.$$

The normalizing coefficients can be computed in the following way:

$$\int_0^\infty e^{-x}\mathscr{L}_n^2(x)\,dx = \int_0^\infty (-1)^n x^n\frac{d^n}{dx^n}(x^ne^{-x})\,dx = n!\int_0^\infty x^ne^{-x}dx = (n!)^2.$$

Therefore the functions

$$\varphi_n(x) = \frac{e^{-\frac{x}{2}} \mathcal{L}_n(x)}{n!} \qquad (n = 0, 1, 2, \ldots)$$

form an orthonormal system on the halfline $(0, \infty)$, and are called the orthonormal Čebyšev-Laguerre functions.

2. Let us consider the equation

$$u'' + \left\{\lambda - \left(\frac{t^2}{16} - \frac{1}{4t^2} - \frac{1}{2}\right)\right\} u = 0 \quad (0 \leqslant t < \infty). \tag{6.3}$$

By means of the substitutions $x = t^2/4$, $y(x) = u(t) t^{-1/2} \exp(t^2/8)$, equation (6.3) reduces to the form

$$xy'' + (1 - x)y' + \lambda y = 0 \qquad (0 \leqq x < \infty). \tag{6.4}$$

Since, in equation (6.3), $q(t) = t^2/16 - 1/4t^2 - 1/2$ tends monotonically to $+\infty$ as $t \to \infty$, the spectrum of (6.3) is discrete. On the other hand, equation (6.4) coincides with equation (6.2), for which the eigenvalues are $\lambda_n = n$ and the eigenfunctions are the Čebyšev-Laguerre polynomials. That the equation (6.4) does not have any eigenfunctions in addition to those just indicated can be proved in the same way as for the Hermite polynomials. Consequently the eigenvalues of (6.3) are $\lambda_n = n$, and the eigenfunctions are

$$u_n(t) = \sqrt{t}\, e^{-\frac{t^2}{8}} \mathcal{L}_n\left(\frac{t^2}{4}\right).$$

3. As for the Legendre functions in §5, the process of differentiation and multiplication by a suitable factor leads here to the higher-order Laguerre functions (generalized Laguerre functions), which satisfy similar differential equations. First of all, differentiating equation (6.4) m times shows that the functions

$$u = \mathcal{L}_n^m(x) = \frac{d^m}{dx^m} \mathcal{L}_n(x)$$

satisfy the differential equation

$$xu'' + (m + 1 - x)u' + (\lambda - m)u = 0, \tag{6.5}$$

which can be written in the following selfadjoint form:

$$(x^{m+1} e^{-x} u')' + x^m e^{-x} (\lambda - m) u = 0.$$

The corresponding orthogonal functions

$$v = x^{m/2} e^{-x/2} \mathcal{L}_n^m(x)$$

satisfy the following equation of Sturm-Liouville type:

$$(xv')' + \left(\frac{1-m}{2} - \frac{x}{4} - \frac{m^2}{4x}\right)v + \lambda v = 0.$$

In order to show that these equations do not have any other eigenvalues and eigenfunctions, we insert the power series $u = \sum_{\nu=0}^{\infty} a_\nu x^\nu$ in equation (6.5) and obtain, by means of recursion relations, the following expression for the coefficients:

$$a_\nu = \frac{a_0}{\nu!} \cdot \frac{(m-\lambda)\ldots(m-\lambda+\nu-1)}{(m+1)\ldots(m+\nu)}.$$

We note that for an arbitrary given value of λ the coefficients a_ν have fixed sign, starting with some value of ν, and the series converges for all values of x; consequently it does actually represent a solution of (6.5) which is regular for $0 \leqq x < \infty$. For $\lambda = n$ a positive integer $(n > m)$ the series terminates and is therefore a polynomial. For every other value of λ it is easy to obtain a bound of the form

$$|a_\nu| > c/\nu!\nu^r,$$

where c is a suitable constant, and r is a positive integer. But from this it follows that as $x \to \infty$ our solution tends to infinity not slower than e^x/x^r. We have thus proved that it cannot be an eigenfunction of the problem being considered.

§7. The "hydrogen atom"

1. We will consider the equation

$$\frac{d^2R}{dr^2} + \frac{2}{r} \cdot \frac{dR}{dr} - \frac{l(l+1)}{r^2} R + \left(E + \frac{2}{r}\right) R = 0 \quad (0 < r < \infty), \tag{7.1}$$

where l is a positive integer or zero.

In quantum mechanics the study of the energy levels of the hydrogen atom leads to this equation.[6] This substitution $R = y/r$ reduces equation (7.1) to the form

$$\frac{d^2y}{dr^2} + \left\{E + \frac{2}{r} - \frac{l(l+1)}{r^2}\right\} y = 0. \tag{7.2}$$

Just as in the case of Bessel's equation (cf. §2), one can show that in a finite interval $[0, b]$ the spectrum is discrete.

Let us consider the infinite interval $(0, \infty)$. We will show that for negative E the spectrum remains discrete. Let $E < 0$. Obviously there exists a number $r = r_E$ such that

[6] See, for example, Blohincev [1], §50.

$$E+\frac{2}{r}-\frac{l\,(l+1)}{r^2}<0$$

outside the interval $(0, r_E)$. Therefore every solution of (7.2) has at most one zero in the interval (r_E, ∞). The number of zeros in the interval $(0, r_E)$ is finite. From the finiteness, for $E<0$, of the number of zeros of any solution follows the discreteness of the spectrum.

To find the negative eigenvalues we will seek a solution of (7.2) in the form $y(r) = e^{-\lambda r}v(r)$, $\lambda = \sqrt{-E}$, $E<0$.

Then from (7.2) we find the following equation for the function $v(r)$:

$$(7.2') \qquad \frac{d^2v}{dr^2}-2\lambda\frac{dv}{dr}+\left\{\frac{2}{r}-\frac{l\,(l+1)}{r^2}\right\}v=0.$$

We now represent the solution $v(r)$ of (7.2') in the form of a power series

$$(7.3) \qquad v\,(r)=r^{l+1}\sum_{\nu=0}^{\infty}a_\nu r^\nu$$

and, inserting it in equation (7.2'), we obtain for the coefficients the recursion relation

$$(7.4) \qquad a_{\nu+1}=\frac{2\lambda\,(\nu+l+1)-2}{(\nu+l+2)\,(\nu+l+1)-l\,(l+1)}\,a_\nu \quad (\nu=0,\,1,\,\ldots).$$

It is not difficult to see that the series (7.3) will converge for all values of r, but that for $r\to\infty$ the solution of (7.1),

$$(7.5) \qquad R(r) = e^{-\lambda r}v(r)/r,$$

tends to infinity. In fact, putting $\mu = 1/\lambda$, $s = 2l+1$, we rewrite (7.4) in the form

$$a_{\nu+1}=\frac{2\lambda}{\nu+1}\,\frac{\left(\nu+\dfrac{s+1}{2}\right)-\mu}{(\nu+s+1)}\,a_\nu.$$

Hence it is evident that $a_{\nu+1}/a_\nu \to 2\lambda/(\nu+1)$ as $\nu\to\infty$. Further, we can choose $\nu=\nu'$ such that

$$\frac{\nu'+\dfrac{s+1}{2}-\mu}{\nu'+s+1}>\frac{1}{2}(1+\varepsilon),$$

where $\epsilon>0$, $\frac{1}{2}(1+\epsilon)<1$. Starting with this value of ν, the coefficients a_ν grow faster than the coefficients of a series defined by the recursion relation

$$b_{\nu+1}=\frac{\lambda\,(1+\varepsilon)}{\nu+1}\,b_\nu.$$

The series with these coefficients is $v_1(r) = e^{\lambda(1+\epsilon)r}$.

Therefore $v(r)$ grows faster than $v_1(r)$, and consequently the function (7.5) tends to infinity as $r \to \infty$.

Thus, by virtue of (7.5) the solution of equation (7.1) can be square integrable over the halfline $(0, \infty)$ only if the series (7.3) terminates. Then $v(r)$ will be a polynomial. Obviously the series (7.3) can terminate at some point only for particular values of the parameter λ (cf. (7.4)). Suppose that a given coefficient a_n is not zero. In order that the next coefficient a_{n+1} be zero, it is necessary that $\lambda = 1/(n+l+1)$. It is clear that when this is the case not only a_{n+1}, but all subsequent coefficients will equal zero, since all of them are proportional to a_{n+1}. Putting $k = n+l+1$ and taking into account that $E = -\lambda^2$, we conclude that the negative eigenvalues are the numbers $E_k = -1/k^2$ $(k = 1, 2, \cdots)$.

Let us consider the form of the eigenfunctions corresponding to the eigenvalues $E_k = -1/k^2$. Since $\lambda_k = \sqrt{-E_k} = 1/k$, we obtain from (7.4) the following relation for the coefficients a_ν:

$$a_{\nu+1} = -\frac{2}{k} \cdot \frac{k-(\nu+l+1)}{(\nu+1)(2l+\nu+2)} a_\nu .$$

Computing one coefficient from the other (a_0 is arbitrary) and inserting their values in (7.3), we obtain

$$v(r) = a_0 r^{l+1} \Big\{ 1 - \frac{k-l-1}{1!\,(2l+2)} \cdot \frac{2r}{k} + \frac{(k-l-1)(k-l-2)}{2!\,(2l+2)(2l+3)} \left(\frac{2r}{k}\right)^2 + \dots + (-1)^n \times \frac{(k-l-1)(k-l-2)\dots 3 \cdot 2 \cdot 1}{(k-1)!\,(2l+2)(2l+3)\dots(2l+n)} \left(\frac{2r}{k}\right)^n \Big\}.$$

Putting $\xi = 2r/k$, by virtue of (7.5) and the preceding result we obtain the following definitive expression for the function $R(r)$:

$$R_{kl}(r) = C_{kl} e^{-\frac{1}{2}\xi} \xi^l \mathscr{L}_{k+l}^{2l+1}(\xi),$$

where C_{kl} is some constant, and $\mathscr{L}_{k+l}^{2l+1}(\xi)$ is a generalized Laguerre function of order $2l+1$ (cf. §6.3).

3. We now consider positive values of E; let $\sqrt{E} = s$.

As is known,[7] for a solution of (7.2) which is bounded at zero one has the following asymptotic formula for $s \to \infty$:

$$(7.6) \qquad y = \frac{e^{\frac{\pi}{2s}}}{\left|\Gamma\left(l+1+\frac{i}{s}\right)\right|} \frac{1}{s} \cos\left[sr + \frac{1}{s}\ln sr - (l+1)\frac{\pi}{2} + \alpha\right] + o(1),$$

where $\alpha = \arg \Gamma(l+1+i/s)$.

[7] Cf. Fok [1], p. 155.

It easily follows from (7.6) that as $b \to \infty$ the distance between successive zeros of the function $y(b,s)$ —let us denote two of them by s_1 and s_2—tends to zero. We remark that, as is not hard to show, $y(b,s)$ does not have multiple roots. Using these facts, one can obtain a more precise estimate for $s_2 - s_1$:

$$(s_2 - s_1)b + \frac{1}{s_2}\ln 2bs_2 - \frac{1}{s}\ln 2bs_1 = \pi + o(1), \tag{7.7}$$

$$s_2 - s_1 = \frac{\pi}{b} + o\left(\frac{\ln b}{b}\right). \tag{7.8}$$

Inserting (7.8) in (7.7), we obtain

$$(s_2 - s_1)b + \left(\frac{s_2 - s_1}{s_1 s_2}\right)\ln b = (s_2 - s_1)\,b + \frac{\pi}{b s_1 s_2}\ln b + o\left(\frac{ln^2 b}{b}\right) = \pi + o(1).$$

The final estimate for $s_2 - s_1$ is

$$s_2 - s_1 = \frac{\pi}{b} + o\left(\frac{1}{b}\right). \tag{7.9}$$

We also need the following estimate for $b \to \infty$:

$$\int_a^b \cos\left(2sr + \frac{2}{3}\ln 2sr + 2\alpha\right)dr = O(1). \tag{7.10}$$

Let $a > 0$ be arbitrary. We have

$$\int_a^b \cos\left(2sr + \frac{2}{s}\ln 2sr + 2\alpha\right)dr$$
$$= \int_a^b \cos 2sr \cdot \cos\left(\frac{2}{s}\ln 2sr + 2\alpha\right)dr$$
$$- \int_a^b \sin 2sr \cdot \sin\left(\frac{2}{s}\ln 2sr + 2\alpha\right)dr.$$

Both integrals on the right side can be estimated in the same way. Let us consider, for example, the first integral. Integrating by parts twice, we obtain

$$\int_a^b \cos 2sr \cdot \cos\left(\frac{2}{s}\ln 2sr + 2\alpha\right)dr$$
$$= O(1) + \frac{1}{2s^3}\int_a^b \cos 2sr \left\{-\frac{1}{r^2}\sin\left(\frac{2}{s}\ln 2sr + \alpha\right)\right.$$
$$\left. + \frac{1}{2sr^2}\cos\left(\frac{2}{s}\ln 2sr + \alpha\right)\right\} dr = O(1).$$

From the estimates (7.9) and (7.10) follows

$$\Delta\rho_b(s)=\rho_b(s+\Delta)-\rho_b(s)=\sum_{s<s_n\leqslant s+\Delta}\frac{1}{\int_a^b y^2(r,s_n)\,dr}$$

$$=\frac{1}{\pi}\sum_{s<s_n\leqslant s+\Delta}(s_{n+1}-s_n)\,s_n^2e^{\frac{\pi}{s_n}}\left|\Gamma\left(l+1+\frac{i}{s_n}\right)\right|^2$$

$$\text{(7.11)}\qquad\times\left\{\frac{1}{\frac{1}{b}\int_0^b\cos^2\left[s_nr+\frac{1}{s_n}\ln 2s_nr-(l+1)\frac{\pi}{2}+\alpha\right]dr}+o(1)\right\}$$

$$=\frac{2}{\pi}\sum_{s<s_n\leqslant s+\Delta}s_n^2e^{\frac{\pi}{s_n}}\left|\Gamma\left(l+1+\frac{i}{s_n}\right)\right|^2\{1+o(1)\}\{s_{n+1}-s_n\}$$

$$\to\frac{2}{\pi}\int_s^{s+\Delta}s^2e^{\frac{\pi}{s}}\left|\Gamma\left(l+1+\frac{i}{s}\right)\right|^2ds.$$

Using the well-known properties of the Γ-function

$$\Gamma(z+1)=z\Gamma(z),\qquad \Gamma(z)\Gamma(1-z)=\pi/\sin\pi z,$$

we obtain

$$\left|\Gamma\left(l+1+\frac{i}{s}\right)\right|^2=\Gamma\left(l+1+\frac{i}{2}\right)\Gamma\left(l+1-\frac{i}{s}\right)$$

$$=\left(l+\frac{i}{s}\right)\cdots\left(1+\frac{i}{s}\right)\frac{i}{s}\left(l-\frac{i}{s}\right)\cdots\left(1-\frac{i}{s}\right)1\left(\frac{i}{s}\right)\Gamma\left(1-\frac{i}{s}\right)$$

$$=\left(l^2+\frac{1}{s^2}\right)\cdots\left(1+\frac{1}{s^2}\right)\frac{1}{s}\frac{\pi}{\operatorname{sh}\frac{\pi}{3}}=\frac{\pi}{s}\frac{2}{e^{\pi/s}-e^{-\pi/s}}\prod_{n=1}^{l}\left(n^2+\frac{1}{s^2}\right).$$

Therefore we finally obtain from (7.11)

$$\Delta\rho(s)=\int_s^{s+\Delta}\frac{4s\prod_{n=1}^{l}\left(n^2+\frac{1}{s^2}\right)}{\left(1-e^{-\frac{2\pi}{s}}\right)},$$

i.e. the spectrum is continuous.

§8. A Dirac system

We will consider the equation

$$\text{(8.1)}\qquad\begin{pmatrix}p(x) & q(x)+\frac{d}{dx}\\ q(x)-\frac{d}{dx} & r(x)\end{pmatrix}\begin{pmatrix}y_1(x)\\ y_2(x)\end{pmatrix}=\lambda\begin{pmatrix}y_1(x)\\ y_2(x)\end{pmatrix}.\qquad(0\leqslant x<\infty).$$

$$(0\leqslant x<\infty).$$

Let $p(x) = a$, $q(x) = 0$, $r(x) = -a$, where a is an arbitrary positive number. Then equation (8.1) assumes the form

$$y_1' + (a+\lambda)y_2 = 0, \qquad y_2' + (a-\lambda)y_1 = 0. \tag{8.2}$$

Let us denote by

$$\varphi(x,\lambda) = \begin{pmatrix} \varphi_1 \\ \varphi_2 \end{pmatrix}$$

the solution of the system (8.2) satisfying the initial conditions

$$\varphi_1(0,\lambda) = -\sin\alpha, \qquad \varphi_2(0,\lambda) = \cos\alpha, \tag{8.3}$$

where α is an arbitrary real number. It follows from the system (8.2) that the functions $\varphi_1(x,\lambda)$ and $\varphi_2(x,\lambda)$ satisfy the same second-order equation, namely

$$y'' = (a^2 - \lambda^2)y. \tag{8.4}$$

Further, from the system (8.2) and the conditions (8.3) follows

$$\varphi_1'(0,\lambda) = -(a+\lambda)\cos\alpha, \qquad \varphi_2'(0,\lambda) = (a-\lambda)\sin\alpha. \tag{8.5}$$

Now, solving equation (8.4) for the initial conditions (8.3) and (8.5), we obtain the following explicit expressions for $\varphi_1(x,\lambda)$ and $\varphi_2(x,\lambda)$:

$$\begin{aligned} \varphi_1(x,\lambda) &= -\sin\alpha\cdot\cos\{x\sqrt{\lambda^2-a^2}\} \\ &\quad - \sqrt{\frac{\lambda+a}{\lambda-a}}\cos\alpha\cdot\sin\{x\sqrt{\lambda^2-a^2}\}, \\ \varphi_2(x,\lambda) &= \cos\alpha\cdot\cos\{x\sqrt{\lambda^2-a^2}\} - \\ &\quad - \sqrt{\frac{\lambda-a}{\lambda+a}}\sin\alpha\cdot\sin\{x\sqrt{\lambda^2-a^2}\}. \end{aligned} \tag{8.6}$$

Now let

$$\theta(x,\lambda) = \begin{pmatrix} \theta_1(x,\lambda) \\ \theta_2(x,\lambda) \end{pmatrix}$$

be the solution of the system (8.2) satisfying the initial conditions

$$\theta_1(0,\lambda) = \cos\alpha, \qquad \theta_2(0,\lambda) = \sin\alpha.$$

Proceeding as above, we obtain the following explicit expressions for $\theta_1(x,\lambda)$ and $\theta_2(x,\lambda)$:

$$\begin{aligned} \theta_1(x,\lambda) &= \cos\alpha\cdot\cos\{x\sqrt{\lambda^2-a^2}\} - \sqrt{\frac{\lambda+a}{\lambda-a}}\sin\alpha\cdot\sin\{x\sqrt{\lambda^2-a^2}\}, \\ \theta_2(x,\lambda) &= \sin\alpha\cdot\cos\{x\sqrt{\lambda^2-a^2}\} + \sqrt{\frac{\lambda-a}{\lambda+a}}\cos\alpha\cdot\sin\{x\sqrt{\lambda^2-a^2}\}. \end{aligned} \tag{8.7}$$

All the square roots in (8.6) and (8.7) are real and positive if λ is real and greater than a.

From the general theory (cf. Chapter 3, §2) it follows that the function $\psi(x,\lambda)=\theta(x,\lambda)+m(\lambda)\varphi(x,\lambda)$ belongs to $\mathcal{L}^2(0,\infty)$ for $\operatorname{Im}\lambda>0$. Consequently, the function $\theta_1(x,\lambda)+m(\lambda)\varphi_1(x,\lambda)$ also belongs to $\mathcal{L}^2(0,\infty)$ for $\operatorname{Im}\lambda>0$, and so it is a multiple of $\exp\{ix\sqrt{\lambda^2-a^2}\}$, i.e. we have

$$\{\cos\alpha-m(\lambda)\sin\alpha\}\cos\{x\sqrt{\lambda^2-a^2}\}$$
$$-\sqrt{\frac{\lambda+a}{\lambda-a}}\{\sin\alpha+m(\lambda)\cos\alpha\}\sin\{x\sqrt{\lambda^2-a^2}\}$$
$$=C\left[\cos\{x\sqrt{\lambda^2-a^2}\}+i\sin\{x\sqrt{\lambda^2-a^2}\}\right].$$

Hence we obtain

$$-\sqrt{\frac{\lambda+a}{\lambda-a}}\{\sin\alpha+m(\lambda)\cos\alpha\}=i\{\cos\alpha-m(\lambda)\sin\alpha\}$$

or

$$m(\lambda)=-\frac{\sqrt{\lambda+a}\sin\alpha+i\sqrt{\lambda-a}\cos\alpha}{\sqrt{\lambda+a}\cos\alpha-i\sqrt{\lambda-a}\sin\alpha}. \tag{8.8}$$

If $\lambda>a$, it follows from (8.8) that

$$\operatorname{Im} m(\lambda)=-\frac{\sqrt{\lambda^2-a^2}}{\lambda+a\cos 2\alpha}.$$

If $-a<\lambda<a$, then, as follows from (8.8), the function $m(\lambda)$ is real, while if $\lambda<-a$, say $\lambda=-\mu$ $(\mu>a)$, then $\sqrt{\lambda+a}=-\sqrt{\mu-a}$, $\sqrt{\lambda-a}=-\sqrt{\mu+a}$ and

$$\operatorname{Im} m(\lambda)=-\frac{\sqrt{\mu^2-a^2}}{\mu-a\cos 2\alpha}.$$

If $\cos\alpha$ and $\sin\alpha$ have the same sign, then $m(\lambda)$ has no poles on the real axis between $-a$ and a, and therefore in this case the expansion formula has the form

$$f(x)=\frac{1}{\pi}\int_a^\infty\frac{\sqrt{\mu^2-a^2}}{\mu-a\cos 2\alpha}\varphi(x,-\mu)F(-\mu)\,d\mu$$
$$+\frac{1}{\pi}\int_a^\infty\frac{\sqrt{\lambda^2-a^2}}{\lambda+a\cos 2\alpha}\varphi(x,\lambda)F(\lambda)\,d\lambda, \tag{8.9}$$

where $F(x)$ is the Fourier transform of $f(x)$ with respect to the function $\varphi(x,\lambda)$.

If $\cos\alpha$ and $\sin\alpha$ have opposite signs, then $m(\lambda)$ has a pole at the point $\lambda=-a\cos 2\alpha$ with residue $-2a\sin 2\alpha$. Since

$$\varphi_1(x, -a\cos 2\alpha) = -\sin\alpha \cdot e^{ax\sin 2\alpha}, \quad \varphi_2(x, -a\cos 2\alpha) = \cos\alpha \cdot e^{ax\sin 2\alpha},$$

one has to add to the right of the expansion (8.9) the term

$$-2a\sin 2\alpha\varphi(x, -a\cos 2\alpha)\int_0^\infty \{-f_1(t)\sin\alpha + f_2(t)\cos\alpha\}\, e^{at\sin 2\alpha}dt.$$

Bibliographical references

§1. The results of this section are classical.

§2. Cf. Watson [**1**], Chapter 18, Titchmarsh [**1**], Chapter 4. The method discussed here is due to Levitan.

§3. Cf. Weyl [**3**], Watson [**1**], Chapter 14, Titchmarsh [**1**], Chapter 4, Levitan [**5**].

§4. Szegö [**1**], Titchmarsh [**1**], Chapter 4, Levitan [**1**].

§5. Szegö [**1**], Titchmarsh [**1**], Chapter 4, Courant and Hilbert [**1**].

§6. Courant and Hilbert [**1**].

§7. Schrödinger [**1**], Titchmarsh [**1**], Chapter 4. The method discussed here is due to Levitan [**1**].

§8. This example is taken from a paper by Titchmarsh [**5**].

CHAPTER 6

SOLUTION OF THE CAUCHY PROBLEM FOR THE ONE-DIMENSIONAL WAVE EQUATION

§1. Application of the method of successive approximations

Let $q(x)$ be a real function on the line $(-\infty, \infty)$ which is summable over every finite interval, and let $f(x)$ be a twice-differentiable function, also defined on the line $(-\infty, \infty)$.

Under these assumptions on the functions $q(x)$ and $f(x)$, we will consider the following Cauchy problem:

$$\partial^2 u/\partial t^2 = \partial^2 u/\partial x^2 - q(x)u, \tag{1.1}$$

$$u|_{t=0} = f(x), \qquad \partial u/\partial t|_{t=0} = 0. \tag{1.2}$$

Let us construct an integral equation for the solution of this Cauchy problem. To this end we rewrite equation (1.1) in the form

$$\partial^2 u/\partial t^2 - \partial^2 u/\partial x^2 = -q(x)u. \tag{1.1'}$$

If $q(x) \equiv 0$, then (1.1′) assumes the form

$$\partial^2 u/\partial t^2 - \partial^2 u/\partial x^2 = 0. \tag{1.3}$$

As is known, the solution of the problem (1.3) + (1.2) has the form

$$u_0(x,t) = \tfrac{1}{2}[f(x+t) + f(x-t)]. \tag{1.4}$$

Let us now consider the inhomogeneous equation

$$\partial^2 u/\partial t^2 - \partial^2 u/\partial x^2 = g(x,t) \tag{1.5}$$

with known right side $g(x,t)$. We denote by $\tilde{u}(x,t)$ the solution of equation (1.5) which satisfies the initial conditions

$$\tilde{u}|_{t=0} = 0, \qquad \partial\tilde{u}/\partial t|_{t=0} = 0. \tag{1.6}$$

It is well known that

$$\tilde{u}(x, t) = \frac{1}{2}\iint_{\Delta_{x,t}} g(y, \tau)\, dy\, d\tau, \tag{1.7}$$

where $\Delta_{x,t}$ is the triangle in the (y,τ) plane whose vertices are the points $(x-t,0)$, (x,t) and $(x+t,0)$.

Using formulas (1.4) and (1.7), it is not hard to construct an integral equation for the solution of the problem (1.1′) + (1.2). In fact, considering the right side of (1.1′) as a known function and applying formulas (1.4) and (1.7), we obtain the following integral equation, which is equivalent to the problem (1.1′) + (1.2):

$$u(x, t) = \frac{1}{2}\{f(x+t) + f(x-t)\} - \frac{1}{2}\int_0^t d\tau \int_{x-(t-\tau)}^{x+(t-\tau)} q(y)\, u(y, \tau)\, dy$$

$$= u_0(x, t) - \frac{1}{2}\int_0^t d\tau \int_{x-(t-\tau)}^{x+(t-\tau)} q(y)\, u(y, \tau)\, dy. \tag{1.8}$$

The integral equation (1.8) is an equation of Volterra type and can therefore be solved by the method of successive approximations.

Let the function $u_0(x,t)$ be defined by (1.4). For $k \geqq 1$ we put

$$u_k(x, t) = \frac{1}{2}\int_0^t d\tau \int_{x-(t-\tau)}^{x+(t-\tau)} q(y)\, u_{k-1}(y, \tau)\, dy \qquad (k = 1, 2, \ldots).$$

Then from the uniform convergence of the successive approximations, whose proof we omit, assuming it to be well known, it follows that the solution $u(x,t)$ of the integral equation (1.8) is given by

$$u(x,t) = u_0(x,t) - u_1(x,t) + u_2(x,t) - \cdots. \tag{1.9}$$

We will show that each of the functions $u_k(x,t)$ $(k = 1, 2, \cdots)$ can be represented in the form

$$u_k(x, t) = \frac{1}{2}\int_{x-t}^{x+t} f(s)\, w_k(x, t, s)\, ds. \tag{1.10}$$

For $k = 1$ we have

$$u_1(x, t) = \frac{1}{2}\int_0^t d\tau \left\{ \int_{x-(t-\tau)}^{x+(t-\tau)} q(z)\, \frac{1}{2}[f(z+\tau) + f(z-\tau)]\, dz \right\}$$

$$= \frac{1}{4}\int_0^t d\tau \int_{x-(t-\tau)}^{x+(t-\tau)} q(z) f(z+\tau)\, dz + \frac{1}{4}\int_0^t d\tau \int_{x-(t-\tau)}^{x+(t-\tau)} q(z) f(z-\tau)\, dz$$

$$= \frac{1}{4}\int_0^t d\tau \int_{x-t+2\tau}^{x+t} q(s-\tau) f(s)\, ds + \frac{1}{4}\int_0^t d\tau \int_{x-t}^{x+t-2\tau} q(s+\tau) f(s)\, ds.$$

Interchanging the order of integration in both integrals, we obtain

$$u_1(x, t) = \frac{1}{4} \int_{x-t}^{x+t} f(s)\,ds \int_0^{\frac{1}{2}(s-x+t)} q(s-\tau)\,d\tau$$

$$+ \frac{1}{4} \int_{x-t}^{x+t} f(s)\,ds \int_0^{\frac{1}{2}(x-s+t)} q(s+\tau)\,d\tau$$

$$= \frac{1}{2} \int_{x-t}^{x+t} f(s) \left\{ \frac{1}{2} \int_{\frac{1}{2}(x+s-t)}^{\frac{1}{2}(x+s+t)} q(\tau)\,d\tau \right\} ds.$$

Thus we obtain formula (1.10), if we put

$$w_1(x, t, s) = \frac{1}{2} \int_{\frac{1}{2}(x+s-t)}^{\frac{1}{2}(x+s+t)} q(\tau)\,d\tau. \tag{1.11}$$

Let us now assume that for $k = 1, 2, \cdots, n-1$ formula (1.10) has already been proved. We will prove it for $k = n$. We have

$$u_n(x, t) = \frac{1}{2} \int_0^t d\tau \left\{ \int_{x-(t-\tau)}^{x+(t-\tau)} q(z) \left[\frac{1}{2} \int_{z-\tau}^{z+\tau} f(s)\, w_{n-1}(z, \tau, s)\,ds \right] dz \right\}. \tag{1.12}$$

Change the order of integration so that the outer integral is with respect to s. Since $z - \tau \leqq s \leqq z + \tau$ and $x - t + \tau \leqq z \leqq x + t - \tau$, we have $x - t \leqq s \leqq x + t$. Therefore the limits of integration with respect to s are $x - t$ and $x + t$. We thus obtain

$$u_n(x, t) = \frac{1}{2} \int_{x-t}^{x+t} f(s) \left\{ \frac{1}{2} \iint_{\sigma_{\tau z}} w_{n-1}(z, \tau, s)\, q(z)\,d\tau\,dz \right\} ds,$$

where $\sigma_{\tau z}$ is some region in the (τ, z) plane (depending on x and t), whose exact form will not be needed in what follows. Thus, formula (1.10) has been proved for $k = n$, where

$$w_n(x, t, s) = \frac{1}{2} \iint_{\sigma_{\tau z}} q(z)\, w_{n-1}(z, \tau, s)\,d\tau\,dz. \tag{1.13}$$

From (1.8) and (1.9) follows the formula

$$u(x, t) = \frac{1}{2}\{f(x+t) + f(x-t)\} + \frac{1}{2} \int_{x-t}^{x+t} w(x, t, s) f(s)\,ds, \tag{1.14}$$

where we have put

(1.15) $w(x,t,s) = -w_1(x,t,s) + w_2(x,t,s) - \cdots + (-1)^n w_n(x,t,s) + \cdots.$

We now derive various estimates for the function $w(x,t,s)$, valid for small t.

Lemma 1.1. *Suppose that $q(x)$ is summable over every finite interval. Then for $t > 0$ we have the estimate*

$$|w(x, t, s)| \leqslant \frac{1}{2} \int_{x-t}^{x+t} |q(\tau)| \, d\tau \exp\left(\frac{1}{2} t \int_{x-t}^{x+t} |q(\tau)| \, d\tau\right). \tag{1.16}$$

Proof. We will show that under the stated assumptions one has the inequalities

$$|w_k(x, t, s)| \leqslant \frac{t^{k-1}}{(k-1)!} \left\{\frac{1}{2} \int_{x-t}^{x+t} |q(\tau)| \, d\tau\right\}^k \qquad (k = 1, 2, \cdots). \tag{1.17}$$

$(k = 1, 2, \ldots)$.

Let us first consider the case $k = 1$. We find from formula (1.11) that

$$|w_1(x, t, s)| \leqslant \frac{1}{2} \int_{\frac{1}{2}(x+s-t)}^{\frac{1}{2}(x+s+t)} |q(\tau)| \, d\tau.$$

Since $x - t \leqq s \leqq x + t$ (cf. the formula preceding (1.11)), the preceding formula implies the estimate

$$|w_1(x, t, s)| \leqslant \frac{1}{2} \int_{x-t}^{x+t} |q(\tau)| \, d\tau,$$

i.e. (1.17).

Let us now assume that (1.17) has been proved for all $k = 1, 2, \cdots, n-1$. We will prove it for $k = n$.

From (1.13) and the estimate (1.17) follows the estimate

$$|w_n(x, t, s)| \leqslant \frac{1}{2} \iint_{\sigma_{\tau z}} |w_{n-1}(z, \tau, s)| \, |q(z)| \, d\tau \, dz$$

$$\leqslant \frac{1}{2} \iint_{\sigma_{\tau z}} |q(z)| \frac{\tau^{n-2}}{(n-2)!} \left\{\frac{1}{2} \int_{z-\tau}^{z+\tau} |q(r)| \, dr\right\}^{n-1} d\tau \, dz.$$

Since (cf. (1.12)) $x - t + \tau \leqq z \leqq x + t - \tau$, we have $x - t \leqq z - \tau \leqq z + \tau \leqq x + t$. Therefore the last estimate implies the inequality

$$(1.18) \qquad |w_n(x,\ t,\ s)| \leqslant \frac{1}{2\,(n-2)\,!}\left\{\frac{1}{2}\int_{x-t}^{x+t}|q(r)|\,dr\right\}^{n-1}\iint_{\sigma_{\tau z}}|q(z)|\,\tau^{n-2}dz\,d\tau.$$

Further, since $0 \leqq \tau \leqq t$ and $x - t + \tau \leqq z \leqq x + t - \tau$, we have $x - t \leqq z \leqq x + t$, and consequently the region $\sigma_{\tau z}$ lies in the rectangle $0 \leqq \tau \leqq t$; $x - t \leqq z \leqq x + t$. Therefore from the inequality (1.18) follows the estimate

$$|w_n(x,\ t,\ s)| \leqslant \frac{1}{2\,(n-2)\,!}\left\{\frac{1}{2}\int_{x-t}^{x+t}|q(r)|\,dr\right\}^{n-1}\int_0^t \tau^{n-2}d\tau\int_{x-t}^{x+t}|q(z)|\,dz$$

$$= \frac{t^{n-1}}{(n-1)\,!}\left\{\frac{1}{2}\int_{x-t}^{x+t}|q(\tau)|\,d\tau\right\}^{n},$$

i.e. (1.17).

The inequality (1.16) follows at once from (1.17). In fact, by virtue of formula (1.15)

$$|w(x,\ t,\ s)| \leqslant \sum_{k=1}^{\infty}|w_k(x,\ t,\ s)|$$

$$\leqslant \sum_{k=1}^{\infty}\frac{t^{k-1}}{(k-1)\,!}\left\{\frac{1}{2}\int_{x-t}^{x+t}|q(\tau)|\,d\tau\right\}^{k}$$

$$= \frac{1}{2}\int_{x-t}^{x+t}|q(\tau)|\,d\tau\,\cdot\exp\left(\frac{1}{2}t\int_{x-t}^{x+t}|q(\tau)|\,d\tau\right),$$

as was to be proved.

By making various assumptions concerning the behaviour of the function $q(x)$, we can obtain various estimates for the function $w(x, t, s)$ for small t.

Theorem 1.1. *If the function $q(x)$ is summable over every finite interval, then for every fixed finite interval (x_0, x_1) and any fixed number $t_0 > 0$ there exists a constant $C = C(x_0, x_1; t_0)$ such that for $x \in (x_0, x_1)$ and $0 < t < t_0$ we have $|w(x, t, s)| < C$.*

The proof follows at once from (1.16).

Theorem 1.2. *Suppose that in some interval the function $q(x)$ satisfies the following condition: there exist positive constants $C = C(x_0, x_1)$ and $\alpha = \alpha(x_0, x_1)$ such that for $t \leqq t_0$ (t_0 is a fixed number) and $x \in (x_0, x_1)$ one has the inequality*

$$(1.19) \qquad \int_{x-t}^{x+t}|q(\tau)|\,d\tau < Ct^{\alpha}.$$

Then there exists a constant C which depends upon the constants in the inequality (1.19) *but not upon t, such that for* $x \in (x_0, x_1)$ *and* $t \leqq t_0$ *we have*

$$|w(x,t,s)| < Ct^{\alpha}. \tag{1.20}$$

The proof follows at once from (1.16).

REMARK. For the bound (1.19) it is sufficient that there exist constants $C(x_0, x_1; t_0) = C$ and $p = p(x_0, x_1; t_0) > 1$ such that for $x \in (x_0 - t_0, x_1 + t_0)$ one has the inequality

$$\left(\int_{x-t_0}^{x+t_0} |q(\tau)|^p \, d\tau \right)^{1/p} < C.$$

In fact, by Hölder's inequality $(1/p + 1/p' = 1)$ we have for $t < t_0$

$$\int_{x-t}^{x+t} |q(\tau)| \, d\tau \leqslant \left(\int_{x-t_0}^{x+t_0} |q(\tau)|^p \, d\tau \right)^{p} \left(\int_{x-t}^{x+t} d\tau \right)^{1/p'} < Ct^{1/p'}.$$

From the differentiability of the functions $w_k(x,t,s)$ with respect to x and from the uniform convergence of the series obtained by termwise differentiation of the series (1.15), which results from the uniform convergence of the series obtained by termwise (multiple) differentiation of (1.15) with respect to x, follows:

LEMMA 1.2. *If the function* $q(x)$ *has a derivative of order* $k-1$, *then the function* $w(x,t,s)$ *is k times differentiable with respect to x and s.*

§2. Reduction to the Goursat problem

In this section we will show that the function $w(x,t,s)$ is the solution of the Goursat problem. We will carry out the argument under the assumption that formula (1.14) has already been established, and will derive the Goursat problem for the function $w(x,t,s)$. By a modification of the argument we could at the same time obtain a new justification of the formula (1.14), but we will not do this.

Thus, we will consider the Cauchy problem

$$\partial^2 u/\partial t^2 = \partial^2 u/\partial x^2 - q(x)u, \tag{2.1}$$

$$u(x,t)|_{t=0} = f(x), \tag{2.2}$$

$$\partial u(x,t)/\partial t|_{t=0} = 0. \tag{2.3}$$

Suppose that the solution of this problem is given by (1.14), i.e.

$$u(x, t) = \frac{1}{2} \{f(x+t) + f(x-t)\} + \frac{1}{2} \int_{x-t}^{x+t} w(x, t, s) f(s) \, ds. \tag{2.4}$$

We wish to clarify the conditions that the function $w(x,t,s)$ must satisfy in order that the function $u(x,t)$, defined by (2.4), be a solution of the problem (2.1) + (2.2) + (2.3). It follows from (2.4) that the condition (2.2) is automatically fulfilled. Further, differentiating (2.4) with respect to t and then putting $t=0$ and using the condition (2.3), we obtain

$$0 = \tfrac{1}{2}\{ w(x,t,x+t)f(x+t) + w(x,t,x-t)f(x-t) \}|_{t=0} = w(x,0,x)f(x).$$

Therefore

$$w(x,0,x) = 0. \tag{2.5}$$

The condition (2.5) will be used further on.

We will now derive a differential equation and supplementary conditions for the function $w(x,t,s)$ (we will see that the Goursat problem is obtained for the function $w(x,t,s)$).

Let us denote by $u(x,t;f)$ the solution of the problem (2.1) + (2.2) + (2.3), and suppose we already know that $u(x,t;f)$ is represented by (2.4). Denote by L_t the operator d^2/dt^2, and by M_x the operator $d^2/dx^2 - q(x)$. We will show that

$$M_x u(x,t;f) = u(x,t;M_x f).$$

In fact, the left and right sides of this equality satisfy the same Cauchy problem

$$M_x v = L_t v, \qquad v|_{t=0} = M_x f(x), \qquad \partial v/\partial t|_{t=0} = 0.$$

Therefore the required equality follows from the uniqueness of the solution of the Cauchy problem.

Thus equation (2.1) can be written in the form

$$L_t u = u(x,t;M_x f). \tag{2.6}$$

Using (2.4), we obtain

$$\begin{aligned} L_t u = {} & \frac{1}{2}\{f''(x+t) + f''(x-t)\} \\ & + \frac{1}{2}\Big\{f(x+t)\frac{d}{dt}w(x,t,x+t) + w(x,t,x+t)f'(x+t) \\ & + f(x-t)\frac{d}{dt}w(x,t,x-t) - w(x,t,x-t)f'(x-t)\Big\} \\ & + \frac{1}{2}\Big\{f(x+t)\frac{\partial w(x,t,s)}{\partial t}\Big|_{s=x+t} + f(x-t)\frac{\partial w(x,t,s)}{\partial t}\Big|_{s=x-t}\Big\} \\ & + \frac{1}{2}\int_{x-t}^{x+t} f(s)\frac{\partial^2 w(x,t,s)}{\partial t^2}\,ds. \end{aligned} \tag{2.7}$$

Further, again using (2.4) and integrating by parts twice, we obtain

$$u(x, t; M_x f) = \frac{1}{2}\{f''(x+t) + f''(x-t) - q(x+t)f(x+t) - q(x-t)f(x-t)\} - \frac{1}{2}\int_{x-t}^{x+t} q(s)w(x, t, s)f(s)\,ds + \frac{1}{2}\{w(x, t, x+t)f'(x+t) - w(x, t, x-t)f'(x-t)\} - \frac{1}{2}\left\{f(x+t)\frac{\partial w(x, t, s)}{\partial s}\bigg|_{s=x+t} - f(x-t)\frac{\partial w(x, t, s)}{\partial s}\bigg|_{s=x-t}\right\} + \frac{1}{2}\int_{x-t}^{x+t} f(s)\frac{\partial^2 w(x, t, s)}{\partial s^2}\,ds. \tag{2.8}$$

Since the function $f(x)$ is arbitrary, the coefficients in (2.7) and (2.8) of $f(x+t)$ and $f(x-t)$ and the integrands must coincide (the coefficients of the derivatives of $f(x)$ cancel each other). Equating the coefficients of $f(x+t)$ in (2.7) and (2.8), we obtain

$$\frac{dw(x, t, x+t)}{dt} + \frac{\partial w(x, t, s)}{\partial t}\bigg|_{s=x+t} = -q(x+t) - \frac{\partial w(x, t, s)}{\partial s}\bigg|_{s=x+t},$$

from which

$$\frac{dw(x, t, x+t)}{dt} = -\frac{1}{2}q(x+t).$$

Integrating with respect to t from 0 to t and using the condition (2.5), we obtain

$$w(x, t, x+t) = -\frac{1}{2}\int_0^t q(x+\tau)\,d\tau. \tag{2.9}$$

In the same way, equating the coefficients of $f(x-t)$ leads to

$$w(x, t, x-t) = -\frac{1}{2}\int_0^t q(x-\tau)\,d\tau. \tag{2.10}$$

Finally, equating the integrands in (2.7) and (2.8), we obtain for the function $w(x,t,s)$, as a function of the variables t and s, the following differential equation:

$$\partial^2 w/\partial t^2 = \partial^2 w/\partial s^2 - q(s)w. \tag{2.11}$$

The equation (2.11), together with the conditions (2.9) and (2.10), defines the Goursat problem for the function $w(x,t,s)$.

§3. The solution of a mixed problem on the halfline

Let $q(x)$ be a real function, defined on the halfline $[0, \infty)$ and summable over every finite interval, and let $f(x)$ be a twice differentiable function on the same halfline. We extend the function $q(x)$ to the negative halfline so that it is still summable over every finite interval, but otherwise arbitrary, and we assume $f(x)$ to be extended to the negative halfline in an as yet unspecified way (further on we will make precise the way in which $f(x)$ is to be extended).

With these assumptions concerning the functions $q(x)$ and $f(x)$, we consider the following problem:

$$\partial^2 u/\partial t^2 = \partial^2 u/\partial x^2 - q(x)u, \tag{3.1}$$

$$u|_{t=0} = f(x), \quad \partial u/\partial t|_{t=0} = 0, \tag{3.2}$$

$$u|_{x=0} = \varphi(t), \tag{3.3}$$

where $\varphi(t)$ is some real function defined for all positive t. The conditions imposed upon $\varphi(t)$ will be clarified later.

We will denote the solution of the problem (3.1) + (3.2) + (3.3) by $u(x,t)$.

As we saw in §1, under the stated assumptions regarding the functions $q(x)$ and $f(x)$ the solution of the Cauchy problem (3.1) + (3.2) can be represented in the form

$$u(x, t) = \frac{1}{2}\{f(x+t) + f(x-t)\} + \frac{1}{2}\int_{x-t}^{x+t} w(x, t, s) f(s)\, ds. \tag{3.4}$$

Using the condition (3.3), we obtain from (3.4)

$$f(t) + f(-t) + \int_{-t}^{t} w(0, t, s) f(s)\, ds = 2\varphi(t). \tag{3.5}$$

If we assume that $f(t)$ and $\varphi(t)$ are given for $t \geqq 0$, then equation (3.5) can be taken to define $f(t)$ for $t < 0$. In fact, let us rewrite (3.5) in the following form:

$$\begin{aligned} f(-t) + \int_{-t}^{0} w(0, t, s) f(s)\, ds \\ = 2\varphi(t) - f(t) - \int_{0}^{t} w(0, t, s) f(s)\, ds = g(t). \end{aligned} \tag{3.6}$$

We obviously know $g(t)$ for $t \geqq 0$. Further,

$$\int_{-t}^{0} w(0, t, s) f(s)\, ds = \int_{0}^{t} K(t, s) f(-s)\, ds,$$

where $K(t,s) = w(0,t,-s)$ is a known function. Thus equation (3.6) assumes the form

$$f(-t) + \int_0^t K(t, s) f(-s)\, ds = g(t). \tag{3.7}$$

Equation (3.7) is an integral equation of Volterra type. Solving it for $f(-t)$, we find that

$$f(-t) = g(t) + \int_0^t K_1(t, s)\, g(s)\, ds. \tag{3.8}$$

Suppose that $f(t)$ and $g(t)$ have continuous second derivatives on the halfline $0 \leq t < \infty$. We extend $f(t)$ to the negative halfline in accordance with (3.8). Let us clarify under what conditions the extended function also has a continuous second derivative on the entire real line. It follows from (3.8) that for $t < 0$ the function $f(t)$ has a continuous second derivative. Thus we only have to study the behaviour of the extended function and its first two derivatives at the point $t = 0$.

Obviously, in order that the problem (3.1)–(3.3) have a solution which is continuous at zero, it is necessary that the condition

$$f(+0) = \varphi(+0) \tag{3.9}$$

be fulfilled. Therefore if this condition is fulfilled, then passing to the limit $t \to +0$ in equation (3.6), we obtain $f(-0) = f(+0)$, i.e. the extended function is continuous at zero.

Further, differentiating (3.6) with respect to t, we obtain

$$\begin{aligned} -f'(-t) + w(0, t, -t) f(-t) + \int_{-t}^{0} \frac{\partial w(0, t, s)}{\partial t} f(s)\, ds \\ = 2\varphi'(t) - f'(t) - f(t) w(0, t, t) - \int_0^t \frac{\partial w(0, t, s)}{\partial t} f(s)\, ds. \end{aligned} \tag{3.10}$$

Letting $t \to +0$ in (3.10), we obtain

$$\begin{aligned} -f'(-0) + f(-0) w(0, 0, 0) \\ = 2\varphi'(+0) - f'(+0) - f(+0) w(0, 0, 0). \end{aligned} \tag{3.10$'$}$$

Since $w(0,0,0) = 0$ (cf. (2.5)), from (3.10′) we obtain

$$f'(+0) - f'(-0) = 2\varphi'(+0). \tag{3.11}$$

Further, it follows from the conditions (3.2) and (3.3) that

$$\varphi'(+0)=0. \tag{3.12}$$

Therefore from (3.11) follows $f'(+0)=f'(-0)$, and so the first derivative is continuous.

Finally, let us consider the second derivative. To this end we differentiate (3.10) with respect to t, obtaining

$$f''(-t)+f(-t)\frac{dw(0,t,-t)}{dt}-f'(-t)w(0,t,-t)$$

$$+f(-t)\frac{\partial w(0,t,s)}{\partial t}\bigg|_{s=-t}+\int_{-t}^{0}\frac{\partial^2 w(0,t,s)}{\partial t^2}f(s)\,ds=2\varphi''(t)$$

$$=f''(t)-f(t)\frac{dw(0,t,t)}{dt}-f'(t)w(0,t,t)$$

$$-f(t)\frac{\partial w(0,t,s)}{\partial t}\bigg|_{s=t}-\int_{0}^{t}\frac{\partial^2 w(0,t,s)}{\partial t^2}f(s)\,ds.$$

Letting $t\to+0$ and taking the preceding results concerning $f(0)$ and $f'(0)$ into account, we obtain

$$2\varphi''(+0)-f''(+0)-f''(-0) \tag{3.13}$$

$$=f(0)\left[\frac{dw(0,t,t)}{dt}+\frac{dw(0,t,-t)}{dt}+\frac{\partial w(0,t,s)}{\partial t}\bigg|_{s=t}+\frac{\partial w(0,t,s)}{\partial t}\bigg|_{s=-t}\right]_{t=0}.$$

Let us compute the term in the brackets on the right side of (3.13). To this end we differentiate (3.4) twice with respect to t, obtaining

$$\frac{\partial^2 u}{\partial t^2}=\frac{1}{2}\{f''(x+t)+f''(x-t)\}+\frac{1}{2}\Big\{f'(x+t)w(x,t,x+t)$$

$$+f(x+t)\frac{dw(x,t,x+t)}{dt}-f'(x-t)w(x,t,x-t)$$

$$+f(x-t)\frac{dw(x,t,x-t)}{dt}+f(x+t)\frac{\partial w(x,t,s)}{\partial t}\bigg|_{s=x+t}$$

$$+f(x-t)\frac{\partial w(x,t,s)}{\partial t}\bigg|_{s=x-t}\Big\}+\frac{1}{2}\int_{x-t}^{x+t}\frac{\partial^2 w(x,t,s)}{\partial t^2}f(s)\,ds.$$

Putting $t=0$, by virtue of (2.5) we obtain

$$\frac{\partial^2 u}{\partial t^2}\bigg|_{t=0}-f''(x)=\frac{1}{2}f(x)\Big\{\frac{dw(x,t,x+t)}{dt}+\frac{dw(x,t,x-t)}{dt}$$
$$+\frac{\partial w(x,t,s)}{\partial t}\bigg|_{s=x+t}+\frac{\partial w(x,t,s)}{\partial t}\bigg|_{s=x-t}\Big\}_{t=0}. \tag{3.14}$$

Further, using the first of conditions (3.2), we obtain

$$\partial^2 u/\partial t^2|_{t=0}=f''(x)-q(x)f(x).$$

Hence

$$\partial^2 u/\partial t^2|_{t=0} - f''(x) = -q(x)f(x). \tag{3.15}$$

From (3.14) and (3.15) follows

$$f(x)\left[\frac{dw(x,t,x+t)}{dt} + \frac{dw(x,t,x-t)}{dt} + \frac{\partial w(x,t,s)}{\partial t}\bigg|_{s=x+t} + \frac{\partial w(x,t,s)}{\partial t}\bigg|_{s=x-t}\right]_{t=0} = -2q(x)f(x). \tag{3.16}$$

Putting $x=0$ in (3.16) and inserting the result in (3.13), we obtain

$$2\varphi''(+0) - f''(+0) - f''(-0) = -2q(0)f(0). \tag{3.17}$$

On the other hand, differentiating the condition (3.3) twice with respect to t, inserting the result in (3.1) and then putting $t=+0$, we obtain

$$\varphi''(+0) = f''(+0) - q(0)f(0). \tag{3.18}$$

From (3.17) and (3.18) follows $f''(+0) = f''(-0)$, i.e. the second derivative of the extended function is continuous.

We have thus proved the following:

THEOREM 3.1. *Suppose that the functions $f(x)$ and $\varphi(x)$ belong to the class $C^2(0,\infty)$. In order that the extension of the function $f(x)$ to the negative halfline, given by*

$$f(x) + f(-x) - \int_{-x}^{x} w(0,x,s)f(s)\,ds = 2\varphi(x)$$

belong to the class $C^2(-\infty,\infty)$, it is necessary and sufficient that the conditions (3.9) *and* (3.12) *hold, i.e.* $\varphi(+0) = f(+0)$, $\varphi'(+0) = 0$.

We will now consider a more general problem, namely:

$$\partial^2 u/\partial t^2 = \partial^2 u/\partial x^2 - q(x)u, \tag{3.1}$$

$$u|_{t=0} = f(x), \quad \partial u/\partial t|_{t=0} = 0, \tag{3.2}$$

$$\partial u/\partial x|_{x=0} - hu|_{x=0} = \varphi(t). \tag{3.3$'$}$$

Let us assume that the function $q(x)$ satisfies the previous requirements and extend it to the entire real line, preserving the condition that it be locally summable. We will seek a solution of the problem (3.1) + (3.2) + (3.3′) in the form

$$u(x,t) = \tfrac{1}{2}\{f(x+t) + f(x-t)\} + \tfrac{1}{2}\int_{x-t}^{x+t} w(x,t,s)f(s)\,ds. \tag{3.4}$$

Let $f(x) \in C^2(0, \infty)$ and $\varphi(x) \in C^2(0, \infty)$. It follows from the conditions (3.2) and (3.3′) that in order for the problem (3.1) – (3.3′) to have a solution which is continuous at zero, it is necessary that

$$\varphi(+0) = f'(+0) - hf(+0), \qquad \varphi'(+0) = 0. \tag{3.19}$$

If we determine $\partial u(x,t)/\partial x|_{x=0}$ and $u(0,t)$ from (3.4) and insert their values in the boundary condition (3.3′), we obtain

$$\begin{aligned} &f'(t) + f'(-t) + f(t)\, w(0, t, t) - f(-t)\, w(0, t, -t) - h\{f(t) + f(-t)\} \\ &+ \int_{-t}^{t} \frac{\partial w(x, t, s)}{\partial x}\Big|_{x=0} f(s)\, ds - h \int_{-t}^{t} w(0, t, s) f(s)\, ds = 2\varphi(t). \end{aligned} \tag{3.20}$$

Let us rewrite equation (3.20) in the form

$$\begin{aligned} &f'(-t) - [w(0, t, -t) + h] f(-t) \\ &\quad + \int_{-t}^{0} \left[\frac{\partial w(x, t, s)}{\partial x}\Big|_{x=0} - hw(0, t, s) \right] f(s)\, ds \\ &= 2\varphi(t) - f'(t) - [w(0, t, t) - h] f(t) \\ &\quad + \int_{0}^{t} \left[-\frac{\partial w(x, t, s)}{\partial x}\Big|_{x=0} + hu(0, t, s) \right] f(s)\, ds. \end{aligned} \tag{3:20′}$$

The right side of (3.20′) is a known function; denote it by $g(t)$. By a change of variables in the integral on the left, we can put (3.20′) in the form

$$f'(-t) - [w(0, t, -t) + h] f(-t) + \int_0^t K(t, s) f(-s)\, ds = g(t), \tag{3.21}$$

where

$$K(t, s) = -\frac{\partial w(x, t, -s)}{\partial x}\Big|_{x=0} + hw(0, t, -s).$$

Equation (3.21) is an integro-differential equation in the unknown function $f(-t)$. If we integrate it once, we obtain an integral equation of Volterra type for $f(-t)$. Integrating (3.21) from 0 to t with respect to t, we obtain

$$\begin{aligned} &-f(-t) + f(-0) - \int_0^t \{w(0, t, -t) + h\} f(-t)\, dt \\ &\qquad + \int_0^t dt \int_0^t K(t, s) f(-s)\, ds = \int_0^t g(t)\, dt. \end{aligned}$$

After some simple transformations, this relation can be rewritten in the form

$$f(-t)-f(-0)+\int_0^t K_1(t,s)f(-s)\,ds=g_1(t). \tag{3.22}$$

We put $f(-0)=C$ and write equation (3.22) in the form

$$f(-t)+\int_0^t K_1(t,s)f(-s)\,ds=g_1(t)+C. \tag{3.23}$$

Equation (3.23) has a solution for any C, and so it will have a solution for $C=f(+0)$. On the other hand, for any C the solution of equation (3.23) is also a solution of the integro-differential equation (3.21). Therefore, putting $C=f(+0)$ in (3.23) and then $t=0$, we obtain

$$f(-0)=C=f(+0), \tag{3.24}$$

i.e. the extended function is continuous at $x=0$.

Further, putting $t=0$ in (3.20′) and taking into account (2.5), i.e. $w(x,0,x)=0$ for any x, we obtain

$$f'(-0)-hf(-0)=2\varphi(+0)-f'(+0)+hf(+0).$$

Hence by virtue of (3.24) and the first of conditions (3.19) there follows

$$f'(-0)=f'(+0), \tag{3.25}$$

i.e. the first derivative of the extended function is also continuous at $x=0$.

Finally, let us consider the second derivative of the extension. To this end we differentiate (3.20′) with respect to t, obtaining

$$\begin{aligned}
&-f''(-t)-f(-t)\frac{dw(0,t,-t)}{dt}+f'(-t)[w(0,t,-t)+h]\\
&\quad+f(-t)\left[\frac{\partial w(x,t,s)}{\partial x}\Big|_{x=0}-h(0,t,s)\right]\Big|_{s=-t}\\
&\quad+\int_{-t}^{0} f(s)\frac{\partial}{\partial t}\left[\frac{\partial w(x,t,s)}{\partial x}\Big|_{x=0}-hw(0,t,s)\right]ds
\end{aligned} \tag{3.26}$$

$$\begin{aligned}
&=2\varphi'(t)-f''(t)-f(t)\frac{dw(0,t,t)}{dt}+f'(t)[-w(0,t,t)+h]\\
&\quad-f(t)\left[\frac{\partial w(x,t,s)}{\partial x}\Big|_{x=0}-hw(0,t,s)\right]\Big|_{s=t}\\
&\quad-\int_0^t f(s)\frac{\partial}{\partial t}\left[\frac{\partial w(x,t,s)}{\partial x}\Big|_{x=0}-hw(0,t,s)\right]ds.
\end{aligned}$$

Putting $t=0$ in (3.26) and taking into account (2.5), (3.24), (3.25) and the second of the equalities (3.19), we obtain

$$f''(-0)-f''(+0)=f(0)\left\{\frac{dw(0,t,t)}{dt}-\frac{dw(0,t-t)}{dt}\right.$$

(3.27)

$$\left.+\frac{\partial w(x,t,s)}{\partial x}\bigg|_{\substack{x=0\\ s=-t}}+\frac{\partial w(x,t,s)}{\partial x}\bigg|_{\substack{x=0\\ s=t}}\right\}_{t=0}.$$

Further, since (cf. formulas (2.9) and (2.10))

$$\frac{dw(0,t,t)}{dt}=-\frac{1}{2}q(t),\qquad \frac{dw(0,t,-t)}{dt}=-\frac{1}{2}q(-t),$$

we have

$$\left[\frac{dw(0,t,t)}{dt}-\frac{dw(0,t,-t)}{dt}\right]_{t=0}=-\frac{1}{2}\{q(0)-q(0)\}=0.$$

Therefore (3.27) assumes the form

$$(3.27')\qquad f''(-0)-f''(+0)=f(0)\left\{\frac{\partial w(x,t,s)}{\partial x}\bigg|_{\substack{x=0\\ s=t}}+\frac{\partial w(x,t,s)}{\partial x}\bigg|_{\substack{x=0\\ s=-t}}\right\}_{t=0}.$$

Let us compute the term in brackets on the right side of (3.27′). To this end, differentiating (3.4) once with respect to x and then once with respect to t, we obtain

$$\frac{\partial^2 u}{\partial x\partial t}=\frac{1}{2}\{f''(x+t)-f''(x-t)\}$$
$$+\frac{1}{2}\{f'(x+t)w(x,t,x+t)+f'(x-t)w(x,t,x-t)\}$$
$$+\frac{1}{2}\left\{f(x+t)\frac{dw(x,t,x+t)}{dt}-f(x-t)\frac{dw(x,t,x-t)}{dt}\right\}$$
$$+\frac{1}{2}\left\{f(x-t)\frac{\partial w(x,t,s)}{\partial x}\bigg|_{s=x+t}+f(x-t)\frac{\partial w(x,t,s)}{\partial x}\bigg|_{s=x-t}\right\}$$
$$+\frac{1}{2}\int_{x-t}^{x+t}\frac{\partial^2 w(x,t,s)}{\partial x\partial t}f(s)\,ds.$$

Putting $t=0$ and taking account of the condition (3.2) and (2.5), (2.9) and (2.10), we obtain

$$(3.28)\qquad \left\{\frac{\partial w(x,t,s)}{\partial x}\bigg|_{s=x+t}+\frac{\partial w(x,t,s)}{\partial x}\bigg|_{s=x-t}\right\}_{t=0}=0,$$

which holds for all x. Therefore from (3.27) follows $f''(-0)=f''(+0)$, which gives the continuity of the second derivative of the extended function at $x=0$.

We have thus proved the following:

THEOREM 3.2. *Let* $f(t) \in C^2(0, \infty)$. *In order that the extension of* $f(t)$ *to the negative halfline, defined by* (3.20′), *belong to* $C^2(-\infty, \infty)$, *it is necessary and sufficient that condition* (3.19) *hold, i.e.* $\varphi(+0) = f'(+0) - hf(+0)$, $\varphi'(+0) = 0$.

§4. Solution of a mixed problem on a finite interval

We now consider a mixed problem for the one-dimensional wave equation on a finite interval of the real line which, without loss of generality, we may assume to be the interval $[0, \pi]$:

$$\partial^2 u/\partial t^2 = \partial^2 u/\partial x^2 - q(x)u, \tag{4.1}$$

$$u|_{t=0} = f(x), \qquad \partial u/\partial t|_{t=0} = 0, \tag{4.2}$$

$$\partial u/\partial x|_{x=0} - hu|_{x=0} = \varphi(t), \tag{4.3}$$

$$\partial u/\partial x|_{x=\pi} + Hu|_{x=\pi} = \psi(t). \tag{4.4}$$

We will show that the problem (4.1)–(4.4) can be solved by the extension method which we already used in the preceding section, without resorting to an eigenfunction expansion. However, in contrast with the considerations of the preceding section for the case of the halfline, for the case of a finite interval the extension has to be carried out both to the left of the point $x = 0$ and to the right of the point $x = \pi$.

Let the function $q(x)$ be continuous on $[0, \pi]$. We extend $q(x)$ to the entire real line, preserving its continuity but otherwise arbitrarily, and we seek a solution of the problem (4.1)–(4.4), as before, in the form

$$u(x, t) = \frac{1}{2}\{f(x+t) + f(x-t)\} + \frac{1}{2}\int_{x-t}^{x+t} w(x, t, s) f(s)\, ds, \tag{4.5}$$

where the function $f(t)$ has to be extended outside the interval $[0, \pi]$, using the boundary conditions (4.3) and (4.4).

We will first derive certain necessary conditions which the functions $\varphi(t)$ and $\psi(t)$ have to satisfy.

It follows from the condition (4.3) that

$$\varphi(+0) = f'(+0) - hf(+0). \tag{4.6}$$

From condition (4.3) and the second of conditions (4.2) it follows that

$$\varphi'(+0) = 0. \tag{4.7}$$

Similarly, from condition (4.4) and the second of conditions (4.2) it follows that

$$\psi(+0) = f'(\pi - 0) + Hf(\pi - 0), \tag{4.8}$$

$$\psi'(+0) = 0. \tag{4.9}$$

Using the boundary condition (4.3), we extend the function $f(t)$ from the interval $(0, \pi)$ to the interval $(-\pi, 0)$. Suppose that $f(t)$, $\varphi(t) \in C^2(0, \pi)$. We saw in the preceding section that the conditions (4.6) and (4.7) are necessary and sufficient for the extended function $f(t)$ to belong to $C^2(-\pi, \pi)$. To extend the function $f(t)$ into the interval $(\pi, 2\pi)$ we will use the boundary condition (4.4), from which, as we saw above, follow the conditions (4.8) and (4.9). We will show that for the function $f(t)$, after extension to $(\pi, 2\pi)$, to belong to $C^2(0, 2\pi)$, conditions (4.8) and (4.9) are not only necessary, but also sufficient. In fact, inserting (4.5) in the boundary condition (4.4), we obtain

$$\begin{aligned} &f'(\pi+t) + f'(\pi-t) + H\{f(\pi+t) + f(\pi-t)\} \\ &+ w(\pi, t, \pi+t) f(\pi+t) - w(\pi, t, \pi-t) f(\pi-t) \\ &+ \int_{\pi-t}^{\pi+t} \frac{\partial w(x, t, s)}{\partial x}\Big|_{x=\pi} f(s)\, ds + H \int_{\pi-t}^{\pi+t} w(\pi, t, s) f(s)\, ds = 2\psi(t). \end{aligned} \tag{4.10}$$

We have obtained an integro-differential equation for the determination of $f(\pi + t)$ $(0 \leq t \leq \pi)$. If this equation is integrated once, then we obtain an integral equation of Volterra type for $f(\pi + t)$. Let us first rewrite (4.10) in the form

$$\begin{aligned} &f'(\pi+t) + \{H + w(\pi, t, \pi+t)\} f(\pi+t) \\ &\quad + \int_{\pi}^{\pi+t} \left\{\frac{\partial w(x, t, s)}{\partial x}\Big|_{x=\pi} + Hw(\pi, t, s)\right\} f(s)\, ds \\ &= 2\psi(t) - f'(\pi-t) - \{H + w(\pi, t, \pi-t)\} f(\pi-t) \\ &\quad - \int_{\pi-t}^{\pi} \left\{\frac{\partial w(x, t, s)}{\partial x}\Big|_{x=\pi} + Hw(\pi, t, s)\right\} f(s)\, ds \equiv g_1(t), \end{aligned} \tag{4.11}$$

or after a change of variables in the integral on the left side

$$\begin{aligned} &f'(\pi+t) + \{H + w(\pi, t, \pi+t)\} f(\pi+t) \\ &+ \int_0^t \left\{\frac{\partial w(x, t, s)}{\partial x}\Big|_{\substack{x=\pi \\ s=\pi+\sigma}} + Hw(\pi, t, \sigma+\pi)\right\} f(\pi+\sigma)\, d\sigma = g_1(t). \end{aligned} \tag{4.11'}$$

Putting

$$K_1(t, \sigma) = \left.\frac{\partial w(x, t, s)}{\partial x}\right|_{\substack{x=\pi \\ s=\pi+\sigma}} + Hw(\pi, t, \pi+\sigma)$$

and integrating equation (4.11′) from 0 to t with respect to t, we obtain

$$f(\pi+t) - f(\pi+0) + \int_0^t \{H + w(\pi, t, \pi+t)\} f(\pi+t)\,dt$$

$$+ \int_0^t dt \int_0^t K_1(t, \sigma) f(\pi+\sigma)\,d\sigma = g(t),$$

or after simple transformations

$$f(\pi+t) - f(\pi+0) + \int_0^t K(t, s) f(\pi+s)\,ds = g(t). \tag{4.12}$$

Let us consider the integral equation

$$f(\pi+t) + \int_0^t K(t, s) f(\pi+s)\,ds = g(t) + C, \tag{4.13}$$

where C is an arbitrary constant. It is obvious that for any C a solution of the integral equation (4.13) is also a solution of the integro-differential equation (4.11). Therefore for $C = f(\pi - 0)$ equation (4.13) also has a solution, which is in turn a solution of equation (4.11). Putting $C = f(\pi - 0)$ and $t = 0$ in (4.13), we obtain

$$f(\pi+0) = f(\pi-0), \tag{4.14}$$

i.e. the extended function $f(t)$ is continuous at the point $t = \pi$.

Further, putting $t = 0$ in (4.11) and taking into account (2.5), (4.8) and (4.14), we find that

$$f'(\pi+0) = f'(\pi-0), \tag{4.15}$$

i.e. the first derivative ot the function $f(t)$ extended according to (4.11) is continuous at the point $t = \pi$.

Finally, let us examine the second derivative of the extended function $f(t)$. To this end, we differentiate equation (4.11) with respect to t, obtaining

$$f''(\pi+t) + f'(\pi+t)\{H + w(\pi, t, \pi+t)\} + f(\pi+t)\frac{dw(\pi, t, \pi+t)}{dt}$$

$$+f(\pi+t)\left\{\frac{\partial w(x,\ t,\ s)}{\partial x}\bigg|_{\substack{x=\pi\\ s=\pi+t}}+Hw(\pi,\ t,\ \pi+t)\right\}$$

$$+\int_{\pi}^{\pi+t}\left\{\frac{\partial^2 w(x,\ t,\ s)}{\partial x\partial t}\bigg|_{x=\pi}+H\frac{\partial w(x,\ t,\ s)}{\partial t}\right\}f(s)\,ds$$

$$(4.16)\qquad =2\psi'(t)+f''(\pi-t)+f'(\pi-t)\{H+w(\pi,\ t,\ \pi-t)\}$$

$$-f(\pi-t)\frac{dw(\pi,\ t,\ \pi-t)}{dt}$$

$$-f(\pi-t).\left\{\frac{\partial w(x,\ t,\ s)}{\partial x}\bigg|_{\substack{x=\pi\\ s=\pi-t}}+Hw(\pi,\ t,\ \pi-t)\right\}$$

$$-\int_{\pi-t}^{\pi}\left\{\frac{\partial^2 w(x,\ t,\ s)}{\partial x\partial t}\bigg|_{x=\pi}+H\frac{\partial w(x,\ t,\ s)}{\partial t}\bigg|_{x=\pi}\right\}f(s)\,ds.$$

Putting $t=0$ and taking account of (2.5), (4.9), (4.14) and (4.15), we obtain

$$(4.17)\quad f''(\pi+0)-f''(\pi-0)=f(\pi)\left\{\frac{dw(\pi,\ t,\ \pi-t)}{dt}-\frac{dw(\pi,\ t,\ \pi+t)}{dt}\right\}_{t=0}.$$

Further, by virtue of (2.9) and (2.10) we have

$$\frac{dw(\pi,\ t,\ \pi+t)}{dt}\bigg|_{t=0}=-\frac{1}{2}q(\pi),$$

$$\frac{dw(\pi,\ t,\ \pi-t)}{dt}\bigg|_{t=0}=-\frac{1}{2}q(\pi).$$

Therefore from (4.17) it finally follows that $f''(\pi-0)=f''(\pi+0)$, i.e. the second derivative of the function $f(t)$, extended according to (4.11), is continuous at the point $t=\pi$.

Thus we have extended the original function $f(t)$ from the interval $(0,\pi)$ to the intervals $(-\pi,0)$ and $(\pi,2\pi)$. This extension is of class $C^2(-\pi,2\pi)$. The integral equations (3.23) and (4.13) make it possible to extend $f(t)$ to the entire real line. Let us show, for example, how to obtain an extension of $f(t)$ to the interval $(-2\pi,-\pi)$. To do this we have to assume that t varies in equation (3.23) in the interval $(0,2\pi)$. Since the solution of a Volterra equation is differentiable as many times as is the free term (assuming the kernel is sufficiently differentiable, which is the case here), our extension is of class $C^2(-2\pi,0)$. Since equation (3.23) is a Volterra equation, the original function, which is undefined in $(-\pi,0)$, does not appear. Therefore the extension which we obtain is of class $C^2(-2\pi,2\pi)$.

If we now assume that t varies from 0 to 3π in equation (4.13), then we obtain an extension of the original function $f(t)$ to the interval

$(2\pi, 4\pi)$. This extension is of class $C^2(-2\pi, 4\pi)$. Thus step by step we extend the original function to the entire real axis. This extension is of class $C^2(-\infty, \infty)$. If the function thus obtained is inserted in (4.5), then by virtue of the very process used to extend it we obtain a solution of the problem (4.1) – (4.4).

Bibliographical references

§1. Formula (1.14) can be obtained by Riemann's method. Indeed it is derived in this way in a paper by Povzner [**1**]. The method which we have presented here is also, in essence, close to that of Riemann. As concerns the estimate (1.16) and what follows, see the paper by Levitan [**14**]. See also the joint paper by the authors [**3**].

§2. A similar method, but in another connection, was applied in the paper by Gel′fand and Levitan [**1**].

§§3, 4. So far as we know, the extension method has been previously applied only for the case of mixed problems for the simple wave equation $u_{tt} = u_{xx}$.

CHAPTER 7

EIGENFUNCTION EXPANSION OF A STURM-LIOUVILLE OPERATOR

§1. Derivation of auxiliary formulas

We will study the equation

$$y'' + [\lambda - q(x)]y = 0, \tag{1.1}$$

defined on the entire real line. Concerning the function $q(x)$ we will assume that it is real and summable over every finite interval. Denote by $\varphi(x,\lambda)$ the solution of (1.1) satisfying the initial conditions

$$\varphi(0,\lambda) = 1, \qquad \varphi'(0,\lambda) = 0, \tag{1.2}$$

and by $\psi(x,\lambda)$ the solution of (1.1) satisfying the conditions

$$\psi(0,\lambda) = 0, \qquad \psi'(0,\lambda) = 1. \tag{1.2'}$$

Together with equation (1.1) we will consider the partial differential equation

$$\partial^2 u/\partial t^2 = \partial^2 u/\partial x^2 - q(x)u. \tag{1.3}$$

Suppose that the function $f(x)$ has a continuous second derivative. We denote by $u(x,t;f)$ the solution of equation (1.3) satisfying the initial conditions

$$u|_{t=0} = f(x), \qquad \partial u/\partial t|_{t=0} = 0. \tag{1.3'}$$

As was shown in the preceding chapter, the function $u(x,t;f)$ can be represented in the form

$$u(x,\ t;\ f) = \frac{1}{2}\{f(x+t) + f(x-t)\} + \frac{1}{2}\int_{x-t}^{x+t} w(x,\ t,\ s)\, f(s)\, ds. \tag{1.4}$$

Putting[1] $f(x)$ in (1.4) equal successively to $\varphi(x,\lambda)$ and $\psi(x,\lambda)$, we obtain the equalities

---

[1] More precisely, putting $f(x)$ in the conditions (1.3′) equal successively to $\varphi(x,\lambda)$ and $\psi(x,\lambda)$, then solving the problem (1.3) – (1.3′) by Fourier's method, from (1.4) we obtain, by the uniqueness of the solution of this problem, (1.5) and (1.5′).

$$\varphi(x, \lambda)\cos\sqrt{\lambda}t = \frac{1}{2}\{\varphi(x+t, \lambda) + \varphi(x-t, \lambda)\} + \frac{1}{2}\int_{x-t}^{x+t} w(x, t, s)\varphi(s, \lambda)\,ds, \tag{1.5}$$

$$\psi(x, \lambda)\cos\sqrt{\lambda}t = \frac{1}{2}\{\psi(x+t, \lambda) + \psi(x-t, \lambda)\} + \frac{1}{2}\int_{x-t}^{x+t} w(x, t, s)\psi(s, \lambda)\,ds. \tag{1.5'}$$

Let us denote by $g_\epsilon(t)$ a function satisfying the following conditions:

1. $g_\epsilon(t)$ is even, i.e. $g_\epsilon(t) = g_\epsilon(-t)$.
2. $g_\epsilon(t)$ vanishes outside the interval $(-\epsilon, \epsilon)$.
3. $g_\epsilon(t)$ has a piecewise continuous and piecewise monotonic first derivative.

Further, let us denote by $\psi_\epsilon(\mu)$ the Fourier cosine transform of the function $g_\epsilon(t)$, i.e. we put

$$\psi_\varepsilon(\mu) = \int_0^\varepsilon g_\varepsilon(t)\cos\mu t\,dt. \tag{1.6}$$

It follows from condition 3 that for large μ one has the estimate

$$\psi_\epsilon(\mu) = O(1/\mu^2). \tag{1.7}$$

Multiplying both sides of (1.5) by $g_\epsilon(t)$ and integrating with respect to t from 0 to ϵ, we obtain

$$\varphi(x,\lambda)\psi_\epsilon(\sqrt{\lambda}) = \frac{1}{2}\int_{x-\varepsilon}^{x+\varepsilon}\{g_\varepsilon(x-s) + \chi_\varepsilon(x, s)\}\varphi(s, \lambda)\,ds, \tag{1.8}$$

where

$$\chi_\varepsilon(x, s) = \int_{|x-s|}^{\varepsilon} w(x, t, s)g_\varepsilon(t)\,dt. \tag{1.9}$$

In the same way, it follows from (1.5′) that

$$\psi(x,\lambda)\psi_\epsilon(\sqrt{\lambda}) = \frac{1}{2}\int_{x-\varepsilon}^{x+\varepsilon}\{g_\varepsilon(x-s) + \chi_\varepsilon(x, s)\}\psi(s, \lambda)\,ds, \tag{1.8'}$$

where $\chi_\epsilon(x, s)$ is defined by (1.9).

Let $f(x), g(x) \in \mathscr{L}^2(-\infty, \infty)$. We put

$$F_1(\lambda)=\int_{-\infty}^{\infty} f(x)\varphi(x,\lambda)\,dx,$$

$$F_2(\lambda)=\int_{-\infty}^{\infty} f(x)\psi(x,\lambda)\,dx,$$

(1.10)

$$G_1(\lambda)=\int_{-\infty}^{\infty} g(x)\varphi(x,\lambda)\,dx,$$

$$G_2(\lambda)=\int_{-\infty}^{\infty} g(x)\psi(x,\lambda)\,dx.$$

The functions $F_1(\lambda)$, $F_2(\lambda)$, $G_1(\lambda)$ and $G_2(\lambda)$ are called the Fourier transforms of the functions $f(x)$ and $g(x)$. If $f(x)$ and $g(x)$ vanish outside some finite interval, then the integrals in formulas (1.10) exist in the ordinary sense. In the contrary case (1.10) is to be understood in a generalized sense (cf. §6, Chapter 2).

Further, as is known, there exist monotone functions $\xi(\lambda)$ and $\zeta(\lambda)$ which are bounded in every finite interval, and a function $\eta(\lambda)$ which is of bounded variation over every finite interval, such that if $f(x)$, $g(x)\in\mathscr{L}^2(-\infty,\infty)$, then one has the Parseval equality

$$\int_{-\infty}^{\infty} f(x)g(x)\,dx=\int_{-\infty}^{\infty} F_1(\lambda)G_1(\lambda)\,d\xi(\lambda)$$

(1.11)

$$+\int_{-\infty}^{\infty}\{F_1(\lambda)G_2(\lambda)+F_2(\lambda)G_1(\lambda)\}\,d\eta(\lambda)+\int_{-\infty}^{\infty}F_2(\lambda)G_2(\lambda)\,d\zeta(\lambda).$$

Formulas (1.8) and (1.8′) show that for fixed x the functions $\varphi(x,\lambda)\psi_\epsilon(\sqrt{\lambda})$ and $\psi(x,\lambda)\psi_\epsilon(\sqrt{\lambda})$ are the Fourier transforms of the function which is equal to $\frac{1}{2}[g_\epsilon(x-s)+\chi_\epsilon(x,s)]$ in the interval $(x-\epsilon, x+\epsilon)$ and equal to zero outside this interval. Therefore, replacing x by y in (1.8′) and applying the Parseval equality (1.11), we obtain

$$\int_{-\infty}^{\infty}\psi_\epsilon^2(\sqrt{\lambda})\{\varphi(x,\lambda)\varphi(y,\lambda)\,d\xi(\lambda)+[\varphi(x,\lambda)\psi(y,\lambda)$$

$$+\varphi(y,\lambda)\psi(x,\lambda]\,d\eta(\lambda)+\psi(x,\lambda)\psi(y,\lambda)\,d\zeta(\lambda)\}$$

(1.12)

$$=\frac{1}{4}\int_{\Delta_{xy}}\{g_\epsilon(x-s)+\chi_\epsilon(x,s)\}\{g_\epsilon(y-s)+\chi_\epsilon(y,s)\}\,ds,$$

where Δ_{xy} is the intersection of the intervals $(x-\epsilon, x+\epsilon)$ and $(y-\epsilon, y+\epsilon)$.

Formula (1.12) assumes a simpler form if we introduce the spectral function of equation (1.1):

$$\theta(x, y, \lambda) = \begin{cases} \int_0^\lambda \{\varphi(x, \lambda)\varphi(y, \lambda)\, d\xi(\lambda) + [\varphi(x, \lambda)\psi(y, \lambda) \\ \quad + \varphi(y, \lambda)\psi(x, \lambda)]\, d\eta(\lambda) + \psi(x, \lambda)\psi(y, \lambda)\, d\zeta(\lambda)\} \quad (\lambda > 0), \\ -\int_\lambda^0 \{\varphi(x, \lambda)\varphi(y, \lambda)\, d\xi(\lambda) + [\varphi(x, \lambda)\psi(y, \lambda) \\ \quad + \varphi(y, \lambda)\psi(x, \lambda)]\, d\eta(\lambda) + \psi(x, \lambda)\psi(y, \lambda)\, d\zeta(\lambda)\} \quad (\lambda < 0). \end{cases} \tag{1.13}$$

Using the spectral function $\theta(x, y, \lambda)$, we can write (1.12) in the form

$$\int_{-\infty}^{\infty} \psi_\varepsilon^2(\sqrt{\lambda})\, d_\lambda \theta(x, y, \lambda) = \frac{1}{4}\int_{\Delta_{xy}} \{g_\varepsilon(x-s) + \chi_\varepsilon(x, s)\}\{g_\varepsilon(y-s) + \chi_\varepsilon(y, s)\}\, ds. \tag{1.12'}$$

In particular, if $x = y$, then

$$\int_{-\infty}^{\infty} \psi_\varepsilon^2(\sqrt{\lambda})\, d_\lambda \theta(x, x, \lambda) = \frac{1}{4}\int_{x-\varepsilon}^{x+\varepsilon} \{g_\varepsilon(x-s) + \chi_\varepsilon(x, s)\}^2\, ds. \tag{1.14}$$

The function $\theta(x, y, \lambda)$ yields the resolution of the identity of the operator (1.1) and therefore has the following properties (cf., for example, Chapter 13):

1. Positivity. Let Δ denote the interval $(\lambda, \lambda+\Delta)$ and put

$$\theta(x, y, \Delta) = \theta(x, y, \lambda+\Delta) - \theta(x, y, \lambda).$$

For $\Delta > 0$

$$\theta(x, x, \Delta) \geq 0.$$

2. Orthogonality. For any two intervals $\Delta = (\lambda, \lambda+\Delta)$ and $\Delta' = (\mu, \mu+\Delta')$ one has

$$\int_{-\infty}^{\infty} \theta(x, t, \Delta)\, \theta(t, y, \Delta')\, dt = \theta(x, y, \Delta\cdot\Delta'), \tag{1.15}$$

where $\Delta\cdot\Delta'$ denotes the intersection of Δ and Δ'.

If we put $\Delta = \Delta'$, then from (1.15) and the Cauchy-Bunjakovskiĭ inequality there follows the bound

$$|\theta(x, y, \Delta)| \leqslant \left(\int_{-\infty}^{\infty} \theta^2(x, t, \Delta)\,dt\right)^{1/2} \left(\int_{-\infty}^{\infty} \theta^2(t, y, \Delta)\,dt\right)^{1/2}$$

(1.16)

$$= [\theta(x, x, \Delta)\theta(y, y, \Delta)]^{1/2} \leqslant \frac{1}{2}[\theta(x, x, \Delta) + \theta(y, y, \Delta)].$$

§2. Preliminary bound on the spectral function. The case of the entire line

As we already mentioned, the function $\theta(x, y, \lambda)$ is called the spectral function of equation (1.1). In this section, making use of (1.14), we will obtain several important estimates for the spectral function. Subsequently, using these preliminary estimates, we will obtain the asymptotic behaviour of the spectral function of equation (1.1) for $\lambda \to +\infty$.

Lemma 2.1. *Let ϵ_0 be an arbitrary positive number and (x_0, x_1) an arbitrary finite interval on the real line. There exists a constant $C = C(x_0, x_1, \epsilon_0)$ such that if x and y belong to the interval (x_0, x_1), then one has the inequality*

$$\int_{-\infty}^{0} e^{\varepsilon_0\sqrt{|\lambda|}}\,|d_\lambda\theta(x, y, \lambda)| < C.$$

In particular, for any finite ϵ, x and y

$$\int_{-\infty}^{0} e^{\varepsilon\sqrt{|\lambda|}}\,|d_\lambda\theta(x, y, \lambda)| < \infty.$$

Proof. We take for the function $g_\epsilon(t)$ the function

$$g_\varepsilon(t) = \begin{cases} \frac{1}{\varepsilon^2}(\varepsilon - |t|) & \text{for } |t| \leqq \epsilon, \\ 0 & \text{for } |t| \geqq \epsilon. \end{cases}$$

Then as is easily seen,

$$\psi_\varepsilon(\sqrt{\lambda}) = \int_0^\varepsilon g_\varepsilon(t)\cos\sqrt{\lambda}\,t\,dt = \left\{\frac{\sin\frac{1}{2}\varepsilon\sqrt{\lambda}}{\frac{1}{2}\varepsilon\sqrt{\lambda}}\right\}^2. \tag{2.1}$$

Inserting the value of $\psi_\epsilon(\sqrt{\lambda})$ from (2.1) into (1.14), we obtain

$$\int_{-\infty}^{0}\left[\frac{\operatorname{sh}\frac{1}{2}\varepsilon\sqrt{|\lambda|}}{\frac{1}{2}\varepsilon\sqrt{|\lambda|}}\right]^4 d_\lambda\theta(x, x, \lambda) < \int_{x-\varepsilon}^{x+\varepsilon}\{g_\varepsilon(x-s) + \chi_\varepsilon(x, s)\}^2\,ds. \tag{2.2}$$

Put $\epsilon\sqrt{|\lambda|} = u$. It follows from the mean value theorem that $\operatorname{sh} u = \operatorname{sh} u - \operatorname{sh} 0 = u \operatorname{ch} \theta u > u$. We therefore deduce from (2.2) that

$$\int_{-\infty}^{0} \frac{\operatorname{sh} \frac{1}{2}\varepsilon\sqrt{|\lambda|}}{\sqrt{|\lambda|}} d_\lambda \theta(x, x, \lambda) < \frac{1}{2}\varepsilon \int_{x-\varepsilon}^{x+\varepsilon} \{g_\varepsilon(x-s) + \chi_\varepsilon(x, s)\}^2 ds. \tag{2.3}$$

Let $\epsilon = \epsilon_0 + 1$. It is easily seen that for $|\lambda| > 1$ we have the inequality

$$\frac{\operatorname{sh} \varepsilon\sqrt{|\lambda|}}{\sqrt{|\lambda|}} > e^{\varepsilon_0\sqrt{|\lambda|}},$$

and therefore the inequality (2.3) implies the bound

$$\int_{-\infty}^{0} e^{\frac{1}{2}\varepsilon_0\sqrt{|\lambda|}} d_\lambda \theta(x, x, \lambda)$$

$$< \frac{1}{2}\varepsilon \int_{x-\varepsilon}^{x+\varepsilon} \{g_\varepsilon(x-s) + \chi_\varepsilon(x, s)\}^2 ds + \int_{-1}^{0} e^{\varepsilon_0\sqrt{|\lambda|}} d_\lambda \theta(x, x, \lambda). \tag{2.4}$$

The bound (2.4) proves the lemma for $x = y$. The case $x \neq y$ reduces to the case $x = y$ by means of the inequality (1.16). This completely proves the lemma.

Let us denote by a an arbitrary real number and put $g_\epsilon(t, a) = g_\epsilon(t) \times \cos at$. Since the function $g_\epsilon(t, a)$ satisfies all the requirements placed on the function $g_\epsilon(t)$, replacing $g_\epsilon(t)$ by $g_\epsilon(t, a)$ in (1.14) we obtain

$$\int_{-\infty}^{\infty} \psi_\varepsilon^2(\sqrt{\lambda}, a) d_\lambda \theta(x, x, \lambda) = \frac{1}{4} \int_{x-\varepsilon}^{x+\varepsilon} \{g_\varepsilon(x-s, a) + \chi_\varepsilon(x, s, a)\}^2 ds, \tag{2.5}$$

where

$$\chi_\varepsilon(x, s, a) = \int_{|x-\varepsilon|}^{\varepsilon} w(x, t, s) g_\varepsilon(t, a) dt, \tag{2.6}$$

$$\psi_\varepsilon(\sqrt{\lambda}, a) = \int_0^\varepsilon g_\varepsilon(t, a) \cos\sqrt{\lambda}\, t\, dt = \frac{1}{2}\{\psi_\varepsilon(\sqrt{\lambda} + a) + \psi_\varepsilon(\sqrt{\lambda} - a)\}. \tag{2.7}$$

Put

$$H(x, y, a, \varepsilon) = \int_{-\infty}^{0} \psi_\varepsilon^2(\sqrt{\lambda}, a) d_\lambda \theta(x, y, \lambda). \tag{2.8}$$

If the function $g_\epsilon(t)$ is chosen as before, then from

$$\psi_\varepsilon(\sqrt{\lambda}, a) = \frac{1}{\varepsilon^2}\int_0^\varepsilon (\varepsilon - |t|)\cos at \cdot \cos\sqrt{\lambda}\, t\, dt$$

there follows for $\lambda < 0$ the bound

$$|\psi_\epsilon(\sqrt{\lambda}, a)| \leqq \operatorname{ch}\sqrt{|\lambda|}\,\epsilon;$$

therefore

$$|H(x, y, a, \varepsilon)| < \int_{-\infty}^{0} \operatorname{ch}^2 \sqrt{|\lambda|}\,\varepsilon\, |d_\lambda \theta(x, y, \lambda)|.$$

From this bound and the preceding lemma there at once follows:

Lemma 2.2. *If the coefficient $q(x)$ in equation* (1.1) *is summable over every finite interval and if (x_0, x_1) is an arbitrary finite interval of the real line and ϵ_0 is an arbitrary positive number, then there exists a constant $C = C(x_0, x_1, \epsilon_0)$, which does not depend upon a, such that for $x, y \in (x_0, x_1)$ and $\epsilon \leqq \epsilon_0$ we have the inequality*

$$|H(x, y, a, \epsilon)| < C. \tag{2.9}$$

If $x \neq y$, then by (2.8) the equality (2.5) can be written in the form

$$\begin{aligned}\int_0^\infty \psi_\varepsilon^2(\sqrt{\lambda}, a)\, d_\lambda\theta(x, y, \lambda) &= \frac{1}{4}\int_{\Delta_{xy}} \{g_\varepsilon(x - s, a) + \chi_\varepsilon(x, s, a)\}\{g_\varepsilon(y - s, a) + \chi_\varepsilon(y, s, a)\}\, ds \\ &\quad - H(x, y, a, \varepsilon),\end{aligned} \tag{2.10}$$

where $\chi_\epsilon(x, s, a)$ is defined by (2.6).

Let us put $\lambda = \mu^2$, $\theta(x, y, \lambda) = \theta(x, y, \mu^2) = \theta_1(x, y, \mu)$ for $\lambda > 0$ and for fixed x and y extend the function $\theta_1(x, y, \mu)$ to the negative semiaxis as an odd function. Then (2.10) assumes the form

$$\begin{aligned}\int_{-\infty}^\infty \psi_\varepsilon^2(\mu, a)\, d_\mu\theta_1(x, y, \mu) &= \frac{1}{2}\int_{\Delta_{xy}} \{g_\varepsilon(x - s, a) + \chi_\varepsilon(x, s, a)\}\{g_\varepsilon(y - s, a) + \chi_\varepsilon(y, s, a)\}\, ds \\ &\quad - 2H(x, y, a, \varepsilon).\end{aligned} \tag{2.10'}$$

In particular, for $x = y$ we obtain

$$\int_{-\infty}^{\infty} \psi_\varepsilon^2(\mu, a)\, d_\mu \theta_1(x, x, \mu)$$

$$= \frac{1}{2} \int_{x-\varepsilon}^{x+\varepsilon} \{g_\varepsilon(x-s, a) + \chi_\varepsilon(x, s, a\}^2\, ds - 2H(x, x, a, \varepsilon).$$

LEMMA 2.3. *If the coefficient* $q(x)$ *in equation* (1.1) *is summable over every finite interval and* (x_0, x_1) *is an arbitrary finite interval of the real line, then there exists a constant* $C = C(x_0, x_1)$ *such that for* $x, y \in (x_0, x_1)$ *one has the inequality*

$$\bigvee_{\mu}^{\mu+1} \{\theta_1(x, y, \mu)\} < C. \tag{2.12}$$

PROOF. Put $\epsilon = 2$ in (2.11). Then we obtain (by virtue of (2.1))

$$\begin{aligned}&\int_{-\infty}^{\infty} \left\{\frac{\sin^2(\mu+a)}{(\mu+a)^2} + \frac{\sin^2(\mu-a)}{(\mu-a)^2}\right\}^2 d_\mu \theta_1(x, x, \mu)\\ &= \frac{1}{2} \int_{x-1}^{x+1} \{g_1(x-s, a) + \chi_1(x, s, a\}^2\, ds - 2H(x, x, a, 1).\end{aligned} \tag{2.13}$$

By Theorem 1.1 of Chapter 6 and Lemma 2.2 there exists a constant $C = C(x_0, x_1)$ such that if $x \in (x_0, x_1)$, then the right side of (2.13) does not exceed C. Therefore from (2.13) follows the bound

$$\int_{-\infty}^{\infty} \frac{\sin^4(\mu-a)}{(\mu-a)^4}\, d_\mu \theta_1(x, x, \mu) < C,$$

or

$$\int_{-\infty}^{\infty} \frac{\sin^4\mu}{\mu^4}\, d_\mu \theta_1(x, x, \mu+a) < C,$$

and so a fortiori

$$\int_0^1 \frac{\sin^4\mu}{\mu^4}\, d_\mu \theta_1(x, x, \mu+a) < C. \tag{2.14}$$

As is known, for $0 \leqq \mu \leqq \pi/2$ one has the inequality $\sin\mu/\mu \geqq 2/\pi$. Therefore

$$\int_0^1 \frac{\sin^4\mu}{\mu^4}\, d_\mu \theta_1(x, x, \mu+a) \geqslant \frac{16}{\pi^4}\{\theta_1(x, x, a+1) - \theta_1(x, x, a)\}. \tag{2.14'}$$

The inequalities (2.14) and (2.14′) prove the lemma for $x = y$. With the help of (1.16) the case $x \neq y$ reduces to the case $x = y$.

Suppose that $f(s) \in \mathcal{L}^2(-\infty, \infty)$, and let $\nu_\epsilon(x,s) = g_\epsilon(x-\epsilon) + \chi_\epsilon(x,s)$ for $s \in (x-\epsilon, x+\epsilon)$ and zero otherwise. Then by the Parseval equality (1.11) it follows from (1.8), (1.8′) and the first two of equalities (1.10) that

$$\begin{aligned}\int_{-\infty}^{\infty} \psi_\varepsilon(\sqrt{\lambda})\{F_1(\lambda)\varphi(x,\lambda)\,d\xi(\lambda)\\ + [F_1(\lambda)\psi(x,\lambda) + F_2(\lambda)\varphi(x,\lambda)]\,d\eta(\lambda)\\ + F_2(\lambda)\psi(x,\lambda)\,d\zeta(\lambda)\}\\ = \frac{1}{2}\int_{x-\varepsilon}^{x+\varepsilon}\{g_\varepsilon(x-s) + \chi_\varepsilon(x,s)\}f(s)\,ds.\end{aligned} \tag{2.15}$$

Using the spectral function $\theta(x,y,\lambda)$ (cf. (1.13)) and the definition of the functions $F_1(\lambda)$ and $F_2(\lambda)$ (cf. (1.10)), we can write (2.15) in the form

$$\begin{aligned}\int_{-\infty}^{\infty} \psi_\varepsilon(\sqrt{\lambda})\,d_\lambda \int_{-\infty}^{\infty} f(s)\,\theta(x,s,\lambda)\,ds\\ = \frac{1}{2}\int_{x-\varepsilon}^{x+\varepsilon} f(s)\{g_\varepsilon(x-s) + \chi_\varepsilon(x,s)\}\,ds.\end{aligned} \tag{2.16}$$

Since the function $f(s)$ is arbitrary, it follows from (2.16) that

$$2\int_{-\infty}^{\infty} \psi_\varepsilon(\sqrt{\lambda})\,d_\lambda\theta(x,s,\lambda) = \begin{cases} g_\varepsilon(x-s) + \chi_\varepsilon(x,s), & |x-s| \leqslant \varepsilon,\\ 0, & |x-s| \geqslant \varepsilon.\end{cases} \tag{2.17}$$

The convergence of the integral on the left follows from Lemmas 2.1 and 2.2, and also from the bound (1.7). The formula (2.16) can be written in the form

$$\begin{aligned}\int_{-\infty}^{\infty} \psi_\varepsilon(\mu)\,d_\mu S_1(x,\mu)\\ = \int_{x-\varepsilon}^{x+\varepsilon} f(s)\{g_\varepsilon(x-s) + \chi_\varepsilon(x,s)\}\,ds - 2\int_{-\infty}^{0} \psi_\varepsilon(\sqrt{\lambda})\,d_\lambda S(x,\lambda),\end{aligned} \tag{2.18}$$

where we have put

$$S(x, \lambda) = \int_{-\infty}^{\infty} f(s)\,\theta(x, s, \lambda)\,ds,$$

$$S_1(x, \mu) = S(x, \mu^2) \qquad (\mu > 0),$$
$$S_1(x, \mu) = -S_1(x, -\mu) \quad (\mu < 0).$$

Similarly, (2.17) can be written in the form

$$(2.19)\qquad \int_{-\infty}^{\infty} \psi_\varepsilon(\mu)\,d_\mu\theta_1(x, s, \mu) = \begin{cases} g_\varepsilon(x-s) + \chi_\varepsilon(x, s) \\ \quad -2\int_{-\infty}^{0} \psi_\varepsilon(\sqrt{\lambda})\,d_\lambda\theta(x, s, \lambda), & |x-s| \leqslant \varepsilon, \\ -2\int_{\infty}^{0} \psi_\varepsilon(\sqrt{\lambda})\,d_\lambda\theta(x, s, \lambda), & |x-s| \geqslant \varepsilon. \end{cases}$$

If $q(x) \equiv 0$ in (1.1), then $w(x,t,s) \equiv 0$, and consequently $\chi_\epsilon(x,s,a) \equiv 0$. Therefore (2.18) and (2.19) assume the form

$$(2.20)\qquad \int_{-\infty}^{\infty} \psi_\varepsilon(\mu)\,d_\mu S_1^*(x, \mu) = \int_{x-\varepsilon}^{x+\varepsilon} f(s)\,g_\varepsilon(x-s)\,ds,$$

$$(2.21)\qquad \int_{-\infty}^{\infty} \psi_\varepsilon(\mu)\,d_\mu\theta_1^*(x, s, \mu) = \begin{cases} g_\varepsilon(x-s), & |x-s| \leqslant \varepsilon, \\ 0, & |x-s| \geqslant \varepsilon, \end{cases}$$

where we have put

$$\theta_1^*(x, s, \mu) = \frac{1}{\pi}\frac{\sin\mu(x-s)}{x-s}, \quad S_1^*(x, \mu) = \int_{-\infty}^{\infty} f(s)\,\theta_1^*(x, s, \mu)\,ds.$$

Subtracting (2.20) from (2.18), we obtain

$$(2.22)\qquad \int_{-\infty}^{\infty} \psi_\varepsilon(\mu)\,d_\mu R(x, \mu, f) = \int_{x-\varepsilon}^{x+\varepsilon} f(s)\,\chi_\varepsilon(x, s)\,ds - 2\int_{-\infty}^{0} \psi_\varepsilon(\sqrt{\lambda})\,d_\lambda S(x, \lambda),$$

where we have put $R(x,\mu,f) = S_1(x,\mu) - S_1^*(x,\mu)$. Subtracting (2.21) from (2.19), we obtain

$$\int_{-\infty}^{\infty} \psi_\varepsilon(\mu)\, d_\mu \Phi(x, s, \mu)$$

$$(2.23)\qquad = \begin{cases} \chi_\varepsilon(x, s) - 2\displaystyle\int_{-\infty}^{0} \psi_\varepsilon(\sqrt{\lambda})\, d_\lambda \theta(x, s, \lambda), & |x - s| \leqslant \varepsilon, \\ -2\displaystyle\int_{-\infty}^{0} \psi_\varepsilon(\sqrt{\lambda})\, d_\lambda \theta(x, s, \lambda), & |x - s| \geqslant \varepsilon, \end{cases}$$

where $\Phi(x,s,\mu) = \theta_1(x,s,\mu) - \theta_1^*(x,s,\mu)$. Formulas (2.22) and (2.23) play a fundamental role in what follows.

§3. Asymptotic behaviour of the spectral function. Case of the entire line

In this section we will study the asymptotic behaviour of the spectral function $\theta(x,y,\lambda)$ of equation (1.1).

We will here place upon the function $g_\epsilon(t)$ more rigid restrictions than were imposed earlier. We will, namely, assume that $g_\epsilon(t)$ satisfies the following conditions:

1. $g_\epsilon(t) = g_\epsilon(-t)$.
2. $g_\epsilon(t) = 1$ if $|t| \leqq \epsilon/2$.
3. $g_\epsilon(t) = 0$ if $|t| \geqq \epsilon$.
4. $g_\epsilon(t)$ is bounded as a function of t and ϵ.
5. $g_\epsilon(t)$ has a continuous fourth derivative.

LEMMA 3.1. *Let us assume that the coefficient $q(x)$ in equation (1.1) is summable over every finite interval. Let (x_0, x_1) denote an arbitrary finite interval. There exists a constant $C = C(x_0, x_1)$ such that if $x, s \in (x_0, x_1)$, then*

$$(3.1)\qquad \int_{-\infty}^{0} |\psi_\varepsilon(\sqrt{\lambda})\| d_\lambda \theta(x, s, \lambda)| < C.$$

PROOF. Since

$$|\psi_\varepsilon(\sqrt{\lambda})| = O\left(\int_0^\varepsilon \cos\sqrt{\lambda}\, t\, dt\right) = O\left(\varepsilon e^{\varepsilon\sqrt{|\lambda|}}\right) = O\left(e^{(\sqrt{|\lambda|}+1)\varepsilon}\right),$$

the bound (3.1) follows at once from Lemma 2.1.

Let us now proceed to the investigation of the asymptotic behaviour of the spectral function $\theta(x,s,\mu)$. We will first transform the right side of (2.23).

Let us first consider the term

$$\chi_\varepsilon(x, s) = \int_{|x-s|}^{\varepsilon} w(x, t, s)\, g_\varepsilon(t)\, dt.$$

Put

$$\alpha(x, s, \mu) = \int_{|x-s|}^{1} w(x, t, s) \cos \mu t\, dt.$$

Using the Fourier inversion formula, we have

$$g_\varepsilon(t) = \frac{2}{\pi} \int_0^\infty \psi_\varepsilon(\mu) \cos \mu t\, d\mu.$$

It therefore follows from the Parseval equality for the ordinary Fourier integral that

$$\int_0^\infty \psi_\varepsilon(\mu)\, \alpha(x, s, \mu)\, d\mu$$

$$= \begin{cases} \dfrac{\pi}{2} \displaystyle\int_{|x-s|}^{\varepsilon} w(x, t, s)\, g_\varepsilon(t)\, dt = \dfrac{\pi}{2} \chi_\varepsilon(x, s), & |x-s| \leqslant \varepsilon, \\ 0. & |x-s| \geqslant \varepsilon. \end{cases}$$

Using the evenness of the function $\alpha(x, s, \mu)$, this last formula can be transformed to the form

$$\frac{1}{2} \int_{-\infty}^{\infty} \psi_\varepsilon(\mu)\, \alpha(x, s, \mu)\, d\mu = \begin{cases} \dfrac{\pi}{2} \chi_\varepsilon(x, s), & |x-s| \leqslant \varepsilon, \\ 0, & |x-s| \geqslant \varepsilon. \end{cases} \tag{3.2}$$

Consider the term corresponding to the negative spectrum. We have (changing the order of integration)

$$\int_{-\infty}^{0} \left\{ \int_0^\varepsilon g_\varepsilon(t) \cos \sqrt{\lambda}\, t\, dt \right\} d_\lambda \theta(x, s, \lambda) = \int_0^\varepsilon g_\varepsilon(t)\, h(x, s, t)\, dt,$$

where

$$h(x, s, t) = \int_{-\infty}^{0} \cos \sqrt{\lambda}\, t\, d_\lambda \theta(x, s, \lambda).$$

Put

$$\beta(x, s, \mu) = \int_0^1 h(x, s, t) \cos \mu t\, dt.$$

It follows from the Parseval equality for the ordinary Fourier integral that (cf. the derivation of (3.2))

$$\frac{1}{2}\int_{-\infty}^{\infty} \psi_\varepsilon(\mu)\,\beta(x, s, \mu)\,d\mu = \frac{\pi}{2}\int_{0}^{\varepsilon} g_\varepsilon(t)\,h(x, s, t)\,dt. \tag{3.3}$$

From (2.23), (3.2) and (3.3) follows the important equality

$$\int_{-\infty}^{\infty} \psi_\varepsilon(\mu)\,d_\mu\Phi^*(x, s, \mu) = 0, \tag{3.4}$$

where

$$\Phi^*(x, s, \mu) = \begin{cases} \Phi(x, s, \mu) - \frac{1}{2}\int_0^\mu \alpha(x, s, \nu)\,d\nu + \\ + \frac{2}{\pi}\int_0^\mu \beta(x, s, \nu)\,d\nu, & |x - s| \leqslant \varepsilon, \\ \frac{2}{\pi}\int_0^\mu \beta(x, s; \nu)\,d\nu, & |x - s| \geqslant \varepsilon. \end{cases} \tag{3.5}$$

By virtue of the estimates $(\mu \to +\infty)$

$$\bigvee_{\mu}^{\mu+1}\left\{\int_0^\mu \alpha(x, s, \nu)\,d\nu\right\}, \quad \bigvee_{\mu}^{\mu+1}\left\{\int_0^\mu \beta(x, s, \nu)\,d\nu\right\} = o(1),$$

and Lemma 2.3, the following estimate (for $\mu \to +\infty$) holds for the function $\Phi^*(x, s, \mu)$:

$$\bigvee_{\mu}^{\mu+1}\{\Phi^*(x, s, \mu)\} < C, \tag{3.6}$$

where the constant C depends only upon the interval of variation of x and s. Therefore by virtue of (3.4) a Tauberian theorem of Levitan (cf. Chapter 14, Theorem 4.1) is applicable, from which it follows that as $\mu \to \infty$ we have, uniformly for x and s in any finite interval,

$$\Phi^*(x, s, \mu) = O(1). \tag{3.7}$$

From (3.7) it is not hard to obtain the asymptotic behaviour of the spectral function.

THEOREM 3.1. *If* $q(x)$ *satisfies the conditions* (1.19), *then for every fixed* x *and* s

$$\theta_1(x, s, \mu) - \frac{1}{\pi}\frac{\sin\mu(x - s)}{x - s} = O(1). \tag{3.8}$$

The estimate (3.8) *holds uniformly as x and s vary over any finite interval.*

PROOF. By virtue of the asymptotic formula (3.7) and formula (3.5), to prove (3.8) it suffices to show that as $\mu \to +\infty$

$$\int_0^\mu \alpha(x, s, \nu)\, d\nu = o(1), \tag{3.9}$$

$$\frac{2}{\pi}\int_0^\mu \beta(x, s, \nu)\, d\nu = -\theta(x, s, -\infty) + o(1), \tag{3.10}$$

uniformly in every finite region.

We will first prove (3.9). We have

$$\int_0^\mu \alpha(x, s, \nu)\, d\nu = \int_{|x-s|}^{\varepsilon} w(x, t, s) \frac{\sin \mu t}{t}\, dt.$$

Let η denote an arbitrary positive number not exceeding ϵ and unity, and put

$$\int_{|x-s|}^{\varepsilon} w(x, t, s) \frac{\sin \mu t}{t}\, dt = \left\{ \int_{|x-s|}^{\eta} + \int_{\eta}^{\varepsilon} \right\} w(x, t, s) \frac{\sin \mu t}{t}\, dt = I_1 + I_2.$$

By virtue of the bound (1.20)

$$|I_1| \leqslant C \int_0^\eta t^{\alpha-1} dt = C\eta^\alpha.$$

Therefore for any positive number ζ one can choose η so small that for all μ we will have

$$|I_1| < \zeta/2. \tag{3.11}$$

Further, it follows from the Riemann-Lebesgue lemma that for fixed η one can choose μ_0 so large that for $\mu > \mu_0$ we have

$$|I_2| < \zeta/2. \tag{3.12}$$

From the bounds (3.11) and (3.12) it follows that for $\mu > \mu_0$

$$\left| \int_{|x-s|}^{\varepsilon} w(x, t, s) \frac{\sin \mu t}{t}\, dt \right| < \zeta,$$

which proves (3.9) since the number ζ is arbitrary.

Let us prove (3.10). We have

$$\frac{2}{\pi}\int_0^{\mu} \beta(x, s, \nu)\, d\nu = \frac{2}{\pi}\int_0^{\varepsilon} h(x, s, t)\, \frac{\sin \mu t}{t}\, dt.$$

Since

$$\int_{-\infty}^{0} \cos \sqrt{\lambda}\, t\, |d_{\lambda}\theta(x, s, \lambda)| < \infty,$$

the function $h(x,s,t)$ is an entire analytic function of t, and so, in particular, it is differentiable in a neighborhood of $t=0$. Therefore by a well-known theorem of the theory of Fourier series it follows that as $\mu \to \infty$

$$\frac{2}{\pi}\int_0^{\mu} \beta(x, s, \nu)\, d\nu = \frac{2}{\pi}\int_0^{\varepsilon} h(x, s, t)\frac{\sin \mu t}{t}\, dt$$

$$= h(x, s, 0) + o(1) = -\theta(x, s, -\infty) + o(1),$$

as was to be proved. The uniformity of (3.10) follows from the boundedness of the functions $h(x,s,t)$ and $\partial h(x,s,t)/\partial t$.

REMARK. From a Tauberian theorem of Marčenko (cf. Chapter 14, Theorem 4.2, special case) there follows a result which is more precise than (3.8), namely,

$$\lim_{\mu\to\infty}\left\{\theta_1(x, s; \mu) - \frac{1}{\pi}\,\frac{\sin \mu (x-s)}{x-s}\right\} = \theta(x, s; -\infty). \tag{3.13}$$

For $x=s$ the equality (3.13) follows directly from the identity (2.23) and the indicated theorem of Marčenko. For $x \neq s$ one has to consider the monotonic function

$$P(x,s;\mu) = \theta_1(x,x;\mu) + 2\theta_1(x,s;\mu) + \theta_1(s,s;\mu)$$

and apply Marčenko's theorem to it (again using the identity (2.23)).

§4. Asymptotic behaviour of the spectral function. The case of the halfline

Let the equation

$$y'' + [\lambda - q(x)]y = 0 \tag{4.1}$$

be defined on the halfline $[0, \infty)$. We will assume the function $q(x)$ summable over every finite interval. We adjoin to equation (4.1) a boundary condition at $x=0$:

$$y'(0) - hy(0) = 0, \tag{4.2}$$

where h is an arbitrary real number. The case $h = \infty$ will not be considered here. It is rather completely treated in the paper [**15**] by Levitan.

Let $\varphi_h(x, \lambda)$ denote the solution of equation (4.1) satisfying the initial conditions

(4.3) $$\varphi_h(0, \lambda) = 1, \qquad \varphi_h'(0, \lambda) = h.$$

As is known, for a given h there exists at least one function $\rho(\lambda)$ which is monotonic, bounded in every finite interval, continuous from the left, and which generates an isometric mapping of the space $\mathscr{L}^2(0, \infty)$ onto the space $\mathscr{L}^2_{\{\rho(\lambda)\}}(-\infty, \infty)$ according to the formula

$$F(\lambda) = \lim_{A\to\infty} \int_0^A f(x)\, \varphi_h(x, \lambda)\, dx, \quad f(x) = \lim_{A\to\infty} \int_{-A}^{A} F(\lambda)\, \varphi_h(x, \lambda)\, d\rho(\lambda),$$

where the integrals converge in the metrics of $\mathscr{L}^2_{\{\rho(\lambda)\}}(-\infty, \infty)$ and $\mathscr{L}^2(0, \infty)$ respectively, and one has the Parseval equality

$$\int_0^\infty f^2(x)\, dx = \int_{-\infty}^{\infty} F^2(\lambda)\, d\rho(\lambda).$$

We will call the function

$$\theta_h(x, y, \lambda) = \int_0^\lambda \varphi_h(x, \lambda)\, \varphi_h(y, \lambda)\, d\rho(\lambda).$$

the spectral function of equation (4.1) (for the initial conditions (4.3)).

Further, we denote by $\theta_h^*(x, y, \lambda)$ the spectral function of the same problem for $q(x) \equiv 0$. Let us first assume that $h > 0$. In this case the equation

(4.4) $$y'' + \lambda y = 0$$

does not have a negative spectrum. For $\lambda > 0$ put $\lambda = \mu^2$. Then

$$\theta_h(x, y, \lambda) = \theta_h(x, y, \mu^2) = \theta_{h,1}(x, y, \mu),$$
$$\theta_h^*(x, y, \lambda) = \theta_h^*(x, y, \mu^2) = \theta_{h,1}^*(x, y, \mu).$$

Repeating the arguments of the preceding section,[2)] we obtain the asymptotic formula ($\mu \to \infty$)

(4.5) $$\theta_{h,1}(x, y, \mu) - \theta_{h,1}^*(x, y, \mu) = \theta_h(x, y, -\infty) + o(1).$$

[2)] Cf. the Remark following the proof of Theorem 3.1.

Let us find the principal term of $\theta_{h,1}^*(x,y,\mu)$. As is known (cf. §1, Chapter 5),

$$\theta^*_{h,1}(x,y,\mu)=\frac{2}{\pi}\int_0^\mu\left(\cos\nu x+\frac{h}{\nu}\sin\nu x\right)\left(\cos\nu y+\frac{h}{\nu}\sin\nu y\right)\frac{\nu^2}{\nu^2+h^2}\,d\nu$$

$$=\frac{2}{\pi}\int_0^\mu\cos\nu x\cdot\cos\nu y\,\frac{\nu^2}{\nu^2+h^2}\,d\nu$$

$$+\frac{2h}{\pi}\int_0^\mu\frac{\sin(x+y)\nu}{\nu}\,\frac{\nu^2}{\nu^2+h^2}\,d\nu+\frac{2h^2}{\pi}\int_0^\mu\frac{\sin\nu x\cdot\sin\nu y}{\nu^2+h^2}\,d\nu$$

$$(4.6)\qquad =\frac{2}{\pi}\int_0^\mu\cos\nu x\cdot\cos\nu y\,d\nu-\frac{2h^2}{\pi}\int_0^\mu\frac{\cos\nu x\cdot\cos\nu y}{\nu^2+h^2}\,d\nu$$

$$+\frac{2h^2}{\pi}\int_0^\mu\frac{\sin\nu x\cdot\sin\nu y}{\nu^2+h^2}\,d\nu+\frac{2h}{\pi}\int_0^\mu\frac{\sin(x+y)\nu}{\nu^2+h^2}\,\nu\,d\nu$$

$$=\frac{2}{\pi}\int_0^\mu\cos\nu x\cdot\cos\nu y\,d\nu+\frac{2h}{\pi}\int_0^\mu\frac{\sin\nu(x+y)}{\nu^2+h^2}\,\nu\,d\nu$$

$$-\frac{2h^2}{\pi}\int_0^\mu\frac{\cos\nu(x-y)}{\nu^2+h^2}\,d\nu=\theta_1^*(x,y,\mu)+\frac{2h}{\pi}\int_0^\infty\frac{\sin\nu(x+y)}{\nu^2+h^2}\,\nu\,d\nu$$

$$-\frac{2h^2}{\pi}\int_0^\infty\frac{\cos\nu(x-y)}{\nu^2+h^2}\,d\nu+o(1)\qquad(x+y>0),$$

where

$$\theta_1^*(x,y,\mu)=\frac{2}{\pi}\int_0^\mu\cos\nu x\cdot\cos\nu y\,d\nu$$

is the spectral function of the problem (4.1) + (4.2) for $q(x)\equiv 0$ and $h=0$.

From the asymptotic formula (4.5) and the relation (4.6) there follows at once:

THEOREM 4.1. *Let the function $q(x)$ satisfy the condition* (1.19) *of Chapter* 6. *For every fixed $x\neq 0$ and $y\neq 0$*

$$(4.7)\qquad \lim_{\mu\to\infty}\{\theta_{h,1}(x,y,\mu)-\theta_1^*(x,y,\mu)\}=\theta_h(x,y,-\infty)$$

$$+\frac{2h}{\pi}\int_0^\infty\frac{\sin\nu(x+y)}{\nu^2+h^2}\,\nu\,d\nu-\frac{2h^2}{\pi}\int_0^\infty\frac{\cos\nu(x-y)}{\nu^2+h^2}\,d\nu.$$

The relation (4.7) *holds uniformly as x and y vary in any bounded region* $(x+y \geqq \epsilon > 0)$.

For $x = y = 0$ we obtain from (4.5)

$$\lim_{\mu\to\infty}\left\{\rho(\mu)-\frac{2}{\pi}\mu\right\}=\rho(-\infty)-h. \tag{4.8}$$

Let us consider the case $h < 0$. In this case equation (4.4) has one negative eigenvalue $\lambda_0 = -h^2$. Indeed, the eigenfunctions corresponding to the negative eigenvalues are square integrable. For $\lambda < 0$ the solutions of (4.4) have the form

$$\begin{aligned}\omega_h^*(x,\lambda) &= \cos h\sqrt{|\lambda|}\,x + h\,\frac{\sin h\sqrt{|\lambda|}\,x}{\sqrt{|\lambda|}}\\ &= e^{-\sqrt{|\lambda|}\,x}\left(1-\frac{h}{\sqrt{|\lambda|}}\right)+e^{\sqrt{|\lambda|}\,x}\left(1+\frac{h}{\sqrt{|\lambda|}}\right).\end{aligned}$$

Therefore if $\omega_h^*(x,\lambda) \in \mathscr{L}^2(0,\infty)$, then $1+h/\sqrt{|\lambda|}=0$, i.e. $\lambda=\lambda_0=-h^2$. The corresponding normalized eigenfunction has the form $\sqrt{2h}\,e^{hx}$. Thus, upon comparing the spectral function $\theta_h(x,y,\lambda)$ with the spectral function $\theta_h^*(x,y,\lambda)$ an additional term $-2he^{h(x+y)}$, corresponding to the negative eigenvalue of the problem (4.4) – (4.2), appears.

Consequently, instead of formula (4.7) we obtain $(x+y>0)$

$$\begin{aligned}\lim_{\mu\to\infty}\{\theta_{h,1}(x,y,\mu)-\theta_1^*(x,y,\mu)\} &= \theta_h(x,y,-\infty)\\ -2he^{h(x+y)}+\frac{2h}{\pi}\int_0^\infty \frac{\sin\nu(x+y)}{\nu^2+h^2}\,\nu\,d\nu &- \frac{2h^2}{\pi}\int_0^\infty\frac{\cos\nu(x-y)}{\nu^2+h^2}\,d\nu.\end{aligned}$$

Let us now consider the case $x+y=0$. Instead of (4.5) we now obtain

$$\theta_{h,1}(x,y,\mu)-\theta_{h,1}^*(x,y,\mu)=\theta_h(x,y,-\infty)-2h+o(1),$$

from which for $x=y=0$ we again obtain (4.8).

§5. Riesz summability of the spectral function

The results obtained in §3 enable us to study the asymptotic behaviour of the Riesz means of spectral functions. We will consider the case of the entire line. The case of the halfline and also the case of a finite interval can be studied analogously.

By virtue of (3.7) and (3.9), there follows from the Tauberian Theorem 4.1 (cf. Chapter 14) the bound ($s \geqq 0$ is an arbitrary number)

$$\left| \frac{1}{\Gamma(s+1)} \int_0^{\mu} \left(1-\frac{\nu^2}{\mu^2}\right)^s d_\nu \Phi^*(x, y, \nu) \right| < \frac{C}{\mu^s}, \tag{5.1}$$

where the constant C depends only upon the domain of variation of the point (x, y).

Using the bound (5.1), it is not hard to obtain the asymptotic behaviour as $\mu \to \infty$ of the integral

$$\frac{1}{\Gamma(s+1)} \int^{\mu} \left(1-\frac{\nu^2}{\mu^2}\right) d_\nu \theta_1(x, y, \nu).$$

In fact, by (3.8) and from the definition of the functions $\Phi^*(x, y, \nu)$ and $\theta_1^*(x, y, \nu)$ we have

$$\begin{aligned} \frac{1}{\Gamma(s+1)} \int_0^{\mu} \left(1-\frac{\nu^2}{\mu^2}\right)^s d_\nu \Phi^*(x, y, \nu) &= \frac{1}{\Gamma(s+1)} \int_0^{\mu} \left(1-\frac{\nu^2}{\mu^2}\right)^s d_\nu \theta_1(x, y, \nu) \\ &- \frac{1}{\Gamma(s+1)} \frac{1}{\pi} \int_0^{\mu} \left(1-\frac{\nu^2}{\mu^2}\right)^s \cos \nu(x-y)\, d\nu \\ - \frac{1}{\pi} \frac{1}{\Gamma(s+1)} \int_0^{\mu} \left(1-\frac{\nu^2}{\mu^2}\right)^s \alpha(x, y, \nu)\, d\nu &+ \frac{2}{\pi} \frac{1}{\Gamma(s+1)} \int_0^{\mu} \left(1-\frac{\nu^2}{\mu^2}\right)^s \beta(x, y, \nu)\, d\nu \\ &= \frac{1}{\Gamma(s+1)} \int_0^{\mu} \left(1-\frac{\nu^2}{\mu^2}\right)^s d_\nu \theta_1(x, y, \nu) - I_1 - I_2 + I_3. \end{aligned} \tag{5.2}$$

By a well-known formula (cf. Titchmarsh [**3**], §7.1)

$$\frac{1}{\Gamma(s+1)} \int_0^{\mu} \left(1-\frac{\nu^2}{\mu^2}\right)^s \cos \nu t\, d\nu = \sqrt{\pi}\, \mu 2^{s-\frac{1}{2}} \frac{J_{s+\frac{1}{2}}(\mu t)}{(\mu t)^{s+\frac{1}{2}}}. \tag{5.3}$$

Therefore

$$\begin{aligned} I_1 &= \frac{1}{\pi \Gamma(s+1)} \int_0^{\mu} \left(1-\frac{\nu^2}{\mu^2}\right) \cos \nu(x-y)\, d\nu \\ &= \frac{1}{\sqrt{\pi}} \mu 2^{s-\frac{1}{2}} \frac{J_{s+\frac{1}{2}}[\mu(x-y)]}{[\mu(x-y)]^{s+\frac{1}{2}}}, \end{aligned}$$

$$\begin{aligned} I_2 &= \frac{1}{\pi \Gamma(s+1)} \int_0^{\mu} \left(1-\frac{\nu^2}{\mu^2}\right)^s \alpha(x, y, \nu)\, d\nu \\ &= \frac{1}{\pi \Gamma(s+1)} \int_{|x-y|}^{1} w(x, t, y) \left[\int_0^{\mu} \left(1-\frac{\nu^2}{\mu^2}\right)^s \cos \nu t\, d\nu \right] dt \\ &= \mu \frac{2^{s-\frac{1}{2}}}{\sqrt{\pi}} \int_{|x-y|}^{1} w(x, t, y) \frac{J_{s+\frac{1}{2}}(\mu t)}{(\mu t)^{s+\frac{1}{2}}}\, dt. \end{aligned} \tag{5.4}$$

From the last formula, by Theorem 1.2, Chapter 6, i.e. by virtue of the bound (1.20) of Chapter 6, there follows the bound

$$|I_2| \leqslant C\mu \int_0^1 t^\alpha \frac{\left|J_{s+\frac{1}{2}}(\mu t)\right|}{(\mu t)^{s+\frac{1}{2}}}\, dt = \frac{C}{\mu^\alpha} \int_0^\mu t^\alpha \frac{\left|J_{s+\frac{1}{2}}(t)\right|}{t^{s+\frac{1}{2}}}\, dt$$

(5.5)

$$\leqslant \frac{C}{\mu^\alpha}\left[\int_0^1 dt + \int_1^\mu \frac{dt}{t^{s+1-\alpha}}\right] \leqslant \begin{cases} \dfrac{C}{\mu^\nu}, & \nu = \min(\alpha, s),\ \alpha \neq s, \\ \dfrac{C \ln \mu}{\mu^\alpha}, & \alpha = s. \end{cases}$$

Further, we have

$$I_3 = \frac{2}{\pi \Gamma(s+1)} \int_0^\mu \left(1 - \frac{\nu^2}{\mu^2}\right)^s \beta(x, y, \nu)\, d\nu = \frac{2^{s+\frac{1}{2}}}{\sqrt{\pi}} \mu \int_0^1 h(x, y, t) \frac{J_{s+\frac{1}{2}}(\mu t)}{(\mu t)^{s+\frac{1}{2}}}\, dt$$

(5.6)

$$= -\frac{2^{s+\frac{1}{2}}}{\sqrt{\pi}} \mu \theta(x, y, -\infty) \int_0^\mu \frac{J_{s+\frac{1}{2}}(\mu t)}{(\mu t)^{s+\frac{1}{2}}}\, dt + \int_0^1 O(t) \frac{J_{s+\frac{1}{2}}(\mu t)}{(\mu t)^{s+\frac{1}{2}}}\, dt$$

$$= -\frac{2^{s+\frac{1}{2}}}{\sqrt{\pi}} \theta(x, y, -\infty) \int_0^\infty \frac{J_{s+\frac{1}{2}}(t)}{t^{s+\frac{1}{2}}}\, dt + o(1).$$

From formulas (5.2), (5.4), (5.5), (5.6) and the bound (5.1) follows:

THEOREM 5.1. *Let the function $q(x)$ satisfy the condition (1.19) of Chapter 6. Then for every fixed x and y, and $\mu \to \infty$*

$$\frac{1}{\Gamma(s+1)} \int_0^\mu \left(1 - \frac{\nu^2}{\mu^2}\right)^s d_\nu \theta_1(x, y, \nu) - \frac{2^{s-\frac{t}{2}}}{\sqrt{\pi}} \mu \frac{J_{s+\frac{1}{2}}[\mu(x-y)]}{[\mu(x-y)]^{s+\frac{1}{2}}}$$

(5.7)

$$= \frac{1}{\Gamma(s+1)} \theta(x, y, -\infty) + \begin{cases} \dfrac{C}{\mu^\nu}, & \nu = \min(\alpha, s),\ \alpha \neq s, \\ \dfrac{C \ln \mu}{\mu^\alpha}, & \alpha = s. \end{cases}$$

This relation holds uniformly as x and y vary over any bounded region.

§6. Proof of a theorem on equiconvergence

Making use of formula (2.22) and the Tauberian Theorem 4.1 of Chapter 14, we will in this section give a complete derivation of a theorem concerning the equiconvergence of an expansion in a generalized Fourier integral and an expansion in an ordinary Fourier integral under minimal requirements on the coefficient $q(x)$ —its summability over every finite interval.

We will consider the case of the entire line. The case of the halfline and the case of a finite interval can be studied analogously.

Let us turn first to the transformation of the right side of (2.22). Changing the order of integration, we obtain

$$\int_{x-\varepsilon}^{x+\varepsilon} f(s)\left\{\int_{|x-s|}^{\varepsilon} w(x, t, s)\, g_\varepsilon(t)\, dt\right\} ds = \int_0^{\varepsilon} g_\varepsilon(t)\left\{\int_{x-t}^{x+t} w(x, t, s) f(s)\, ds\right\} dt.$$

Put

$$h(x, t) = \int_{x-t}^{x+t} w(x, t, s) f(s)\, ds,$$

$$\alpha(x, \mu) = \int_0^1 h(x, t)\cos \mu t\, dt.$$

From the Parseval equality for the ordinary Fourier integral follows

$$\begin{aligned}\frac{1}{2}\int_{-\infty}^{\infty} \psi_\varepsilon(\mu)\, \alpha(x, \mu)\, d\mu &= \frac{\pi}{2}\int_0^{\varepsilon} g_\varepsilon(t)\, h(x, t)\, dt \\ &= \frac{\pi}{2}\int_{x-\varepsilon}^{x+\varepsilon} f(s)\left\{\int_{|x-s|}^{\varepsilon} w(x, t, s)\, g_\varepsilon(t)\, dt\right\} ds.\end{aligned} \tag{6.1}$$

Further, we have (as a consequence of changing the order of integration)

$$\int_{-\infty}^{0}\left\{\int_0^{\varepsilon} g_\varepsilon(t)\cos\sqrt{\lambda}\, t\, dt\right\} d_\lambda \int_{-\infty}^{\infty} \theta(x, s, \lambda) f(s)\, ds = \int_0^{\varepsilon} g_\varepsilon(t)\, k(x, t)\, dt,$$

where

$$k(x, t) = \int_{-\infty}^{0} \cos\sqrt{\lambda}\, d_\lambda \int_{-\infty}^{\infty} \theta(x, s, \lambda) f(s)\, ds.$$

Put

$$\beta(x, \mu) = \int_0^1 k(x, t) \cos \mu t \, dt.$$

From the Parseval equality for the ordinary Fourier integral follows

$$\frac{1}{2} \int_{-\infty}^{\infty} \psi_\varepsilon(\mu) \beta(x, \mu) \, d\mu = \frac{\pi}{2} \int_0^\varepsilon k(x, t) g_\varepsilon(t) \, dt. \tag{6.2}$$

From (2.22), (6.1) and (6.2) we finally obtain

$$\int_{-\infty}^{\infty} \psi_\varepsilon(\mu) \, d_\mu R^*(x, \mu) = 0 \qquad (0 < \varepsilon \leqslant 1), \tag{6.3}$$

where

$$R^*(x, \mu) = R(x, \mu, f) - \frac{1}{\pi} \int_0^\mu \alpha(x, \nu) \, d\nu + \frac{2}{\pi} \int_0^\mu \beta(x, \nu) \, d\nu. \tag{6.4}$$

LEMMA 6.1. *For every fixed* x

$$\lim_{\mu \to \infty} \bigvee_{\mu}^{\mu+1} \{R^*(x, \mu)\} = 0. \tag{6.5}$$

This relation holds uniformly with respect to x *over every finite interval.*

PROOF. It is sufficient to prove a relation analogous to (6.5) for each term in the right side of (6.4) separately. We have

$$R(x, \mu, f) = S_1(x, \mu) - S_1^*(x, \mu).$$

We will prove (6.5) for $S_1(x, \mu)$. Let $[a, b]$ denote an arbitrary finite interval and consider the problem

$$y'' + \{\lambda - q(x)\} y = 0 \qquad (a \leqslant x \leqslant b),$$
$$y(a) \cos \alpha + y'(a) \sin \alpha = 0,$$
$$y(b) \cos \beta + y'(b) \sin \beta = 0,$$

in which α and β are arbitrary real numbers. We denote by $\lambda_n = \mu_n^2$ the positive eigenvalues of this problem, arranged in increasing order, and by $\psi_n(x)$ corresponding normalized eigenfunctions. Further, put

$$\theta_{1, a, b}(x, s, \mu) = \sum_{\mu_n < \mu} \psi_n(x) \psi_n(s),$$

$$S_{1, a, b}(x, \mu) = \int_a^b f(s) \theta_{1, a, b}(x, s, \mu) \, ds = \sum_{\mu_n < \mu} c_n \psi_n(x).$$

It is known (cf., for example, Chapter 2) that as $(a,b)\to(-\infty,\infty)$,[3)]

$$\theta_{1,a,b}(x,s,\mu)\to\theta_1(x,s,\mu),\qquad S_{1,a,b}(x,\mu)\to S_1(x,\mu).$$

Further,

$$\bigvee_{\mu}^{\mu+1}\{S_{1,a,b}(x,\mu)\}=\sum_{\mu\leq\mu_n<\mu+1}|c_n||\psi_n(x)|$$

$$\leq\Big(\sum_{\mu\leq\mu_n<\mu+1}|c_n|^2\Big)^{\frac{1}{2}}\Big(\sum_{\mu\leq\mu_n<\mu+1}\psi_n^2(x)\Big)^{\frac{1}{2}} \tag{6.5'}$$

$$=\Big(\sum_{\mu\leq\mu_n<\mu+1}|c_n|^2\Big)^{\frac{1}{2}}[\theta_{1,a,b}(x,x,\mu+1)-\theta_{1,a,b}(x,x,\mu)]^{\frac{1}{2}}.$$

Since[3)]

$$\lim_{(a,b)\to(-\infty,\infty)}\sum_{\mu\leq\mu_n<\mu+1}|c_n|^2$$

$$=\int_{\mu}^{\mu+1}E^2(t)\,d\xi(t)+2E(t)F(t)\,d\eta(t)+F^2(t)\,d\zeta(t),$$

(6.5) follows from the bound (6.5′) by passing to the limit, first with respect to (a,b), and then with respect to μ.

For the function $S_1^*(x,\mu)$ the relation (6.5) can be obtained in similar fashion or, even simpler, starting from the explicit form of this function, by an application of the Cauchy-Bunjakovskiĭ inequality.

Further, we have

$$\bigvee_{\mu}^{\mu+1}\left\{\int_0^{\mu}\alpha(x,t)\,dt\right\}=\int_{\mu}^{\mu+1}|\alpha(x,t)|\,dt=o(1)$$

and similarly for the integral with $\beta(x,\nu)$. Thus the lemma is entirely proved. From this lemma and from the Tauberian Theorem 4.1 of Chapter 14 for Fourier integrals there follows at once:

LEMMA 6.2. *For every fixed* x

$$\lim_{\mu\to\infty}R^*(x,\mu)=0. \tag{6.6}$$

This relation holds uniformly with respect to x *over every finite interval.*

---

3) For the case in which the spectral function is nonunique, one has to pass to the limit along some sequence $\{(a_k,b_k)\}$.

To prove the theorem on equiconvergence, it remains to study the behaviour of the functions $\int_0^\mu \alpha(x,\nu)\,d\nu$ and $\int_0^\mu \beta(x,\nu)\,d\nu$ as $\mu \to \infty$.

LEMMA 6.3. *Suppose that the coefficient $q(x)$ is summable over every finite interval. Then for every fixed x we have*

$$\lim_{\mu\to\infty}\int_0^\mu \alpha(x,\nu)\,d\nu = 0. \tag{6.7}$$

This relation holds uniformly over every finite interval.

PROOF. We have

$$\int_0^\mu \alpha(x,\nu)\,d\nu = \int_0^1 h(x,t)\frac{\sin\mu t}{t}\,dt.$$

Further, from the bound $|w(x,t,s)| < C$ (cf. Theorem 1.1, Chapter 6) follows

$$\begin{aligned} |h(x,t)| &\leqslant \int_{x-t}^{x+t} |w(x,t,s)|\cdot|f(s)|\,ds \leqslant C\int_{x-t}^{x+t}|f(s)|\,ds \\ &\leqslant C\left(\int_{-\infty}^{\infty} f^2(s)\,ds\right)^{\frac{1}{2}}\left(\int_{x-t}^{x+t} ds\right)^{\frac{1}{2}} \leqslant Ct^{\frac{1}{2}}. \end{aligned} \tag{6.8}$$

Let η denote an arbitrary positive number and put

$$I = \int_0^t h(x,t)\frac{\sin\mu t}{t}\,dt = \left\{\int_0^\eta + \int_\eta^t\right\} h(x,t)\frac{\sin\mu t}{t}\,dt = I_1 + I_2.$$

By virtue of the bound (6.8)

$$|I_1| \leqslant C\int_0^\eta \frac{dt}{\sqrt{t}} = C\eta^{\frac{1}{2}}.$$

Therefore for any number $\zeta > 0$ we can choose η so small that for all μ we will have

$$|I_1| < \zeta/2. \tag{6.9}$$

Choosing η, we then take μ_0 so large that for all $\mu > \mu_0$ we will have[4)]

[4)] From the proof of the Riemann-Lebesgue lemma it follows that the bound (6.10) holds uniformly over every finite interval.

$$|I_2| < \zeta/2, \tag{6.10}$$

which is possible by the Riemann-Lebesgue lemma. The relation (6.7) follows from the bounds (6.9) and (6.10) and the arbitrariness of the number ζ.

LEMMA 6.5. *For every fixed* x

$$\lim_{\mu\to\infty} \frac{2}{\pi} \int_0^\infty \beta(x, \nu)\, d\nu = -S(x, -\infty). \tag{6.11}$$

This relation holds uniformly over every finite interval.

PROOF. The function $k(x,t)$ (for whose definition see the beginning of this section) is analytic (and, in particular, differentiable) with respect to t. Therefore it follows from a well-known theorem of the theory of Fourier series that

$$\lim_{\mu\to\infty} \frac{2}{\pi} \int_0^\mu \beta(x, \nu)\, d\nu = \lim_{\mu\to\infty} \frac{2}{\pi} \int_0^1 k(x, t) \frac{\sin \mu t}{t}\, dt = k(x, 0) = -S(x, -\infty),$$

as was to be proved.[5]

From Lemmas 6.2, 6.3 and 6.4 follows:

THEOREM 6.1 (ON EQUICONVERGENCE). *Suppose that the coefficient* $q(x)$ *is summable over every finite interval and that* $f(x) \in \mathcal{L}^2(-\infty, \infty)$. *Then we have, uniformly in every finite interval,*

$$\lim_{\mu\to\infty} \left\{ \int_{-\infty}^{\infty} f(s)\, \theta_1(x, s; \mu)\, ds - \frac{1}{\pi} \int_{-\infty}^{\infty} f(s) \frac{\sin \mu(x-s)}{x-s}\, ds \right\}$$

$$= \int_{-\infty}^{\infty} f(s)\, \theta(x, s; -\infty)\, ds,$$

i.e. the difference between an expansion with respect to the eigenfunctions of a Sturm-Liouville operator and the ordinary Fourier integral tends to zero uniformly in every finite interval.

[5] That (6.11) holds uniformly in every finite interval follows from the boundedness of the function $\partial k(x,t)/\partial t$.

§7. Convergence and summability of an eigenfunction expansion

From the theorem on equiconvergence, proved in §6, there immediately follows a theorem on the convergence of an expansion in the eigenfunctions of a Sturm-Liouville operator.

THEOREM 7.1. *If the coefficient $q(x)$ is summable over every finite interval and $f(x) \in \mathscr{L}^2(-\infty, \infty)$, then at every point at which $f(x)$ satisfies local conditions for its expandibility in an ordinary Fourier integral, we have*

$$\lim_{\lambda\to\infty} \int_{-\infty}^{\infty} [\theta(x, s, \lambda) - \theta(x, s, -\lambda)] f(s)\, ds = f(x).$$

In this section we will study the Riesz summability of the generalized Fourier integral.

By virtue of (6.3) and the bound (6.5), it follows from the Tauberian Theorem 4.1 of Chapter 14 that

$$\left| \frac{1}{\Gamma(s+1)} \int_0^{\mu} \left(1 - \frac{\nu^2}{\mu^2}\right)^s d_\nu R^*(x, \nu) \right| < \frac{C}{\mu^s} \qquad (0 < s \leq 1),$$

where the constant C depends upon the interval in which x varies but not upon μ, and the function $R^*(x, \nu)$ is defined by (6.4).

Therefore to study the question of the Riesz summability of an expansion, i.e. to estimate

$$\frac{1}{\Gamma(s+1)} \int_0^{\mu} \left(1 - \frac{\nu^2}{\mu^2}\right)^s d_\nu R(x, \nu),$$

we have to estimate the integrals

$$I_1 = \frac{1}{\pi} \frac{1}{\Gamma(s+1)} \int_0^{\mu} \left(1 - \frac{\nu^2}{\mu^2}\right)^s \alpha(x, \nu)\, d\nu,$$

$$I_2 = \frac{2}{\pi} \frac{1}{\Gamma(s+1)} \int_0^{\mu} \left(1 - \frac{\nu^2}{\mu^2}\right)^s \beta(x, \nu)\, d\nu.$$

Let us start with the first integral. We have

$$I_1 = \frac{\mu}{\sqrt{\pi}} 2^{s-\frac{1}{2}} \int_0^1 h(x, t) \frac{J_{s+\frac{1}{2}}(\mu t)}{(\mu t)^{s+\frac{1}{2}}}\, dt.$$

Suppose that the bound (1.20) of Chapter 6 holds. Then

$$|h(x,t)| \leqslant Ct^{\alpha} \int_{x-t}^{x+t} |f(s)|\, ds < Ct^{\alpha+\frac{1}{2}}.$$

Therefore

$$(7.1) \qquad |I_1| \leqslant \frac{C}{\mu^{\alpha+\frac{1}{2}}} \int_0^{\mu} t^{\alpha+\frac{1}{2}} \frac{\left|J_{s+\frac{1}{2}}(t)\right|}{t^{s+\frac{1}{2}}}\, dt$$

$$\leqslant \frac{C}{\mu^{\alpha+\frac{1}{2}}} \left\{ \int_0^1 dt + \int_1^{\mu} \frac{dt}{t^{s-\alpha+\frac{1}{2}}} \right\} \leqslant \frac{C}{\mu^{\alpha+\frac{1}{2}}} + \frac{C}{\mu^{s}}.$$

Further, we have $k(x,t) = k(x,0) + O(t) = -S(x,-\infty) + O(t)$. Therefore

$$I_2 = \frac{2^{s+\frac{1}{2}}}{\sqrt{\pi}} \mu \int_0^1 k(x,t) \frac{J_{s+\frac{1}{2}}(\mu t)}{(\mu t)^{s+\frac{1}{2}}}\, dt + \int_0^1 O(t) \frac{J_{s+\frac{1}{2}}(\mu t)}{(\mu t)^{s+\frac{1}{2}}}\, dt$$

$$(7.2) \qquad = -\frac{2^{s+\frac{1}{2}}}{\sqrt{\pi}} S(x,-\infty) \int_0^{\infty} \frac{J_{s+\frac{1}{2}}(t)}{t^{s+\frac{1}{2}}}\, dt + O(\mu^{-s})$$

$$= -\frac{1}{\Gamma(s+1)} S(x,-\infty) + O(\mu^{-s}).$$

From the estimates (7.1) and (7.2) follows:

Theorem 7.2. *Suppose that the coefficient $q(x)$ is summable over every finite interval, and suppose that condition* (1.19) *of Chapter* 6 *is satisfied. If $f(x) \in \mathcal{L}^2(-\infty,\infty)$, then we have, uniformly in every finite interval,*

$$\frac{1}{\Gamma(s+1)} \int_0^{\mu} \left(1 - \frac{\nu^2}{\mu^2}\right)^{s} d_{\nu} R(x,\nu) = \frac{1}{\Gamma(s+1)} S(x,-\infty) + O\left(\frac{1}{\mu^{\gamma}}\right),$$

where $\gamma = \min(s, \alpha + \frac{1}{2})$.

§8. Asymptotic behaviour of the spectral function of an equation of order $2n$

We will consider a formally symmetric differential expression $l(y)$, defined on the real line or the halfline, of the form

$$l(y) = (-1)^n y^{(2n)} + p_{2n-2}(x) y^{(2n-2)} + \cdots + p_0(x) y.$$

We assume the coefficients $p_k(x)$ to be real.

Let L be one of its selfadjoint extensions, which exist since the coefficients are real and therefore the deficiency indices are equal. The spectral family E_λ of the operator L is in this case a family of integral operators with kernel $\theta(x, \xi; \lambda)$, which is called the spectral function. Another, equivalent definition of the spectral function can be given.

Let $u_j(x, \lambda)$ be the solutions of the equation

$$l(y) = \lambda y \tag{8.1}$$

which satisfy the conditions

$$u_j^{(s-1)}(x, \lambda)\,|_{x=0} = \delta_{js} \qquad (j, s = 1, 2, \cdots, 2n). \tag{8.2}$$

Then for the case of the entire line the spectral function $\theta(x, \xi; \lambda)$ is given by

$$\begin{aligned} \theta(x, \xi; \lambda_2) - \theta(x, \xi; \lambda_1) &= \theta(x, \xi; \Delta) \\ &= \int_\Delta \sum_{j,k=1}^{2n} u_j(x, \lambda) u_k(\xi, \lambda)\, d\sigma_{jk}(\lambda), \end{aligned} \tag{8.3}$$

where $\| \sigma_{jk}(\lambda) \|$ is the matrix introduced in §16, Chapter 1, i.e. the spectral matrix of the operator L.

If the operator L is defined on the halfline $[0, \infty)$, then

$$\theta(x, \xi; \Delta) = \int_\Delta \sum_{j,k=1}^{n} u_j(x, \lambda) u_k(\xi, \lambda)\, d\sigma_{jk}(\lambda), \tag{8.3'}$$

where the $u_j(x, \lambda)$ are certain solutions of equation (8.1) which satisfy the initial conditions

$$u_j^{(s-1)}(x, \lambda)\,|_{x=0} = \delta_{js} \qquad (j, s = 1, 2, \cdots, n). \tag{8.2'}$$

A proof of the equivalence of these two definitions of the spectral function $\theta(x, \xi; \lambda)$ can be found in the monograph by Naĭmark [1].

It follows from the representations (8.3) and (8.3′) that

$$\sigma_{jk}(\lambda) = D_x^{j-1} D_\xi^{k-1} \theta(x, \xi; \lambda)\,|_{\substack{x=0 \\ \xi=0}} = \theta^{j-1, k-1}(0, 0; \lambda)$$

$$(j, k = 1, 2, \cdots, 2n \text{ if } -\infty < x < \infty;\ j, k = 1, 2, \cdots, n \text{ if } 0 \leq x < \infty).$$

Thus the question of the asymptotic behaviour of the spectral matrix $\| \sigma_{jk}(\lambda) \|$ can be reduced to the question of finding the asymptotic behaviour of the spectral function $\theta(x, \xi; \lambda)$ and its derivatives.

We now formulate a result concerning the asymptotic behaviour of the spectral function of the operator L on the entire line and the halfline.

THEOREM 8.1. *Let the operator L be defined on the entire line* $-\infty < x < \infty$ *and be semibounded, and suppose that its coefficients* $p_k(x)$ *are such that* $p_{2n-2}(x)$ *is a piecewise smooth function, while the remaining coefficients are locally summable.*

Then in every bounded region of the (x, ξ) *plane, and as* $\lambda \to \infty$, *we have for the spectral function* $\theta(x, \xi; \lambda)$ *of any semibounded selfadjoint extension, i.e. for any* $\theta(x, \xi; \lambda)$ *generated by an arbitrary spectral matrix* $\| \sigma_{jk}(\lambda) \|$ $(\sigma_{jk}(\lambda) = 0, \lambda < \lambda_0)$, *the asymptotic formula*

$$\theta(x, \xi; \lambda) = \theta_0(x, \xi; \lambda) + o(1), \tag{8.4}$$

where $\theta_0(x, \xi; \lambda)$ *is the spectral function of the operator* $(-1)^n y^{(2n)}$ *on the entire line, and is given by*

$$\theta_0(x, \xi; \lambda) = \frac{1}{\pi} \frac{\sin \lambda^{\frac{1}{2n}} (x - \xi)}{x - \xi}.$$

If the operator L is defined on the halfline $[0, \infty)$ with the selfadjoint boundary conditions

$$B_j(y) = \sum_{i=1}^{2n} b_{ij} y^{(i-1)}(0) \qquad (j = 1, 2, \ldots, n), \tag{8.5}$$

then a theorem analogous to that just stated holds, if the boundary conditions satisfy certain specific conditions.

Namely, let λ_i^+ be the roots of the equation $\tau^{2n} - \lambda = 0$ lying in the upper halfplane and let

$$\mu^+(\tau, \lambda) = \prod_{i=1}^{n} (\tau - \lambda_i^+).$$

Let us denote by $B_j'(\tau, \lambda) = \sum_{i=1}^{n} b_{ij}' \tau^{i-1}$ the remainder resulting from the division of $B_j(\tau) = \sum_{i=1}^{2n} b_{ij} \tau^{i-1}$ by $\mu^+(\tau, \lambda)$. We require that

1. $\det \| b_{ij}' \| \neq 0$ (condition of solvability).
2. The elements c_{ji} of the inverse of $\| b_{ij}' \|$ satisfy the inequality $|c_{ji}| \leq c |\lambda|^{-r_j + i - 1}$, where r_j is the highest order of differentiation in the jth boundary condition.

We then have the following result:

THEOREM 8.2. *For the spectral function* $\theta(x, \xi; \lambda)$ *of any semibounded selfadjoint extension of the operator L on the halfline* $[0, \infty)$, *if the conditions 1 and 2 are satisfied, we have the following asymptotic formula as* $\lambda \to \infty$:

$$\theta(x, \xi; \lambda) = \theta_0(x, \xi; \lambda) + o(1), \tag{8.6}$$

uniformly with respect to x and ξ *in any bounded region. Here* $\theta_0(x, \xi; \lambda)$ *is the spectral function of the operator* $(-1)^n y^{(2n)}$ *on the halfline* $[0, \infty)$ *with the same boundary conditions* (8.5).

For the case $n = 1$ Theorems 8.1 and 8.2 imply Theorems 3.1 and 4.1, respectively, of the present chapter.

For the case $n > 1$ the scheme for deriving the asymptotic formulas (8.4) and (8.6), based on a consideration of the Cauchy problem for the wave equation, cannot be carried through here. This is connected with the fact that the equation $\partial^2 u/\partial t^2 = Lu$ is hyperbolic only for the case $n = 1$, and this hyperbolicity is the determining factor in the consideration of an operator L with arbitrary (increasing) coefficients. It is also not hard to show that for $n > 1$ there does not exist a polynomial $P(\tau, \lambda)$ in the two variables τ and λ such that $P(\partial/\partial t, L)u = 0$ is a hyperbolic equation.

We will now indicate the scheme for proving Theorem 8.1.

One first studies the spectral function $\tilde{\theta}_0(x, \xi; \lambda)$ of an operator $\tilde{L}$ whose coefficients have compact support, i.e. vanish outside some finite interval. The asymptotic behaviour of $\tilde{\theta}_0(x, \xi; \lambda)$ can be established in a direct way. Namely, let us consider the equation for the kernel $\tilde{K}(x, \xi; \lambda)$ of the resolvent operator of $\tilde{L}$:

$$\tilde{K}(x, \xi; \lambda) = K_0(x, \xi; \lambda) + \int_{\infty}^{\infty} K_0(x, s; \lambda)\, \tilde{P}\left(s, \frac{d}{ds}\right) \tilde{K}(s, \xi; \lambda)\, ds, \tag{8.7}$$

where $K_0(x, \xi; \lambda)$ is the kernel of the resolvent of the operator $(-1)^n y^{(2n)}$ on the entire line, and $\tilde{P}(x, d/dx)$ is the operator containing all lower terms of the operator $\tilde{L}$ (its coefficients have compact support), i.e.

$$\tilde{P}(x, d/dx) = \tilde{L} - (-1)^n d^{2n}/dx^{2n}.$$

It is not hard to show that for large $|\lambda|$ equation (8.7) can be solved by the method of successive approximations and that as $\lambda \to \infty$ one has, uniformly with respect to (x, ξ),

$$\tilde{K}(x, \xi; \lambda) = K_0(x, \xi; \lambda) + o(1)/|\lambda|. \tag{8.8}$$

Integrating the left and right sides of (8.8) along a closed contour Γ_η in the complex plane $\lambda = \eta + i\tau$ which passes through the point η on the real axis, we obtain as $\eta \to \infty$

$$\tilde{\theta}_0(x, \xi; \eta) = \theta_0(x, \xi; \eta) + o(1), \tag{8.9}$$

where $\theta_0(x, \xi; \eta)$ is the spectral function of the operator $(-1)^n y^{(2n)}$.

Suppose now that we have to establish (8.4) in the region S ($a \leq x \leq b$, $a_1 \leq \xi \leq b_1$). We choose some region $u = \{x : a - h \leq x \leq b + h\}$ with $h > 0$ and consider the "reduced" operator $\tilde{L}$, whose coefficients coincide with the coefficients $p_i(x)$ of the operator L in the region u, and vanish outside $u_\epsilon = \{x : a - h - \epsilon \leqq x \leqq b + h + \epsilon\}$.

Let $\widetilde{G}_0(x, \xi; t)$ denote the Green's function of the Cauchy problem for the parabolic equation $\partial u / \partial t = -\widetilde{L}u$.

If now $G(x, \xi; t)$ is the kernel of the integral operator e^{-Lt}, i.e. the Green's function of the Cauchy problem for the operator equation $\partial u / \partial t = -Lu$, then as can be seen by means of Green's formula, we have in the region S

$$G(x, \xi; t) - \widetilde{G}_0(x, \xi; t) = O(1)\, e^{-ch/t^{\frac{1}{2n-1}}}, \tag{8.10}$$

where c is a constant which does not depend upon ϵ, h or t.

On the other hand, as is not difficult to see,

$$G(x, \xi; t) = \int_0^\infty e^{-\lambda t}\, d\theta(x, \xi; \lambda).$$

Taking account of this representation, (8.10) can be rewritten in the form

$$\int_0^\infty e^{-\lambda t} d\theta(x, \xi; \lambda) = \int_0^\infty e^{-\lambda t} d\tilde{\theta}_0(x, \xi; \lambda) + O(1)\, e^{-ch/t^{\frac{1}{2n-1}}}, \tag{8.11}$$

where $\tilde{\theta}_0(x, \xi; \lambda)$ is the spectral function of the "reduced" operator $\widetilde{L}$. After multiplying the left and right sides of (8.11) by $e^{-\mu t}$ and integrating with respect to t from 0 to ∞, we obtain

$$\int_0^\infty \frac{d\theta(x, \xi; \lambda)}{\lambda + \mu} = \int_0^\infty \frac{d\tilde{\theta}_0(x, \xi; \lambda)}{\lambda + \mu} + O(1)\, e^{-c^* h^* \mu^{\frac{1}{2n}}},$$

where c^* is a constant which depends only upon c, and $h^* = h^{(2n-1)/2n}$. Using this last relation, we can establish the following useful relation:

$$\int_0^\infty e^{-s\lambda^{1/2n}} d\theta(x, \xi; \lambda) = \int_0^\infty e^{-s\lambda^{1/2n}} d\tilde{\theta}_0(x, \xi; \lambda) + G(s), \tag{8.12}$$

where $G(s)$ is analytic with respect to $s = \eta + i\tau$ in the strip $|\tau| < \tilde{h} = h^* c^* / \cos(\pi/2n)$ and $G(0) = 0$.

Putting $\rho(\lambda) = \theta(x, x; \lambda^{2n})$, $\sigma(\lambda) = \tilde{\theta}_0(x, x; \lambda^{2n})$, on the basis of (8.12) we obtain

$$\int_0^\infty e^{-s\lambda} d\rho(\lambda) = \int_0^\infty e^{-s\lambda} d\sigma(\lambda) + G(s). \tag{8.13}$$

From (8.9) we find that

$$\sigma(\lambda) = \tilde{\theta}_0(x, x; \lambda^{2n}) = \frac{1}{\pi}\lambda + o(1).$$

Further, on the basis of Marčenko's Tauberian theorem (§4, Chapter 14), which makes possible a sharper estimate of the remainder in Ikehara's Tauberian theorem, we obtain from (8.13)

$$\overline{\lim_{\lambda\to\infty}} \left| \frac{\rho(\lambda+0)-\rho(\lambda)}{2} - \sigma(\lambda) - G(0) \right| \leqslant \bigvee_{\lambda}^{\lambda+\tilde{h}} \sigma(\xi) = \frac{A}{\tilde{h}\pi},$$

where A is an absolute constant.

Since the number h can be chosen arbitrarily large, and $\tilde{h}$, along with h, is arbitrarily large, taking into account (8.7) and the fact that $G(0)=0$, we obtain the assertion of Theorem 8.1 for the case $\xi = x$. Following this, the proof for the case $\xi \neq x$ is carried out by consideration of the nondecreasing function

$$F(x,\xi;\lambda) = \theta(x,x;\lambda) + 2\theta(x,\xi;\lambda) + \theta(\xi,\xi;\lambda).$$

The proof of Theorem 8.2 is technically rather complicated, but can be carried out according to the same scheme.

We remark that applying results of Nilsson [1] on estimates for the spectral function of general elliptic operators on the negative spectrum, one can also establish Theorems 8.1 and 8.2 for operators which are not semibounded. At least for operators with sufficiently smooth coefficients, this can be done rather easily. Namely, one has to consider the square of the non-semibounded operator L which is now a positive operator. One further has to use Theorem 8.1 or 8.2.

From Theorem 8.1 follows:

THEOREM 8.3 (ON EQUICONVERGENCE). *Every function $f(x)$ with compact support which can be expanded in a Fourier integral can also be expanded as an integral with respect to eigenfunctions, and conversely.*

We are concerned here with the pointwise convergence of the expansions.

This theorem follows immediately from Theorem 8.1, since the partial sums of an eigenfunction expansion and a Fourier integral expansion are equal respectively to

$$S(x, f; \lambda) = \int_a^b \theta(x, \xi; \lambda) f(\xi)\, d\xi,$$

$$S_0(x, f; \lambda) = \int_a^b \theta_0(x, \xi; \lambda) f(\xi)\, d\xi,$$

and consequently

$$S(x, f; \lambda) - \frac{1}{\pi}\int_a^b \frac{\sin \lambda^{1/2n}(x-\xi)}{x-\xi} f(\xi)\, d\xi \to 0, \qquad \lambda \to \infty.$$

By this method one can also obtain the corresponding asymptotic behaviour of the derivatives $D_x^i D_\xi^j \theta(x, \xi; \lambda) = \theta^{(i+j)}(x, \xi; \lambda)$, which enables one to write down the asymptotic behaviour of the spectral matrix.

Bibliographical references

§1. The basic results concerning eigenfunction expansions are presented in the monographs by Titchmarsh [**1**], Levitan [**1**] and others. The identities (1.12) and (1.14) were obtained in a paper by Levitan [**14**].

§2. Lemmas 2.1 and 2.2 for the case of the halfline $[0, \infty)$ are in essential part due to Marčenko [**1, 2**]. The proof of Lemma 2.1 given here is due to Levitan [**14**].

§3. The asymptotic formula (3.8) was obtained in papers by Levitan [**14, 15**]. Cf. also Marčenko [**3**].

§4. Formula (4.5) was derived in a paper by Levitan [**15**]; cf. also a paper by Marčenko [**3**].

§5. The Riesz summability of the spectral function was considered in papers by Levitan [**14, 15**].

§§6, 7. The theorem on equiconvergence was first proved by Levitan [**14**]; cf. also his paper [**15**] and Marčenko [**1, 2**]. The Riesz summability of an eigenfunction expansion was considered by Levitan [**14, 15**].

§8. The results of this section are due to Kostjučenko [**1, 2**].

CHAPTER 8

DIFFERENTIATION OF AN EIGENFUNCTION EXPANSION

§1. Preliminary estimates of the derivatives of the spectral function

1. In this chapter we will study the question of the differentiation of an expansion with respect to the eigenfunctions of the selfadjoint second-order differential operator

$$y'' + [\lambda - q(x)]y = 0, \tag{1.1}$$

defined on the halfline $[0, \infty)$ with the initial conditions

$$y(0) = 1, \qquad y'(0) = 0. \tag{1.2}$$

We assume that $q(x)$ is a real function defined on the halfline $[0, \infty)$ and summable over every finite interval.

Throughout the chapter we will assume that the spectrum of the problem (1.1) – (1.2) is nonnegative or, if a negative spectrum exists, that it is bounded from below. The latter case can easily be reduced to the former. Thus, let $\lambda = \mu^2$ $(\lambda > 0)$. We denote by $\varphi(x, \mu)$ the solution of the problem (1.1) – (1.2).

As is known, there exists a monotonic function $\rho(\mu)$ which is bounded on every finite interval, such that for every function $f(x) \in \mathcal{L}^2(0, \infty)$ one has the Parseval equality

$$\int_0^\infty f^2(x)\,dx = \int_0^\infty F^2(\mu)\,d\rho(\mu), \tag{1.3}$$

where

$$F(\mu) = \underset{A\to\infty}{\mathrm{l.i.m}} \int_0^A f(x)\,\varphi(s, \mu)\,ds.$$

Let us denote by $\theta(x, s, \mu)$ the spectral function of equation (1.1) (for the initial data (1.2)), i.e. we put

$$\theta(x, s, \mu) = \begin{cases} \int_0^\mu \varphi(x, \nu)\,\varphi(s, \nu)\,d\rho(\nu), & \mu > 0, \\ 0, & \mu = 0. \end{cases} \tag{1.4}$$

In this section we will obtain various preliminary estimates for the derivatives of the spectral function $\theta(x,s,\mu)$.

As we have seen in the preceding chapter, for arbitrary real t we have

$$\varphi(x,\mu)\cos\mu t=\frac{1}{2}\{\varphi(x+t,\mu)+\varphi(x-t,\mu)\}+\frac{1}{2}\int_{x-t}^{x+t} w(x,t,s)\varphi(s,\mu)\,ds. \tag{1.5}$$

Differentiating (1.5) with respect to x, we obtain

$$\begin{aligned}\varphi'_x(x,\mu)\cos\mu t=&\frac{1}{2}\{\varphi'_x(x+t,\mu)+\varphi'_x(x-t,\mu)\}\\&+\frac{1}{2}\{w(x,t,x+t)\varphi(x+t,\mu)\\&-w(x,t,x-t)\varphi(x-t,\mu)\}\\&+\frac{1}{2}\int_{x-t}^{x+t}\frac{\partial w(x,t,s)}{\partial x}\varphi(s,\mu)\,ds.\end{aligned} \tag{1.6}$$

Suppose that the function $g_\epsilon(t)$ satisfies the conditions indicated on p. 296. If we multiply both sides of (1.6) by $g_\epsilon(t)$ and then integrate with respect to t from 0 to ϵ, we obtain

$$\begin{aligned}\varphi'_x(x,\mu)\psi_\varepsilon(\mu)=&\frac{1}{2}\int_0^\varepsilon\{\varphi'_x(x+t,\mu)+\varphi'_x(x-t,\mu)\}g_\varepsilon(t)\,dt\\&+\frac{1}{2}\int_0^\varepsilon\{w(x,t,x+t)\varphi(x+t,\mu)\\&-w(x,t,x-t)\varphi(x-t,\mu)\}g_\varepsilon(t)\,dt\\&+\frac{1}{2}\int_0^\varepsilon g_\varepsilon(t)\left\{\int_{x-t}^{x+t}\frac{\partial w(x,t,s)}{\partial x}\varphi(s,\mu)\,ds\right\}dt\equiv I_1+I_2+I_3.\end{aligned} \tag{1.7}$$

Let us transform the right side of (1.7). Consider the first term. Integrating by parts, we find that

$$\begin{aligned}\int_0^\varepsilon\varphi'_x(x+t,\mu)g_\varepsilon(t)\,dt=&\int_0^\varepsilon\varphi'_t(x+t,\mu)g_\varepsilon(t)\,dt\\=&-\varphi(x,\mu)g_\varepsilon(0)-\int_0^\varepsilon\varphi(x+t,\mu)g'_\varepsilon(t)\,dt,\end{aligned}$$

$$\int_0^\varepsilon \varphi'_x(x-t,\ \mu)\, g_\varepsilon(t)\, dt = -\int_0^\varepsilon \varphi'_t(x-t,\ \mu)\, g_\varepsilon(t)\, dt$$

$$= \varphi(x,\ \mu)\, g_\varepsilon(0) + \int_0^\varepsilon \varphi(x-t,\ \mu) g'_\varepsilon(t)\, dt.$$

Consequently, using the evenness of $g_\epsilon(t)$ with respect to t, we obtain

$$I_1 = -\int_0^\varepsilon \varphi(x+t,\ \mu) g'_\varepsilon(t)\, dt + \int_0^\varepsilon \varphi(x-t,\ \mu)\, g'_\varepsilon(t)\, dt$$

$$= -\int_x^{x+\varepsilon} \varphi(s,\ \mu) \frac{\partial g_\varepsilon(s-x)}{\partial s}\, ds - \int_{x-\varepsilon}^{x} \varphi(s,\ \mu) \frac{\partial g_\varepsilon(s-x)}{\partial s}\, ds \tag{1.8}$$

$$= -\int_{x-\varepsilon}^{x+\varepsilon} \varphi(s,\ \mu)\, \frac{\partial g_\varepsilon(s-x)}{\partial s}\, ds.$$

We now transform the second term of (1.7). By virtue of (2.9) and (2.10) of Chapter 6 we obtain

$$\int_0^\varepsilon w(x,\ t,\ x+t)\, \varphi(x+t,\ \mu)\, g_\varepsilon(t)\, dt$$

$$= -\frac{1}{2} \int_0^\varepsilon \varphi(x+t,\ \mu)\, g_\varepsilon(t) \left\{ \int_0^t q(x+\tau)\, d\tau \right\} dt$$

$$= -\frac{1}{2} \int_x^{x+\varepsilon} \varphi(s,\ \mu)\, g_\varepsilon(s-x) \left\{ \int_0^{s-x} q(x+\tau) d\tau \right\} ds$$

$$= -\frac{1}{2} \int_x^{x+\varepsilon} \left\{ g_\varepsilon(s-x) \int_x^s q(\tau) d\tau \right\} \varphi(s,\ \mu)\, ds,$$

$$-\int_0^\varepsilon w(x,\ t,\ x-t)\, \varphi(x-t,\ \mu)\, g_\varepsilon(t)\, dt$$

$$= \frac{1}{2} \int_0^\varepsilon \varphi(x-t,\ \mu)\, g_\varepsilon(t) \left\{ \int_0^t q(x-\tau)\, d\tau \right\} dt$$

$$= \frac{1}{2} \int_x^{x-\varepsilon} \varphi(s,\ \mu)\, g_\varepsilon(x-s) \left\{ \int_0^{x-s} q(x-\tau)\, d\tau \right\} ds$$

$$= -\frac{1}{2} \int_{x-\varepsilon}^{x} \left\{ g_\varepsilon(s-x) \int_x^s q(\tau)\, d\tau \right\} \varphi(s,\ \mu)\, ds.$$

Consequently,

$$I_2 = -\frac{1}{2}\int_{x-\varepsilon}^{x+\varepsilon}\left\{g_\varepsilon(s-x)\int_x^s q(\tau)\,d\tau\right\}\varphi(s,\mu)\,ds. \tag{1.9}$$

Finally, changing the order of integration in the last integral in (1.7), we obtain

$$I_3 = \int_{x-\epsilon}^{x+\varepsilon}\varphi(s,\mu)\lambda_\varepsilon(x,s)\,ds, \tag{1.10}$$

where

$$\lambda_\varepsilon(x,s) = \int_{|x-s|}^{\varepsilon} g_\varepsilon(t)\frac{\partial w(x,t,s)}{\partial x}\,dt. \tag{1.11}$$

By virtue of (1.8), (1.9) and (1.10), the relation (1.7) assumes the form

$$\psi'_x(x,\mu)\psi_\varepsilon(\mu) = -\frac{1}{2}\int_{x-\varepsilon}^{x+\varepsilon}\varphi(s,\mu)\frac{\partial g_\varepsilon(s-x)}{\partial s}\,ds$$

$$-\frac{1}{4}\int_{x-\varepsilon}^{x+}\left\{g_\varepsilon(s-x)\int_x^s q(\tau)\,d\tau\right\}\varphi(s,\mu)\,ds + \frac{1}{2}\int_{x-\varepsilon}^{x+\varepsilon}\lambda_\varepsilon(x,s)\varphi(s,\mu)\,ds \tag{1.12}$$

$$= \frac{1}{2}\int_{x-\varepsilon}^{x+\varepsilon}\left\{\lambda_\varepsilon(x,s) - \frac{\partial g_\varepsilon(s-x)}{\partial s} - \frac{1}{2}g_\varepsilon(s-x)\int_x^s q(\tau)\,d\tau\right\}\varphi(s,\mu)\,ds.$$

It follows from (1.12) that for fixed x the function $\varphi'_x(x,\mu)\psi_\epsilon(\mu)$ can be regarded as the generalized Fourier transform (with respect to the eigenfunctions $\varphi(x,\mu)$) of the function which is equal to

$$\frac{1}{2}\left\{\lambda_\varepsilon(x,s) - \frac{\partial g_\varepsilon(s-x)}{\partial s} - \frac{1}{2}g_\varepsilon(s-x)\int_x^s q(\tau)\,d\tau\right\} = h_\varepsilon(x,s) \tag{1.13}$$

for $s \in (x-\epsilon, x+\epsilon)$, and is equal to zero outside this interval.

Therefore by the Parseval equality we obtain from (1.12)

$$\int_0^\infty F(\mu)\varphi'_x(x,\mu)\psi_\varepsilon(\mu)\,d\rho(\mu) = \int_{x-\varepsilon}^{x+\varepsilon} f(s)h_\varepsilon(x,s)\,ds. \tag{1.13'}$$

Hence, by the arbitrariness of $f(s)$ and by (1.13) we obtain

$$\int_0^\infty \psi_\varepsilon(\mu)\varphi(s,\mu)\varphi'_x(x,\mu)\,d\rho(\mu) =$$

$$
(1.14)\qquad = \begin{cases} \lambda_\varepsilon(x,s) - \dfrac{\partial g_\varepsilon(s-x)}{\partial s} - \dfrac{1}{2} g_\varepsilon(s-x) \displaystyle\int_x^s q(\tau)\,d\tau, & |x-s| \leqslant \varepsilon, \\ 0, & |x-s| \geqslant \varepsilon. \end{cases}
$$

Differentiating the identity (1.14) with respect to s [1] and using (1.11), we obtain for $x > s$

$$
\int_0^\infty \psi_\varepsilon(\mu)\,\varphi_s'(s,\mu)\,\varphi_x'(x,\mu)\,d\rho(\mu)
$$

$$
(1.15)\qquad = \begin{cases} g_\varepsilon(s-x) \left.\dfrac{\partial w(x,t,s)}{\partial x}\right|_{t=x-s} \\ \quad + \displaystyle\int_{x-\varepsilon}^{\varepsilon} g_\varepsilon(t) \dfrac{\partial^2 w(x,t,s)}{\partial x\,\partial s}\,dt - \dfrac{\partial^2 g_\varepsilon(s-x)}{\partial s^2} \\ \quad - \dfrac{1}{2}\dfrac{\partial g_\varepsilon(s-x)}{\partial s} \displaystyle\int_x^s q(\tau)\,d\tau - \dfrac{1}{2} g_\varepsilon(s-x)\,q(s), & |x-s| \leqslant \varepsilon, \\ 0, & |x-s| \geqslant \varepsilon. \end{cases}
$$

By the definition of the spectral function $\theta(x,s,\mu)$, i.e. (1.4), the identity (1.15) assumes the form

$$
\int_0^\infty \psi_\varepsilon(\mu)\,d_\mu \frac{\partial^2 \theta(x,s,\mu)}{\partial x\,\partial s}
$$

$$
(1.15')\qquad = \begin{cases} g_\varepsilon(s-x) \left.\dfrac{\partial w(x,t,s)}{\partial x}\right|_{t=x-s} \\ \quad + \displaystyle\int_{x-s}^{\varepsilon} g_\varepsilon(t) \dfrac{\partial^2 w(x,t,s)}{\partial x\,\partial s}\,dt - \dfrac{\partial^2 g_\varepsilon(s-x)}{\partial s^2} \\ \quad - \dfrac{1}{2}\dfrac{\partial g_\varepsilon(s-x)}{\partial s} \displaystyle\int_x^s q(\tau)\,d\tau - \dfrac{1}{2} g_\varepsilon(s-x)\,q(s), & |x-s| \leqslant \varepsilon, \\ 0, & |x-s| \geqslant \varepsilon. \end{cases}
$$

[1] We assume here that the function $q(x)$ is absolutely continuous.

Extending the function $\theta(x, s, \mu)$ in the variable μ so as to obtain an odd function, for $s = x$ we obtain from (1.15)

$$\int_{-\infty}^{\infty} \psi_\varepsilon(\mu)\, d_\mu \frac{\partial^2\theta(x, s, \mu)}{\partial x\, \partial s}\bigg|_{s=x} = g_\varepsilon(0) \frac{\partial w(x, t, s)}{\partial x}\bigg|_{\substack{t=x-s \\ s=x}} \tag{1.16}$$
$$+ \int_0^\varepsilon g_\varepsilon(t) \frac{\partial^2 w(x, t, s)}{\partial x\, \partial s}\bigg|_{s=x} dt - \frac{\partial^2 g_\varepsilon(s-x)}{\partial s^2}\bigg|_{s=x} - \frac{1}{2} g_\varepsilon(0)\, q(x).$$

2. We will now prove some simple lemmas concerning the behaviour of the derivatives of the eigenfunctions.

LEMMA 1.1. *If the function $q(x)$ is absolutely continuous in every finite interval, then as $a \to \infty$ we have the following estimate:*

$$\bigvee_a^{a+1} \left\{\frac{\partial^2\theta(x, s, \mu)}{\partial x\, \partial s}\right\}_{s=x} = \int_a^{a+1} [\varphi_x'(x, \mu)]^2\, d\rho(\mu) = O(a^2). \tag{1.17}$$

This estimate holds uniformly as x varies in any finite interval.

PROOF. From well-known asymptotic formulas for the eigenfunctions it follows that for x in a finite interval and $\mu \to \infty$

$$\varphi_x'(x, \mu) = O(\mu). \tag{1.18}$$

Hence (1.17) follows from Lemma 2.3 of Chapter 7.

One can prove the following lemma in a similar way.

LEMMA 1.2. *If the function $q(x)$ has a derivative of order $2k$ which is summable over every finite interval, then for every fixed x we have, as $a \to \infty$,*

$$\bigvee_a^{a+1} \left\{\frac{\partial^{2k}(x, s, \mu)}{\partial x^k \partial s^k}\right\}_{s=x} = \int_a^{a+1} \{\varphi_x^{(k)}(x, \mu)\}^2\, d\rho(\mu) = O(a^{2k}) \tag{1.19}$$
$$(k = 1, 2, \ldots).$$

This estimate holds uniformly as x varies over any finite interval.

From Lemma 2.3 of Chapter 7, Lemma 1.1 and the Cauchy-Bunjakovskiĭ inequality follows:

LEMMA 1.3. *If $q(x)$ has a first derivative which is summable over every finite interval, then for every fixed x and s we have, as $a \to \infty$,*

$$\bigvee_a^{a+1} \left\{\frac{\partial\theta(x, s, \mu)}{\partial x}\right\} = \int_a^{a+1} |\varphi(s, \mu)| \cdot |\varphi_x'(x, \mu)|\, d\rho(\mu) = O(a). \tag{1.20}$$

This estimate holds uniformly as x and s vary over any finite intervals.

From Lemma 1.2 and the Cauchy-Bunjakovskiĭ inequality follows:

LEMMA 1.4. *Under the conditions of Lemma 1.2 we have the estimate*

$$\bigvee_a^{a+1}\left\{\frac{\partial^{2k}\theta(x,s,\mu)}{\partial x^i\partial s^j}\right\}=\int_a^{a+1}|\varphi_x^{(i)}(x,\mu)|\cdot|\varphi_s^{(j)}(s,\mu)|\,d\rho(\mu)=O(a^{2k}) \tag{1.21}$$
$$(i+j=2k).$$

This estimate holds uniformly as x and s vary over any finite intervals.

Let $\theta^*(x,s,\mu)$ denote the spectral function of the problem (1.1) – (1.2) for $q(x)\equiv 0$, i.e. we put

$$\theta^*(x,s,\mu)=\frac{2}{\pi}\int_0^\mu \cos\nu x\cdot\cos\nu s\,d\nu. \tag{1.22}$$

From the definition of $\theta^*(x,s,\mu)$ there immediately follows:

LEMMA 1.5. *As $a\to\infty$ we have the estimate*

$$\bigvee_a^{a+1}\left\{\frac{\partial^n\theta^*(x,s,\mu)}{\partial x^k\partial s^j}\right\}=O(a^n)\qquad(k+j=n=0,1,\ldots). \tag{1.23}$$

This estimate holds uniformly over the entire halfline $(0,\infty)$.

§2. Asymptotic behaviour of the derivatives of the spectral function

Formula (1.14) enables us to study the asymptotic behaviour of the derivatives of the spectral functions $\theta(x,s,\mu)$. In order to study the asymptotic behaviour of the derivatives of the spectral function of the problem (1.1) + (1.2), we will compare them with the derivatives of the spectral function of a simpler problem, namely the problem

$$y''+\lambda y=0\qquad(\lambda>0), \tag{2.1}$$

$$y(0)=1,\qquad y'(0)=0. \tag{2.2}$$

To this end we will transform the right side of (1.14). We have

$$\begin{aligned}&\int_{-\infty}^{\infty}\psi_\varepsilon(\mu)\varphi(s,\mu)\varphi_x'(x,\mu)\,d\rho(\mu)\\&=\begin{cases}\lambda_\varepsilon(x,s)-\frac{1}{2}g_\varepsilon(s-x)\int_x^\varepsilon q(\tau)\,d\tau-\frac{\partial g_\varepsilon(s-x)}{\partial s}, & |x-s|\leqslant\varepsilon,\\ 0, & |x-s|\geqslant\varepsilon.\end{cases}\end{aligned} \tag{2.3}$$

Using the spectral function $\theta(x,s,\mu)$, we can give (2.3) the form

$$\int_{-\infty}^{\infty} \psi_\varepsilon(\mu)\, d_\mu \frac{\partial\theta(x,s,\mu)}{\partial x} = \begin{cases} \lambda_\varepsilon(x,s) - \frac{1}{2} g_\varepsilon(s-x)\int_x^s q(\tau)\,d\tau - \frac{\partial g_\varepsilon(s-x)}{\partial s}, & |x-s| \leqslant \varepsilon, \\ 0, & |x-s| \geqslant \varepsilon. \end{cases} \tag{2.3'}$$

Writing down for the function $\theta^*(x,s,\mu)$, i.e. for the spectral function of the problem (2.1) + (2.2), a formula analogous to (2.3′), subtracting it from (2.3′) and taking into account that for the problem (2.1) + (2.2) one has $w(x,t,s) \equiv 0$, we obtain

$$\int_{-\infty}^{\infty} \psi_\varepsilon(\mu)\, d_\mu \left\{\frac{\partial\theta(x,s,\mu)}{\partial x} - \frac{\partial\theta^*(x,s,\mu)}{\partial x}\right\} = \begin{cases} \lambda_\varepsilon(x,s) - \frac{1}{2} g_\varepsilon(s-x)\int_x^s q(\tau)\,d\tau, & |x-s| \leqslant \varepsilon, \\ 0, & |x-s| \geqslant \varepsilon, \end{cases} \tag{2.4}$$

where

$$\lambda_\varepsilon(x,s) = \int_{|x-s|}^{\varepsilon} g_\varepsilon(t) \frac{\partial w(x,t,s)}{\partial x}\, dt. \tag{2.5}$$

Put

$$\alpha(x,s,\mu) = \int_{|x-s|}^{1} \frac{\partial w(x,t,s)}{\partial x} \cos\mu t\, dt. \tag{2.6}$$

Then by the Parseval equality for the ordinary Fourier integral it follows from (1.6), Chapter 7, and (2.6) that

$$\int_0^{\infty} \psi_\varepsilon(\mu)\, \alpha(x,s,\mu)\, d\mu = \frac{\pi}{2} \int_{|x-s|}^{\varepsilon} \frac{\partial w(x,t,s)}{\partial x} g_\varepsilon(t)\, dt. \tag{2.7}$$

Further, by virtue of the inversion formula we find from (1.6), Chapter 7, that

$$\begin{aligned} g_\varepsilon(s-x) &= \frac{1}{\pi} \int_{-\infty}^{\infty} \psi_\varepsilon(\mu) \cos(s-x)\mu\, d\mu \\ &= \frac{1}{\pi} \int_{-\infty}^{\infty} \psi_\varepsilon(\mu)\, d_\mu \frac{\sin(s-x)\mu}{s-x}. \end{aligned} \tag{2.8}$$

Therefore since the functions $\psi_\epsilon(\mu)$ and $\alpha(x,s,\mu)$ are even with respect to μ, it follows from (2.7) and (2.8) that (2.4) can be rewritten in the following form:

$$\begin{aligned}\int_{-\infty}^{\infty} \psi_s(\mu)\, d_\mu \Bigg\{ & \frac{\partial\theta(x, s, \mu)}{\partial x} - \frac{\partial\theta^*(x, s, \mu)}{\partial x} - \frac{1}{\pi}\int_0^\mu \alpha(x, s, \nu)\, d\nu \\ & + \frac{1}{4\pi}\frac{\sin(s-x)\mu}{s-x}\int_x^s q(\tau)\, d\tau\Bigg\} = 0.\end{aligned} \tag{2.9}$$

Theorem 2.1. *If $q(x)$ is absolutely continuous in every finite interval, then for every fixed x and s we have $(\mu\to\infty)$*

$$\int_0^\mu \left(1-\frac{\nu^2}{\mu^2}\right) d_\nu \left[\frac{\partial\theta(x, s, \nu)}{\partial x} - \frac{\partial\theta^*(x, s, \nu)}{\partial x}\right] = O(1) \tag{2.10}$$

This asymptotic relation holds uniformly as x and s vary over any finite intervals.

Proof. Since

$$\bigvee_\mu^{\mu+1}\left\{\int_0^\mu \alpha(x, s, \nu)\, d\nu\right\} = o(\mu), \qquad \bigvee_\mu^{\mu+1}\left\{\int_0^\mu \cos\nu(s-x)\, d\nu\right\} = O(\mu),$$

then by (1.20) and (2.9) the Tauberian Theorem 4.1 of Chapter 14 is applicable, from which it follows that for every fixed x and s the Riesz means of order 1 of the function

$$\begin{aligned}\frac{\partial\theta(x, s, \mu)}{\partial x} & - \frac{\partial\theta^*(x, s, \mu)}{\partial x} - \frac{1}{\pi}\int_0^\mu \alpha(x, s, \nu)\, d\nu \\ & + \frac{1}{4\pi}\int_x^s q(\tau)\, d\tau \int_0^\mu \cos\nu(s-x)\, d\nu\end{aligned}$$

remain bounded as $\mu\to\infty$, i.e. we have

$$\begin{aligned}\int_0^\mu \left(1-\frac{\nu^2}{\mu^2}\right) d_\nu \Bigg\{ & \frac{\partial\theta(x, s, \nu)}{\partial x} - \frac{\partial\theta^*(x, s, \nu)}{\partial x} - \frac{1}{\pi}\int_0^\nu \alpha(x, s, \sigma)\, d\sigma \\ & + \frac{1}{4\pi}\int_x^s q(\tau)\, d\tau \int_0^\nu \cos\sigma(s-x)\, d\sigma\Bigg\} = O(1).\end{aligned} \tag{2.11}$$

To prove the theorem it remains to compute the integrals

$$I_1 = \frac{1}{\pi}\int_0^\mu \left(1-\frac{\nu^2}{\mu^2}\right)\alpha(x, s, \nu)\, d\nu,$$

$$I_2 = \frac{1}{4\pi} \int_x^s q(\tau)\, d\tau \int_0^\mu \left(1 - \frac{\nu^2}{\mu^2}\right) \cos \nu (s - x)\, d\nu.$$

Let us compute the first one. By the definition of the function $\alpha(x, s, \nu)$, i.e. (2.6), we have

$$\pi I_1 = \int_0^\mu \left(1 - \frac{\nu^2}{\mu^2}\right) \left\{ \int_{|x-s|}^1 \frac{\partial w(x, t, s)}{\partial x} \cos \nu t\, dt \right\} d\nu.$$

Changing the order of integration, we obtain

$$\pi I_1 = \int_{|x-s|}^1 \frac{\partial w(x, t, s)}{\partial x} \left\{ \int_0^\mu \left(1 - \frac{\nu^2}{\mu^2}\right) \cos \nu t\, d\nu \right\} dt. \tag{2.12}$$

Further, by a well-known theorem (cf. Titchmarsh [**3**], §7.1)

$$\int_0^\mu \left(1 - \frac{\nu^2}{\mu^2}\right) \cos \nu t\, d\nu = \sqrt{\frac{\pi}{2}}\, \mu \frac{J_{3/2}(\mu t)}{(\mu t)^{3/2}}. \tag{2.13}$$

From the last two formulas we finally obtain

$$I_1 = \frac{1}{\sqrt{2\pi}}\, \mu \int_{|x-s|}^1 \frac{\partial w(x, t, s)}{\partial x} \frac{J_{3/2}(\mu t)}{(\mu t)^{3/2}}\, dt. \tag{2.14}$$

For the second integral we obtain, by (2.13),

$$I_2 = \frac{1}{4\sqrt{2\pi}}\, \mu \frac{J_{3/2}[\mu(s - x)]}{[\mu(s - x)]^{3/2}} \int_x^s q(\tau)\, d\tau. \tag{2.15}$$

Therefore by (2.14) and (2.15) the theorem follows from (2.11).

Similar asymptotic formulas can be obtained for the higher-order derivatives of the spectral function $\theta(x, s, \mu)$. However we will not do this here, restricting ourselves to stating a theorem for the second derivatives.

THEOREM 2.2. *If $q(x)$ has a first derivative which is absolutely continuous in every finite interval, then for every fixed x and s we have the asymptotic formula*

$$\int_0^\mu \left(1 - \frac{\nu^2}{\mu^2}\right)^2 d_\nu \left[\frac{\partial^2 \theta(x, s, \nu)}{\partial x\, \partial s} - \frac{\partial^2 \theta^*(x, s, \nu)}{\partial x\, \partial s} \right] = O(1). \tag{2.16}$$

This formula holds uniformly as x and s vary over any finite intervals.

§3. Equisummability of differentiated eigenfunction expansions

In this section, using the preliminary estimates obtained in §1 for the derivatives of the spectral function $\theta(x, s, \mu)$, we will prove a theorem on the equisummability of differentiated expansions in the eigenfunctions of a Sturm-Liouville operator and in an ordinary Fourier cosine integral of functions in $\mathcal{L}^2(0, \infty)$.

We will first prove some lemmas.

From formula (1.13′) follows

$$\int_{-\infty}^{\infty} \psi_\varepsilon(\mu) F(\mu) \varphi'_x(x, \mu)\, d\rho(\mu) = \int_{x-\varepsilon}^{x+\varepsilon} f(s) h_\varepsilon(x, s)\, ds, \tag{3.1}$$

where $f(s) \in \mathcal{L}^2(0, \infty)$ and

$$F(\mu) = \underset{A\to\infty}{\mathrm{l \cdot i \cdot m}} \int_0^A f(s) \varphi(s, \mu)\, ds, \tag{3.2}$$

$$\begin{aligned} h_\varepsilon(x, s) = \int_{|x-s|}^{\varepsilon} \frac{\partial w(x, t, s)}{\partial x} g_\varepsilon(t)\, dt + \frac{\partial g_\varepsilon(s-x)}{\partial x} - \\ - \frac{1}{2} g_\varepsilon(s-x) \int_x^s q(\tau)\, d\tau. \end{aligned} \tag{3.3}$$

We introduce the notation

$$S(x, \mu) = \int_0^\infty f(s) \theta(x, s, \mu)\, ds = \int_0^\mu F(\nu) \varphi(x, \nu)\, d\rho(\nu), \tag{3.4}$$

$$S^*(x, \mu) = \int_0^\infty f(s) \theta^*(x, s, \mu)\, ds. \tag{3.5}$$

By the definition of the function $S(x, \mu)$, (3.1) assumes the form

$$\begin{aligned} \int_{-\infty}^{\infty} \psi_\varepsilon(\mu)\, d_\mu \frac{\partial S(x, \mu)}{\partial x} = \int_{x-\varepsilon}^{x+\varepsilon} f(s) \Bigg\{ \int_{|x-s|}^{\varepsilon} \frac{\partial w(x, t, s)}{\partial x} g_\varepsilon(t)\, dt \\ + \frac{\partial g_\varepsilon(s-x)}{\partial x} - \frac{1}{2} g_\varepsilon(s-x) \int_x^s q(\tau)\, d\tau \Bigg\} ds. \end{aligned} \tag{3.6}$$

The function $S(x, \mu)$ is a segment of the expansion in the eigenfunctions of a Sturm-Liouville operator, and $S^*(x, \mu)$ is a segment of the expansion in an ordinary Fourier cosine integral, of the function $f(x) \in \mathcal{L}^2(0, \infty)$.

Writing for $S^*(x, \mu)$ a formula analogous to (3.6), taking into account that for this case $q(x) \equiv 0$, $w(x, t, s) \equiv 0$, and subtracting it from (3.6), we obtain

$$
\begin{aligned}
(3.7)\quad &\int_{-\infty}^{\infty} \psi_\varepsilon(\mu)\, d_\mu \left\{\frac{\partial S(x,\mu)}{\partial x} - \frac{\partial S^*(x,\mu)}{\partial x}\right\} \\
&= \int_{x-\varepsilon}^{x+\varepsilon} f(s) \left\{ \int_{|x-s|}^{\varepsilon} \frac{\partial w(x,t,s)}{\partial x} g_\varepsilon(t)\, dt - \frac{1}{2} g_\varepsilon(s-x) \int_x^s q(\tau)\, d\tau \right\} ds.
\end{aligned}
$$

LEMMA 3.1. *Let $f(x) \in \mathscr{L}^2(0,\infty)$. If $q(x)$ is absolutely continuous in every finite interval, then for each fixed x we have, as $\mu \to \infty$, the estimate*

$$
(3.8)\qquad \bigvee_{\mu}^{\mu+1} \left\{\frac{\partial S(x,\mu)}{\partial x}\right\} = \int_{\mu}^{\mu+1} |\varphi'_x(x,\mu)|\,|F(\mu)|\, d\rho(\mu) = o(\mu).
$$

This estimate holds uniformly as x varies in any finite interval.

PROOF. By the definition of $S(x,\mu)$, i.e. by (3.4) we have

$$
\frac{\partial S(x,\mu)}{\partial x} = \int_0^{\mu} \varphi'_x(x,\mu) F(\mu)\, d\rho(\mu).
$$

Therefore from the Cauchy-Bunjakovskiĭ inequality it follows that

$$
\begin{aligned}
(3.9)\qquad \bigvee_{\mu}^{\mu+1} \left\{\frac{\partial S(x,\mu)}{\partial x}\right\} &= \int_{\mu}^{\mu+1} |\varphi'_x(x,\mu) F(\mu)|\, d\rho(\mu) \\
&\leqslant \left(\int_{\mu}^{\mu+1} [\varphi'_x(x,\mu)]^2\, d\rho(\mu)\right)^{1/2} \left(\int_{\mu}^{\mu+1} F^2(\mu)\, d\rho(\mu)\right)^{1/2}
\end{aligned}
$$

By Lemma 1.1 we have

$$
(3.10)\qquad \int_{\mu}^{\mu+1} [\varphi'_x(x,\mu)]^2\, d\rho(\mu) = O(\mu^2).
$$

Further, since $\int_{-\infty}^{\infty} F^2(\mu)\, d\rho(\mu) < \infty$, as $\mu \to \infty$ we have

$$
(3.11)\qquad \int_{\mu}^{\mu+1} F^2(\mu)\, d\rho(\mu) = o(1).
$$

Therefore the lemma follows from the inequality (3.9) by virtue of the estimates (3.10) and (3.11).

One can similarly prove:

LEMMA 3.2. *Let $f(x) \in \mathscr{L}^2(0,\infty)$. If $q(x)$ has a kth derivative which is summable over every finite interval, then for every fixed x we have, as $\mu \to \infty$, the estimate*

$$(3.12)\qquad \bigvee_{\mu}^{\mu+1}\left\{\frac{\partial^k S(x,\mu)}{\partial x^k}\right\}=o(\mu^k)\qquad (k=1,\ 2,\ \ldots).$$

This estimate holds uniformly as x varies over every finite interval.

From the definition of $S^*(x,\mu)$ there immediately follows:

Lemma 3.3. *Let $f(s)\in\mathscr{L}^2(0,\infty)$. Then as $\mu\to\infty$*

$$(3.13)\qquad \bigvee_{\mu}^{\mu+1}\left\{\frac{\partial^k S^*(x,\mu)}{\partial x^k}\right\}=o(\mu^k)\qquad (k=1,\ 2,\ \ldots).$$

This estimate holds uniformly over the halfline $[0,\infty)$.

Theorem 3.1. *Let $f(x)\in\mathscr{L}^2(0,\infty)$. If $q(x)$ is absolutely continuous in every finite interval, then we have, uniformly in every finite interval, the relation*

$$\lim_{\mu\to\infty}\int_0^{\mu}\left(1-\frac{\nu^2}{\mu^2}\right)d_\nu\left\{\frac{\partial S(x,\nu)}{\partial x}-\frac{\partial S^*(x,\nu)}{\partial x}\right\}=0,$$

i.e. the difference between the Riesz means of order one of the first derivatives of the expansion of $f(x)\in\mathscr{L}^2(0,\infty)$ in the eigenfunctions of a Sturm-Liouville operator and of the expansion of $f(x)$ in an ordinary Fourier cosine integral tends to zero uniformly in every finite interval.

Proof. We put

$$(3.14)\qquad h(x,t)=\int_{x-t}^{x+t}\frac{\partial w(x,t,s)}{\partial x}f(s)\,ds,$$

$$(3.15)\qquad a(x,\mu)=\int_0^1 h(x,t)\cos\mu t\,dt,$$

$$(3.16)\qquad k(x,t)=f(x+t)\int_x^{x+t}q(\tau)\,d\tau,$$

$$(3.17)\qquad \beta(x,\mu)=\int_{-1}^{1}f(x+t)\left\{\int_x^{x+t}q(\tau)\,d\tau\right\}\cos\mu t\,dt=\int_{-1}^{1}k(x,t)\cos\mu t\,dt,$$

$$(3.18)\qquad \psi_\varepsilon(\mu)=\frac{1}{2}\int_{-\varepsilon}^{\varepsilon}g_\varepsilon(t)\cos\mu t\,dt\quad (0<\varepsilon\leqslant 1).$$

By the Parseval equality for ordinary Fourier integrals we obtain from (3.15)–(3.16) and (3.17)–(3.18) respectively

$$\begin{aligned}\int_{-\infty}^{\infty} \psi_{\varepsilon}(\mu)\, \alpha(x, \mu)\, d\mu &= \pi \int_0^{\varepsilon} g_{\varepsilon}(t)\, h(x, t)\, dt \\ &= \pi \int_0^{\varepsilon} g_{\varepsilon}(t) \left\{ \int_{x-t}^{x+t} \frac{\partial w(x, t, s)}{\partial x} f(s)\, ds \right\} dt \\ &= \pi \int_{x-\varepsilon}^{x+\varepsilon} f(t) \left\{ \int_{|x-s|}^{\varepsilon} \frac{\partial w(x, t, s)}{\partial x} g_{\varepsilon}(t)\, dt \right\} ds,\end{aligned} \tag{3.19}$$

$$\begin{aligned}\int_{-\infty}^{\infty} \psi_{\varepsilon}(\mu)\, \beta(x, \mu)\, d\mu &= \int_{-\varepsilon}^{\varepsilon} g_{\varepsilon}(t)\, k(x, t)\, dt \\ &= \int_{-\varepsilon}^{\varepsilon} g_{\varepsilon}(t)\, f(x+t) \left\{ \int_x^{x+t} q(\tau)\, d\tau \right\} dt \\ &= \int_{x-\varepsilon}^{x+\varepsilon} f(s)\, g_{\varepsilon}(s-x) \left\{ \int_x^{s} q(\tau)\, dt \right\} ds.\end{aligned} \tag{3.20}$$

Therefore by virtue of (3.19) and (3.20) the identity (3.7) can be rewritten in the form

$$\int_{-\infty}^{\infty} \psi_{\varepsilon}(\mu)\, d_{\mu} \left\{ \frac{\partial S(x, \mu)}{\partial x} - \frac{\partial S^*(x, \mu)}{\partial x} - \frac{1}{\pi} \int_0^{\mu} \alpha(x, \nu)\, d\nu + \frac{1}{\pi} \int_0^{\mu} \beta(x, \nu)\, d\nu \right\} = 0. \tag{3.21}$$

For $\mu \to \infty$ the validity of the following estimates is obvious:

$$\bigvee_{\mu}^{\mu+1} \left\{ \int_0^{\mu} \alpha(x, \nu)\, d\nu \right\} = o(1), \quad \bigvee_{\mu}^{\mu+1} \left\{ \int_0^{\mu} \beta(x, \nu)\, d\nu \right\} = o(1). \tag{3.22}$$

Therefore by (3.8), (3.13), (3.22) and (3.21) the Tauberian Theorem 4.1 of Chapter 14 is applicable, on the basis of which the first-order Riesz mean of the function

$$\frac{\partial S(x, \mu)}{\partial x} - \frac{\partial S^*(x, \mu)}{\partial x} - \frac{1}{\pi} \int_0^{\mu} \{\alpha(x, \nu) - \beta(x, \nu)\}\, d\nu$$

tends to zero uniformly in every finite interval, i.e.

$$\lim_{\mu\to\infty}\int_0^\mu\left(1-\frac{\nu^2}{\mu^2}\right)d_\nu\left\{\frac{\partial S(x,\nu)}{\partial x}-\frac{\partial S^*(x,\nu)}{\partial x}-\frac{1}{\pi}\int_0^\nu[\alpha(x,\nu)-\beta(x,\nu)]\,d\nu\right\}=0$$

or

$$\lim_{\mu\to\infty}\int_0^\mu\left(1-\frac{\nu^2}{\mu^2}\right)d_\nu\left\{\frac{\partial S(x,\nu)}{\partial x}-\frac{\partial S^*(x,\nu)}{\partial x}\right\}$$

$$(3.23)\qquad =\frac{1}{\pi}\lim_{\mu\to\infty}\int_0^\mu\left(1-\frac{\nu^2}{\mu^2}\right)\{\alpha(x,\mu)-\beta(x,\nu)\}\,d\nu.$$

To prove the theorem it suffices to show that

$$(3.24)\qquad \lim_{\mu\to\infty}\int_0^\mu\left(1-\frac{\nu^2}{\mu^2}\right)\{\alpha(x,\nu)-\beta(x,\nu)\}\,d\nu=0$$

holds uniformly in every finite interval. By the definition of the function $\alpha(x,\mu)$, i.e. (3.15), we have

$$\int_0^\mu\left(1-\frac{\nu^2}{\mu^2}\right)\alpha(x,\nu)\,d\nu=\int_0^\mu\left(1-\frac{\nu^2}{\mu^2}\right)\left[\int_0^1 h(x,t)\cos\nu t\,dt\right]d\nu.$$

Changing the order of integration, we obtain

$$(3.25)\qquad \int_0^\mu\left(1-\frac{\nu^2}{\mu^2}\right)\alpha(x,\nu)\,d\nu=\int_0^1 h(x,t)\left\{\int_0^\mu\left(1-\frac{\nu^2}{\mu^2}\right)\cos\nu t\,dt\right\}dt.$$

Since by hypothesis $q(x)$ is absolutely continuous in every finite interval, it is bounded in every finite interval. It then follows from Lemma 1.2, Chapter 7, that the function $\partial w(x,t,s)/\partial x$ is bounded in every finite interval, so that from (3.14) we find that

$$|h(x,t)|<C\int_{x-t}^{x+t}|f(s)|\,ds,$$

from which by the Cauchy-Bunjakovskiĭ inequality we obtain

$$(3.26)\qquad |h(x,t)|<C\left(\int_{x-t}^{x+t}f^2(s)\,ds\right)^{1/2}\left(\int_{x-t}^{x+t}ds\right)^{1/2}\leqslant C't^{1/2},$$

since $f(s)$ is square-integrable. Then from (3.25), (3.26) and (2.14) there follows for $\mu \to \infty$ the estimate

$$\begin{aligned}\left|\int_0^{\mu}\left(1-\frac{\nu^2}{\mu^2}\right)\alpha(x,\nu)\,d\nu\right| &< C\int_0^1 \mu t^{1/2}\frac{|J_{3/2}(\mu t)|}{(\mu t)^{3/2}}\,dt \\ &= \frac{C}{\mu^{1/2}}\int_0^{\mu}\frac{|J_{3/2}(u)|}{|u|}\,du = O(\mu^{-1/2}).\end{aligned} \tag{3.27}$$

Further, arguing analogously, we obtain from (3.15), (3.16) and (2.14)

$$\begin{aligned}&\int_0^{\mu}\left(1-\frac{\nu^2}{\mu^2}\right)\beta(x,\nu)\,d\nu \\ &= \sqrt{\frac{\pi}{2}}\,\mu\int_{-1}^{1} f(x,t)\frac{J_{3/2}(\mu t)}{(\mu t)^{3/2}}\left\{\int_x^{x+t} q(\tau)\,d\tau\right\}dt = O(\mu^{-1/2}).\end{aligned} \tag{3.28}$$

The estimates (3.27) and (3.28) prove (3.24), and thus the theorem.

Starting from Lemmas 3.2 and 3.3, we can prove in a completely analogous way:

THEOREM 3.2. *Let $f(x) \in \mathscr{L}^2(0,\infty)$. If $q(x)$ has a derivative of order $k-1$ which is absolutely continuous in every finite interval, then we have, uniformly in every finite interval,*

$$\lim_{\mu\to\infty}\int_0^{\mu}\left(1-\frac{\nu^2}{\mu^2}\right)^k d_\nu\left\{\frac{\partial^k S(x,\nu)}{\partial x^k}-\frac{\partial^k S^*(x,\nu)}{\partial x^k}\right\}=0,$$

i.e. the difference between the Riesz means of order k of the kth derivatives of the expansion of $f(x)$ with respect to the eigenfunctions of a Sturm-Liouville operator and as an ordinary Fourier cosine integral tends to zero uniformly in every finite interval.

§4. Summation of differentiated ordinary and generalized Fourier integral expansions

From the theorem on the equisummability of a differentiated expansion in the eigenfunctions of a Sturm-Liouville operator and an expansion as an ordinary Fourier cosine integral, which was proved in the preceding section, it is easy to obtain theorems concerning the summation of differentiated expansions in the eigenfunctions of a Sturm-Liouville operator to the values of the corresponding derivatives of the function which is being expanded, if one has such theorems for the summation of the derivatives of the ordinary Fourier integral. Concerning the summation of the first derivative of an ordinary Fourier series, there is the well-known theorem of Fejér-Lebesgue (cf., for example, Natanson [**1**]). In this section, using our method, we will prove a theorem analogous to the Fejér-Lebesgue theorem.

THEOREM 4.1. *Let* $F\{f(x)\}$ *denote the ordinary Fourier series of the function* $f(x)$. *The series* $F'\{f(x)\}$ *is summable at the point* x_0 *by the first-order Riesz method to the value* $f'(x_0)$ *if the following conditions are satisfied at the point* x_0.

1. $f(x)$ *is absolutely continuous in some neighborhood of* x_0.
2. $f'(x)$ *is continuous at* x_0.

PROOF. Suppose that $f(x)$ is absolutely continuous in the interval $(x_0-\delta, x_0+\delta)$ $(\delta>0)$. Further, let $\{f_n(x)\}$ be a sequence of smooth functions, defined on the interval $(0, 2\pi)$, which converges to $f(x)$ in the metric of $\mathscr{L}^1(0, 2\pi)$. By virtue of the smoothness of $f_n(x)$ we have

$$f_n(x)=\sum_{k=0}^{\infty}\{a_k^{(n)}\cos kx+b_k^{(n)}\sin kx\} \quad (n=1, 2, \ldots), \tag{4.1}$$

where $a_k^{(n)}$ and $b_k^{(n)}$ are the Fourier coefficients of $f_n(x)$, and the series (4.1) converges uniformly.

Let t be an arbitrary real number satisfying $|t|<\delta/2$. From (4.1) follows

$$\frac{1}{2}\{f_n(x+t)+f_n(x-t)\}=\sum_{k=0}^{\infty}\{a_k^{(n)}\cos kx+b_k^{(n)}\sin kx\}\cos kt \quad (n=1, 2, \ldots). \tag{4.2}$$

Multiplying both sides of (4.2) by $g_\epsilon(t)$ and integrating with respect to t from 0 to ϵ, by (3.18) we obtain

$$\frac{1}{2}\int_0^{\varepsilon}\{f_n(x+t)+f_n(x-t)\}g_\varepsilon(t)\,dt=\sum_{k=0}^{\infty}\{a_k^{(n)}\cos kx+b_k^{(n)}\sin kx\}\psi_\varepsilon(k) \quad (n=1, 2, \ldots). \tag{4.3}$$

Let us introduce the notation

$$S_n(x, \mu)=\sum_{k<\mu}\{a_k^{(n)}\cos kx+b_k^{(n)}\sin kx\} \quad (n=1, 2, \ldots).$$

Then (4.3) assumes the form

$$\int_{-\infty}^{\infty} \psi_\varepsilon(\mu)\, d_\mu S_n(x, \mu) = \int_0^\varepsilon \{f_n(x+t) + f_n(x-t)\} g_\varepsilon(t)\, dt$$

$$(4.4) \qquad\qquad (n = 1, 2, \cdots).$$

The convergence of the sequence $\{f_n(x)\}$ to $f(x)$ allows us to pass to the limit under the integral sign in the right side of (4.4), and the bound (1.7) allows us to pass to the limit in the left side of (4.4).

As a consequence we obtain

$$(4.5) \qquad \int_{-\infty}^{\infty} \psi_\varepsilon(\mu)\, d_\mu S(x, \mu) = \int_0^\varepsilon \{f(x+t) + f(x-t)\}\, g_\varepsilon(t)\, dt,$$

where

$$S(x, \mu) = \sum_{k<\mu} (a_k \cos kx + b_k \sin kx),$$

and a_k and b_k are the Fourier coefficients of $f(x)$.

Differentiating (4.5) with respect to x for $x \in (x_0 - \delta/2, x_0 + \delta/2)$, we find that

$$(4.6) \qquad \int_{-\infty}^{\infty} \psi_\varepsilon(\mu)\, d_\mu \frac{\partial S(x, \mu)}{\partial x} = \int_0^\varepsilon \{f'_x(x+t) + f'_x(x-t)\}\, g_\varepsilon(t)\, dt.$$

Put

$$(4.7) \qquad \alpha(x, \mu) = \int_0^\varepsilon \{f'_x(x+t) + f'_x(x-t)\} \cos \mu t\, dt.$$

By the Parseval equality for the ordinary Fourier integral, it follows from (4.6) and (4.7) that

$$(4.8) \qquad \int_0^\varepsilon \{f'_x(x+t) + f'_x(x-t)\}\, g_\varepsilon(t)\, dt = \frac{2}{\pi} \int_0^\infty \alpha(x, \mu)\, \psi_\varepsilon(\mu)\, d\mu.$$

Since $\psi_\varepsilon(\mu)$ and $\alpha(x, \mu)$ are even with respect to μ, it follows from (4.6) and (4.8) that

$$(4.9) \qquad \int_{-\infty}^{\infty} \psi_\varepsilon(\mu)\, d_\mu \left\{ \frac{\partial S(x, \mu)}{\partial x} - \frac{1}{\pi} \int_0^\mu \alpha(x, \nu)\, d\nu \right\} = 0.$$

It further follows from the definition of $S(x, \mu)$ and $\alpha(x, \mu)$ that for $\mu \to \infty$ we have the estimates

(4.10) $$\bigvee_{\mu}^{\mu+1}\left\{\frac{\partial S(x,\mu)}{\partial x}\right\}=o(\mu),\qquad \bigvee_{\mu}^{\mu+1}\left\{\int_0^{\mu}\alpha(x,\nu)\,d\nu\right\}=o(1).$$

From (4.9), (4.10) and the Tauberian Theorem 4.1, Chapter 14, it follows that the Riesz mean of order one of the function

$$\frac{\partial S(x,\mu)}{\partial x}-\frac{1}{\pi}\int_0^{\mu}\alpha(x,\nu)\,d\nu$$

tends to zero uniformly in the interval $(x_0-\delta/2, x_0+\delta/2)$, i.e.

(4.11) $$\lim_{\mu\to\infty}\int_0^{\mu}\left(1-\frac{\nu^2}{\mu^2}\right)d_{\nu}\left\{\frac{\partial S(x,\nu)}{\partial x}-\frac{1}{\pi}\int_0^{\nu}\alpha(x,\nu)\,d\nu\right\}=0.$$

It follows from (4.11) that to prove the theorem it suffices to prove the validity of the following equality.

(4.12) $$\lim_{\mu\to\infty} I=\frac{1}{\pi}\lim_{\mu\to\infty}\int_0^{\mu}\left(1-\frac{\nu^2}{\mu^2}\right)\alpha(x_0,\nu)\,d\nu=f'(x_0).$$

From the definition of the function $\alpha(x,\nu)$ we have

$$I=\frac{1}{\pi}\int_0^{\mu}\left(1-\frac{\nu^2}{\mu^2}\right)\left\{\int_0^{\varepsilon}[f'_x(x_0+t)+f'_x(x_0-t)]\cos\nu t\,dt\right\}d\nu.$$

Changing the order of integration, we obtain

(4.13) $$I=\frac{1}{\pi}\int_0^{\varepsilon}[f'_x(x_0+t)+f'_x(x_0-t)]\left\{\int_0^{\mu}\left(1-\frac{\nu^2}{\mu^2}\right)\cos\nu t\,d\nu\right\}dt.$$

Therefore from (2.14) and (4.13) follows

(4.14) $$I=\sqrt{\frac{2}{\pi}}\,\mu\int_0^{\varepsilon}[f'_x(x_0+t)+f'_x(x_0-t)]\frac{J_{3/2}(\mu t)}{(\mu t)^{3/2}}\,dt.$$

We now transform the integral in the right side of (4.14) as follows:

(4.15) $$\begin{aligned}I=&\sqrt{\frac{2}{\pi}}\,\mu\int_0^{\varepsilon}[f'_x(x_0+t)+f'_x(x_0-t)-2f'(x_0)]\frac{J_{3/2}(\mu t)}{(\mu t)^{3/2}}\,dt\\&+2\sqrt{\frac{2}{\pi}}\,\mu f'(x_0)\int_0^{\varepsilon}\frac{J_{3/2}(\mu t)}{(\mu t)^{3/2}}\,dt.\end{aligned}$$

By a change of variables in the integrals in the right side of (4.15) we can transform it to the form

$$I = \sqrt{\frac{2}{\pi}} \int_0^{\varepsilon\mu} \left[f'_x\left(x_0 + \frac{z}{\mu}\right) + f'_x\left(x_0 - \frac{z}{\mu}\right) - 2f'(x_0) \right] \frac{J_{3/2}(z)}{z^{3/2}}\, dz \tag{4.16}$$
$$+ 2\sqrt{\frac{2}{\pi}} f'(x_0) \int_0^{\varepsilon\mu} \frac{J_{3/2}(z)}{z^{3/2}}\, dz.$$

We now wish to estimate the first integral in the right side of (4.16) for large μ.

Let η denote an arbitrary positive number and put

$$\int_0^{\varepsilon\mu} \left[f'_x\left(x_0 + \frac{z}{\mu}\right) + f'_x\left(x_0 - \frac{z}{\mu}\right) - 2f'(x_0) \right] \frac{J_{3/2}(z)}{z^{3/2}}\, dz$$
$$= \left\{ \int_0^{\eta} + \int_{\eta}^{\varepsilon\mu} \right\} \left[f'_x\left(x_0 + \frac{z}{\mu}\right) + f'_x\left(x_0 - \frac{z}{\mu}\right) - 2f'(x_0) \right] \tag{4.17}$$
$$\times \frac{J_{3/2}(z)}{z^{3/2}}\, dz = I_1 + I_2.$$

Since $f'(x)$ is continuous at x_0, for any $\delta > 0$ and fixed η we can choose μ_0 so large that for all $\mu > \mu_0$ we will have

$$|I_1| < \delta/2. \tag{4.18}$$

On the other hand, choosing η sufficiently large, we obtain the inequality

$$|I_2| < \delta/2. \tag{4.19}$$

From (4.18), (4.19) and the arbitrariness of δ we obtain, by virtue of (4.17),

$$\lim_{\mu \to \infty} \int_0^{\varepsilon\mu} \left\{ f'_x\left(x_0 + \frac{z}{\mu}\right) + f'_x\left(x_0 - \frac{z}{\mu}\right) - 2f'(x_0) \right\} \frac{J_{3/2}(z)}{z^{3/2}}\, dz = 0. \tag{4.20}$$

Finally, passing to the limit in (4.16), we obtain by (4.20)

$$\lim_{\mu \to \infty} I = 2\sqrt{\frac{2}{\pi}} f'(x_0) \int_0^{\infty} \frac{J_{3/2}(z)}{z^{3/2}}\, dz. \tag{4.21}$$

Further, by a well-known formula (cf. Watson [1], §13.24)

$$\int_0^{\infty} \frac{J_{3/2}(z)}{z^{3/2}}\, dz = \frac{1}{2}\sqrt{\frac{\pi}{2}},$$

which together with (4.21) proves (4.12) and thus the theorem. By the equivalence of the first-order Cesàro and Riesz summation methods, the theorem of Fejér-Lebesgue follows from the theorem just proved. It is not difficult to see from the proof that the preceding theorem is true for ordinary Fourier integrals under the same assumptions, i.e. we have:

THEOREM 4.2. *Suppose that the following conditions are satisfied:*
1. $f(x) \in \mathcal{L}^2(0, \infty)$.
2. $f(x)$ *is absolutely continuous in some neighborhood of* x_0.
3. $f'(x)$ *is continuous at* x_0.
Then the first derivative of the expansion of $f(x)$ *as an ordinary Fourier integral is summable at the point* x_0 *by the first-order Riesz method to the value* $f'(x_0)$.

From Theorems 3.1 and 4.2 follows:

THEOREM 4.3. *Let* $q(x)$ *be absolutely continuous in every finite interval, and suppose that the following conditions are satisfied:*
1. $f(x) \in \mathcal{L}^2(0, \infty)$.
2. $f(x)$ *is absolutely continuous in some neighborhood of* x_0.
3. $f'(x)$ *is continuous at* x_0.
Then

$$\lim_{\mu\to\infty} \int_0^\mu \left(1 - \frac{\nu^2}{\mu^2}\right) d_\nu \frac{\partial S(x,\nu)}{\partial x}\bigg|_{x=x_0} = f'(x_0),$$

i.e. the first derivative of the expansion of $f(x)$ *with respect to the eigenfunctions of the Sturm-Liouville operator is summable at the point* x_0 *by the first-order Riesz method to the value* $f'(x_0)$.

A similar theorem holds for the higher-order derivatives.

§5. Convergence of a differentiated eigenfunction expansion

THEOREM 5.1. *Suppose that* $q(x)$ *is summable over every finite interval, and suppose that* $f(x)$ *satisfies the following conditions for* $x \geqq 0$:
1. $f(x)$ *and* $\mathcal{L}\{f(x)\} = f''(x) - q(x)f(x)$ *belong to* $\mathcal{L}^2(0, \infty)$.
2. $f(0)\cos\alpha + f'(0)\sin\alpha = 0$.
3. $\lim_{x\to\infty} W\{\varphi(x,\lambda), f(x)\} = 0$ *for all* λ *from the spectrum.*
Then the eigenfunction expansion of $f(x)$ *converges absolutely.*

PROOF. From the hypotheses of the theorem and from Green's identity follows

$$(5.1) \qquad F(\lambda) = \underset{b\to\infty}{\mathrm{l.i.m}} \int_0^b f(x)\varphi(x,\lambda)\,dx =$$

$$= -\frac{1}{\lambda}\, \mathrm{l\cdot i\cdot m}_{b\to\infty} \int_0^b f(x)\,\mathscr{L}\{\varphi(x,\lambda)\}\,dx = -\frac{1}{\lambda}\,\mathscr{E}(\lambda),$$

where

$$\mathscr{E}(\lambda) = \mathrm{l\cdot i\cdot m}_{b\to\infty} \int_0^b \mathscr{L}\{f(x)\}\,\varphi(x,\lambda)\,dx.$$

From Lemma 2.1, Chapter 7 and the Cauchy-Bunjakovskiĭ inequality it follows that

$$\int_{-\infty}^{0} |F(\lambda)|\,|\varphi(x,\lambda)|\,d\rho(\lambda) < \infty. \tag{5.2}$$

Let us consider the integral

$$\int_0^\infty |F(\lambda)|\,|\varphi(x,\lambda)|\,d\rho(\lambda).$$

If we replace λ by μ^2 and put

$$F(\lambda) = F_1(\mu), \quad \varphi(x,\lambda) = \varphi_1(x,\mu), \quad \mathscr{E}(\lambda) = \mathscr{E}_1(\mu), \quad \rho(\lambda) = \rho_1(\mu),$$

this integral assumes the form

$$\int_0^\infty |F_1(\mu)|\,|\varphi_1(x,\mu)|\,d\rho_1(\mu)$$

$$= \int_0^1 |F_1(\mu)|\,|\varphi_1(x,\mu)|\,d\rho_1(\mu) + \sum_{a=1}^{\infty} \int_a^{a+1} |F_1(\mu)|\,|\varphi_1(x,\mu)|\,d\rho_1(\mu). \tag{5.3}$$

It further follows from (5.1), Lemma 2.3 of Chapter 7 and the Cauchy-Bunjakovskiĭ inequality that

$$\int_a^{a+1} |F_1(\mu)|\,|\varphi_1(x,\mu)|\,d\rho_1(\mu)$$

$$\leq \frac{1}{a^2}\left(\int_a^{a+1} \mathscr{E}_1^2(\mu)\,d\rho_1(\mu)\right)(\theta_1(x,x,a+1) - \theta_1(x,x,a)) \leq \frac{C}{a^2},$$

where the constant C does not **depend** upon a. Therefore from (5.3) follows

$$\int_0^\infty |F_1(\mu)|\,|\varphi_1(x,\mu)|\,d\rho_1(\mu) < \int_0^1 |F_1(\mu)|\,|\varphi_1(x,\mu)|\,d\rho_1(\mu) + C\sum_{a=1}^{\infty}\frac{1}{a^2} < \infty,$$

which together with (5.2) proves the theorem.

The purpose of this section is the proof of the following theorem.

THEOREM 5.2. *Suppose that $f(x)$ satisfies the conditions of the previous theorem and, moreover, that the function $g(x)=f''(x)-q(x)f(x)$ can be expanded with respect to the eigenfunctions $\varphi(x,\lambda)$ at the point $x_0 \geqq 0$.*[2] *Then*

$$f''(x_0)=\lim_{\lambda\to\infty}\int_{-\infty}^{\lambda} F(\lambda)\varphi''_{xx}(x_0,\lambda)\,d\rho(\lambda). \tag{5.4}$$

PROOF. By hypothesis

$$\begin{aligned} g(x_0) &= \lim_{\lambda\to\infty}\int_{-\infty}^{\lambda} \mathcal{E}(\lambda)\varphi(x_0,\lambda)\,d\rho(\lambda) = -\lim_{\lambda\to\infty}\int_{-\infty}^{\lambda} \lambda F(\lambda)\varphi(x_0,\lambda)\,d\rho(\lambda) \\ &= \lim_{\lambda\to\infty}\int_{-\infty}^{\lambda} F(\lambda)\varphi''_{xx}(x_0,\lambda)\,d\rho(\lambda) - q(x_0)\lim_{\lambda\to\infty}\int_{-\infty}^{\lambda} F(\lambda)\varphi(x_0,\lambda)\,d\rho(\lambda). \end{aligned} \tag{5.5}$$

Since $g(x_0)=f''(x_0)-q(x_0)f(x_0)$ and

$$\lim_{\lambda\to\infty}\int_{-\infty}^{\lambda} F(\lambda)\varphi(x_0,\lambda)\,d\rho(\lambda)=f(x_0),$$

the relation (5.4) follows from (5.5).

§6. Justification of Fourier's method for the one-dimensional wave equation

In this section we will show that the results of §4 of the preceding chapter, in which we investigated the solution of a mixed problem for the one-dimensional wave equation on a finite interval, together with Theorem 5.2, enable us to justify Fourier's method for the one-dimensional wave equation under weaker conditions on the initial function than the usual ones.

We will consider the following mixed problem on an interval:

$$\partial^2 u/\partial t^2 = \partial^2 u/\partial x^2 - q(x)u, \tag{6.1}$$

$$u|_{t=0}=f(x), \tag{6.2}$$

---

[2] As follows from Theorem 7.1, Chapter 7, for this it is sufficient that the function $g(x)$ satisfy any local conditions which suffice for it to be expandable in a trigonometric Fourier series. Since, by hypothesis, $f(x)$ has a second derivative for $x \geqq 0$, it is for example sufficient that $f''(x)$ and $q(x)$ be of bounded variation in a neighborhood of the point x_0.

(6.3) $$\partial u/\partial t|_{t=0} = 0,$$

(6.4) $$[\partial u/\partial x - hu]_{x=0} = 0,$$

(6.5) $$[\partial u/\partial x + Hu]_{x=\pi} = 0.$$

As was proved (cf. §4, Chapter 6), if $f(x)$ has a continuous second derivative and satisfies the conditions

(6.6) $$f'(+0) - hf(+0) = 0,$$

(6.7) $$f'(\pi - 0) + Hf(\pi - 0) = 0,$$

then the mixed problem (6.1) – (6.5) has a solution for all $t > 0$. Let us denote this solution by $u_0(x,t)$. For fixed t this solution satisfies the boundary conditions and has a continuous second derivative. Therefore by a classical theorem $u_0(x,t)$ can for any fixed t be expanded in an absolutely and uniformly convergent series with respect to the eigenfunctions of the Sturm-Liouville operator

(6.8) $$y'' + \{\lambda - q(x)\}y = 0.$$

(6.9) $$y'(0) - hy(0) = 0, \qquad y'(\pi) + Hy(\pi) = 0,$$

i.e.

(6.10) $$u_0(x, t) = \sum_{n=1}^{\infty} c_n(t)\varphi_n(x),$$

where

(6.11) $$c_n(t) = \int_0^{\pi} u_0(x, t)\varphi_n(x)\,dx,$$

and the $\varphi_n(x)$ are the eigenfunctions of the problem (6.8) + (6.9). To determine the coefficients $c_n(t)$ we will first determine $c_n(0)$ and $c_n'(0)$. To this end we use the initial conditions (6.2) – (6.3). Putting $t = 0$ in (6.11), by virtue of (6.2) we obtain

(6.12) $$c_n(0) = \int_0^{\pi} f(x)\varphi_n(x)\,dx = a_n.$$

Differentiating (6.11) with respect to t and then putting $t = 0$, by (6.3) we obtain

(6.13) $$c_n'(0) = 0.$$

Further, differentiating (6.11) twice with respect to t, by (6.1) we obtain

$$c_n''(t) = \int_0^\pi \frac{\partial^2 u_0(x, t)}{\partial t^2} \varphi_n(x)\,dx = \int_0^\pi \left\{\frac{\partial^2 u_0(x, t)}{\partial x^2} - q(x)\,u_0(x, t)\right\} \varphi_n(x)\,dx.$$

Integrating the last integral by parts twice and using the boundary conditions (6.4) – (6.5) and (6.9), we obtain

$$c_n''(t) = \int_0^\pi u_0(x, t)\,\{\varphi_n''(x) - q(x)\,\varphi_n(x)\}\,dx,$$

from which by (6.8) follows

(6.14) $$c_n''(t) = -\lambda_n \int_0^\pi u_0(x, t)\,\varphi_n(x)\,dx = -\lambda_n c_n(t).$$

Thus, by virtue of (6.12), (6.13) and (6.14) we obtain the following Cauchy problem for the determination of $c_n(t)$:

$$c_n''(t) + \lambda_n c_n(t) = 0, \qquad c_n(0) = a_n, \qquad c_n'(0) = 0.$$

The solution of this problem has the form

(6.15) $$c_n(t) = a_n \cos\sqrt{\lambda_n}\,t.$$

Inserting this value of $c_n(t)$ in (6.10), we obtain

(6.10′) $$u_0(x, t) = \sum_{n=0}^{\infty} a_n \cos\sqrt{\lambda_n}\,t\,\varphi_n(x).$$

If the function $f''(x)$ satisfies a Lipschitz condition with exponent $\alpha > 0$ on the the interval $[0, \pi]$, then it follows from (4.5), Chapter 6, and Theorem 5.2[3] that the series (6.10′) can be differentiated twice with respect to x, i.e.

(6.16) $$\frac{\partial^2 u_0}{\partial x^2} = \sum_{n=0}^{\infty} a_n \cos\sqrt{\lambda_n}\,t\,\varphi_n''(x)$$

where the series (6.16) converges uniformly.

We will prove that the series obtained from (6.10′) by differentiating once with respect to t converges absolutely. In fact, put

[3] Theorem 5.2 considers the halfline $[0, \infty)$. However it is easily seen that this theorem can be proved by the same methods for the case of a regular Sturm-Liouville problem. Of course, as is evident from the proof, for the regular case Theorem 5.2 follows from a classical theorem on the equiconvergence of the expansion of a function in a Sturm-Liouville series and the expansion of this function in a Fourier cosine series (cf. §9, Chapter 1).

$$\alpha_n = \int_0^\pi \{f''(x) - q(x) f(x)\} \varphi_n(x)\, dx.$$

Integrating the last integral by parts twice, by the conditions (6.6) – (6.7) and (6.9) we obtain

$$\alpha_n = \int_0^\pi f(x) \{\varphi''_x(x) - q(x) \varphi_n(x)\}\, dx$$

$$= -\lambda_n \int_0^\pi f(x) \varphi_n(x)\, dx = -\lambda_n a_n,$$

from which

$$a_n = -\alpha_n / \lambda_n. \tag{6.17}$$

Differentiating the series (6.10′) with respect to t, we obtain

$$\frac{\partial u_0}{\partial t} = -\sum_{n=1}^{\infty} a_n \sqrt{\lambda_n} \sin \sqrt{\lambda_n}\, t \varphi_n(x). \tag{6.18}$$

Further, by virtue of (6.17) and the Cauchy-Bunjakovskiĭ inequality we obtain

$$\sum_{n=1}^{\infty} |\sqrt{\lambda_n}\, a_n| = \sum_{n=1}^{\infty} \left| \frac{\alpha_n}{\sqrt{\lambda_n}} \right| \leqslant \left(\sum_{n=1}^{\infty} \alpha_n^2 \right)^{1/2} \left(\sum_{n=1}^{\infty} \frac{1}{\lambda_n} \right)^{1/2} < \infty, \tag{6.19}$$

which follows from the fact that $f''(x) - q(x) f(x) \in \mathscr{L}^2(0, \pi)$ and $\lambda_n = n^2 + O(1/n)$.

It follows from the bound (6.19) that the series (6.18) is absolutely convergent.

Let us now clarify the convergence of the second derivative with respect to t. By virtue of the equation (6.8) we have

$$\frac{\partial^2 u}{\partial t^2} = -\sum_{n=1}^{\infty} \lambda_n a_n \cos \sqrt{\lambda_n}\, t \varphi_n(x) = \sum_{n=1}^{\infty} a_n \cos \sqrt{\lambda_n}\, t \cdot \{\varphi''_n(x) - q(x) \varphi_n(x)\}$$

$$= \sum_{n=1}^{\infty} a_n \varphi''_n(x) \cos \sqrt{\lambda_n}\, t - q(x) \sum_{n=1}^{\infty} a_n \varphi_n(x) \cos \sqrt{\lambda_n}\, t.$$

Hence from the convergence of the series (6.10) and (6.16) follows the convergence of the series

$$\frac{\partial^2 u_0}{\partial t^2} = -\sum_{n=1}^{\infty} \lambda_n a_n \varphi(x) \cos \sqrt{\lambda_n}\, t \quad (x = x_0).$$

We now consider the case

(6.20) $$u(x,0)=0, \qquad u_t'(x,0)=f(x).$$

Let $u_1(x,t)$ denote the solution of the problem (6.1) + (6.20) + (6.4) + (6.5). Then

$$u_1(x,t)=\int_0^t u_0(x,\tau)\,d\tau.$$

If we now put

$$u_1(x,t)=\sum_{n=0}^{\infty} c_n(t)\varphi_n(x), \qquad c_n(t)=\int_0^{\pi} u_1(x,t)\varphi_n(x)\,dx,$$

then we obtain for $c_n(t)$ the following Cauchy problem:

$$\begin{aligned} c_n''(t)+\lambda_n c_n(t)&=0,\\ c_n(0)&=0,\\ c_n'(0)=a_n&=\int_0^{\pi} f(x)\varphi_n(x)\,dx. \end{aligned}$$

Thus

$$c_n(t)=\frac{a_n}{\sqrt{\lambda_n}}\sin\sqrt{\lambda_n}\,t.$$

The rest of this investigation can be carried out analogously to the preceding investigation.

Bibliographical references

§§1, 2, 3, 4. The results of these sections are due to Sargsjan [**1**–**4**].

§5. Theorem 5.2 is due to the authors and was published in their joint papers [**1**—**2**]. Another proof is given in the joint paper [**3**] by the authors.

§6. A justification of Fourier's method was given in the paper by Levitan [**17**]. Cf. also the joint paper [**3**] of the authors.

CHAPTER 9

SOLUTION OF THE CAUCHY PROBLEM FOR A ONE-DIMENSIONAL DIRAC SYSTEM

§1. Derivation of formulas for the solution of the Cauchy problem

Let $p(x)$ and $r(x)$ be real functions defined on the entire line and summable over every finite interval,[1] and let the vector-function

$$f(x) = \begin{pmatrix} f_1(x) \\ f_2(x) \end{pmatrix}$$

be real and continuously differentiable. Let

$$u(x,t) = \begin{pmatrix} u_1(x,t) \\ u_2(x,t) \end{pmatrix}.$$

We will consider the Cauchy problem

$$\begin{pmatrix} i & 0 \\ 0 & i \end{pmatrix} \frac{\partial u}{\partial t} + \begin{pmatrix} 0 & 1 \\ -1 & 0 \end{pmatrix} \frac{\partial u}{\partial x} + \begin{pmatrix} p(x) & 0 \\ 0 & r(x) \end{pmatrix} u = 0, \tag{1.1}$$

$$u(x, 0) = f(x). \tag{1.2}$$

Let us first consider the problem (1.1) + (1.2) for $p(x) = r(x) \equiv 0$, i.e. the problem

$$\begin{gathered} iI \frac{\partial u^0}{\partial t} + B \frac{\partial u^0}{\partial x} = 0, \\ u^0(x, 0) = f(x), \end{gathered} \tag{1.3}$$

where we have put

$$I = \begin{pmatrix} 1 & 0 \\ 0 & 1 \end{pmatrix}, \quad B = \begin{pmatrix} 0 & 1 \\ -1 & 0 \end{pmatrix}.$$

The problem (1.3) is equivalent to two independent problems involving the components $u_1^0(x,t)$ and $u_2^0(x,t)$ of the vector-function $u^0(x,t)$:

---

[1] Further on, depending upon the circumstance, other smoothness conditions will be imposed upon the functions $p(x)$ and $r(x)$.

$$\frac{\partial^2 u_1^0}{\partial t^2} - \frac{\partial^2 u_1^0}{\partial x^2} = 0, \tag{1.4}$$
$$u_1^0(x, 0) = f_1(x), \qquad \left.\frac{\partial u_1^0}{\partial t}\right|_{t=0} = if_2'(x),$$

$$\frac{\partial^2 u_2^0}{\partial t^2} - \frac{\partial^2 u_2^0}{\partial x^2} = 0, \tag{1.5}$$
$$u_2^0(x, 0) = f_2(x), \qquad \left.\frac{\partial u_2^0}{\partial t}\right|_{t=0} = -if_1'(x).$$

Solving the problems (1.4) and (1.5) by D'Alembert's method, we obtain

$$u_1^0(x,t) = \tfrac{1}{2}\{ f_1(x+t) + if_2(x+t) \} + \tfrac{1}{2}\{ f_1(x-t) - if_2(x-t) \},$$
$$u_2^0(x,t) = \tfrac{1}{2}\{ -if_1(x+t) + f_2(x+t) \} + \tfrac{1}{2}\{ if_1(x-t) + f_2(x-t) \}.$$

Consequently the solution of the problem (1.3) is given by

$$u^0(x,t) = \tfrac{1}{2}\{ Hf(x+t) + H^T f(x-t) \}, \tag{1.6}$$

where

$$H = \begin{pmatrix} 1 & i \\ -i & 1 \end{pmatrix}, \tag{1.6'}$$

and H^T is the transpose of H.

Let us now consider an inhomogeneous problem with homogeneous initial data

$$iI\frac{\partial \tilde{u}}{\partial t} + B\frac{\partial \tilde{u}}{\partial x} = g(x, t), \tag{1.7}$$
$$\tilde{u}(x, 0) = 0.$$

where

$$g(x,t) = \begin{pmatrix} g_1(x,t) \\ g_2(x,t) \end{pmatrix}$$

is a differentiable vector-function. It is not hard to see that this problem also breaks up into two independent problems involving the components $\tilde{u}_1(x,t)$ and $\tilde{u}_2(x,t)$ of the vector-function $\tilde{u}(x,t)$, namely:

$$\frac{\partial^2 \tilde{u}_1}{\partial t^2} - \frac{\partial^2 \tilde{u}_1}{\partial x^2} = -i\frac{\partial g_1}{\partial t} + \frac{\partial g_2}{\partial x}, \tag{1.8}$$
$$\tilde{u}_1(x, 0) = 0, \quad \left.\frac{\partial \tilde{u}_1}{\partial t}\right|_{t=0} = -ig_1(x, 0),$$

$$\frac{\partial^2 \tilde{u}_2}{\partial t^2} - \frac{\partial^2 \tilde{u}_2}{\partial x^2} = -i\frac{\partial g_2}{\partial t} - \frac{\partial g_1}{\partial x}, \tag{1.9}$$
$$\tilde{u}_2(x, 0) = 0, \quad \left.\frac{\partial \tilde{u}_2}{\partial t}\right|_{t=0} = -ig_2(x, 0).$$

As is known,[2] the solution of the problem

$$\partial^2 v/\partial t^2 - \partial^2 v/\partial x^2 = h(x,t),$$

$$v(x,0) = 0, \qquad \partial v/\partial t|_{t=0} = q(x)$$

is given by

$$v(x, t) = \frac{1}{2} \int_{x-t}^{x+t} q(\tau)\, d\tau + \frac{1}{2} \int_0^t d\tau \int_{x-(t-\tau)}^{x+(t-\tau)} h(s, \tau)\, ds.$$

Therefore, applying this formula to the problems (1.8) and (1.9), we find that the solutions of these problems are given respectively by

$$\tilde{u}_1(x, t) = \frac{1}{2} \int_0^t \{-ig_1(x+t-\tau, \tau) + g(x+t-\tau, \tau) - ig_1(x-t+\tau, \tau) - g_2(x-t+\tau, \tau)\}\, d\tau,$$

$$\tilde{u}_2(x, t) = \frac{1}{2} \int_0^t \{-g_1(x+t-\tau, \tau) - ig_2(x+t-\tau, \tau) + g(x-t+\tau, \tau) - ig_2(x-t+\tau, \tau)\}\, d\tau.$$

Thus the solution of (1.7) is given by

$$\tilde{u}(x, t) = -\frac{i}{2} \int_0^t \{Hg(x+t-\tau, \tau) + H^T g(x-t+\tau, \tau)\}\, d\tau. \tag{1.10}$$

We now turn to the problem (1.1) + (1.2). We will rewrite it in the form

$$iI \frac{\partial u}{\partial t} + B \frac{\partial u}{\partial x} = -Q(x) u(x, t), \tag{1.1'}$$

$$u(x, 0) = f(x), \tag{1.2}$$

where

$$Q(x) = \begin{pmatrix} p(x) & 0 \\ 0 & r(x) \end{pmatrix}.$$

Let us assume that $p(x)$ and $r(x)$ have continuous derivatives.[3]

Putting now $g(x,t) = -Q(x)u(x,t)$, assuming $g(x,t)$ to be known, and then taking into account that the solution of the problem (1.1′) + (1.2) is the sum of the solutions of the problems (1.3) and (1.7), by virtue of (1.6) and (1.10) we obtain

---

[2] Cf. Courant and Hilbert [1], Chapter IV, §2.3.

[3] This condition will subsequently be weakened.

$$u(x, t) = \frac{1}{2}\{Hf(x+t) + H^T f(x-t)\} + \frac{i}{2}\int_0^t \{HQ(x+t-\tau)u(x+t-\tau, \tau) + H^T Q(x-t+\tau)u(x-t+\tau, \tau)\}\,d\tau$$

or, changing the variable of integration,

$$u(x, t) = \frac{1}{2}\{Hf(x+t) + H^T f(x-t)\} + \frac{i}{2}\int_x^{x+t} HQ(s)u(s, x+t-s)\,ds + \frac{i}{2}\int_{x-t}^{x} H^T Q(s)u(s, s-x+t)\,ds. \tag{1.11}$$

Thus, we have constructed for the problem (1.1) + (1.2) an equivalent system of integral equations (1.11). These equations are of Volterra type and can therefore be solved by the method of successive approximations.

Let $u_0(x,t)$ be defined by

$$u_0(x,t) = \tfrac{1}{2}\{Hf(x+t) + H^T f(x-t)\}. \tag{1.6}$$

For $p \geqq 1$, $p = 1, 2, 3, \cdots$, we put

$$u_p(x, t) = \frac{1}{2}i\int_x^{x+t} HQ(s)u_{p-1}(s, x+t-s)\,ds + \frac{1}{2}i\int_{x-t}^{x} H^T Q(s)u_{p-1}(s, s-x+t)\,ds. \tag{1.12}$$

Then from the uniform convergence of the successive approximations, whose proof we omit, it follows that the solution of equation (1.11) is given by

$$u(x,t) = u_0(x,t) + u_1(x,t) + \cdots + u_p(x,t) + \cdots. \tag{1.13}$$

We will show that each of the vector-functions $u_p(x,t)$, $p = 1, 2, 3, \cdots$, can be represented in the form

$$u_p(x, t) = K_p(x, t)f(x+t) + L_p(x, t)f(x-t) + \frac{1}{2}\int_{x-t}^{x+t} H_p(x, t, s)f(s)\,ds, \tag{1.14}$$

where $K_p(x,t)$, $L_p(x,t)$ and $H_p(x,t,s)$ are 2×2 matrices.

Indeed, for $p=1$ it follows from (1.12), by virtue of (1.6), that

$$u_1(x,t)=\frac{1}{4}i\int_x^{x+t} HQ(s)\{Hf(x+t)+H^Tf(2s-x-t)\}\,ds$$
$$+\frac{1}{4}i\int_{x-t}^{x} H^TQ(s)\{Hf(2s-x+t)+H^Tf(x-t)\}\,ds,$$

or, changing variables,

$$u_1(x,t)=\left(\frac{1}{4}i\int_x^{x+t} HQ(s)H\,ds\right)f(x+t)$$
$$(1.15)\qquad +\left(\frac{1}{4}i\int_{x-t}^{x} H^TQ(s)H^T\,ds\right)f(x-t)$$
$$+\frac{1}{2}\int_{x-t}^{x+t}\frac{1}{4}i\left\{HQ\left(\frac{x+t+s}{2}\right)H^T+H^TQ\left(\frac{x-t+s}{2}\right)H\right\}f(s)\,ds,$$

from which (1.14) follows if we put

$$K_1(x,t)=\frac{1}{4}i\int_x^{x+t} HQ(s)H\,ds,\qquad L_1(x,t)=\frac{1}{4}i\int_{x-t}^{x} H^TQ(s)H^T\,ds.$$
$$(1.16)$$
$$H_1(x,t,s)\frac{1}{4}i\left\{HQ\left(\frac{x+t+s}{2}\right)H^T+H^TQ\left(\frac{x-t+s}{2}\right)H\right\}.$$

Let us now assume that (1.14) has been proved for $p=1,2,\cdots,q-1$. We will prove it for $p=q$. Putting $p=q-1$ in (1.14) and inserting the expression for $u_{q-1}(x,t)$ in (1.12), we obtain

$$u_q(x,t)=\frac{1}{2}i\int_x^{x+t} HQ(s)\{K_{q-1}(s,x+t-s)f(x+t)$$
$$+L_{q-1}(s,x+t-s)f(2s-x-t)\}\,ds$$
$$+\frac{1}{2}i\int_{x-t}^{x} H^TQ(s)\{K_{q-1}(s,s-x+t)f(2s-x+t)$$
$$(1.17)\qquad +L_{q-1}(s,s-x+t)f(x-t)\}\,ds$$

$$+\frac{1}{2} i \int_{x}^{x+t} HQ(s)\left\{\frac{1}{2} \int_{2s-x-t}^{x+t} H_{q-1}(s,\ x+t-s,\ \tau) f(\tau)\, d\tau\right\} ds$$

$$+\frac{1}{2} i \int_{x-t}^{x} H^T Q(s)\left\{\frac{1}{2} \int_{x-t}^{2s-x+t} H_{q-1}(s,\ s-x+t,\ \tau) f(\tau)\, d\tau\right\} ds$$

Put

$$u_p(x,t) = I_1 + I_2 + I_3 + I_4. \tag{1.17'}$$

We will transform each of the integrals I_k, $k = 1, 2, 3, 4$. Making a change of variables in the second part of the integral I_1, we obtain

$$I_1 = \left(\frac{1}{2} i \int_{x}^{x+t} HQ(s) K_{q-1}(s,\ x+t-s)\, ds\right) f(x+t)$$

$$+\frac{1}{4} i \int_{x-t}^{x+t} HQ\left(\frac{x+t+\tau}{2}\right) L_{q-1}\left(\frac{x+t+\tau}{2},\quad \frac{x+t-\tau}{2}\right) f(\tau)\, d\tau.$$

Similarly, we obtain

$$I_2 = \left(\frac{1}{2} i \int_{x-t}^{x} H^T Q(s) L_{q-1}(s,\ s-x+t)\, ds\right) f(x-t) +$$

$$+\frac{1}{4} i \int_{x-t}^{x+t} H^T Q\left(\frac{x-t+\tau}{2}\right) K_{q-1}\left(\frac{x-t+\tau}{2},\quad \frac{\tau-x+t}{2}\right) f(\tau)\, d\tau.$$

Finally, changing the order of integration in the integrals I_3 and I_4, we obtain

$$I_3 + I_4 = \frac{1}{2} i \int_{x-t}^{x+t} \frac{1}{2}\left\{\int_{x}^{\frac{1}{2}(x+t+\tau)} HQ(s) H_{q-1}(s,\ x+t-s,\ \tau)\, ds\right.$$

$$\left. + \int_{x-t}^{\frac{1}{2}(\tau+x-t)} H^T Q(s) H_{q-1}(s,\ s-x+t,\ \tau)\, ds\right\} f(\tau)\, d\tau.$$

If we now put

$$K_q(x, t) = \frac{1}{2} i \int_x^{x+t} HQ(s) K_{q-1}(s, x+t-s)\, ds, \tag{1.18}$$

$$L_q(x, t) = \frac{1}{2} i \int_{x-t}^{x} H^T Q(s) L_{q-1}(s, s-x+t)\, ds, \tag{1.19}$$

$$\begin{aligned} H_q(x, t, s) = {} & \frac{1}{2} iHQ\left(\frac{x+t+s}{2}\right) L_{q-1}\left(\frac{x+t+s}{2}, \frac{x+t-s}{2}\right) \\ & + \frac{1}{2} iH^T Q\left(\frac{x-t+s}{2}\right) K_{q-1}\left(\frac{x-t+s}{2}, \frac{s-x+t}{2}\right) + \\ & + \frac{1}{2} i \int_x^{\frac{1}{2}(x+t+s)} HQ(\tau) H_{q-1}(\tau, x+t-\tau, s)\, ds + \\ & + \frac{1}{2} i \int_{x-t}^{\frac{1}{2}(s+x-t)} H^T Q(\tau) H_{q-1}(\tau, \tau-x+t, s)\, d\tau, \end{aligned}$$

then (1.17) can be rewritten in the form

$$u_q(x, t) = K_q(x, t) f(x+t) + L_q(x, t) f(x-t) + \frac{1}{2} \int_{x-t}^{x+t} H_q(x, t, s) f(s)\, ds,$$

which is (1.14) for $p = q$. Thus (1.14) is proved for any $p \geqq 1$.

Now, inserting the expression for $u_p(x,t)$, $p = 0, 1, 2, \cdots$, in (1.13), we find that the solution of the problem (1.1) + (1.2) is given by

$$u(x, t) = K(x, t) f(x+t) + L(x, t) f(x-t) + \frac{1}{2} \int_{x-t}^{x+t} H(x, t, s) f(s)\, ds, \tag{1.20}$$

where we have put

$$K(x, t) = \sum_{p=0}^{\infty} K_p(x, t), \quad L(x, t) = \sum_{p=0}^{\infty} L_p(x, t), \quad H(x, t, s) = \sum_{p=1}^{\infty} H_p(x, t, s). \tag{1.20'}$$

The proof of the uniform convergence of these series is analogous to the proof of the uniform convergence of the successive approximations, and will be omitted.

Let us point out certain properties of the matrices $K_p(x,t)$ and $L_p(x,t)$ which will subsequently enable us to explicitly determine them. From the definition of the matrices $Q(s)$ and H it follows that

$$\begin{aligned} HQ(s)H &= \begin{pmatrix} p(s)+r(s), & i[p(s)+r(s)] \\ -i[p(s)+r(s)], & p(s)+r(s) \end{pmatrix} \\ &= [p(s)+r(s)]\begin{pmatrix} 1 & i \\ -i & 1 \end{pmatrix} = [p(s)+r(s)]H. \end{aligned} \tag{1.21}$$

Therefore by the definition of the matrix $K_1(x,t)$, i.e. (1.16), we have

$$K_1(x,t) = k_1(x,t)H, \tag{1.22}$$

where

$$k_1(x,\ t) = \frac{i}{4}\int_x^{x+t}[p(s)+r(s)]\,ds.$$

Further,

$$\begin{aligned} H^TQ(s)H^T &= \begin{pmatrix} p(s)+r(s), & -i[p(s)+r(s)] \\ i[p(s)+r(s)], & p(s)+r(s) \end{pmatrix} \\ &= [p(s)+r(s)]\begin{pmatrix} 1 & -i \\ i & 1 \end{pmatrix} = [p(s)+r(s)]H^T. \end{aligned} \tag{1.23}$$

Consequently, by virtue of (1.16) we obtain for the matrix $L_1(x,t)$ the expression

$$L_1(x,t) = l_1(x,t)H^T, \tag{1.24}$$

where

$$l_1(x,\ t) = \frac{i}{4}\int_{x-t}^{x}[p(s)+r(s)]\,ds.$$

We will show that each of the matrices $K_p(x,t)$ and $L_p(x,t)$ can be represented in the form

$$\begin{aligned} K_p(x,t) &= k_p(x,t)H, \\ L_p(x,t) &= l_p(x,t)H^T \qquad (p=1,2,\cdots). \end{aligned} \tag{1.25}$$

Indeed, for $p=1$ these formulas follow from (1.22) and (1.24). Let us assume that the formulas (1.25) have been proved for $p=1,2,\cdots,q-1$. We will prove them for $p=q$. Putting $p=q-1$ in (1.25) and inserting the expression for $K_{q-1}(x,t)$ in (1.18), we obtain

$$K_q(x, t) = \frac{1}{2} i \int_x^{x+t} HQ(s) k_{q-1}(s, x+t-s) H \, ds$$
$$= \frac{1}{2} i \int_x^{x+t} k_{q-1}(s, x+t-s) HQ(s) H \, ds.$$

Inserting the expression for $HQ(s)H$ from (1.21), we find that

$$K_q(x, t) = \frac{i}{2} \int_x^{x+t} k_{q-1}(s, x+t-s)[p(s)+r(s)] \, ds \cdot H,$$

which proves (1.25) for the matrix $K_q(x,t)$, if we put

$$k_q(x,t) = \frac{i}{2} \int_x^{x+t} k_{q-1}(s, x+t-s)[p(s)+r(s)] \, ds.$$

The formula (1.25) for $L_q(x,t)$ can be proved analogously by means of (1.19) and (1.23).

Now, inserting the expressions for $K_p(x,t)$ and $L_p(x,t)$ from (1.25) in (1.20) and putting

$$k(x, t) = \sum_{p=0}^{\infty} k_p(x, t), \qquad l(x, t) = \sum_{p=0}^{\infty} l_p(x, t),$$

the formula (1.20) can be rewritten in the form

$$u(x, t) = k(x, t) Hf(x+t) + l(x, t) H^T f(x-t) + \frac{1}{2} \int_{x-t}^{x+t} H(x, t, s) f(s) \, ds. \tag{1.26}$$

Formula (1.26) yields the solution of the problem (1.1) + (1.2) and will play an important role in what follows.

§2. Reduction to the Goursat problem

We again consider the problem (1.1) + (1.2). We will clarify the conditions that the functions $k(x,t)$, $l(x,t)$ and the matrix $H(x,t,s)$ must satisfy in order that the vector-function $u(x,t)$, defined by (1.26), shall be a solution of the problem (1.1) + (1.2). In this section we will prove that the matrix $H(x,t,s)$ can be obtained from a Goursat problem, and that we are able to explicitly determine the functions $k(x,t)$ and $l(x,t)$ (cf. (2.8), (2.10) below).

Suppose that the solution of (1.1) + (1.2) is given by (1.26). By virtue of (1.6), $k_0(x,t) = \frac{1}{2} = l_0(x,t)$, and from (1.22), (1.24) and (1.25) it follows from the expressions for $k_p(x,t)$ and $l_p(x,t)$ that $k_p(x,0)=l_p(x,0)=0$ $(p=1,2,\cdots)$. Therefore

$$k(x,0)=l(x,0)=\tfrac{1}{2}. \tag{2.1}$$

It follows from (1.26) and (2.1) that the initial condition (1.2) is automatically satisfied.

Let $u(x,t;f)$ denote the solution of (1.1) + (1.2), and suppose we know that the vector-function $u(x,t;f)$ can be represented by (1.26). Let T_t denote the operator $T_t=-iI\,\partial/\partial t$, and B_x denote the operator

$$B_x=\begin{pmatrix}0 & 1\\ -1 & 0\end{pmatrix}\frac{\partial}{\partial x}+Q(x),\quad Q(x)=\begin{pmatrix}p(x) & 0\\ 0 & r(x)\end{pmatrix}.$$

We will show that $B_x u(x,t;f)\equiv u(x,t;B_x f)$. In fact, it is not hard to see that the left and right sides of this relation satisfy the same Cauchy problem

$$B_x v(x,t)=T_t v(x,t),\qquad v(x,0)=B_x f(x).$$

Therefore the required equality follows from the uniqueness of the solution of a Cauchy problem.

Consequently, equation (1.1) can be written in the form

$$T_t u(x,t;f)=u(x,t;B_x f). \tag{2.2}$$

By the definition of the operator T_t and formula (1.26) we have

$$\begin{aligned}T_t u(x,t;f)&=-iI\frac{\partial}{\partial t}[u(x,t;f)]\\&=-ik'_t(x,t)Hf(x+t)-ik(x,t)Hf'(x+t)\\&\quad-il'_t(x,t)H^Tf(x-t)+il(x,t)H^Tf'(x-t)\\&\quad-\frac{i}{2}H(x,t,x+t)f(x+t)-\frac{i}{2}H(x,t,x-t)f(x-t)\\&\quad-\frac{i}{2}\int_{x-t}^{x+t}H'_t(x,t,s)f(s)\,ds.\end{aligned} \tag{2.3}$$

On the other hand, by the definition of the operator B_x and formula (1.26) we have

$$\begin{aligned}u(x,t;B_xf)&=k(x,t)HBf'(x+t)\\&\quad+k(x,t)HQ(x+t)f(x+t)+l(x,t)H^TBf'(x-t)+\end{aligned}$$

$$+ l(x, t) H^T Q(x-t) f(x-t) + \frac{1}{2} \int_{x-t}^{x+t} H(x, t, s) B f'(s)\, ds$$

$$+ \frac{1}{2} \int_{x-t}^{x+t} H(x, t, s) Q(s) f(s)\, ds.$$

Integrating the first integral in the right side of this relation by parts and taking into account that $HB = -iH$, $H^T B = iH^T$, this relation can be rewritten in the form

$$\begin{aligned} u(x, t; B_x f) = {} & -ik(x, t) H f'(x+t) \\ & + k(x, t) HQ(x+t) f(x+t) \\ & + l(x, t) H^T Q(x-t) f(x-t) + il(x, t) H^T f'(x-t) \\ & + \frac{1}{2}\{H(x, t, x+t) B f(x+t) - H(x, t, x-t) B f(x-t)\} \\ & - \frac{1}{2} \int_{x-t}^{x+t} \{H'_s(x, t, s) B - H(x, t, s) Q(s)\} f(s)\, ds. \end{aligned} \tag{2.4}$$

Since the vector-function $f(x)$ is arbitrary, it follows by virtue of (2.2) that the coefficients of $f(x+t)$ and $f(x-t)$, as well as the integrands, in the expressions (2.3) and (2.4), must coincide (the coefficients of the derivatives $f'(x+t)$ and $f'(x-t)$ cancel each other). Thus, equating the coefficients of $f(x+t)$ and $f(x-t)$ respectively, we obtain

$$\begin{aligned} -ik'_t(x, t) H - \frac{i}{2} H(x, t, x+t) \\ = k(x, t) HQ(x+t) + \frac{1}{2} H(x, t, x+t) B, \end{aligned} \tag{2.5}$$

$$\begin{aligned} -il'_t(x, t) H^T - \frac{i}{2} H(x, t, x-t) \\ = l(x, t) H^T Q(x-t) - \frac{1}{2} H(x, t, x-t) B. \end{aligned} \tag{2.6}$$

We will rewrite (2.5) in the form

$$k'_t(x,t) H - ik(x,t) HQ(x+t) = \tfrac{1}{2} H(x,t,x+t)\,[iB - I]. \tag{2.5$'$}$$

The left and right sides of (2.5′) are 2×2 matrices, and their equality signifies the equality of their corresponding elements.

Let

$$H(x, t, x+t) = \begin{pmatrix} H_{11}(x, t, x+t) & H_{12}(x, t, x+t) \\ H_{21}(x, t, x+t) & H_{22}(x, t, x+t) \end{pmatrix}.$$

Inserting the values of the matrices $H, Q(x+t)$, B and $H(x,t,x+t)$ in (2.5′), we obtain

$$\begin{pmatrix} k_t'(x,t) - ik(x,t)\,p(x+t), & ik_t'(x,t) + k(x,t)\,r(x+t) \\ -ik_t'(x,t) - k(x,t)\,p(x+t), & k_t'(x,t) - ik(x,t)\,r(x+t) \end{pmatrix}$$

$$(2.5'') \quad = \begin{pmatrix} -iH_{12}(x,t,x+t) - H_{11}(x,t,x+t), & iH_{11}(x,t,x+t) - H_{12}(x,t,x+t) \\ -iH_{22}(x,t,x+t) - H_{21}(x,t,x+t), & iH_{21}(x,t,x+t) - H_{22}(x,t,x+t) \end{pmatrix}.$$

Since, in the matrix on the right, the first element of the first row, when multiplied by i, is equal to the negative of the second element of the first row, the matrix on the left must have the same property. Therefore

$$i\{k_t'(x,t) - ik(x,t)p(x+t)\} = -\{ik_t'(x,t) + k(x,t)r(x+t)\}.$$

i.e.

$$(2.7) \qquad k_t'(x,t) = \frac{i}{2}\{p(x+t) + r(x+t)\}\,k(x,t).$$

Solving this equation with due regard for the initial condition (2.1), we obtain

$$(2.8) \qquad k(x,t) = \frac{1}{2}\exp\left\{\frac{i}{2}\int_0^t [p(x+\tau) + r(x+\tau)]\,d\tau\right\}.$$

We will now transform equation (2.6). We first write it in the form

$$(2.6') \qquad l_t'(x,t)H^T - il_t(x,t)H^TQ(x-t) = -\tfrac{1}{2}H(x,t,x-t)\,[iB+I]$$

or, in greater detail,

$$\begin{pmatrix} l_t'(x,t) - ip(x-t)\,l(x,t), & -il_t'(x,t) - r(x-t)\,l(x,t) \\ il_t'(x,t) + p(x-t)\,l(x,t), & l_t'(x,t) - ir(x-t)\,l(x,t) \end{pmatrix}$$

$$(2.6'') \quad = \begin{pmatrix} -iH_{12}(x,t,x-t) + H_{11}(x,t,x-t), & iH_{11}(x,t,x-t) + H_{12}(x,t,x-t), \\ -iH_{22}(x,t,x-t) + H_{21}(x,t,x-t), & iH_{21}(x,t,x-t) + H_{22}(x,t,x-t) \end{pmatrix}.$$

Since the elements of the first column of the matrix on the right, when multiplied by i, are equal to the corresponding elements of the second column, the matrix on the left must have the same property. Therefore

$$i\{l_t'(x,t) - ip(x-t)l(x,t)\} = -il_t'(x,t) - r(x-t)l(x,t),$$

i.e.

$$l'_t(x,\ t)=\frac{i}{2}\{p(x-t)+r(x-t)\}\,l(x,\ t). \tag{2.9}$$

Integrating this equation with due regard for the initial condition (2.1), we obtain the following explicit expression for the function $l(x,t)$:

$$l(x,\ t)=\frac{1}{2}\exp\left\{\frac{i}{2}\int_0^t [p(x-\tau)+r(x-\tau)\,d\tau\right\}. \tag{2.10}$$

Further, by virtue of equation (2.7) we obtain from (2.5′)

$$\begin{aligned} H(x,\ t,\ x+t)[iB-I] \\ =ik(x,\ t)\{p(x+t)+r(x+t)]\,H-2HQ(x+t)\}. \end{aligned} \tag{2.11}$$

Since

$$\begin{aligned} iB-I=-H^T,\ [p(x+t)+r(x+t)]\,H-2HQ(x+t) \\ =i\,[p(x+t)-r(x+t)]E, \end{aligned}$$

where

$$E=\begin{pmatrix} i & 1 \\ 1 & -i \end{pmatrix},$$

it follows from (2.11) that

$$H(x,t,x+t)\,H^T=k(x,t)\,[p(x+t)\ -r(x+t)\,]E. \tag{2.12}$$

By (2.9) we obtain from (2.6′)

$$\begin{aligned} H(x,\ t,\ x-t)[iB+I] \\ =-il(x,\ t)\{[p(x-t)+r(x-t)]\,H^T-2H^TQ(x-t)\}. \end{aligned} \tag{2.13}$$

But since by the definition of B, H^T and $Q(x-t)$

$$\begin{aligned} iB+I=H,\ [p(x-t)+r(x-t)]\,H^T-2H^TQ(x-t) \\ =i\,[r(x-t)-p(x-t)]\,E', \end{aligned}$$

$$E'=\begin{pmatrix} -i & 1 \\ 1 & i \end{pmatrix},$$

(2.13) assumes the form

$$H(x,t,x-t)\,H=l(x,t)\,[r(x-t)\ -p(x-t)\,]E'. \tag{2.14}$$

Finally, equating the integrands in (2.3) and (2.4), we obtain for the matrix $H(x,t,s)$ the equation

$$iH'_t(x,t,s)\ -H'_s(x,t,s)\,B+H(x,t,s)\,Q(s)\ =0. \tag{2.15}$$

This equation, together with (2.12) and (2.14), defines the Goursat problem for the matrix $H(x,t,s)$.

From (2.5″) and (2.6″) we obtain another relation, which we will need further on. Equating, in each of these equalities, the elements of the first row and column, taking (2.7) and (2.9) into consideration, and then putting $t=0$, by virtue of (2.1) we obtain

$$H_{11}(x,\ 0,\ x)+iH_{12}(x,\ 0,\ x)=\frac{i}{2}[p(x)-r(x)],$$

$$H_{11}(x,\ 0,\ x)-iH_{12}(x,\ 0,\ x)=\frac{i}{2}[p(x)-r(x)],$$

from which

$$H_{11}(x,\ 0,\ x)=\frac{i}{2}[p(x)-r(x)],\qquad H_{12}(x,\ 0,\ x)=0. \tag{2.16}$$

Similarly, equating the elements of the second row and first column, we obtain

$$H_{21}(x,\ 0,\ x)+iH_{22}(x,\ 0,\ x)=\frac{1}{2}[p(x)-r(x)],$$

$$H_{21}(x,\ 0,\ x)-iH_{22}(x,\ 0,\ x)=-\frac{i}{2}[p(x)-r(x)],$$

i.e.

$$H_{21}(x,\ 0,\ x)=0,\ H_{22}(x,\ 0,\ x)=-\frac{i}{2}[p(x)-r(x)]. \tag{2.17}$$

§3. Operator-matrix transformations

Let A_1 and A_2 be two linear differential operators, and L_1 and L_2, two linear function spaces.

DEFINITION. *A continuous linear operator X, which maps L_1 into L_2, is called an* operator transformation *if it satisfies the following two conditions*:

(3.1)
1. $A_1X=XA_2$.
2. *The inverse operator X^{-1} exists.*

Let

$$A_1\equiv\begin{pmatrix} p_1(x) & \frac{d}{dx} \\ -\frac{d}{dx} & r_1(x)\end{pmatrix}=\begin{pmatrix} 0 & 1 \\ -1 & 0\end{pmatrix}\frac{d}{dx}+\begin{pmatrix} p_1(x) & 0 \\ 0 & r_1(x)\end{pmatrix} \tag{3.2}$$

$$\equiv B\frac{d}{dx}+Q_1(x),$$

$$A_2 \equiv \begin{pmatrix} p_2(x) & \frac{d}{dx} \\ -\frac{d}{dx} & r_2(x) \end{pmatrix} = \begin{pmatrix} 0 & 1 \\ -1 & 0 \end{pmatrix} \frac{d}{dx} + \begin{pmatrix} p_2(x) & 0 \\ 0 & r_2(x) \end{pmatrix} \equiv B \frac{d}{dx} + Q_2(x), \tag{3.3}$$

where $p_1(x)$, $r_1(x)$, $p_2(x)$ and $r_2(x)$ are real functions which are summable over every finite interval $(0 \leqq x \leqq b < \infty)$.

We take for L_1 the set of all continuously differentiable vector-functions

$$f(x) = \begin{pmatrix} f_1(x) \\ f_2(x) \end{pmatrix}$$

defined on the interval $[0, b)$ and satisfying the boundary condition

$$f_2(0) - h_1 f_1(0) = 0, \tag{3.4}$$

where h_1 is an arbitrary finite real number.

We take for L_2 the set of all continuously differentiable vector-functions

$$g(x) = \begin{pmatrix} g_1(x) \\ g_2(x) \end{pmatrix}$$

which are defined on the same interval $[0, b)$ and satisfy the boundary condition

$$g_2(0) - h_2 g_1(0) = 0, \tag{3.5}$$

where h_2 is an arbirary finite real number.

We will seek the operator-matrix X in the form $(f(x) \in L_1)$

$$X[f(x)] = R(x) f(x) + \int_0^x K(x, s) f(s)\, ds. \tag{3.6}$$

where $R(x)$ and $K(x, s)$ are second-order square matrices.

By virtue of (3.2) and (3.6) we have

$$A_1 X[f(x)] = BR'(x) f(x) + BR(x) f'(x) + Q_1(x) R(x) f(x) + BK(x, x) f(x) + \int_0^x \{BK'_x(x, s) + Q_1(x) K(x, s)\} f(s)\, ds. \tag{3.7}$$

On the other hand, by virtue of (3.3) and (3.6)

$$XA_2[f(x)] = R(x) Bf'(x) + R(x) Q_2(x) f(x) + \int_0^x K(x, s) \{Bf'(s) + Q_2(s) f(s)\}\, ds.$$

Integrating by parts, this relation can be rewritten in the form

$$
\begin{aligned}
XA_2[f(x)] = {} & R(x)Bf'(x) + R(x)Q_2(x)f(x) \\
& + K(x,x)Bf(x) - K(x,0)Bf(0) \\
& + \int_0^x \{K(x,s)Q_2(s) - K'_s(x,s)B\} f(s)\,ds.
\end{aligned} \tag{3.8}
$$

Since $f(x)$ is an arbitrary vector-function from L_1, by virtue of (3.1) the coefficients of $f(x)$ and $f'(x)$, and the integrands in the expressions in (3.7) and (3.8), must be equal. Therefore, equating the coefficients of $f'(x)$, we obtain

$$BR(x) = R(x)B. \tag{3.9}$$

Then if we put

$$R(x) = \begin{pmatrix} \alpha(x) & \beta(x) \\ \gamma(x) & \delta(x) \end{pmatrix},$$

we obtain from (3.9) $\delta(x) = \alpha(x)$ and $\gamma(x) = -\beta(x)$, i.e. the matrix $R(x)$ has the form

$$R(x) = \begin{pmatrix} \alpha(x) & \beta(x) \\ -\beta(x) & \alpha(x) \end{pmatrix}. \tag{3.10}$$

Let us determine the functions $\alpha(x)$ and $\beta(x)$. To do this, by equating the coefficients of $f(x)$ in (3.7) and (3.8), we obtain the following equation for the determination of the matrix $R(x)$:

$$BR'(x) + Q_1(x)R(x) - R(x)Q_2(x) = K(x,x)B - BK(x,x). \tag{3.11}$$

Put

$$K(x,s) = \begin{pmatrix} K_{11}(x,s), & K_{12}(x,s) \\ K_{21}(x,s), & K_{22}(x,s) \end{pmatrix}.$$

Using the explicit forms of the matrices $Q_1(x)$, $Q_2(x)$, B and $R(x)$ (cf. (3.10)), we can write (3.11) in the following form:

$$
\begin{aligned}
&\begin{pmatrix} -\beta'(x) + [p_1(x) - p_2(x)]\alpha(x), & \alpha'(x) + [p_1(x) - r_2(x)]\beta(x) \\ -\alpha'(x) + [p_2(x) - r_1(x)]\beta(x), & -\beta'(x) + [r_1(x) - r_2(x)]\alpha(x) \end{pmatrix} \\
&= \begin{pmatrix} -[K_{12}(x,x) + K_{21}(x,x)], & K_{11}(x,x) - K_{22}(x,x) \\ K_{11}(x,x) - K_{22}(x,x), & K_{12}(x,x) + K_{21}(x,x) \end{pmatrix}.
\end{aligned} \tag{3.12}
$$

In the matrix on the right side of (3.12) the elements on the principal diagonal are equal except for having opposite signs, while the two off-diagonal elements are equal. The same is therefore true of the matrix on the left side of (3.12), and so it follows that

$$\alpha'(x) + [p_1(x) - r_2(x)]\beta(x) = -\alpha'(x) + [p_2(x) - r_1(x)]\beta(x),$$
$$\beta'(x) - [p_1(x) - p_2(x)]\alpha(x) = -\beta'(x) + [r_1(x) - r_2(x)]\beta(x),$$

i.e.

$$(3.13) \qquad 2\alpha'(x) + q(x)\beta(x) = 0, \qquad 2\beta'(x) - q(x)\alpha(x) = 0,$$

where

$$(3.14) \qquad q(x) = p_1(x) - p_2(x) + r_1(x) - r_2(x).$$

From the system (3.13) follows $2\alpha(x)\alpha'(x) + 2\beta(x)\beta'(x) = 0$, i.e. $\{\alpha^2(x) + \beta^2(x)\}' = 0$, from which we obtain

$$(3.15) \qquad \alpha^2(x) + \beta^2(x) = \alpha^2(0) + \beta^2(0).$$

Suppose that the vector-function

$$f(x) = \begin{pmatrix} f_1(x) \\ f_2(x) \end{pmatrix}$$

is continuously differentiable and satisfies the conditions

$$(3.16) \qquad f_1(0) = 1, \qquad f_2(0) = h_1.$$

Then $f(x)$ obviously satisfies the boundary condition (3.4), and therefore $f(x) \in L_1$. Further, let

$$(3.17) \qquad X[f(x)] = g(x),$$

where the vector-function

$$g(x) = \begin{pmatrix} g_1(x) \\ g_2(x) \end{pmatrix}$$

is an element of the space L_2 and so satisfies the boundary condition (3.5). Then for $x = 0$ there follows from (3.17) by the definition of the operator-matrix X, i.e. by virtue of (3.6) and (3.10),

$$g_1(0) = \alpha(0)f_1(0) + \beta(0)f_2(0), \qquad g_2(0) = -\beta(0)f_1(0) + \alpha(0)f_2(0).$$

Multiplying the first of these equations by h_2, subtracting it from the second and taking into account the boundary condition (3.5) and the condition (3.16), we find that

$$\beta(0) = \frac{h_1 - h_2}{1 + h_1h_2}\alpha(0)$$

Put

(3.18) $$\alpha(0) = 1.$$

Then

(3.19) $$\beta(0) = (h_1 - h_2)/(1 + h_1 h_2).$$

Therefore

(3.20) $$\alpha^2(0) + \beta^2(0) = \frac{(1+h_1)^2(1+h_2)^2}{(1+h_1h_2)^2} = \varkappa^2.$$

Now solving the system (3.13) and taking account of the expressions (3.15), (3.18)–(3.20), we obtain the following explicit expressions for $\alpha(x)$ and $\beta(x)$:

(3.21) $$\alpha(x) = \varkappa \sin\left\{-\frac{1}{2}\int_0^x q(\tau)\,d\tau + \arcsin\frac{1}{\varkappa}\right\},$$

(3.22) $$\beta(x) = \varkappa \cos\left\{-\frac{1}{2}\int_0^x q(\tau)\,d\tau + \arcsin\frac{1}{\varkappa}\right\},$$

where the function $q(x)$ is defined by (3.14), and the number κ is defined by (3.20).

Now equating the integrands in (3.7) and (3.8), we obtain the following equation for the matrix-kernel $K(x,s)$:

(3.23) $$K_s'(x,s)B + BK_x'(x,s) = K(x,s)Q_2(s) - Q_1(x)K(x,s).$$

Finally, the term in (3.8) which contains $f(0)$ must be zero in view of the absence of a similar term in (3.7). Thus $K(x,0)Bf(0) = 0$, i.e.

$$\begin{pmatrix} -K_{12}(x,0), & K_{11}(x,0) \\ -K_{22}(x,0), & K_{21}(x,0) \end{pmatrix} f(0) = 0,$$

which is equivalent to the system of equations

$$K_{12}(x,0)f_1(0) = K_{11}(x,0)f_2(0),$$
$$K_{22}(x,0)f_1(0) = K_{21}(x,0)f_2(0).$$

Hence by virtue of the boundary condition (3.4) we finally obtain

(3.24) $$K_{12}(x,0) = h_1 K_{11}(x,0), \qquad K_{22}(x,0) = h_1 K_{21}(x,0).$$

Let us further put

(3.25) $$K_{11}(x,0) = \varphi(x), \qquad K_{21}(x,0) = \psi(x),$$

where $\varphi(x)$ and $\psi(x)$ are for the present arbitrary continuously differentiable functions. The conditions (3.24) and (3.25) for the matrix-kernel $K(x,s)$ combine to give the condition

$$K(x, s)|_{s=0} = \begin{pmatrix} \varphi(x), & h_1\varphi(x) \\ \psi(x), & h_1\psi(x) \end{pmatrix}, \tag{3.26}$$

and this condition together with equation (3.23) defines a Cauchy problem for $K(x,s)$. This problem is solvable.[4]

Expressing the solution of the problem (3.23) + (3.26) by means of (1.26) and inserting it in (3.11), we obtain an integral equation for the determination of the vector-function $\begin{pmatrix} \varphi(2x) \\ \psi(2x) \end{pmatrix}$. The quantities $\varphi(0)$ and $\psi(0)$, which also appear in this equation, can be determined from the equation (3.12) for $x = 0$. Therefore, carrying out the calculations in the opposite order, we can prove that the operator-matrix X, defined by (3.6), in which the matrix-kernel $K(x,s)$ is the solution of the Cauchy problem (3.23) + (3.26), satisfies the relation (3.1). The existence of the inverse X^{-1} follows from the form of the operator X.

§4. Solution of a mixed problem on the halfline

Let $p(x)$, $r(x)$ and the vector-function $f(x) = \begin{pmatrix} f_1(x) \\ f_2(x) \end{pmatrix}$ satisfy the previous requirements. We extend the functions $p(x)$ and $r(x)$ to the negative halfline in such a way that they are still summable over every finite interval, but otherwise arbitrary, and extend $f_1(x)$ and $f_2(x)$ in a manner which for the present we leave unspecified (the manner in which they are to be extended will be made precise further on).

Under these assumptions we will consider the mixed problem

$$-iI\frac{\partial u}{\partial t} = B\frac{\partial u}{\partial x} + Q(x)u, \tag{4.1}$$

$$u(x, 0) = f(x), \tag{4.2}$$

$$u_2(0, t) - hu_1(0, t) = 0, \tag{4.3}$$

where, as before, I is the unit matrix,

$$B = \begin{pmatrix} 0 & 1 \\ -1 & 0 \end{pmatrix}, \quad Q(x) = \begin{pmatrix} p(x) & 0 \\ 0 & r(x) \end{pmatrix},$$

and h is an arbitrary finite real number.

[4] Cf. formula (1.26).

Let

$$u(x,t) = \begin{pmatrix} u_1(x,t) \\ u_2(x,t) \end{pmatrix}$$

denote the solution of the problem (4.1)–(4.3). According to the formula (1.26) the solution of the problem (4.1)+(4.2) can be represented in the form

$$u(x,t) = k(x,t)Hf(x+t) + l(x,t)H^T f(x-t) + \frac{1}{2}\int_{x-t}^{x+t} H(x,t,s)f(s)\,ds, \tag{4.4}$$

where by the results of §2

$$k(x,t) = \frac{1}{2}\exp\left\{\frac{i}{2}\int_0^t [p(x+\tau)+r(x+\tau)]\,d\tau\right\},$$
$$l(x,t) = \frac{1}{2}\exp\left\{\frac{i}{2}\int_0^t [p(x-\tau)+r(x-\tau)]\,d\tau\right\}. \tag{4.4'}$$

Let us now consider the application of operator transformations in order to extend the solution of (4.1) to the negative halfline. We will show how to express the solution $u(x,t)$ at the point $-x$ as the result of a linear operator acting on $u(s,t)$ $(0 \leq s \leq x)$. Let the matrices A_1 and A_2 have the form

$$A_1 = B\frac{d}{dx} + Q(-x), \quad A_2 = B\frac{d}{dx} + Q(x) \quad (x \geqslant 0). \tag{4.5}$$

We will assume that the functions $p(x)$ and $r(x)$ have been extended to the negative halfline in such a way that $p(-0) = p(+0)$, $r(-0) = r(+0)$, i.e.

$$Q(-0) = Q(+0). \tag{4.6}$$

In other words, we will assume that the extension of the matrix $Q(x)$ is continuous at zero. Further, let us denote by X an operator-matrix transformation which maps the space L_h of continuously differentiable vector-functions $f(x)$ satisfying the boundary condition (4.3), i.e. $f_2(0) - hf_1(0) = 0$, onto the space L_h.

Suppose that the vector-function

$$u^+(x,t) = \begin{pmatrix} u_1^+(x,t) \\ u_2^+(x,t) \end{pmatrix}$$

is the solution of the mixed problem (4.1) – (4.3) on the halfline $[0, \infty)$. By virtue of (4.3), for every fixed t the vector-function $u^+(x,t)$ belongs to the space L_h; therefore the operator X can be applied to $u^+(x,t)$.

The extension of the solution $u^+(x,t)$ to the negative halfline will be defined by the formula

$$u(-x,t) \equiv u^-(x,t) = X[u^+(x,t)], \tag{4.7}$$

i.e. by the definition of X (cf. (3.6))

$$X[u^+(x, t)] = u^-(x, t) = R(x)u^+(x, t) + \int_0^x K(x, s)u^+(s, t)\,ds. \tag{4.8}$$

We will show that the vector-function $u^-(x,t)$ satisfies the equation (4.1) on the negative halfline. By virtue of the definition of the matrix A_1 we have to show that

$$A_1[u^-(x, t)] = -iI\frac{\partial u^-}{\partial t}. \tag{4.9}$$

Since for every fixed t the solution $u^+(x,t)$ belongs to the space L_h, by the definition of the operator-matrix transformation X and (3.1) we have

$$A_1X[u^+(x,t)] = XA_2[u^+(x,t)]. \tag{4.10}$$

According to (4.7) the left side of (4.10) coincides with the left side of (4.9). Let us calculate the right side of (4.10). By virtue of (4.1), which can be rewritten in the form

$$A_2[u^+(x, t)] = -iI\frac{\partial u^+}{\partial t},$$

we have

$$XA_2[u^+(x, t)] = X\left[-iI\frac{\partial u^+}{\partial t}\right] = -iIX\left[\frac{\partial u^+}{\partial t}\right].$$

It follows in obvious fashion from (4.8) that the operator-matrix X and the operator of differentiation $\partial/\partial t$ commute; therefore by (4.7) the above relation implies that

$$XA_2[u^+(x, t)] = -iI\frac{\partial}{\partial t}X[u^+(x, t)] = -iI\frac{\partial u^-}{\partial t},$$

i.e. the right sides of (4.9) and (4.10) coincide, which proves our assertion.

We will now show that the extension of the solution $u^+(x,t)$ to the negative halfline is continuous together with its first derivative with respect to x.

In fact, putting $x=0$ in (4.8), we obtain

$$u^-(-0,t)=R(0)u^+(+0,t). \tag{4.11}$$

By definition (cf. (3.10))

$$R(0)=\begin{pmatrix}\alpha(0) & \beta(0)\\ -\beta(0) & \alpha(0)\end{pmatrix},$$

where $\alpha(0)=1$, and $\beta(0)$ can be computed from (3.19). Since in the present instance the spaces L_1 and L_2 which were considered in the previous section coincide, $L_1=L_2=L_h$, it follows that $h_1=h_2=h$, and therefore $\beta(0)=0$. Hence $R(0)=I$, and the continuity at zero of the extension of $u^+(x,t)$ follows from (4.11).

Further, using the explicit form of the operators A_1, A_2 and X, by virtue of (4.7) the relation (4.10) can be rewritten in the form

$$B\frac{\partial u^-}{\partial x}+Q(-x)u^-=R(x)\Big[B\frac{\partial u^+}{\partial x}+Q(x)u^+\Big] + \int_0^x K(x,s)\Big[B\frac{\partial u^+}{\partial s}+Q(s)u(s,t)\Big]ds.$$

Putting $x=0$, we obtain

$$B\frac{\partial u(-x,t)}{\partial x}\Big|_{x=0}+Q(-0)u(-0,t) = R(0)\Big[B\frac{\partial u(x,t)}{\partial x}\Big|_{x=0}+Q(+0)u(+0,t)\Big].$$

Since $R(0)=I$, it follows by (4.6) and (4.11) that

$$\frac{\partial u(-x,t)}{\partial x}\Big|_{x=0}=\frac{\partial u(x,t)}{\partial x}\Big|_{x=0}, \tag{4.12}$$

i.e. the first derivative is continuous at zero.

Formula (4.8) can be used to extend the original vector-function $f(x)$. In fact, putting $t=0$ in (4.8) and taking account of the initial condition (4.2), we obtain

$$f(-x)=R(x)f(x)+\int_0^x K(x,s)f(s)\,ds. \tag{4.13}$$

The continuity at zero of the extension of $f(x)$ and its derivative follows from (4.11) and (4.12), if we put $t=+0$ in them.

Let us now return to the problem (4.1)–(4.3).

If $x>t>0$, then the solution of this problem coincides with the solution of the problem (4.1) + (4.2), and consequently is given by (4.4). Now if $0 < x < t$, then we have to use the extension of the vector-function $f(x)$ to the negative halfline according to (4.13), in the following way: we break up the interval of integration $(x-t, x+t)$ in (4.4) into two intervals, $(x-t, 0)$ and $(0, x+t)$, and then in the integral over $(x-t, 0)$, $x-t<0$, we replace the variable of integration s by $-s$, and use (4.13) for the value of the function $f(-s)$. We similarly change the value of the function $f(x-t)$ in (4.4). Then following simple transformations we obtain the following formula for the solution of the problem (4.1) – (4.3) for $0 < x < t$:

$$\begin{aligned} u(x, t) &= k(x, t) Hf(x+t) + l(x, t) H^T R(t-x) f(t-x) \\ &+ \frac{1}{2}\int_0^{x+t} H(x, t, s) f(s)\, ds + \frac{1}{2}\int_0^{t-x} H_1(x, t, s) f(s)\, ds, \end{aligned} \tag{4.14}$$

where

$$\begin{aligned} H_1(x, t, s) = 2l(x, t) H^T K(t-x, s) + \\ + H(x, t, -s) R(s) + \int_s^{t-x} H(x, t, -\tau) K(\tau, s)\, d\tau. \end{aligned}$$

§5. Solution of the problem (1.1) + (1.2) for $t < 0$

We will again consider the problem (1.1) + (1.2), i.e. the problem

$$-iI\frac{\partial u}{\partial t} = B\frac{\partial u}{\partial x} + Q(x)u, \tag{5.1}$$

$$u(x, 0) = f(x), \tag{5.2}$$

assuming now, however, that $t < 0$. Put $t = -\tau$. Then equation (5.1) assumes the form

$$iI\frac{\partial u(x, -\tau)}{\partial \tau} = B\frac{\partial u(x, -\tau)}{\partial x} + Q(x)u(x, -\tau). \tag{5.1'}$$

Let I_1 denote the matrix

$$I_1 = \begin{pmatrix} 0 & 1 \\ 1 & 0 \end{pmatrix}$$

and put $v(x, +\tau) = I_1 u(x, -\tau)$. Since $I_1^2 = I$,

$$u(x, -\tau) = I_1 v(x, \tau). \tag{5.3}$$

Inserting now the value of $u(x, -\tau)$ in equation (5.1′), we obtain

$$iI_1 \frac{\partial v(x, \tau)}{\partial \tau} = BI_1 \frac{\partial v(x, \tau)}{\partial x} + Q(x) I_1 v(x, \tau).$$

Multiplying both sides of this equation from the left by $-I_1$, and taking into account that $I_1^2 = I$, $-I_1 B I_1 = B$, we find that

(5.4) $$-iI \frac{\partial v}{\partial \tau} = B \frac{\partial v}{\partial x} + Q_1(x) v,$$

where

(5.5) $$Q_1(x) = -I_1 Q(x) I_1 = \begin{pmatrix} -r(x) & 0 \\ 0 & -p(x) \end{pmatrix}.$$

By virtue of (5.3), the initial condition (5.2) becomes

(5.6) $$v(x, 0) = I_1 f(x),$$

which together with equation (5.4) defines a Cauchy problem for $v(x, \tau)$. Since the problem (5.4) + (5.6) is of the same form as the problem (5.1) + (5.2), applying (4.4) we obtain

(5.7) $$v(x, \tau) = k_1(x, \tau) H I_1 f(x + \tau) + l_1(x, \tau) H^T I_1 f(x - \tau) + \frac{1}{2} \int_{x-\tau}^{x+\tau} H_1(x, \tau, s) I_1 f(s)\, ds,$$

where according to (4.4)

$$k_1(x, \tau) = \frac{1}{2} \exp\left\{ \frac{i}{2} \int_0^\tau \operatorname{tr} Q_1(x + \sigma)\, d\sigma \right\},$$

$$l_1(x, \tau) = \frac{1}{2} \exp\left\{ \frac{i}{2} \int_0^\tau \operatorname{tr} Q_1(x - \sigma)\, d\sigma \right\}.$$

Since by the definition of the matrix $Q_1(x)$, i.e. (5.5), $\operatorname{tr} Q_1(x) = -\operatorname{tr} Q(x)$, it easily follows from the preceding expressions that

(5.8) $$k_1(x, \tau) = l(x, -\tau), \qquad l_1(x, \tau) = k(x, -\tau),$$

where $k(x, -\tau)$ and $l(x, -\tau)$ are given by (4.4′). Therefore (5.7) can be rewritten in the form

$$v(x, \tau) = l(x, -\tau) H I_1 f(x + \tau) + k(x, -\tau) H^T I_1 f(x - \tau) + \frac{1}{2} \int_{x-\tau}^{x+\tau} H_1(x, \tau, s) I_1 f(s)\, ds.$$

Hence, multiplying both sides from the left by I_1, by virtue of (5.3) we obtain

$$u(x, -\tau) = k(x, -\tau) Hf(x-\tau) + l(x, -\tau) H^T f(x+\tau) + \frac{1}{2}\int_{x-\tau}^{x+\tau} I_1 H_1(x, \tau, s) I_1 f(s)\, ds,$$

since $I_1 H I_1 = H^T$, $I_1 H^T I_1 = H$. Putting now $-\tau = t$, we will have

$$u(x, t) = k(x, t) Hf(x+t) + l(x, t) H^T f(x-t) + \frac{1}{2}\int_{x+t}^{x-t} \tilde{H}(x, t, s) f(s)\, ds, \tag{5.9}$$

where

$$\widetilde{H}(x,t,s) = I_1 H_1(x, -t, s) I_1. \tag{5.10}$$

Let us now take up the derivation of an equation for the matrix-kernel $\widetilde{H}(x,t,s)$. According to (2.15) the matrix-kernel $H_1(x,\tau,s)$ in (5.7) satisfies the equation

$$iH'_\tau(x,\tau,s) - H'_s(x,\tau,s) B + H_1(x,\tau,s) Q_1(s) = 0.$$

Replacing τ by $-t$ and taking account of (5.5), we obtain

$$-iH'_{1t}(x,-t,s) - H'_{1s}(x,-t,s) B - H_1(x,-t,s) I\, Q(s) I_1 = 0,$$

from which, multiplying both sides from the left and right by I_1, we find that

$$iI_1 H'_{1t}(x,-t,s) I_1 + I_1 H'_{1s}(x,-t,s) B I_1 + I_1 H_1(x,-t,s) I_1 Q(s) = 0.$$

It follows from the definition of the matrices I_1 and B that $BI_1 = -I_1 B$; therefore by (5.10) the preceding equation can be rewritten in the form

$$i\widetilde{H}_t(x,t,s) - \widetilde{H}'_s(x,t,s) B + \widetilde{H}(x,t,s) Q(s) = 0. \tag{5.11}$$

We now compute $\widetilde{H}(x,0,x)$. Let

$$H_1(x, \tau, s) = \begin{pmatrix} H_{11}^{(1)}, & H_{12}^{(1)} \\ H_{21}^{(1)}, & H_{22}^{(1)} \end{pmatrix}.$$

Then by the definition of the matrix $Q_1(x)$, i.e. (5.5), and according to (2.16) and (2.17) we have

$$\begin{aligned} H_{11}^{(1)}(x, 0, x) &= -H_{22}^{(1)}(x, 0, x) = \frac{i}{2}[p(x) - r(x)]; \\ H_{12}^{(1)}(x, 0, x) &= H_{21}^{(1)}(x, 0, x) = 0. \end{aligned} \tag{5.12}$$

On the other hand, by (5.10) we have

$$\tilde{H}(x,t,s)\equiv\begin{pmatrix}\tilde{H}_{11}, & \tilde{H}_{12}\\ \tilde{H}_{21}, & \tilde{H}_{22}\end{pmatrix}=I_1H_1(x,-t,s)I_1=\begin{pmatrix}H_{22}^{(1)}, & H_{21}^{(1)}\\ H_{12}^{(1)}, & H_{11}^{(2)}\end{pmatrix}.$$

Therefore, using (5.12) we obtain

$$\begin{gathered}\tilde{H}_{11}(x,0,x)=-\tilde{H}_{22}(x,0,x)=-\frac{i}{2}[p(x)-r(x)],\\ \tilde{H}_{12}(x,0,x)=\tilde{H}_{21}(x,0,x)=0\end{gathered}\tag{5.12'}$$

Finally, by virtue of (5.10), (5.5) and (5.8) it is not hard to obtain from (2.12) and (2.14) the following conditions for the matrix $\widetilde{H}(x,t,s)$ on the characteristics:

$$\left.\begin{aligned}\tilde{H}(x,t,x+t)H^T&=k(x,t)[r(x+t)-p(x+t)]E,\\ \tilde{H}(x,t,x-t)H&=l(x,t)[p(x-t)-r(x-t)]E',\end{aligned}\right\}\tag{5.11'}$$

which together with equation (5.11) define the Goursat problem for the matrix-kernel $\widetilde{H}(x,t,s)$.

REMARK. The formulas obtained in this section for the solution of the Cauchy problem for $t<0$ can also be obtained on the basis of the results in §1.

Summarizing all the results of this chapter, we arrive at the following theorem, which is important for the next two chapters.

THEOREM 5.1. *Let $p(x)$ and $r(x)$ be real functions, defined on the halfline $[0,\infty)$ and continuously differentiable,*[5] *and let $f(x)$ be a real vector-function which is defined and continuously differentiable on $[0,\infty)$ and satisfies the condition $f_2(0)-hf_1(0)=0$, where h is an arbitrary real number. Then the solution of the problem*

$$-iI\frac{\partial u}{\partial t}=B\frac{\partial u}{\partial x}+Q(x)u,$$

$$u(x,0)=f(x),\quad u_2(0,t)-hu_1(0,t)=0,$$

where I is the 2×2 unit matrix, and

$$B\begin{pmatrix}0 & 1\\ -1 & 0\end{pmatrix},\quad Q(x)=\begin{pmatrix}p(x) & 0\\ 0 & r(x)\end{pmatrix}$$

is given by the following formulas:

1. *For $0<t<x$*

[5] This condition will subsequently be weakened.

$$u(x, t) = k(x, t) Hf(x+t) + l(x, t) H^T f(x-t) + \frac{1}{2}\int_{x-t}^{x+t} H(x, t, s) f(s)\, ds; \tag{5.13}$$

2. *for* $t < 0,\ 0 < -t < x$

$$u(x, t) = k(x, t) Hf(x+t) + l(x, t) H^T f(x-t) + \frac{1}{2}\int_{x+t}^{x-t} \tilde{H}(x, t, s) f(s)\, ds. \tag{5.14}$$

where

$$k(x, t) = \frac{1}{2}\exp\left\{\frac{i}{2}\int_0^t \operatorname{tr} Q(x+\tau)\, d\tau\right\},$$

$$l(x, t) = \frac{1}{2}\exp\left\{\frac{i}{2}\int_0^t \operatorname{tr} Q(x-\tau)\, d\tau\right\}$$

$$H = \begin{pmatrix} 1 & i \\ -i & 1 \end{pmatrix},$$

and the 2×2 *matrices* $H(x,t,s)$ *and* $\tilde{H}(x,t,s)$ *are the solutions of the same matrix equation*

$$iW_t'(x,t,s) - W_s'(x,t,s)B + W(x,t,s)Q(s) = 0,$$

satisfying the following conditions along the characteristics:

$$H(x, t, x+t)H^T = k(x, t)\{p(x+t) - r(x+t)\}E,$$

$$H(x, t, x-t)H = l(x, t)\{r(x-t) - p(x-t)\}E',$$

$$\tilde{H}(x, t, x+t)H^T = k(x, t)\{r(x+t) - p(x+t)\}E,$$

$$\tilde{H}(x, t, x-t)H = l(x, t)\{p(x-t) - r(x-t)\}E',$$

where

$$E = \begin{pmatrix} i & 1 \\ 1 & -i \end{pmatrix}, \quad E' = \begin{pmatrix} -i & 1 \\ 1 & i \end{pmatrix};$$

3. *for* $0 < x < t$ *the solution can be constructed in the following way. We extend the functions* $p(x)$ *and* $r(x)$ *to the negative halfline, preserving their differentiability properties, but otherwise arbitrarily, and we extend the vector-function* $f(x)$ *according to the formula*

$$f(-x) = R(x)f(x) + \int_0^x K(x, s)f(s)\, ds, \tag{5.15}$$

where

$$R(x)=\begin{pmatrix} \alpha(x) & \beta(x) \\ -\beta(x) & \alpha(x)\end{pmatrix},$$

with

$$\alpha(x)=\cos\left\{\frac{1}{2}\int_0^x \mathbf{tr}\,[Q(\tau)-Q(-\tau)]\,d\tau\right\},$$

$$\beta(x)=\sin\left\{\frac{1}{2}\int_0^x \mathbf{tr}\,[Q(\tau)-Q(-\tau)]\,d\tau\right\},$$

and the 2×2 *matrix* $K(x,s)$ *is the solution of the problem*

$$K_s'(x,s)B+BK_x'(x,s)+Q(x)K(x,s)-K(x,s)Q(-s)=0,$$

$$K(x,x)B-BK(x,x)=BR'(x)+Q(x)R(x)-R(x)Q(-x),$$

$$K(x,0)Bf(0)=0.$$

Then the solution is given by

$$\begin{aligned} u(x,t)&=k(x,t)Hf(x+t)+l(x,t)H^TR(t-x)f(t-x)\\ &+\frac{1}{2}\int_0^{x+t}H(x,t,s)f(s)\,ds+\frac{1}{2}\int_0^{t-x}H_1(x,t,s)f(s)\,ds, \end{aligned} \tag{5.16}$$

where

$$H_1(x,t,s)=2l(x,t)H^TK(t-x,s)+$$

$$+H(x,t,-s)R(s)+\int_s^{t-x}H(x,t,-\tau)K(\tau,s)\,d\tau.$$

Bibliographical references

The results of this chapter are due to Sargsjan [**8, 11**].

As concerns §3, cf. the paper by Prats and Toll [**1**]; and for §4, the joint paper [**4**] by the authors.

CHAPTER 10

ASYMPTOTIC BEHAVIOUR OF THE SPECTRAL KERNEL AND ITS DERIVATIVES FOR THE CASE OF A DIRAC SYSTEM

§1. Derivation of the basic identities

We will study the problem

$$B\frac{dy}{dx}+Q(x)\,y=\lambda I y, \tag{1.1}$$

$$y_2(0)-hy_1(0)=0, \tag{1.2}$$

where I is the 2×2 unit matrix,

$$B=\begin{pmatrix}0 & 1\\ -1 & 0\end{pmatrix},\qquad Q(x)=\begin{pmatrix}p(x) & 0\\ 0 & r(x)\end{pmatrix},$$

the functions $p(x)$ and $r(x)$ satisfy the conditions of the preceding chapter, i.e. are continuously differentiable real functions defined on the halfline $[0,\infty)$, and h is an arbitrary real number.

Together with the problem (1.1) – (1.2) we will consider the Cauchy problem

$$-iI\frac{\partial u}{\partial t}=B\frac{\partial u}{\partial x}+Q(x)\,u, \tag{1.3}$$

$$u(x,\,0)=f(x), \tag{1.4}$$

where

$$f(x)=\begin{pmatrix}f_1(x)\\ f_2(x)\end{pmatrix}$$

is a real and continuously differentiable vector-function defined on $[0,\infty)$ which satisfies the condition $f_2(0)-hf_1(0)=0$.

We extend the functions $p(x)$ and $r(x)$ to the negative halfline, preserving their differentiability but otherwise arbitrarily, and extend $f(x)$ according to (5.15) of the preceding chapter, i.e.

$$f(-x)=R(x)\,f(x)+\int_0^x K(x,\,s)\,f(s)\,ds.$$

Let

$$\varphi(x,\lambda) = \begin{pmatrix} \varphi_1(x,\lambda) \\ \varphi_2(x,\lambda) \end{pmatrix}$$

denote the solution of equation (1.1) which satisfies the initial conditions

$$\varphi_1(0,\lambda) = 1, \qquad \varphi_2(0,\lambda) = h. \tag{1.5}$$

Obviously $\varphi(x,\lambda)$ satisfies the boundary condition (1.2).

Let us now put $f(x) = \varphi(x,\lambda)$ and consider the problem (1.3) + (1.4). On the one hand, the solution of this problem is given, for $0 < t < x$, $0 < -t < x$ and $0 < x < t$, by (5.13), (5.14) and (5.16) of the preceding chapter, respectively. On the other hand, solving the problem (1.3) + (1.4) by Fourier's method, it is easy to see by direct verification that the solution of this problem can be given by the formula

$$u(x,t) = \varphi(x,\lambda)e^{i\lambda t}. \tag{1.6}$$

Therefore by the uniqueness of the solution of the problem (1.3) + (1.4), equating the right sides of (5.13), (5.14) and (5.16) of the preceding chapter with the right side of (1.6), respectively, we arrive at the following identities:

1. for $0 < t < x$

$$\begin{aligned} \varphi(x, \lambda) e^{i\lambda t} &= k(x, t) H\varphi(x+t, \lambda) + l(x, \lambda) H^T\varphi(x-t, \lambda) \\ &\quad + \frac{1}{2}\int_{x-t}^{x+t} H(x, t, s)\varphi(s, \lambda)\,ds; \end{aligned} \tag{1.7}$$

2. for $t < 0$, $0 < -t < x$

$$\begin{aligned} \varphi(x, \lambda) e^{i\lambda t} &= k(x, t) H\varphi(x+t, \lambda) + l(x, t) H^T\varphi(x-t, \lambda) \\ &\quad + \frac{1}{2}\int_{x+t}^{x-t} \tilde{H}(x, t, s)\varphi(s, \lambda)\,ds; \end{aligned} \tag{1.8}$$

3. for $0 < x < t$

$$\begin{aligned} \varphi(x, \lambda) e^{i\lambda t} &= k(x, t) H\varphi(x+t, \lambda) \\ &\quad + l(x, t) H^T R(t-x)\varphi(t-x, \lambda) + \frac{1}{2}\int_0^{x+t} H(x, t, s)\varphi(s, \lambda)\,ds \\ &\quad + \frac{1}{2}\int_0^{t-x} H_1(x, t, s)\varphi(s, \lambda)\,ds. \end{aligned} \tag{1.9}$$

Let ϵ denote an arbitrary positive number and let $g_\epsilon(t)$ be a function satisfying the following conditions:

1. $g_\epsilon(t)$ vanishes outside the interval $(-\epsilon, \epsilon)$;
2. $g_\epsilon(t)$ has a piecewise continuous first derivative.

Further, let $\psi_\epsilon(\lambda)$ denote the Fourier transform of $g_\epsilon(t)$, i.e. we put

$$\psi_\varepsilon(\lambda) = \int_{-\varepsilon}^{\varepsilon} g_\varepsilon(t) e^{i\lambda t} dt. \tag{1.10}$$

We denote by x an arbitrary point of the halfline $[0, \infty)$.

Let us assume that $x \geqq \epsilon$.[1] Multiplying both sides of the identity (1.7) by $g_\epsilon(t)$ and integrating with respect to t from 0 to ϵ, we have

$$\begin{aligned} \varphi(x, \lambda) \int_0^\varepsilon g_\varepsilon(t) e^{i\lambda t} dt &= H \int_0^\varepsilon k(x, t) \varphi(x+t, \lambda) g_\varepsilon(t)\, dt \\ &+ H^T \int_0^\varepsilon l(x, t) \varphi(x-t, \lambda) g_\varepsilon(t)\, dt \\ &+ \frac{1}{2} \int_0^\varepsilon \left\{ \int_{x-t}^{x+t} H(x, t, s) \varphi(s, \lambda)\, ds \right\} g_\varepsilon(t)\, dt. \end{aligned} \tag{1.11}$$

Changing variables in the first two integrals on the right and changing the order of integrating in the third integral, we can write the identity (1.11) in the form

$$\begin{aligned} \varphi(x, \lambda) \int_0^\varepsilon g_\varepsilon(t) e^{i\lambda t} dt &= H \int_x^{x+\varepsilon} k(x, s-x) g_\varepsilon(s-x) \varphi(s, \lambda)\, ds \\ &+ H^T \int_{x-\varepsilon}^{x} l(x, x-s) g_\varepsilon(x-s) \varphi(s, \lambda)\, ds \\ &+ \frac{1}{2} \int_{x-\varepsilon}^{x+\varepsilon} \left\{ \int_{|x-s|}^{\varepsilon} H(x, t, s) g_\varepsilon(t)\, dt \right\} \varphi(s, \lambda)\, ds. \end{aligned} \tag{1.11'}$$

Multiplying both sides of the identity (1.8) by $g_\epsilon(t)$ and integrating with respect to t from $-\epsilon$ to 0, and then proceeding as we did in order to obtain the identity (1.11′), we obtain

$$\varphi(x, \lambda) \int_{-\varepsilon}^{0} g_\varepsilon(t) e^{i\lambda t} dt = H \int_{x-\varepsilon}^{x} k(x, s-x) g_\varepsilon(s-x) \varphi(s, \lambda)\, ds\ .$$

[1] Since the study of the case $x < \epsilon$ is analogous, we will not consider it.

$$+H^T\int_x^{x+\varepsilon} l(x,\ x-s)\, g_\varepsilon(x-s)\,\varphi(s,\ \lambda)\,ds \tag{1.12}$$

$$+\frac{1}{2}\int_{x-\varepsilon}^{x+\varepsilon}\left\{\int_{-\varepsilon}^{-|x-s|}\tilde{H}(x,\ t,\ s)\, g_\varepsilon(t)\,dt\right\}\varphi(s,\ \lambda)\,ds.$$

Finally, adding the identities (1.11′) and (1.12) and taking account of (1.10), we obtain the identity

$$\varphi(x,\ \lambda)\,\psi_\varepsilon(\lambda)=\int_{x-\varepsilon}^{x+\varepsilon} W(x,\ s;\ \varepsilon)\,\varphi(s,\ \lambda)\,ds, \tag{1.13}$$

where

$$W(x,s;\epsilon) = Hk(x,s-x)g_\epsilon(s-x) + H^T l(x,x-s)g_\epsilon(x-s)$$

$$+\frac{1}{2}\int_{|x-s|}^{\varepsilon} H(x,\ t,\ s)\, g_\varepsilon(t)\,dt+\frac{1}{2}\int_{-\varepsilon}^{-|x-s|}\tilde{H}(x,\ t,\ s)\, g_\varepsilon(t)\,dt, \tag{1.14}$$

which will be important in what follows.

Further, since by condition 1 on the function $g_\epsilon(t)$ it follows from (1.14) that $W(x,x+\epsilon;\epsilon) = W(x,x-\epsilon;\epsilon) = 0$, differentiating the identity (1.15) with respect to x, we obtain a second identity which will be important in what follows:

$$\varphi'_x(x,\ \lambda)\,\psi_\varepsilon(\lambda)=\int_{x-\varepsilon}^{x+\varepsilon}\frac{\partial W(x,\ s;\ \varepsilon)}{\partial x}\,\varphi(s,\ \lambda)\,ds. \tag{1.15}$$

§2. Preliminary estimation of the spectral kernel and its derivatives

Let

$$f(x) = \begin{pmatrix} f_1(x) \\ f_2(x) \end{pmatrix}$$

be an arbitrary vector-function from $\mathscr{L}^2(0,\infty)$, and let $\varphi(x,\lambda)$, as before, be the solution of the problem (1.1) + (1.5).

As we have already proved in Chapter 3, for a given h there exists a nondecreasing function[2] $\rho(\lambda)$, $-\infty < \lambda < \infty$, which is bounded in every finite interval and continuous from the left, which defines an

[2] It will be proven in §7, Chapter 13, that if the coefficients $p(x)$ and $r(x)$ in equation (1.1) are continuous, then $\rho(\lambda)$ is unique.

isometric mapping of the space $\mathcal{L}^2(0, \infty)$ of vector-functions onto the space $\mathcal{L}^2_{\{\rho(\lambda)\}}(-\infty, \infty)$ according to

$$F(\lambda) = \int_0^\infty f^T(x)\,\varphi(x, \lambda)\,dx, \tag{2.1}$$

$$f(x) = \int_{-\infty}^\infty F(\lambda)\,\varphi(x, \lambda)\,d\rho(\lambda), \tag{2.2}$$

where the integrals (2.1) and (2.2) converge in the metrics of the spaces $\mathcal{L}^2_{\{\rho(\lambda)\}}(-\infty, \infty)$ and $\mathcal{L}^2(0, \infty)$ respectively, and one has the Parseval equality

$$\int_0^\infty |f(x)|^2\,dx = \int_{-\infty}^\infty F^2(\lambda)\,d\rho(\lambda). \tag{2.3}$$

Further, if $f(x), g(x) \in \mathcal{L}^2(0, \infty)$, and $F(\lambda)$, $G(\lambda)$ are their Fourier transforms, then as was proved in Chapter 3, one has the following generalized Parseval equality:

$$\int_0^\infty f^T(x)\,g(x)\,dx = \int_{-\infty}^\infty F(\lambda)\,G(\lambda)\,d\rho(\lambda). \tag{2.4}$$

We introduce the following terminology:

1. *The function* $\rho(\lambda)$ *will be called the spectral function of the problem* (1.1) + (1.2).

2. *The* 2×2 *matrix*

$$\theta(x, s; \lambda) = \int_0^\lambda \varphi(x, \mu)\,\varphi^T(s, \mu)\,d\rho(\mu) \tag{2.5}$$

will be called the spectral kernel of the problem (1.1) + (1.2).

In this section we will obtain various estimates for the spectral kernel and its derivatives which we will need further on.

In the preceding section we obtained the identity (1.13),

$$\varphi(x, \lambda)\,\psi_\varepsilon(\lambda) = \int_{x-\varepsilon}^{x+\varepsilon} W(x, s; \varepsilon)\,\varphi(s, \lambda)\,ds, \tag{2.6}$$

where the 2×2 matrix

$$W(x, s; \varepsilon) = \begin{pmatrix} W_{11}(x, s; \varepsilon) & W_{12}(x, s; \varepsilon) \\ W_{21}(x, s; \varepsilon) & W_{22}(x, s; \varepsilon) \end{pmatrix}$$

is defined by (1.14).

The identity (2.6) says that for any $x \geqq 0$ the functions $\varphi_i(x,\lambda)\psi_\epsilon(\lambda)$ $(i=1,2)$ (the components of the vector-function $\varphi(x,\lambda)\psi_\epsilon(\lambda)$) are the Fourier transforms of the vector-functions which are equal to

$$\begin{pmatrix} W_{i1}(x,s;\epsilon) \\ W_{i2}(x,s;\epsilon) \end{pmatrix}$$

for $|x-s| \leqq \epsilon$ and equal to zero for $|x-s| > \epsilon$. Therefore if

$$f(x) = \begin{pmatrix} f_1(x) \\ f_2(x) \end{pmatrix} \in \mathscr{L}^2(0,\infty),$$

then by the generalized Parseval equality (2.4) we have

$$\int_{-\infty}^{\infty} \varphi_i(x,\lambda)\psi_\varepsilon(\lambda)F(\lambda)\,d\rho(\lambda) = \int_0^\infty \{W_{i1}(x,s;\varepsilon)f_1(s) + W_{i2}(x,s;\varepsilon)f_2(s)\}\,ds \quad (i=1,2), \tag{2.7}$$

or in matrix form

$$\int_{-\infty}^{\infty} \varphi(x,\lambda)F(\lambda)\psi_\varepsilon(\lambda)\,d\rho(\lambda) = \int_0^\infty W(x,s;\varepsilon)f(s)\,ds. \tag{2.7'}$$

Remark. The identity (2.7′), which was derived under the assumption that $p(x)$ and $r(x)$ are continuously differentiable functions, is valid for summable $p(x)$ and $r(x)$. This is easily obtained by a passage to the limit.

Lemma 2.1. *If the elements $p(x)$ and $r(x)$ of the matrix $Q(x)$ of the system (1.1) are summable over every finite interval and if (x_0,x_1) is an arbitrary finite interval of the real line, then there exists a constant $C = C(x_0,x_1)$ such that for any x and s from the interval (x_0,x_1) and for all a we have the bounds*

$$\bigvee_a^{a+1}\{\theta_{ik}(x,s;\lambda)\} = \int_a^{a+1} |\varphi_i(x,\lambda)||\varphi_k(s,\lambda)|\,d\rho(\lambda) < C \quad (i,k=1,2). \tag{2.8}$$

Proof. Let the function $g_\epsilon(t)$ be defined by

$$g_\varepsilon(t) = \begin{cases} \frac{1}{\varepsilon}(\varepsilon - |t|), & |t| \leqslant \varepsilon, \\ 0, & |t| > \varepsilon. \end{cases}$$

Then

$$\psi_\varepsilon(\lambda) = \int_{-\varepsilon}^{\varepsilon} g_\varepsilon(t) e^{i\lambda t} dt = \frac{\varepsilon}{2} \left(\frac{\sin \frac{\varepsilon\lambda}{2}}{\frac{\varepsilon\lambda}{2}} \right)^2 .$$

Let a be an arbitrary positive number, and put $g_\epsilon(t, a) = g_\epsilon(t) e^{-iat}$. Then

$$\psi_\varepsilon(\lambda, a) = \int_{-\varepsilon}^{\varepsilon} g_\varepsilon(t, a) e^{i\lambda t} dt = \frac{\varepsilon}{2} \left(\frac{\sin \frac{\varepsilon(\lambda - a)}{2}}{\frac{\varepsilon(\lambda - a)}{2}} \right)^2 . \tag{2.9}$$

Therefore from (2.6) and (2.9) we have, by the Parseval equality,

$$\frac{\varepsilon^2}{4} \int_{-\infty}^{\infty} \left\{ \frac{\sin [\varepsilon(\lambda - a)]/2}{\varepsilon(\lambda - a)/2} \right\}^4 |\varphi(x, \lambda)|^2 d\rho(\lambda) = \int_0^{\infty} \| W(x, s; \varepsilon, a) \|^2 ds, \tag{2.10}$$

where $\| W(x, s; \epsilon, a) \|^2$ denotes the sum of the squares of the elements of the matrix $W(x, s; \epsilon, a)$, where the latter is defined by (1.14), in which the function $g_\epsilon(t)$ is replaced by $g_\epsilon(t, a) = g_\epsilon(t) e^{-iat}$. But since $g_\epsilon(t)$ has compact support, so does the matrix $W(x, s; \epsilon, a)$, and since the elements $p(x)$ and $r(x)$ of the matrix $Q(x)$ are summable, so is the function $\| W(x, s; \epsilon, a) \|^2$. Therefore the integral in the right side of (2.10) exists. Then from (2.10) for $\epsilon = 2$ there follows

$$\int_{-\infty}^{\infty} \left\{ \frac{\sin(\lambda - a)}{\lambda - a} \right\}^4 |\varphi(x, \lambda)|^2 d\rho(\lambda) < C,$$

and so a fortiori

$$\int_a^{a+1} \frac{\sin^4(\lambda - a)}{(\lambda - a)^4} |\varphi(x, \lambda)|^2 d\rho(\lambda) < C. \tag{2.11}$$

Further, since for $0 \leq x \leq 1$ one has $\sin x / x > 2/\pi$, the bound (2.11) implies the bounds

$$\int_a^{a+1} \varphi_i^2(x, \lambda) d\rho(\lambda) < \frac{\pi^4}{16} C = C_1 \qquad (i = 1, 2), \tag{2.12}$$

which prove the lemma for the functions $\theta_{ii}(x, s; \lambda)$ $(i = 1, 2)$ for $s = x$. The bounds (2.8) for these functions for $s \neq x$ follow from (2.12) by the Cauchy-Bunjakovskiĭ inequality. The completion of the proof for the functions $\theta_{ij}(x, s; \lambda)$ for $i \neq j$ is similar. This proves the lemma.

LEMMA 2.2. *Under the conditions of Lemma* 2.1 *we have, for all* a,

$$\rho(a+1)-\rho(a)<C. \tag{2.8'}$$

The proof of this lemma follows from the bounds (2.12) if we put $i=1$, $x=0$, and take into account that $\varphi_1(0,\lambda)=1$.

We now take up the estimation of the derivatives of the spectral kernel $\theta(x,s;\lambda)$. To this end let us consider the formula (1.15);

$$\varphi_x'(x,\lambda)\psi_\varepsilon(\lambda)=\int_{x-\varepsilon}^{x+\varepsilon}\frac{\partial W(x,s;\varepsilon)}{\partial x}\varphi(s,\lambda)\,ds. \tag{2.13}$$

It follows from this relation that for fixed x the functions $\varphi_i'(x,\lambda)\psi_\epsilon(\lambda)$ $(i=1,2)$ can be regarded as the generalized Fourier transformations of the vector-functions which equal

$$\begin{pmatrix}\dfrac{\partial W_{i1}}{\partial x}(x,s;\varepsilon),\\ \dfrac{\partial W_{i2}}{\partial x}(x,s;\varepsilon)\end{pmatrix}\qquad(i=1,2)$$

for $s\in(x-\epsilon,x+\epsilon)$ and equal zero outside this interval. Therefore by the Parseval equality we have

$$\int_{-\infty}^{\infty}|\psi_\varepsilon(\lambda)|^2|\varphi_x'(x,\lambda)|^2\,d\rho(\lambda)=\int_{x-\varepsilon}^{x+\varepsilon}\left\|\frac{\partial W(x,s;\varepsilon)}{\partial x}\right\|^2ds. \tag{2.14}$$

Let $g_\epsilon(t)$ be defined in the same way as in the proof of Lemma 2.1. Replacing in (2.14) $g_\epsilon(t)$ by $g_\epsilon(t,a)=g_\epsilon(t)e^{-iat}$ and taking account of (2.9), we can rewrite (2.14) in the form

$$\frac{\varepsilon^2}{4}\int_{-\infty}^{\infty}\left\{\frac{\sin\dfrac{\varepsilon(\lambda-a)}{2}}{\dfrac{\varepsilon(\lambda-a)}{2}}\right\}^4|\varphi_x'(x,\lambda)|^2\,d\rho(\lambda)=\int_{x-\varepsilon}^{x+\epsilon}\left\|\frac{\partial W(x,s;\varepsilon)}{\partial x}\right\|^2ds, \tag{2.14'}$$

where the matrix $W(x,s;\epsilon,a)$ is as in the proof of Lemma 2.1. We now estimate the integral in the right side of (2.14) for large a. Since the matrix $W(x,s;\epsilon,a)$ is defined by (1.14) with $g_\epsilon(t)$ replaced by $g_\epsilon(t)e^{-iat}$, we have

$$\begin{aligned}W(x,s;\varepsilon,a)&=Hk(x,s-x)g_\varepsilon(s-x)e^{-ia(s-x)}\\&\quad+H^Tl(x,x-s)g_\varepsilon(x-s)e^{-ia(x-s)}\\&\quad+\frac{1}{2}\int_{|x-s|}^{\varepsilon}H(x,t,s)g_\varepsilon(t)e^{-iat}dt+\frac{1}{2}\int^{-|x-s|}\tilde{H}(x,t,s)g_\varepsilon(t)e^{-iat}dt.\end{aligned}$$

Since $g_\epsilon(t)$ has compact support, so do the matrices $W(x,s;\epsilon,a)$ and $\partial W(x,s;\epsilon,a)/\partial x$. On the other hand, by direct differentiation of the matrix $W(x,s;\epsilon,a)$ with respect to x it is easily seen that for large a its elements are of order $O(a)$. But if the elements $p(x)$ and $r(x)$ of the matrix $Q(x)$ have summable first derivatives, then for $a\to\infty$

$$\int_{x-\varepsilon}^{x+\varepsilon}\left\|\frac{\partial}{\partial x}W(x,s;\varepsilon,a)\right\|^2 ds=O(a^2). \tag{2.15}$$

Lemma 2.3. *If the elements $p(x)$ and $r(x)$ of the matrix $Q(x)$ of the system* (1.1) *are absolutely continuous in every finite interval, then for $a\to\infty$ we have*

$$\bigvee_a^{a+1}\left\{\frac{\partial^2\theta_{ik}(x,s;\lambda)}{\partial x\,\partial s}\right\}=\int_a^{a+1}|\varphi_i'(x,\lambda)||\varphi_k'(s,\lambda)|\,d\rho(\lambda)=O(a^2) \quad (i,k=1,2), \tag{2.16}$$

where $\theta_{ik}(x,s;\lambda)$ $(i,k=1,2)$ are the elements of the matrix $\theta(x,s;\lambda)$ (cf. (2.5)).

The estimates (2.16) *hold uniformly in every finite interval.*

Proof. By virtue of the estimate (2.15) we can obtain from (2.14′), similarly to what was done in the proof of Lemma 2.1,

$$\int_a^{a+1}|\varphi_i'(x,\lambda)|^2\,d\rho(\lambda)=O(a^2) \quad (i=1,2), \tag{2.17}$$

which proves the lemma for $i=k$. For the other cases the lemma follows from (2.17) by the Cauchy-Bunjakovskiĭ inequality.

From Lemmas 2.1 and 2.3 and the Cauchy-Bunjakovskiĭ inequality follows:

Lemma 2.4. *When the conditions of Lemma* 2.3 *are satisfied, we have for $a\to\infty$*

$$\bigvee_a^{a+1}\left\{\frac{\partial\theta_{ik}(x,s;\lambda)}{\partial x}\right\}=\int_a^{a+1}|\varphi_i'(x,\lambda)||\varphi_k(s,\lambda)|\,d\rho(\lambda)=O(a) \quad (i,k=1,2). \tag{2.18}$$

These estimates hold uniformly in every finite interval.

One can similarly prove the following lemma.

Lemma 2.5. *If the elements $p(x)$ and $r(x)$ of the matrix $Q(x)$ have absolutely continuous derivatives of order $l-1$ in every finite interval, then for $a \to \infty$*

$$\bigvee_a^{a+1} \left\{ \frac{\partial^l \theta_{ik}(x, s; \lambda)}{\partial x^m \partial s^n} \right\} = \int_a^{a+1} \left| \frac{\partial^m \varphi_i(x, \lambda)}{\partial x^m} \right| \left| \frac{\partial^n \varphi_k(s, \lambda)}{\partial s^n} \right| d\rho(\lambda) = O(a^l), \tag{2.19}$$

where $m + n = l$, $i, k = 1, 2$. These estimates hold uniformly as x and s vary over any finite intervals.

Let us now consider the problem (1.1) + (1.2) under the conditions $p(x) = r(x) \equiv 0$, i.e. the problem

$$B \frac{dy}{dx} = \lambda I y \quad (0 \leqslant x < \infty), \tag{2.20}$$

$$y_2(0) - h y_1(0) = 0. \tag{2.21}$$

Let

$$\psi(x, \lambda) = \begin{pmatrix} \psi_1(x, \lambda) \\ \psi_2(x, \lambda) \end{pmatrix}$$

denote the solution of equation (2.20) which satisfies the initial conditions

$$\psi_1(0, \lambda) = 1, \qquad \psi_2(0, \lambda) = h. \tag{2.22}$$

As is easily seen, the problem (2.20) + (2.22) breaks up into two independent problems for the functions $\psi_1(x, \lambda)$ and $\psi_2(x, \lambda)$, whose solutions are given by

$$\psi_1(x, \lambda) = \cos \lambda x - h \sin \lambda x, \qquad \psi_2(x, \lambda) = \sin \lambda x + h \cos \lambda x. \tag{2.23}$$

We now determine the spectral function and spectral kernel of the problem (2.20) + (2.21).

Let us adjoin to the boundary condition (2.21) the condition

$$y_2(b) - H y_1(b) = 0, \tag{2.24}$$

where b is an arbitrary positive number, and H is an arbitrary real number, and determine the eigenvalues of the boundary-value problem (2.20) + (2.21) + (2.24). Since $\psi(x, \lambda)$ satisfies the boundary condition (2.21), to determine the eigenvalues we have to insert the expressions for the functions $\psi_1(x, \lambda)$ and $\psi_2(x, \lambda)$ from (2.23) in the boundary condition (2.24). We have

$$\sin \lambda b + h \cos \lambda b - H \cos \lambda b + hH \sin \lambda b = 0,$$

i.e.

(2.25) $$(1+hH)\sin\lambda b+(h-H)\cos\lambda b=0.$$

If we put

$$\cos\omega=\frac{1+hH}{\sqrt{(1+h^2)(1+H^2)}},$$

then equation (2.25) can be rewritten in the form $\sin(\lambda b+\omega)=0$, from which $\lambda_{n,b}b+\omega=n\pi$ $(n=0,\pm 1,\pm 2,\cdots)$. Consequently, we obtain for the eigenvalues the expression

(2.26) $$\lambda_{n,b}=n\cdot\pi/b+\omega/b \qquad (n=0,\pm 1,\pm 2,\cdots).$$

Further, since

$$\alpha_{n,b}^2=\int_0^b\{\psi_1^2(x,\lambda_{n,b})+\psi_2^2(x,\lambda_{n,b})\}\,dx=(1+h^2)\,b,$$

we have by the definition of the spectral function $\rho_b(\lambda)$ (cf. Chapter 3, §1)

$$\rho_b^*(\lambda)=\sum_{\lambda_{n,b}\leqslant\lambda}\frac{1}{(1+h^2)\,b},\quad \lambda\geqslant 0,$$

$$\rho_b^*(\lambda)=-\sum_{\lambda_{n,b}>\lambda}\frac{1}{(1+h^2)\,b},\quad \lambda<0.$$

Let us transform $\rho^*(\lambda)$ for $\lambda\geqq 0$ (for $\lambda<0$ the transformation is analogous). Using (2.26), we obtain

$$\begin{aligned}\rho_b^*(\lambda)&=\frac{1}{1+h^2}\sum_{\lambda_{n,b}\leqslant\lambda}\frac{1}{b}=\frac{1}{\pi(1+h^2)}\sum_{\lambda_{n,b}\leqslant\lambda}\frac{\pi}{b}\\&=\frac{1}{\pi(1+h^2)}\sum_{\lambda_{n,b}\leqslant\lambda}(\lambda_{n+1,b}-\lambda_{n,b})\\&=\frac{1}{\pi(1+h^2)}\sum_{\lambda_{n,b}\leqslant\lambda}\Delta\lambda_{n,b}\to\frac{\lambda}{\pi(1+h^2)},\quad(b\to\infty).\end{aligned}$$

For $\lambda<0$ the result is the same.

Thus the spectral function $\rho^*(\lambda)$ of the boundary-value problem (2.20) + (2.21) has the form

$$\rho^*(\lambda)=\lambda/\pi(1+h^2),$$

and the spectral kernel has the form

(2.27) $$\theta^*(x,s;\lambda)=\frac{1}{\pi(1+h^2)}\int_0^\lambda\psi(x,\mu)\,\psi^T(s,\mu)\,d\mu.$$

From the definition of the matrix $\theta^*(x,s;\lambda)$ there immediately follows:

LEMMA 2.6. *For $a \to \infty$ we have, uniformly for all positive values of x and s,*

$$\bigvee_a^{a+1} \{\theta^*_{ik}(x, s; \lambda)\} < C \quad (i, k = 1, 2), \tag{2.28}$$

$$\bigvee_a^{a+1} \left\{\frac{\partial^l \theta^*_{ik}(x, s; \lambda)}{\partial x^m \partial s^n}\right\} = O(a^l) \quad (l = m+n;\ i, k = 1, 2), \tag{2.29}$$

where $\theta^*_{ik}(x,s;\lambda)$ are the elements of the matrix $\theta^*(x,s;\lambda)$.

§3. Asymptotic behaviour of the spectral kernel and spectral function

Let

$$f(x) = \begin{pmatrix} f_1(x) \\ f_2(x) \end{pmatrix}$$

be an arbitrary vector-function from $\mathscr{L}^2(0, \infty)$ which vanishes outside some finite interval. Then its Fourier transform

$$F(\lambda) = \int_0^\infty f^T(s)\,\varphi(s, \lambda)\,ds \tag{3.1}$$

exists in the ordinary sense.

Now, inserting the expression for $F(\lambda)$ from (3.1) in (2.7), and then changing the order of integration in the left side, which is permissible by virtue of the bound (2.8′) and Fubini's theorem, we obtain

$$\int_0^\infty \left\{\int_{-\infty}^\infty \varphi(x, \lambda)\,\varphi^T(s, \lambda)\,\psi_\varepsilon(\lambda)\,d\rho(\lambda)\right\} f(s)\,ds = \int_0^\infty W(x, s; \varepsilon)\,f(s)\,ds.$$

Due to the arbitrariness of the vector-function $f(x)$, the last relation implies the identity

$$\int_{-\infty}^\infty \varphi(x, \lambda)\,\varphi^T(s, \lambda)\,\psi_\varepsilon(\lambda)\,d\rho(\lambda) = W(x, s; \varepsilon), \tag{3.2}$$

where $W(x,s;\epsilon)$ is given by (1.14).

Now, putting

$$W^*(x, s; \varepsilon) = \int_{|x-s|}^{\varepsilon} H(x, t, s)\,g_\varepsilon(t)\,dt + \int_{-\varepsilon}^{-|x-s|} \tilde{H}(x, t, s)\,g_\varepsilon(t)\,dt \tag{3.3}$$

and taking account of (1.14) and the definition of the spectral kernel $\theta(x,s;\lambda)$, the identity (3.2) can be rewritten in the form

$$\int_{-\infty}^{\infty} \psi_\varepsilon(\lambda)\, d_\lambda \theta(x, s; \lambda) = \tag{3.2'}$$
$$= \begin{cases} Hk(x, s-x)\, g_\varepsilon(s-x) + H^T l(x, x-s)\, g_\varepsilon(x-s) + \\ \qquad\qquad + W^*(x, s; \varepsilon), & |x-s| \leqslant \varepsilon, \\ 0, & |x-s| \geqslant \varepsilon. \end{cases}$$

We will now write down an identity analogous to (3.2′) for the spectral matrix $\theta^*(x,s;\lambda)$. For $p(x) = r(x) \equiv 0$ it follows from the equalities (4.4′) of the preceding chapter that $k(x,t) = l(x,t) \equiv \frac{1}{2}$. On the other hand, as easily follows from the problems (2.15) + (2.12) + (2.14) and (5.11) + (5.11′) of the preceding chapter, for $p(x) = r(x) \equiv 0$ the matrices $H(x,t,s)$ and $\widetilde{H}(x,t,s)$ are the zero matrix, and consequently in this case $W^*(x,s,\epsilon)$ is also the zero matrix. Therefore for the case being considered the identity (3.2′) assumes the form

$$\int_{-\infty}^{\infty} \psi_\varepsilon(\lambda)\, d_\lambda \theta^*(x, s; \lambda) \tag{3.4}$$
$$= \begin{cases} \frac{1}{2} H g_\varepsilon(s-x) + \frac{1}{2} H^T g_\varepsilon(x-s), & |x-s| \leqslant \varepsilon, \\ 0, & |x-s| \geqslant \varepsilon. \end{cases}$$

Now subtracting (3.4) from (3.2′), we obtain

$$\int_{-\infty}^{\infty} \psi_\varepsilon(\lambda)\, d_\lambda \{\theta(x, s; \lambda) - \theta^*(x, s; \lambda)\} \tag{3.5}$$
$$= \begin{cases} H\left\{k(x, s-x) - \frac{1}{2}\right\} g_\varepsilon(s-x) + H^T\left\{l(x, x-s) - \frac{1}{2}\right\} g_\varepsilon(x-s) \\ \qquad\qquad + W^*(x, s; \varepsilon), & |x-s| \leqslant \varepsilon \\ 0, & |x-s| \geqslant \varepsilon. \end{cases}$$

We now take up the transformation of the right side of the identity (3.5). To this end we put

$$\begin{aligned} \alpha^{(1)}(x, s; \lambda) &= \int_{|x-s|}^{1} H(x, t, s)\, e^{i\lambda t}\, dt, \\ \alpha^{(2)}(x, s; \lambda) &= \int_{-1}^{-|x-s|} H(x, t, s)\, e^{i\lambda t}\, dt, \\ \alpha(x, s; \lambda) &= \alpha^{(1)}(x, s; \lambda) + \alpha^{(2)}(x, s; \lambda). \end{aligned} \tag{3.6}$$

On the other hand (cf. (1.10)),

$$\psi_\varepsilon(\lambda) = \int_{-\varepsilon}^{\varepsilon} g_\varepsilon(t)\, e^{i\lambda t} dt. \tag{3.7}$$

Then by the Parseval equality for the classical Fourier transform as well as (3.3), it follows from the last two equalities that ($\epsilon \leqq 1$)

$$\frac{1}{2\pi} \int_{-\infty}^{\infty} \psi_\varepsilon(\lambda)\, \overline{\alpha(x, s; \lambda)}\, d\lambda = W^*(x, s; \varepsilon). \tag{3.8}$$

Further, inverting (3.7), we obtain

$$g_\varepsilon(t) = \frac{1}{2\pi} \int_{-\infty}^{\infty} \psi_\varepsilon(\lambda)\, e^{-i\lambda t} d\lambda. \tag{3.9}$$

On the basis of (3.8) and (3.9) the identity (3.5) assumes the final form

$$\int_{-\infty}^{\infty} \psi_\varepsilon(\lambda)\, d_\lambda \Phi(x, s; \lambda) = 0, \tag{3.10}$$

where we have put

$$\begin{aligned} \Phi(x, s; \lambda) = \theta(x, s; \lambda) - \theta^*(x, s; \lambda) - \frac{1}{\pi} \int_0^\lambda \overline{\alpha(x, s; \nu)}\, d\nu \\ + \frac{1}{2\pi} H \left\{ k(x, s - x) - \frac{1}{2} \right\} \frac{e^{-i\lambda(s-x)} - 1}{i(s-x)} \\ + \frac{1}{2\pi} H^T \left\{ l(x, x - s) - \frac{1}{2} \right\} \frac{e^{-i\lambda(x-s)} - 1}{i(x-s)}. \end{aligned} \tag{3.11}$$

The identity (3.10) and Marčenko's Tauberian theorem for Fourier integrals (cf. Chapter 14) enable us to study the behaviour of the spectral kernel for $|\lambda| \to \infty$. We will first prove the following lemma.

LEMMA 3.1. *If the coefficients $p(x)$ and $r(x)$ of the system* (1.1) *are summable over every finite interval, then for any fixed x, s we have for all a*

$$\bigvee_a^{a+1} \{\Phi_{ik}(x, s; \lambda)\} < C \quad (i, k = 1, 2), \tag{3.12}$$

where $\Phi_{ik}(x,s;\lambda)$ are the elements of the matrix $\Phi(x,s;\lambda)$ defined by (3.11). *The bounds* (3.12) *hold uniformly as x and s vary over any finite intervals.*

PROOF. To prove the lemma it is sufficient to obtain the bound (3.12) for every element of the matrix on the right side of (3.11). For the elements of the matrices $\theta(x,s;\lambda)$ and $\theta^*(x,s;\lambda)$ the bound (3.12) coincides with the bounds (2.8) and (2.28). Further, from the definition of the matrix $\alpha(x,s;\lambda)$, i.e. from (3.6), it follows from the summability of the elements of the matrices $H(x,t,s)$ and $\check{H}(x,t,s)$ and the well-known Riemann-Lebesgue lemma that

$$\lim_{\lambda\to\infty} \alpha_{ik}(x,s;\lambda) = 0 \qquad (i,k=1,2),$$

uniformly for x and s in any bounded regions, where $\alpha_{ik}(x,s;\lambda)$ are the elements of the matrix $\alpha(x,s;\lambda)$. Therefore for $a\to\infty$

$$\bigvee_a^{a+1}\left\{\int_0^\lambda \overline{\alpha_{ik}(x,s;\nu)\,d\nu}\right\} = \int_a^{a+1} |\alpha_{ik}(x,s;\nu)|\,d\nu = o(1).$$

Finally, for the elements of the matrices

$$H\left\{k(x,s-x)-\frac{1}{2}\right\}\frac{e^{-i\lambda(s-x)}-1}{2\pi i(s-x)}, \qquad H^T\left\{l(x,x-s)-\frac{1}{2}\right\}\frac{e^{-i\lambda(x-s)}-1}{2\pi i(x-s)}$$

the bounds (3.12) can be obtained at once from the expressions for these matrices, if we write them in the form

$$\frac{1}{2\pi}H\left\{k(x,s-x)-\frac{1}{2}\right\}\frac{e^{-i\lambda(s-x)}-1}{i(s-x)}$$

$$= H\left\{k(x,s-x)-\frac{1}{2}\right\}e^{i\lambda\frac{x-s}{2}}\frac{\sin\lambda\frac{x-s}{2}}{\pi(x-s)},$$

(3.13)

$$\frac{1}{2\pi}H^T\left\{l(x,x-s)-\frac{1}{2}\right\}\frac{e^{-i\lambda(x-s)}-1}{i(x-s)} =$$

$$= -H^T\{l(x,x-s)\}e^{i\lambda\frac{s-x}{2}}\frac{\sin\lambda\frac{s-x}{2}}{\pi(s-x)}.$$

This proves the lemma.

Lemma 3.2. *Suppose that $p(x)$ and $r(x)$ satisfy a Dini condition uniformly in every finite interval. Then for any fixed x and s we have*

$$\lim_{\lambda\to\infty}\int_{-\lambda}^{\lambda}\overline{\alpha(x,s;\nu)\,d\nu} = 0. \tag{3.14}$$

This relation holds uniformly as x and s vary over any finite intervals.

Proof. By the definition of the matrix $\alpha(x,s;\nu)$, i.e. (3.6), we have

$$\int_{-\lambda}^{\lambda}\alpha(x,s;\nu)\,d\nu = \int_{-\lambda}^{\lambda}\left\{\int_{|x-s|}^{1}H(x,t,s)e^{i\nu t}dt + \int_{-1}^{-|x-s|}\tilde{H}(x,t,s)e^{i\nu t}dt\right\}d\nu$$

$$= 2\int_{|x-s|}^{1}\{H(x,t,s)+\tilde{H}(x,-t,s)\}\frac{\sin\lambda t}{t}\,dt. \tag{3.15}$$

From the form of the matrices $H(x,t,s)$ and $\widetilde{H}(x,t,s)$ it follows that $H(x,0,s)+\widetilde{H}(x,0,s)=0$. Therefore (3.15) can be written in the form

$$\int_{-\lambda}^{\lambda} \alpha(x, s; \nu)\, d\nu = \int_{|x-s|}^{1} \{F(x, t, s) - F(x, 0, s)\} \frac{\sin \lambda t}{t}\, dt, \tag{3.16}$$

where $F(x,t,s)=2[H(x,t,s)+H(x,-t,s)]$. From the conditions of the lemma and the form of the matrices $H(x,t,s)$ and $\widetilde{H}(x,t,s)$ it follows that $F(x,t,s)$, as a function of t, satisfies a Dini condition at the point $t=0$. Therefore

$$\lim_{\lambda\to\infty} \int_{|x-s|}^{1} \{F(x, t, s) - F(x, 0, s)\} \frac{\sin \lambda t}{t}\, dt,$$

whence the proof of the lemma follows by virtue of (3.16).

We will now prove the basic theorem of this section.

Theorem 3.1. *If $p(x)$ and $r(x)$ satisfy a Dini condition uniformly in every finite interval, then for any fixed x and s we have the asymptotic formula*

$$\begin{aligned} &\theta(x, s; \lambda) - \theta(x, s; -\lambda) = \theta^*(x, s; \lambda) - \theta^*(x, s; -\lambda) \\ &+ [Hk(x, s-x) + H^T l(x, x-s) - I] \frac{\sin \lambda (x-s)}{\pi (x-s)} = o(1). \end{aligned} \tag{3.17}$$

This formula holds uniformly as x and s vary over any finite intervals.

Proof. Let (f,g) denote the scalar product of the vectors f and g, i.e. put $(f,g)=f^T\cdot g$. Let f and g be arbitrary constant vectors. It follows from (3.5) that

$$\begin{aligned} &\int_{-\infty}^{\infty} \psi_\varepsilon(\lambda)\, d_\lambda \{\sigma_1(x, s, \lambda; f, g) - \sigma_2(x, s, \lambda; f, g)\} \\ &= \begin{cases} a(x, s; f, g)\, g_\varepsilon(s-x) + b(x, s, \varepsilon; f, g), & |x-s| \leqslant \varepsilon, \\ 0, & |x-s| \geqslant \varepsilon, \end{cases} \end{aligned} \tag{3.18}$$

where we have put

$$\sigma_1(x,s,\lambda; f, g) = (\theta(x,s;\lambda) f, g),$$

$$\sigma_2(x,s,\lambda; f,g) = (\theta^*(x,s;\lambda) f,g),$$

$$a(x,s;f,g) = \left(\left[Hk(x, s-x) - \frac{I}{2}\right] f, g\right) + \left(\left[H^T l(x, x-s) - \frac{I}{2}\right] f, g\right),$$

$$b(x,s;\epsilon;f,g) = (W^*(x,s;\epsilon) f,g).$$

For $s = x$ and $g = f$ it is obvious that σ_1 and σ_2 are monotonic functions of λ, and that $a(x,x;f,f) \equiv 0$. Therefore it follows from the identity (3.18), by virtue of Marčenko's Tauberian theorem (cf. Chapter 14, Theorem 4.2, special case) as well as (3.14), that

$$\lim_{\lambda\to\infty} \{\sigma_1(x, x, \lambda; f, f) - \sigma_1(x, x, -\lambda; f, f) - [\sigma_2(x, x, \lambda; f, f) - \sigma_2(x, x, -\lambda; f, f)]\} = 0. \tag{3.18'}$$

For $s \neq x$ and $g = f$ let us consider the monotonic functions of λ

$$\tau_1(x,s;\lambda) = \sigma_1(x,x,\lambda;f,f) + 2\sigma_1(x,s,\lambda;f,f) + \sigma_1(s,s,\lambda;f,f),$$

$$\tau_2(x,s;\lambda) = \sigma_2(x,x,\lambda;f,f) + 2\sigma_2(x,s,\lambda;f,f) + \sigma_2(s,s,\lambda;f,f)$$

and again apply the aforementioned Tauberian theorem of Marčenko. As a consequence we obtain, by (3.18′),

$$\lim_{\lambda\to\infty} \{\sigma_1(x, s, \lambda; f, f) - \sigma_1(x, s, -\lambda; f, f) - [\sigma_2(x, s, \lambda; f, f) - \sigma_2(x, s, -\lambda; f, f)] - a(r,s;f,f) \frac{\sin \lambda (x-s)}{\pi (x-s)}\} = 0$$

Now, transforming the quadratic form to a bilinear form, we obtain from the preceding formula

$$\lim_{\lambda\to\infty} \Big\{\sigma_1(x, s, \lambda; f, g) - \sigma_1(x, s, -\lambda; f, g) - [\sigma_2(x, s, \lambda; f, g) - \sigma_2(x, s, -\lambda; f, g)] - a(x, s; f, g) \frac{\sin \lambda (x-s)}{\pi (x-s)}\Big\} = 0.$$

Since the vectors f and g are arbitrary, (3.17) follows from this equality. This proves the theorem.

From the asymptotic formula (3.17) there follows:

Theorem 3.2. *Under the conditions of Theorem 3.1 we have*

$$\lim_{\lambda\to\infty} \Big\{\rho(\lambda) - \rho(-\lambda) - \frac{2}{\pi(1+h^2)}\lambda\Big\} = 0. \tag{3.19}$$

In fact, from the definition of the spectral kernel $\theta^*(x,s;\lambda)$, i.e. from (2.27), and from the explicit form of the vector-function

$$\psi(x,\lambda) = \begin{pmatrix} \psi_1(x,\lambda) \\ \psi_2(x,\lambda) \end{pmatrix}$$

(cf. (2.23)) we find that

$$\theta^*_? (0, 0; \lambda) - \theta^*(0, 0; \lambda) = \begin{pmatrix} 1 & h \\ h & h^2 \end{pmatrix} \frac{2}{\pi(1+h^2)} \lambda. \tag{3.20}$$

Further, by virtue of (2.1), Chapter 9,

$$k(x,s-x)\big|_{s=x=0}=l(x,x-s)\big|_{s=x=0}=\tfrac{1}{2}.$$

Consequently, taking account of this and of the definition of the matrix H (cf. (1.6′), Chapter 9), we obtain

$$[Hk(x,s-x)+H^{T}l(x,x-s)-I]\big|_{s=x=0}=0. \tag{3.21}$$

Finally, it follows from the definition of the spectral kernel $\theta(x,s;\lambda)$ (cf. (2.5)) and the condition (1.5) that

$$\theta(0,0;\lambda)-\theta(0,0;-\lambda)=\begin{pmatrix}1 & h\\ h & h^2\end{pmatrix}\{\rho(\lambda)-\rho(-\lambda)\}. \tag{3.22}$$

Now, putting $s=x=0$ in (3.18) and taking account of (3.20), (3.21) and (3.22), we obtain

$$\lim_{\lambda\to\infty}\begin{pmatrix}1 & h\\ h & h^2\end{pmatrix}\left\{\rho(\lambda)-\rho(-\lambda)-\frac{2}{\pi(1+h^2)}\lambda\right\}=0,$$

which proves the asymptotic formula (3.19). This proves the theorem.

Formula (3.19) gives the asymptotic behaviour of the odd component of the spectral function $\rho(\lambda)$. To determine the asymptotic behaviour of $\rho(\lambda)$ itself, we obviously have to find the asymptotic behaviour of its even component. We have the following:

Theorem 3.3. *Suppose that $p(x)$ and $r(x)$ satisfy a Lipschitz condition uniformly in every finite interval. Then for $|\lambda|\to\infty$ we have*

$$\rho(\lambda)+\rho(-\lambda)=o(1). \tag{3.23}$$

Proof. By virtue of (3.3), it follows from (3.5) for $x=s=0$ that

$$\int_{-\infty}^{\infty}\psi_\varepsilon(\lambda)\,d_\lambda\{\theta_{11}(0,0;\lambda)-\theta^*_{11}(0,0;\lambda)\}=\int_{-\varepsilon}^{\varepsilon}G(t)\,g_\varepsilon(t)\,dt, \tag{3.24}$$

where

$$G(t)=\begin{cases}\tilde{H}_{11}(0,t,0), & -\varepsilon\leqslant t<0,\\ \tilde{H}_{11}(0,t,0), & 0<t\leqslant\varepsilon.\end{cases}$$

Since

$$\theta_{11}(0,0;\lambda)=\rho(\lambda),\qquad \theta^*_{11}(0,0;\lambda)=\frac{\lambda}{\pi(1+h^2)},$$

by Marčenko's Tauberian theorem (cf. Chapter 14, Theorem 4.2′) we obtain from (3.24)

$$\rho(\lambda)+\rho(-\lambda)=\frac{\lambda}{|\lambda|}\cdot\frac{G(+0)+G(-0)}{2}+o(1). \tag{3.25}$$

On the other hand, it follows from (2.16) and (5.12′), Chapter 9, that $G(+0)+G(-0)=0$. Then (3.23) follows from (3.25), which proves the theorem.

From this theorem and Theorem 3.2 follows:

Theorem 3.4. *Suppose that the hypotheses of Theorem 3.3 are satisfied. Then for* $|\lambda| \to \infty$ *we have*

$$\rho(\lambda) = \frac{\lambda}{\pi(1+h^2)} + o(1).$$

§4. Asymptotic behaviour of the derivatives of the spectral kernel

In §1 we derived the formula (cf. (1.15))

$$\varphi'_x(x, \lambda)\psi_\varepsilon(\lambda) = \int_{x-\varepsilon}^{x+\varepsilon} \frac{\partial W(x, s; \varepsilon)}{\partial x} \varphi(s, \lambda)\, ds. \tag{4.1}$$

It follows from this formula that for fixed x the components of the vector-function $\varphi'_x(x,\lambda)\psi_\epsilon(\lambda)$, i.e. the functions

$$\psi_\epsilon(\lambda)\frac{\partial}{\partial x}\varphi_1(x,\lambda) \quad \text{and} \quad \psi_\epsilon(\lambda)\frac{\partial}{\partial x}\varphi_2(x,\lambda),$$

can be regarded as the Fourier transforms of the vector-functions which are equal respectively to

$$\begin{pmatrix} \frac{\partial}{\partial x} W_{11}(x, s; \varepsilon) \\ \frac{\partial}{\partial x} W_{12}(x, s; \varepsilon) \end{pmatrix}, \qquad \begin{pmatrix} \frac{\partial}{\partial x} W_{21}(x, s; \varepsilon) \\ \frac{\partial}{\partial x} W_{22}(x, s; \varepsilon) \end{pmatrix}$$

for $|x-s| \leqq \epsilon$, and zero for $|x-s| > \epsilon$. Therefore if

$$f(x) = \begin{pmatrix} f_1(x) \\ f_2(x) \end{pmatrix} \in \mathscr{L}^2(0, \infty)$$

and equals zero outside some finite interval, then by the generalized Parseval equality (2.4) it follows from (3.1) and (4.1) that

$$\int_{-\infty}^{\infty} \varphi'_x(x, \lambda) F(\lambda) \psi_\varepsilon(\lambda)\, d\rho(\lambda) = \int_{x-\varepsilon}^{x+\varepsilon} \frac{\partial W(x, s; \varepsilon)}{\partial x} f(s)\, ds.$$

Hence, since the vector-function $f(x)$ is arbitrary, we obtain the identity

$$\int_{-\infty}^{\infty} \varphi'_x(x, \lambda)\varphi^T(s, \lambda)\psi_\varepsilon(\lambda)\, d\rho(\lambda) = \frac{\partial}{\partial x} W(x, s; \varepsilon), \tag{4.2}$$

where

$$W(x, s; \varepsilon) = Hk(x, s-x) g_\varepsilon(s-x) + H^T l(x, x-s) g_\varepsilon(x-s) + \frac{1}{2} \int_{|x-s|}^{\varepsilon} H(x, t, s) g_\varepsilon(t)\, dt + \frac{1}{2} \int_{-\varepsilon}^{-|x-s|} \tilde{H}(x, t, s) g_\varepsilon(t)\, dt. \tag{4.3}$$

Further, taking account of the definition of the spectral kernel $\theta(x, s; \lambda)$, the identity (4.2) can be rewritten in the form

$$\int_{-\infty}^{\infty} \psi_\varepsilon(\lambda)\, d_\lambda \frac{\partial \theta(x, s; \lambda)}{\partial x} = \begin{cases} \dfrac{\partial}{\partial x} W(x, s; \varepsilon), & |x-s| \leqslant \varepsilon, \\ 0, & |x-s| > \epsilon. \end{cases} \tag{4.2'}$$

Differentiating (3.4) with respect to x and subtracting the result from (4.2′), we obtain

$$\int_{-\infty}^{\infty} \psi_\varepsilon(\lambda)\, d_\lambda \left\{ \frac{\partial \theta(x, s; \lambda)}{\partial x} - \frac{\partial \theta^*(x, s; \lambda)}{\partial x} \right\} = \begin{cases} \dfrac{\partial}{\partial x} W(x, s; \varepsilon) - \dfrac{1}{2} \dfrac{\partial}{\partial x} \{ H g_\varepsilon(s-x) + H^T g_\varepsilon(x-s) \}, & |x-s| \leqslant \varepsilon, \\ 0, & |x-s| \geqslant \varepsilon. \end{cases} \tag{4.4}$$

We will now transform the right side of (4.4). From the definition of the matrix $W(x, s; \epsilon)$ follows

$$\begin{aligned} &\frac{\partial}{\partial x} W(x, s; \varepsilon) - \frac{1}{2} \frac{\partial}{\partial x} \{ H g_\varepsilon(s-x) + H^T g_\varepsilon(x-s) \} \\ &= \{ H k'_x(x, s-x) g_\varepsilon(s-x) + H^T l'_x(x, x-s) g_\varepsilon(x-s) \} \\ &\quad + \left\{ H \left[k(x, s-x) - \frac{1}{2} \right] \frac{\partial g_\varepsilon(s-x)}{\partial x} + H^T \left[l(x, x-s) - \frac{1}{2} \right] \frac{\partial g_\varepsilon(x-s)}{\partial x} \right\} \\ &\quad - \operatorname{sign}(x-s) \frac{1}{2} \{ H(x, |x-s|, s) g_\varepsilon(|x-s|) \\ &\quad \operatorname{sign}(x-s) \frac{1}{2} \tilde{H}(x, -|x-s|, s) g_\varepsilon(-|x-s|) \} \\ &\quad + \frac{1}{2} \left\{ \int_{|x-s|}^{\varepsilon} H'_x(x, t, s) g_\varepsilon(t)\, dt + \int_{-\varepsilon}^{-|x-s|} \tilde{H}'_x(x, t, s) g_\varepsilon(t)\, dt \right\} \\ &= I_1 + I_2 + I_3 + I_4. \end{aligned} \tag{4.5}$$

From the definition of the functions $k(x,t)$ and $l(x,t)$ (cf. (2.8) and (2.10), Chapter 9) it follows that

$$k'_x(x, s-x) = -\frac{i}{4}[p(x)+r(x)]\exp\left\{\frac{i}{2}\int_x^s [p(\tau)+r(\tau)]\,d\tau\right\}, \tag{4.6}$$

$$l'_x(x, x-s) = \frac{i}{4}[p(x)+r(x)]\exp\left\{-\frac{i}{2}\int_x^s [p(\tau)+r(\tau)]\,d\tau\right\} \tag{4.6'}$$

On the other hand, from (3.9) we have

$$g_\varepsilon[\pm(s-x)] = \frac{1}{2\pi}\int_{-\infty}^{\infty} \psi_\varepsilon(\lambda)\, e^{\mp i\lambda(s-x)}\,d\lambda. \tag{4.7}$$

Therefore the term I_1 from (4.5) can be represented in the form

$$\begin{aligned} I_1 &= \frac{1}{2\pi}\int_{-\infty}^{\infty} \psi_\varepsilon(\lambda)\, d_\lambda \left\{Hk'_x(x,\ s-x)\int_0^\lambda e^{-i\nu(s-x)}d\nu \right.\\ &\qquad \left. + H^T l'_x(x,\ x-s)\int_0^\lambda e^{i\nu(s-x)}d\nu\right\} \\ &= \frac{1}{2\pi}\int_{-\infty}^{\infty} \psi_\varepsilon(\lambda)\, d_\lambda \left\{iHk'_x(x,\ s-x)\,\frac{e^{-i\lambda(s-x)}-1}{s-x}\right.\\ &\qquad \left. - iH^T l'_x(x,\ x-s)\,\frac{e^{i\lambda(s-x)}-1}{s-x}\right\}. \end{aligned} \tag{4.8}$$

Since

$$i\,\frac{e^{-i\lambda(s-x)}-1}{s-x} = 2e^{-i\lambda\frac{s-x}{2}}\,\frac{\sin\lambda\frac{s-x}{2}}{s-x},$$

$$-i\,\frac{e^{i\lambda(s-x)}-1}{s-x} = 2e^{i\lambda\frac{s-x}{2}}\,\frac{\sin\lambda\frac{s-x}{2}}{s-x},$$

(4.8) can be rewritten as

$$\begin{aligned} I_1 = \frac{1}{\pi}\int_{-\infty}^{\infty} \psi_\varepsilon(\lambda)\, d_\lambda &\left\{\left[Hk'_x(x,\ s-x)\, e^{-i\lambda\frac{s-x}{2}}\right.\right.\\ &\left.\left. + H^T l'_x(x,\ x-s)\, e^{i\lambda\frac{s-x}{2}}\right]\frac{\sin\lambda\frac{s-x}{2}}{s-x}\right\}. \end{aligned} \tag{4.8'}$$

Put

$$q(x, s; \lambda) = \frac{1}{2} \int_x^s [p(\tau) + r(\tau)] \, d\tau - \lambda \frac{s-x}{2}. \tag{4.9}$$

Now, inserting the expressions for $k'_x(x, s-x)$ and $l'_x(x, x-s)$ from (4.6) and (4.6′) in (4.8′) and using (4.9), we obtain

$$I_1 = \frac{p(x) + r(x)}{2\pi} \int_{-\infty}^{\infty} \psi_\varepsilon(\lambda) \, d_\lambda \left\{ \left[\frac{H}{2i} e^{iq(x, s; \lambda)} - \frac{H^T}{2i} e^{-iq(x, s; \lambda)} \right] \frac{\sin \lambda \frac{s-x}{2}}{s-x} \right\}. \tag{4.8″}$$

Further, taking into account the definition of the matrix

$$H = \begin{pmatrix} 1 & i \\ -i & 1 \end{pmatrix},$$

we have

$$\begin{aligned} &\frac{H}{2i} e^{iq(x, s; \lambda)} - \frac{H^T}{2i} e^{-iq(x, s; \lambda)} = \\ &= \begin{pmatrix} \sin q(x, s; \lambda) & \cos q(x, s; \lambda) \\ -\cos q(x, s; \lambda) & \sin q(x, s; \lambda) \end{pmatrix} \equiv A(x, s; \lambda). \end{aligned} \tag{4.10}$$

Using (4.9), we find that

$$A(x, s; \lambda) = A(x, s) \cos \lambda \frac{s-x}{2} - A(x, s) \, T \sin \lambda \frac{s-x}{2}, \tag{4.10′}$$

where

$$A(x, s) = A(x, s; 0); \qquad T = \begin{pmatrix} 0 & -1 \\ 1 & 0 \end{pmatrix}. \tag{4.11}$$

Inserting the expression for $A(x, s; \lambda)$ from (4.10′) in (4.8″) and putting

$$\begin{aligned} \alpha_1(x, s; \lambda) = {} & \frac{p(x) + r(x)}{4\pi} A(x, s) \\ & \times \left[\frac{\sin \lambda (s-x)}{s-x} - T \frac{1 - \cos \lambda (s-x)}{s-x} \right], \end{aligned} \tag{4.12}$$

we obtain for I_1 the final form

$$I_1 = \int_{-\infty}^{\infty} \psi_\varepsilon(\lambda) \, d_\lambda \alpha_1(x, s; \lambda). \tag{4.13}$$

We will now transform the term I_2 from the expression (4.5).

Let us assume that the function $g_\varepsilon(t)$ has a piecewise continuous and piecewise monotone second derivative. Then it easily follows from (3.7) (integrating by parts and taking account of the conditions imposed upon the function $g_\epsilon(t)$) that for large $|\lambda|$ we have the estimate

$$\psi_\epsilon(\lambda) = O(1/\lambda^3).$$

This estimate enables us to differentiate (3.9) with respect to t, following which we obtain

$$g'_\varepsilon(t) = \frac{1}{2\pi i} \int_{-\infty}^{\infty} \lambda \psi_\varepsilon(\lambda) e^{-i\lambda t} dt. \tag{4.14}$$

Since

$$\frac{\partial}{\partial x} g_\varepsilon(s-x) = -\frac{\partial}{\partial x} g_\varepsilon(x-s),$$

by virtue of (4.14)

$$I_2 = \frac{1}{2\pi i} \int_{-\infty}^{\infty} \psi_\varepsilon(\lambda) d_\lambda \Big\{ \Big[Hk(x, s-x) - H^T l(x, x-s) - \frac{1}{2}(H - H^T) \Big] \int_0^\lambda \nu e^{-i\nu(s-x)} d\nu \Big\}.$$

On the one hand, taking into account the expressions for the functions $k(x,t)$ and $l(x,t)$ and the matrix H, we have

$$Hk(x, s-x) - H^T l(x, x-s) - \tfrac{1}{2}(H - H^T) = i[A(x,s) + T],$$

where the matrices $A(x,s)$ and T have their previous values, given by (4.10), (4.10′) and (4.11).

On the other hand, since

$$\int_0^\lambda \nu e^{-i\nu(s-x)} d\nu = -i \frac{\partial}{\partial x} \int_0^\lambda e^{-i\nu(s-x)} d\nu = \frac{\partial}{\partial x} \left\{ \frac{e^{-i\lambda(s-x)} - 1}{s-x} \right\},$$

putting

$$\alpha_2(x, s; \lambda) = \frac{1}{\pi} [A(x, s) + T] e^{-i\lambda \frac{s-x}{2}} \frac{\sin \lambda \frac{s-x}{2}}{i(s-x)}, \tag{4.15}$$

we obtain the following final form for I_2:

$$I_2 = \int_{-\infty}^{\infty} \psi_\varepsilon(\lambda) d_\lambda \alpha_2(x, s; \lambda). \tag{4.16}$$

Further, by virtue of (3.9) we easily obtain the following representation for I_3:

$$I_3 = \int_{-\infty}^{\infty} \psi_\varepsilon(\lambda) d_\lambda \alpha_3(x, s; \lambda), \tag{4.17}$$

where

$$\begin{aligned}\alpha_3(x, s; \lambda) = &-\frac{\operatorname{sign}(x-s)}{2\pi}\Big\{H(x, |x-s|, s)\, e^{-i\lambda\frac{|x-s|}{2}} \\ &+ \tilde{H}(x, -|x-s|, s)\, e^{i\lambda\frac{|x-s|}{2}}\Big\} \frac{\sin\lambda\frac{|x-s|}{2}}{|x-s|}.\end{aligned} \tag{4.18}$$

Finally, we transform I_4. To do this we put

$$\begin{aligned}\alpha_4^{(1)}(x, s; \lambda) &= \frac{1}{2}\int_{|x-s|}^{1} H'_x(x, t, s)\, e^{i\lambda t}\, dt, \\ \alpha_4^{(2)}(x, s; \lambda) &= \frac{1}{2}\int_{-1}^{-|x-s|} \tilde{H}'_x(x, t, s)\, e^{i\lambda t}\, dt.\end{aligned} \tag{4.19}$$

Then by the classical Parseval equality there follows from (3.7) and (4.19) respectively ($\epsilon \leqq 1$)

$$\frac{1}{2}\int_{|x-s|}^{1} H'_x(x, t, s)\, g_\varepsilon(t)\, dt = \frac{1}{2\pi}\int_{-\infty}^{\infty} \psi_\varepsilon(\lambda)\, \overline{\alpha_4^{(1)}(x, s; \lambda)}\, d\lambda,$$

$$\frac{1}{2}\int_{-1}^{-|x-s|} \tilde{H}'_x(x, t, s)\, g_\varepsilon(t)\, dt = \frac{1}{2\pi}\int_{-\infty}^{\infty} \psi_\varepsilon(\lambda)\, \overline{\alpha_4^{(2)}(x, s; \lambda)}\, d\lambda.$$

Therefore we obtain for I_4 the final form

$$I_4 = \int_{-\infty}^{\infty} \psi_\varepsilon(\lambda)\, d_\lambda \alpha_4(x, s; \lambda), \tag{4.20}$$

where

$$\alpha_4(x, s; \lambda) = \frac{1}{2\pi}\int_0^{\lambda} \left[\overline{\alpha_4^{(1)}(x, s; \nu)} + \overline{\alpha_4^{(2)}(x, s; \nu)}\right] d\nu. \tag{4.21}$$

Thus, taking account of (4.13), (4.16), (4.17) and (4.20), the identity (4.4) can be rewritten in the form

$$\int_{-\infty}^{\infty} \psi_\varepsilon(\lambda)\, d_\lambda R(x, s; \lambda) = 0, \tag{4.22}$$

where we have put

$$R(x,s;\lambda) = \frac{\partial}{\partial x}\{\theta(x, s; \lambda) - \theta^*(x, s; \lambda)\} - \sum_{k=1}^{4} \alpha_k(x, s; \lambda). \tag{4.23}$$

The identity (4.22) and a Tauberian theorem of Levitan make it possible, similarly to what was done in obtaining (3.18), to study the asymptotic behaviour of the derivative with respect to x of the spectral kernel $\theta(x,s;\lambda)$ as $|\lambda| \to \infty$. We first prove the following lemma.

Lemma 4.1. *If the functions $p(x)$ and $r(x)$ have derivatives which are summable over every finite interval, then for $a \to \infty$ we have*

$$\bigvee_a^{a+1} \{R_{ik}(x, s; \lambda)\} = O(a) \quad (i, k = 1, 2), \tag{4.24}$$

where $R_{ik}(x,s;\lambda)$ are the elements of the matrix $R(x,s;\lambda)$ defined by (4.23). *The estimates* (4.24) *hold uniformly as x and s vary over any finite intervals.*

Proof. To prove the lemma it is sufficient to obtain the estimate (4.24) for every element of every term in the expression (4.23). For the elements of the matrices $\partial\theta(x,s;\lambda)/\partial x$ and $\partial\theta^*(x,s;\lambda)/\partial x$ the estimate (4.24) follows from (2.18) and (2.29) upon putting $l=1$ in them. For the elements of the matrices $\alpha_k(x,s;\lambda)$ $(k=1,2,3)$ the estimate (4.24) follows immediately from the explicit form of these matrices. As for the matrix $\alpha_4(x,s;\lambda)$, by virtue of the continuity in t of the matrices $H_x'(x,t;s)$ and $\widetilde{H}_x'(x,t;s)$ (cf. (4.19)) and the Riemann-Lebesgue lemma it follows that

$$\lim_{\lambda\to\infty} \alpha_{4ij}^{(1)}(x, s; \lambda) = \lim_{\lambda\to\infty} \alpha_{4ij}^{(2)}(x, s; \lambda) = 0 \quad (i, j = 1, 2),$$

uniformly for x and s in any finite intervals, where $\alpha_{4ij}^{(1)}(x,s;\lambda)$ and $\alpha_{4ij}^{(2)}(x,s;\lambda)$ are the elements of the matrices $\alpha_4^{(1)}(x,s;\lambda)$ and $\alpha_4^{(2)}(x,s;\lambda)$ respectively. Therefore for $a \to \infty$

$$\bigvee_a^{a+1} \{\alpha_{4ij}(x, s; \lambda)\} = \bigvee_a^{a+1} \left\{ \frac{1}{2\pi} \int_0^\lambda [\overline{\alpha_{4ij}^{(1)}(x, s; \nu)} + \overline{\alpha_{4ij}^{(2)}(x, s; \nu)}] d\nu \right\}$$

$$= \frac{1}{2\pi} \int_a^{a+1} \{|\alpha_{4ij}^{(1)}(x, s; \nu)| + |\alpha_{4ij}^{(2)}(x, s; \nu)|\} d\nu = o(1),$$

which proves the lemma.

Theorem 4.1. *If the functions $p(x)$ and $r(x)$ have derivatives which are summable over every finite interval, then for every fixed x and s we have*

$$\int_{-\lambda}^{\lambda} \left(1 - \frac{\nu^2}{\lambda^2}\right) d_\nu \left[\frac{\partial}{\partial x} \theta(x, s; \nu) - \frac{\partial}{\partial x} \theta^*(x, s; \nu)\right] = O(1). \tag{4.25}$$

This asymptotic formula holds uniformly as x and s vary over any finite intervals.

PROOF. By virtue of the estimates (4.24) and the identity (4.22), Levitan's Tauberian theorem is applicable, according to which the asymptotic formula

$$\int_{-\lambda}^{\lambda}\left(1-\frac{\nu^2}{\lambda^2}\right)d_\nu R(x, s; \nu)=O(1)$$

holds uniformly as x and s vary over any finite intervals, or by virtue of the expression (4.23) for $R(x,s;\nu)$,

$$\int_{-\lambda}^{\lambda}\left(1-\frac{\nu^2}{\lambda^2}\right)d_\nu\left\{\frac{\partial\theta(x, s; \nu)}{\partial x}-\frac{\partial\theta^*(x, s; \nu)}{\partial x}-\sum_{k=1}^{4}\alpha_k(x, s; \nu)\right\}=O(1). \tag{4.26}$$

Let us now calculate each of the integrals

$$A_k=\int_{-\lambda}^{\lambda}\left(1-\frac{\nu^2}{\lambda^2}\right)d_\nu\alpha_k(x, s; \nu) \qquad (k=1, 2, 3, 4).$$

From the definition of $\alpha_1(x,s;\nu)$, i.e. from (4.12), there follows

$$\begin{aligned}A_1&=\int_{-\lambda}^{\lambda}\left(1-\frac{\nu^2}{\lambda^2}\right)d_\nu\alpha_1(x, s; \nu)\\ &=\frac{p(x)+r(x)}{4\pi}A(x, s)\int_{-\lambda}^{\lambda}\left(1-\frac{\nu^2}{\lambda^2}\right)d_\nu\left[\frac{\sin\nu(s-x)}{s-x}-T\frac{1-\cos\lambda(s-x)}{s-x}\right]\\ &=\frac{p(x)+r(x)}{4\pi}A(x, s)\int_{-\lambda}^{\lambda}\left(1-\frac{\nu^2}{\lambda^2}\right)[\cos\nu(s-x)-T\sin\nu(s-x)]\,d\nu.\end{aligned}$$

Further,

$$\int_{-\lambda}^{\lambda}\left(1-\frac{\nu^2}{\lambda^2}\right)[\cos\nu(s-x)-T\sin\nu(s-x)]\,d\nu$$

$$=2\int_{0}^{\lambda}\left(1-\frac{\nu^2}{\lambda^2}\right)\cos\nu(s-x)\,d\nu.$$

By a well-known formula (cf. Titchmarsh [3], §7.1)

$$\int_{0}^{\lambda}\left(1-\frac{\nu^2}{\lambda^2}\right)\cos\nu(s-x)\,d\nu=\sqrt{\frac{\pi}{2}}\,\lambda\frac{I_{3/2}[\lambda(s-x)]}{[\lambda(s-x)]^{3/2}}. \tag{4.27}$$

Therefore we obtain for A_1 the final expression

$$A_1=\frac{p(x)+r(x)}{2\sqrt{2\pi}}\lambda^{-1/2}A(x, s)\frac{I_{3/2}[\lambda(s-x)]}{(s-x)^{3/2}}. \tag{4.28}$$

From (4.15) we obtain for A_2 the expression

$$A_2 = \frac{1}{\pi}[A(x, s) + T]\int_{-\lambda}^{\lambda}\left(1-\frac{\nu^2}{\lambda^2}\right)d_\nu\left[e^{-i\nu\frac{s-x}{2}}\frac{\sin\nu\frac{s-x}{2}}{i(s-x)}\right]$$

$$= \frac{1}{\pi}[A(x, s)]\int_{-\lambda}^{\lambda}\left(1-\frac{\nu^2}{\lambda^2}\right)\left[\frac{e^{-i\nu\frac{s-x}{2}}\cos\nu\frac{s-x}{2}}{2i} - \frac{e^{-i\nu\frac{s-x}{2}}\sin\nu\frac{s-x}{2}}{2}\right]d\nu.$$

Since

$$\frac{1}{2i}\int_{-\lambda}^{\lambda}\left(1-\frac{\nu^2}{\lambda^2}\right)d\nu = \frac{1}{i}\int_{0}^{\lambda}\left(1-\frac{\nu^2}{\lambda^2}\right)\cos^2\nu\frac{s-x}{2}\,d\nu,$$

$$\frac{1}{2}\int_{-\lambda}^{\lambda}\left(1-\frac{\nu^2}{\lambda^2}\right)e^{-i\nu\frac{s-x}{r}}\sin\frac{s-x}{2}\,d\nu = -\frac{1}{i}\int_{0}^{\lambda}\left(1-\frac{\nu^2}{\lambda^2}\right)\sin^2\nu\frac{s-x}{2}\,d\nu,$$

it follows that

$$A_2 = \frac{1}{\pi}[A(x, \lambda) + T]\int^{\lambda}\left(1-\frac{\nu^2}{\lambda^2}\right)d\nu = \frac{4\lambda}{3\pi}[A(x, s) + T]. \tag{4.29}$$

We now determine A_3. Using (4.18), we have

$$A_3 = -\frac{1}{2\pi}H(x, |x-s|, s)\int_{-\lambda}^{\lambda}\left(1-\frac{\nu^2}{\lambda^2}\right)d_\nu\left[e^{-i\nu\frac{|x-s|}{2}}\frac{\sin\nu\frac{|x-s|}{2}}{|x-s|}\right]$$

$$-\frac{1}{2\pi}\tilde{H}(x, -|x-s|, s)\int_{-\lambda}^{\lambda}\left(1-\frac{\nu^2}{\lambda^2}\right)d_\nu\left[e^{i\nu\frac{|x-s|}{2}}\frac{\sin\nu\frac{|x-s|}{2}}{|x-s|}\right],$$

from which after simple computations we obtain for A_3 the expression

$$A_3 = \frac{2\lambda}{3\pi}\{H(x, |x-s|, s) + \tilde{H}(x, -|x-s|, s)\}. \tag{4.30}$$

Finally, we must determine A_4. By (4.19) and (4.21) we have

$$A_4 = \frac{1}{4\pi}\int_{-\lambda}^{\lambda}\left(1-\frac{\nu^2}{\lambda^2}\right)d_\nu\left\{\int_{|x-s|}^{1}\overline{H'_x(x, t, s)e^{i\nu t}\,dt} + \int_{-|x-s|}^{-1}\overline{\tilde{H}'(x, t, s)e^{i\nu t}dt}\right\}.$$

Changing the order of integration and proceeding as above, we obtain

$$A_4 = \frac{1}{2\pi}\int_{|x-s|}^{i}H'_x(x, t, s)\left[\int_0^{\lambda}\left(1-\frac{\nu^2}{\lambda^2}\right)\cos\nu t\,d\nu\right]dt$$

$$+\frac{1}{2\pi}\int_{-1}^{-|x-s|}\tilde{H}'_x(x,t,s)\left[\int_0^{\lambda}\left(1-\frac{\nu^2}{\lambda^2}\right)\cos\nu t\,d\nu\right]dt,$$

from which by virtue of (4.27) we find that

$$\begin{aligned} A_4 &= \frac{\lambda^{-1/2}}{2\sqrt{2\pi}}\int_{|x-s|}^{1} H'_x(x,t,s)\frac{I_{3/2}(\lambda t)}{t^{3/2}}\,dt \\ &+\frac{\lambda^{-1/2}}{2\sqrt{2\pi}}\int_{-1}^{-|x-s|}\tilde{H}'_x(x,t,s)\frac{I_{3/2}(\lambda t)}{t^{3/2}}\,dt. \end{aligned} \tag{4.31}$$

From (2.12), (2.14) and (5.11′), Chapter 9, it easily follows that $A_3 = -A_2$, and from the asymptotic behaviour of the Bessel functions we obtain $A_2 + A_4 = O(1)$. Therefore the asymptotic formula (4.25), and consequently the theorem, follows from (4.26) by virtue of (4.28) $-$(4.31). This proves the theorem.

Bibliographical references

The results of this chapter are due to Sargsjan [**5, 9, 12**].

CHAPTER 11

EXPANSION, AND DIFFERENTIATION OF AN EXPANSION, WITH RESPECT TO THE EIGENFUNCTIONS OF A DIRAC SYSTEM

§1. Some preliminary estimates

We will consider the eigenvalue problem

$$\begin{pmatrix} 0 & 1 \\ -1 & 0 \end{pmatrix} \frac{dy}{dx} + \begin{pmatrix} p(x) & 0 \\ 0 & r(x) \end{pmatrix} y = \lambda I y, \tag{1.1}$$

$$y_2(0) - h y_1(0) = 0. \tag{1.2}$$

where I is the 2×2 unit matrix, and h is an arbitrary real number. The functions $p(x)$ and $r(x)$ are real functions, defined on the halfline $[0, \infty)$ and summable over every finite interval.

In the first section of the preceding chapter we obtained the identity (cf. formula (1.13), Chapter 10)

$$\varphi(x, \lambda) \psi_\varepsilon(\lambda) = \int_{x-\varepsilon}^{x+\varepsilon} W(x, s, \varepsilon) \varphi(s, \lambda) ds, \tag{1.3}$$

where the vector-function

$$\varphi(x, \lambda) = \begin{pmatrix} \varphi_1(x, \lambda) \\ \varphi_2(x, \lambda) \end{pmatrix},$$

as before, denotes the solution of equation (1.1) which satisfies the initial conditions

$$\varphi_1(0, \lambda) = 1, \qquad \varphi_2(0, \lambda) = h, \tag{1.4}$$

and the matrix $W(x, s, \epsilon)$ is defined by (cf. (1.14), Chapter 10)

$$\begin{aligned} W(x, s, \varepsilon) &= A(x, s) g_\varepsilon(x - s) \\ &+ \frac{1}{2} \int_{|x-s|}^{\varepsilon} \{H(x, t, s) + \tilde{H}(x, t, s)\} g_\varepsilon(t) dt. \end{aligned} \tag{1.5}$$

Let

$$f(x) = \begin{pmatrix} f_1(x) \\ f_2(x) \end{pmatrix}$$

be an arbitrary vector-function from $\mathscr{L}^2(0, \infty)$. We will denote by $F(\lambda)$ the generalized Fourier transform of $f(x)$, i.e. we put

$$F(\lambda) = \int_0^\infty f^T(s)\, \varphi(s, \lambda)\, ds. \tag{1.6}$$

It follows from the identity (1.3) that the functions $\varphi_1(x, \lambda)\psi_\epsilon(\lambda)$ and $\varphi_2(x, \lambda)\psi_\epsilon(\lambda)$ are, for each fixed x, the generalized Fourier transforms of the vector-functions which are equal respectively to

$$\begin{pmatrix} W_{11}(x, s, \varepsilon) \\ W_{12}(x, s, \varepsilon) \end{pmatrix} \quad \text{and} \quad \begin{pmatrix} W_{21}(x, s, \varepsilon) \\ W_{22}(x, s, \varepsilon) \end{pmatrix}$$

for $x - \epsilon \leqq s \leqq x + \epsilon$ and equal to zero outside this interval. Therefore by the generalized Parseval equality (2.4) of Chapter 10, it follows from (1.3) and (1.6) that

$$\int_{-\infty}^{\infty} \varphi(x, \lambda) F(\lambda) \psi_\varepsilon(\lambda)\, d\rho(\lambda) = \int_{x-\varepsilon}^{x+\varepsilon} W(x, s, \varepsilon) f(s)\, ds. \tag{1.7}$$

Let us introduce the vector-function

$$S(x, \lambda) \equiv \begin{pmatrix} S_1(x, \lambda) \\ S_2(x, \lambda) \end{pmatrix} = \int_0^\lambda F(\lambda) \varphi(x, \lambda)\, d\rho(\lambda). \tag{1.8}$$

Then (1.7) assumes the form

$$\int_{-\infty}^{\infty} \psi_\varepsilon(\lambda)\, d_\lambda S(x, \lambda) = \int_{x-\varepsilon}^{x+\varepsilon} W(x, s, \varepsilon) f(s)\, ds. \tag{1.9}$$

We will now write down the formula analogous to (1.9) for the problem (1.1) + (1.2) with $p(x) = r(x) \equiv 0$. To this end we will find a new representation for the vector-function $S(x, \lambda)$. Inserting the value of $F(\lambda)$ from (1.6) into the expression (1.8) and then changing the order of integration, we obtain

$$S(x, \lambda) = \int_0^\infty \left\{ \int_0^\lambda \varphi(x, \lambda) \varphi^T(s, \lambda)\, d\rho(\lambda) \right\} f(s)\, ds, \tag{1.10}$$

since

$$\begin{aligned}(f^T(s)\varphi(s, \lambda))\varphi(x, \lambda) &= (f_1(s)\varphi_1(s, \lambda) + f_2(s)\varphi_2(s, \lambda)) \begin{pmatrix} \varphi_1(x, \lambda) \\ \varphi_2(x, \lambda) \end{pmatrix} \\ &= \begin{pmatrix} f_1(s)\varphi_1(s, \lambda)\varphi_1(x, \lambda) + f_2(s)\varphi_2(s, \lambda)\varphi_1(x, \lambda) \\ f_1(s)\varphi_1(s, \lambda)\varphi_2(x, \lambda) + f_2(s)\varphi_2(s, \lambda)\varphi_2(x, \lambda) \end{pmatrix}\end{aligned}$$

$$= \begin{pmatrix} \varphi_1(s, \lambda)\varphi_1(x, \lambda) & \varphi_2(s, \lambda)\varphi_1(x, \lambda) \\ \varphi_1(s, \lambda)\varphi_2(x, \lambda) & \varphi_2(s, \lambda)\varphi_2(x, \lambda) \end{pmatrix} \begin{pmatrix} f_1(s) \\ f_2(s) \end{pmatrix}$$

$$= \begin{pmatrix} \varphi_1(x, \lambda) \\ \varphi_2(x, \lambda) \end{pmatrix} (\varphi_1(s, \lambda), \varphi_2(s, \lambda)) \begin{pmatrix} f_1(s) \\ f_2(s) \end{pmatrix} = (\varphi(x, \lambda)\varphi^T(s, \lambda)) f(s)$$

Therefore by the definition of the spectral kernel $\theta(x, s; \lambda)$ (cf. (2.5), Chapter 10) (1.10) assumes the form

$$S(x, \lambda) = \int_0^\infty \theta(x, s; \lambda) f(s)\, ds. \tag{1.11}$$

Consequently the vector-function analogous to $S(x, \lambda)$ for the problem (1.1) + (1.2) with $p(x) = r(x) \equiv 0$ can be defined by

$$\begin{pmatrix} S_1^*(x, \lambda) \\ S_2^*(x, \lambda) \end{pmatrix} \equiv S^*(x, \lambda) = \int_0^\infty \theta^*(x, s; \lambda) f(s)\, ds, \tag{1.12}$$

where the matrix $\theta^*(x, s; \lambda)$ is given by (2.27), Chapter 10.

Further, for $p(x) = r(x) \equiv 0$ it follows from the definition of the functions $k(x, t)$ and $l(x, t)$ that $k(x, t) = l(x, t) = \frac{1}{2}$, and the matrices $H(x, t, s)$ and $\widetilde{H}(x, t, s)$, as solutions of the Goursat problem which equal zero on the characteristics, are identically equal to the zero matrix. Thus for the case being considered, (1.9) has the form

$$\begin{aligned} \int_{-\infty}^{\infty} \psi_\varepsilon(x)\, d_\lambda S^*(x, \lambda) = \frac{1}{2} H \int_{x-\varepsilon}^{x+\varepsilon} g_\varepsilon(s - x) f(s)\, ds \\ + \frac{1}{2} H^T \int_{x-\varepsilon}^{x+\varepsilon} g_\varepsilon(x - s) f(s)\, ds. \end{aligned} \tag{1.13}$$

Now, subtracting (1.13) from (1.9), we obtain

$$\begin{aligned} &\int_{\infty}^{\infty} \psi_\varepsilon(\lambda)\, d_\lambda \{S(x, \lambda) - S^*(x, \lambda)\} \\ &= \int_{x-\varepsilon}^{x+\varepsilon} \left\{W(x, s; \varepsilon) - \frac{1}{2}[Hg_\varepsilon(s - x) + H^T g_\varepsilon(x - s)]\right\} f(s)\, ds. \end{aligned} \tag{1.14}$$

Lemma 1.1. *If the coefficients $p(x)$ and $r(x)$ in equation (1.1) are continuous, then for any fixed x we have, as $a \to \infty$,*

$$\bigvee_a^{a+1} \{S_k(x, \lambda)\} = \int_a^{a+1} |F(\lambda)| |\varphi_k(x, \lambda)| d\rho(\lambda) = o(1) \tag{1.15}$$

$$(k = 1, 2).$$

These estimates hold uniformly in every finite interval.

Proof. By the definition of the vector-function $S(x,\lambda)$, i.e. (1.8), we have

$$\bigvee_a^{a+1} \{S_k(x, \lambda)\} = \int_a^{a+1} |F(\lambda)| |\varphi_k(x, \lambda)| d\rho(\lambda) \qquad (k = 1, 2).$$

Therefore according to the Cauchy-Bunjakovskiĭ inequality it follows that

$$\bigvee_a^{a+1} \{S_k(x, \lambda)\} \leqslant \left(\int_a^{a+1} |F(\lambda)|^2 d\rho(\lambda) \right)^{1/2} \left(\int_a^{a+1} |\varphi_k(x, \lambda)|^2 d\rho(\lambda) \right)^{1/2} \tag{1.16}$$

$$(k = 1, 2).$$

Now on the one hand, by virtue of Lemma 2.1, Chapter 10,

$$\int_a^{a+1} |\varphi_k(x, \lambda)|^2 d\rho(\lambda) \leqslant C \qquad (k = 1, 2), \tag{1.17}$$

and on the other hand, since by the Parseval equality

$$\int_{-\infty}^{\infty} |F(\lambda)|^2 d\rho(\lambda) = \int_0^{\infty} |f(x)|^2 dx < \infty,$$

it is obvious that as $a \to \infty$

$$\int_a^{a+1} |F(\lambda)|^2 d\rho(\lambda) = o(1). \tag{1.18}$$

The estimates (1.15) follow from the estimates (1.17) and (1.18), which proves the lemma.

From the definition of the vector-function $S^*(x,\lambda)$, i.e. (1.12), and from the explicit form of the matrix $\theta^*(x, s; \lambda)$ (cf. (2.27) and (2.23), Chapter 10) there immediately follows:

Lemma 1.2. *We have, uniformly over the entire halfline $[0, \infty)$ as $a \to \infty$,*

$$\bigvee_a^{a+1} \{S_k^*(x, \lambda)\} = o(1) \qquad (k = 1, 2). \tag{1.19}$$

We will now take up the question of estimating the variation of the derivatives of the vector-functions $S(x,\lambda)$ and $S^*(x,\lambda)$ with respect to x. To this end we will consider the identity (1.15), Chapter 10, i.e. the identity

$$\varphi'_x(x, \lambda)\psi_\varepsilon(\lambda) = \int_{x-\epsilon}^{x+\varepsilon} \frac{\partial W(x, s; \varepsilon)}{\partial x} \varphi(s, \lambda)\, ds. \tag{1.20}$$

It follows from this identity that for fixed x the functions $\varphi'_x(x,\lambda)\psi_\epsilon(\lambda)$ $(k=1,2)$ can be regarded as the generalized Fourier transforms of the vector-functions which are equal respectively to

$$\begin{pmatrix} \frac{\partial}{\partial x} W_{k1}(x, s; \varepsilon) \\ \frac{\partial}{\partial x} W_{k2}(x, s; \varepsilon) \end{pmatrix} \qquad (k=1, 2)$$

for $x-\epsilon \leqq s \leqq x+\epsilon$ and equal to zero outside this interval. Therefore by the generalized Parseval equality (2.4), Chapter 10, it follows from the identity (1.20) and the equality (1.6) that

$$\int_{-\infty}^{\infty} \varphi'_x(x, \lambda) F(\lambda) \psi_\varepsilon(\lambda)\, d\rho(\lambda) = \int_{x-\epsilon}^{x+\varepsilon} \frac{\partial W(x, s; \varepsilon)}{\partial x} f(s)\, ds,$$

which, using the definition of the vector-function $S(x,\lambda)$, assumes the form

$$\int_{-\infty}^{\infty} \psi_\varepsilon(\lambda)\, d_\lambda \frac{\partial S(x, \lambda)}{\partial x} = \int_{x-\varepsilon}^{x+\varepsilon} \frac{\partial W(x, s; \varepsilon)}{\partial x} f(s)\, ds. \tag{1.21}$$

For the case $p(x) = r(x) \equiv 0$ the analogue of (1.21) is, by virtue of (1.13),

$$\int_{-\infty}^{\infty} \psi_\varepsilon(\lambda)\, d_\lambda \frac{\partial S^*(x, \lambda)}{\partial x} = iB \int_{x-\varepsilon}^{x+\varepsilon} \frac{\partial g_\varepsilon(s-x)}{\partial x} f(s)\, ds, \tag{1.22}$$

where

$$B = \begin{pmatrix} 0 & 1 \\ -1 & 0 \end{pmatrix}. \tag{1.23}$$

We have taken into account here that

$$\frac{\partial g_\epsilon(s-x)}{\partial x} = -\frac{\partial g_\epsilon(x-s)}{\partial x} \quad \text{and} \quad \tfrac{1}{2}(H - H^T) = iB.$$

Now, subtracting (1.22) from (1.21), we obtain

$$\int_{-\infty}^{\infty} \psi_{\varepsilon}(\lambda)\, d_{\lambda} \left\{ \frac{\partial S(x, \lambda)}{\partial x} - \frac{\partial S^*(x, \lambda)}{\partial x} \right\}$$

(1.24)

$$= \int_{x-\varepsilon}^{x+\varepsilon} \left\{ \frac{\partial W(x, s; \varepsilon)}{\partial x} - iB \frac{\partial g_{\varepsilon}(s-x)}{\partial x} \right\} f(s)\, ds.$$

Lemma 1.3. *If the coefficients $p(x)$ and $r(x)$ in equation* (1.1) *have derivatives which are absolutely continuous in every finite interval, then for every fixed x we have, as $a \to \infty$,*

$$\bigvee_a^{a+1} \left\{ \frac{\partial S_k(x, \lambda)}{\partial x} \right\} = \int_{-a}^{a+1} |F(\lambda)| \cdot |\varphi_k'(x, \lambda)|\, d\rho(\lambda) = o(a), \tag{1.25}$$

$$(k = 1, 2).$$

These estimates hold uniformly in every finite interval.

Proof. By the Cauchy-Bunjakovskiĭ inequality it follows from the definition of the vector-function $S(x, \lambda)$ that

$$\bigvee_a^{a+1} \left\{ \frac{\partial S_k(x, \lambda)}{\partial x} \right\} \leqslant \left(\int_a^{a+1} |F(\lambda)|^2\, d\rho(\lambda) \right)^{1/2} \left(\int_a^{a+1} |\varphi_k'(x, \lambda)|^2\, d\rho(\lambda) \right)^{1/2} \tag{1.26}$$

$$(k = 1, 2).$$

Then the estimates (1.25) follow from the inequalities (1.26) by virtue of the estimates (1.18) and (2.17), Chapter 10. This proves the lemma.

Proceeding in a similar way and making use of the estimates (2.19), Chapter 10, and (1.18), it is easy to prove the following:

Lemma 1.3. *If the coefficients $p(x)$ and $r(x)$ in equation* (1.1) *have derivatives which are absolutely continuous in every finite interval, then for every fixed x we have, as $a \to \infty$,*

$$\bigvee_a^{a+1} \left\{ \frac{\partial^n S_k(x, \lambda)}{\partial x^n} \right\} = \int_a^{a+1} |F(\lambda)|\, |\varphi_k^{(n)}(x, \lambda)|\, d\rho(\lambda) = o(a^n) \tag{1.27}$$

$$(k = 1, 2).$$

These estimates hold uniformly over every finite interval.

From the definition of the vector-function $S^*(x, \lambda)$ there immediately follows:

LEMMA 1.5. *We have, uniformly over the entire halfline* $[0,\infty)$ *as* $a\to\infty$,

$$\bigvee_a^{a+1}\left\{\frac{\partial^n S_k^*(x,\lambda)}{\partial x^x}\right\}=o(a^n)\qquad(k=1,\ 2). \tag{1.28}$$

§2. Proof of a theorem on equiconvergence

Basing our argument on formula (1.14) and on Levitan's Tauberian theorem (cf. §4, Chapter 14), we will in this section give a proof of a theorem on the equiconvergence of the expansion with respect to the eigenfunctions of the problem (1.1) + (1.2) and of the simpler problem

$$\begin{pmatrix}0&1\\-1&0\end{pmatrix}\frac{dy}{dx}=\lambda Iy,\qquad y_2(0)-hy_1(0)=0 \tag{2.1}$$

for an arbitrary vector-function which is square integrable on $[0,\infty)$.

We remark that if $h=0$ in (2.1), then from (2.23), Chapter 10, we obtain

$$\psi_1(x,\lambda)=\cos\lambda x,\qquad \psi_2(x,\lambda)=\sin\lambda x,$$

and therefore in this case the expansion of a vector-function from $\mathscr{L}^2(0,\infty)$ with respect to the eigenfunctions of the Dirac system is the same as its expansion as an ordinary Fourier integral.

We now take up the problem of transforming the right side of (1.14). Inserting the expression for the matrix $W(x,s;\epsilon)$ in (1.14), we obtain

$$\begin{aligned}\int_{-\infty}^{\infty}\psi_\varepsilon(x)\,d_\lambda\{S(x,\lambda)-S^*(x,\lambda)\}&=\int_{x-\varepsilon}^{x+\varepsilon}\Big\{H\Big[k(x,\ s-x)-\frac{1}{2}\Big]g_\varepsilon(s-x)\\&\quad+H^T\Big[l(x,\ x-s)-\frac{1}{2}\Big]g_\varepsilon(x-s)\Big\}f(s)\,ds\\&\quad+\frac{1}{2}\int_{x-\varepsilon}^{x+\varepsilon}\Big\{\int_{|x-s|}^{\varepsilon}H(x,\ t,\ s)\,g_\varepsilon(t)\,dt\\&\quad+\int_{-\varepsilon}^{-|x-s|}\tilde{H}(x,\ t,\ s)\,g_\varepsilon(t)\,dt\Big\}f(s)\,ds=I_1+I_2.\end{aligned} \tag{2.2}$$

Changing variables, we have

$$I_1 = H\int_{-\varepsilon}^{\varepsilon}\left\{k(x,\ t)-\frac{1}{2}\right\}g_\varepsilon(t)\,f(x+t)\,dt \tag{2.3}$$
$$+H^T\int_{-\varepsilon}^{\varepsilon}\left\{l(x,\ t)-\frac{1}{2}\right\}g_\varepsilon(t)\,f(x-t)\,dt.$$

If we now put

$$\alpha^{(1)}(x,\ \lambda)=\int_{-\varepsilon}^{\varepsilon}H\left\{k(x,\ t)-\frac{1}{2}\right\}f(x+t)\,e^{i\lambda t}dt, \tag{2.4}$$
$$\alpha^{(2)}(x,\ \lambda)=\int_{-\varepsilon}^{\varepsilon}H^T\left\{l(x,\ t)-\frac{1}{2}\right\}f(x-t)\,e^{i\lambda t}dt,$$

$$\alpha(x,\ \lambda)=\alpha^{(1)}(x,\ \lambda)+\alpha^{(2)}(x,\ \lambda), \tag{2.4'}$$

then by the Parseval equality for the classical Fourier transform we find, from (2.4) and the equality (cf. (1.10), Chapter 10)

$$\psi_\varepsilon(\lambda)=\int_{-\varepsilon}^{\varepsilon}g_\varepsilon(t)\,e^{i\lambda t}dt, \tag{2.5}$$

that

$$I_1=\frac{1}{2\pi}\int_{-\infty}^{\infty}\psi_\varepsilon(\lambda)\,\overline{\alpha(x,\ \lambda)}\,d\lambda. \tag{2.6}$$

Changing the order of integration in the second integral in the right side of (2.2), we bring it to the form

$$I_2=\frac{1}{2}\int_0^{\varepsilon}\left\{\int_{x-t}^{x+t}H(x,\ t,\ s)\,f(s)\,ds\right\}g_\varepsilon(t)\,dt \tag{2.7}$$
$$+\frac{1}{2}\int_{-\varepsilon}^{0}\left\{\int_{x+t}^{x-t}\tilde{H}(x,\ t,\ s)\,f(s)\,ds\right\}g_\varepsilon(t)\,dt.$$

Let us further put

$$\mathfrak{b}(x,\ t)=\begin{pmatrix}b_1(x,\ t)\\ b_2(x,\ t)\end{pmatrix}=\begin{cases}\dfrac{1}{2}\displaystyle\int_{x-t}^{x+t}H(x,\ t,\ s)\,f(s)\,ds, & 0\leqslant t\leqslant\varepsilon,\\[2ex] \dfrac{1}{2}\displaystyle\int_{x+t}^{x-t}\tilde{H}(x,\ t,\ s)\,f(s)\,ds, & -\varepsilon\leqslant t\leqslant 0.\end{cases} \tag{2.8}$$

and

$$\beta(x, \lambda) = \int_{-\varepsilon}^{\varepsilon} b(x, t) e^{i\lambda t} dt. \tag{2.8'}$$

Applying the Parseval equality for the classical Fourier transform, we obtain from (2.8′) and (2.5)

$$I_2 = \frac{1}{2\pi} \int_{-\infty}^{\infty} \psi_\varepsilon(\lambda) \overline{\beta(x, \lambda)}\, d\lambda. \tag{2.9}$$

Therefore, taking into account (2.3), (2.6), (2.7) and (2.9), formula (2.2) can be rewritten in the final form

$$\int_{-\infty}^{\infty} \psi_\varepsilon(\lambda)\, d_\lambda R(x, \lambda) = 0, \tag{2.10}$$

where

$$R(x, \lambda) = S(x, \lambda) - S^*(x, \lambda) - \frac{1}{2\pi} \int_0^\lambda \{\overline{\alpha(x, \nu) + \beta(x, \nu)}\}\, d\nu. \tag{2.11}$$

We will now prove several lemmas.

Lemma 2.1. *If the coefficients $p(x)$ and $r(x)$ are summable over every finite interval, then for every fixed x we have, for $a \to \infty$,*

$$\bigvee_a^{a+1} \{R_k(x, \lambda)\} = o(1) \quad (k = 1, 2), \qquad R(x, \lambda) = \begin{pmatrix} R_1(x, \lambda) \\ R_2(x, \lambda) \end{pmatrix}. \tag{2.12}$$

These estimates hold uniformly in every finite interval.

Proof. Since the total variation of a sum does not exceed the sum of the total variations of the summands, to prove the estimates (2.12) it is sufficient to show that these estimates hold for each term of the sum (2.11). For the elements of the vector-functions $S(x, \lambda)$ and $S^*(x, \lambda)$ the estimates (2.12) coincide with the estimates (1.15) and (1.19). Further, since the vector-functions

$$H[k(x,t) - \tfrac{1}{2}]f(x+t) \quad \text{and} \quad H^T[l(x,t) - \tfrac{1}{2}]f(x-t)$$

are summable, the Riemann-Lebesgue lemma is applicable, according to which

$$\lim_{\lambda \to \infty} \alpha_k(x, \lambda) = 0 \quad (k = 1, 2), \qquad \alpha(x, \lambda) = \begin{pmatrix} \alpha_1(x, \lambda) \\ \alpha_2(x, \lambda) \end{pmatrix}.$$

Therefore

$$\bigvee_a^{a+1}\left\{\int_0^\lambda \alpha_k(x,\nu)\,d\nu\right\} = \int_a^{a+1} |\alpha_k(x,\nu)|\,d\nu = o(1) \quad (k=1,2). \tag{2.13}$$

Since the vector-function $b(x,t)$ is even differentiable, the estimate (2.12) for the functions

$$\int_0^\lambda \beta_k(x,\nu)\,d\nu \quad (k=1,2), \quad \beta(x,\nu) = \begin{pmatrix}\beta_1(x,\nu)\\ \beta_2(x,\nu)\end{pmatrix}$$

can be obtained from the Riemann-Lebesgue lemma in the same way that the estimates (2.13) were obtained. This proves the lemma.

Lemma 2.2. *Let $p(x)$ and $r(x)$ be measurable and locally bounded, and let $f(x)$ be an arbitrary vector-function from $\mathscr{L}^2(0,\infty)$. Then for every interval $0 < x \leq x_0$ there exists a constant $C = C(x_0)$ such that for $t \to 0$ we have*

$$|b_k(x,t)| < Ct^{1/2} \quad (k=1,2). \tag{2.14}$$

Proof. We will prove (2.14) for the function $b_1(x,t)$; the proof for $b_2(x,t)$ is analogous. By definition

$$b_1(x,t) = \frac{1}{2}\int_{x-t}^{x+t} \{H_{11}(x,t,s)f_1(s) + H_{12}(x,t,s)f_2(s)\}\,ds. \tag{2.15}$$

Let x be an arbitrary point of the halfline $[0,\infty)$. Since the integral with respect to t in (2.8′) has the limits $-\epsilon, \epsilon$, choosing $\epsilon \leq x$ we obtain $x+t \geq 0$, $x-t \geq 0$. Therefore, applying the Cauchy-Bunjakovskiĭ inequality to the integral in the right side of (2.15), we find that

$$|b_1(x,t)| \leqslant C_1\int_{x-t}^{x+t} |f_1(s)|\,ds + C_2\int_{x-t}^{x+t} |f_2(s)|\,ds$$

$$\leqslant C_1' t^{1/2}\left(\int_0^\infty f_1^2(s)\,ds\right)^{1/2} + C_2' t^{1/2}\left(\int_0^\infty f^2(s)\,ds\right)^{1/2} = Ct^{1/2},$$

where we have put

$$C_k = \max_{x-t\leq s\leq x+t} |H_{1k}(x,t,s)| \quad (k=1,2).$$

Lemma 2.3. *Under the hypotheses of the preceding lemma we have, for every fixed x,*

(2.16)
$$\lim_{\lambda\to\infty}\int_{-\lambda}^{\lambda}\beta_k(x,\nu)\,d\nu=0\qquad(k=1,\ 2).$$

These relations hold uniformly in every finite interval.

PROOF. As in the proof of Lemma 2.2, we will prove (2.16) only for the function $\beta_1(x,\nu)$. From its definition, i.e. (2.8), follows

(2.17)
$$\int_{-\lambda}^{\lambda}\beta_1(x,\nu)\,d\nu=\int_{-\lambda}^{\lambda}\left\{\int_{-\varepsilon}^{\varepsilon}b_1(x,t)\,e^{i\nu t}dt\right\}d\nu$$
$$=\int_{-\varepsilon}^{\varepsilon}b_1(x,t)\left\{\int_{-\lambda}^{\lambda}e^{i\nu t}d\nu\right\}dt=2\int_{-\varepsilon}^{\varepsilon}b_1(x,t)\frac{\sin\lambda t}{t}\,dt.$$

Further, since it follows from (2.15) that $b_1(x,0)=0$, by virtue of (2.14) we have $b_1(x,t)-b_1(x,0)=O(t^\delta)$ $(\delta>0)$. Therefore the assertion of the lemma regarding $\beta_1(x,\nu)$ follows from (2.17), thanks to (3.17), Chapter 10. This proves the lemma.

LEMMA 2.4. *If the coefficients $p(x)$ and $r(x)$ are summable over every finite interval, then for any fixed x we have*

(2.18)
$$\lim_{\lambda\to\infty}\int_{-\lambda}^{\lambda}\alpha_k(x,\nu)\,d\nu=0\qquad(k=1,\ 2).$$

These relations holds uniformly in every finite interval.

PROOF. Once again we will prove the lemma only for $\alpha_1(x,\nu)$. From the definition of $\alpha_1(x,\nu)$, i.e. (2.4) – (2.4′), it follows that

$$\alpha_1(x,\nu)=\alpha_1^{(1)}(x,\nu)+\alpha_1^{(2)}(x,\nu).$$
$$=\int_{-\varepsilon}^{\varepsilon}\left\{\left[k(x,t)-\frac{1}{2}\right]f_1(x+t)+i\left[k(x,t)-\frac{1}{2}\right]f_2(x+t)\right\}e^{i\nu t}dt$$
$$+\int_{-\varepsilon}^{\varepsilon}\left\{\left[l(x,t)-\frac{1}{2}\right]f_1(x-t)-i\left[l(x,t)-\frac{1}{2}\right]f_2(x-t)\right\}e^{i\nu t}dt.$$

Therefore

$$\int_{-\lambda}^{\lambda}\alpha_1(x,\nu)\,d\nu=$$

$$= 2 \int_{-\varepsilon}^{\varepsilon} \left\{ \left[k(x,t) - \frac{1}{2} \right] f_1(x+t) + i \left[k(x,t) - \frac{1}{2} \right] f_2(x+t) \right\} \frac{\sin \lambda t}{t} dt$$

$$+ 2 \int_{-\varepsilon}^{\varepsilon} \left\{ \left[l(x,t) - \frac{1}{2} \right] f_1(x-t) - i \left[l(x,t) - \frac{1}{2} \right] f_2(x-t) \right\} \frac{\sin \lambda t}{t} dt.$$

Changing variables in the second integral in the right side, we obtain

$$\int_{-\lambda}^{\lambda} \alpha_1(x,\nu)\, d\nu = 2 \int_{-\varepsilon}^{\varepsilon} [k(x,t) + l(x,-t) - 1] f_1(x+t) \frac{\sin \lambda t}{t} dt \tag{2.19}$$

$$+ 2 \int_{-\varepsilon}^{\varepsilon} [k(x,t) - l(x,-t)] f_2(x+t) \frac{\sin \lambda t}{t} dt.$$

Since

$$k(x,t) + l(x,-t) = \cos q(x,t), \qquad k(x,t) - l(x,-t) = i \sin q(x,t),$$

where we have put

$$q(x,t) = \frac{1}{2} \int_0^t \{p(x+\tau) + r(x+\tau)\}\, d\tau,$$

(2.19) can be rewritten in the form

$$\int_{-\lambda}^{\lambda} \alpha_1(x,\nu)\, d\nu = -4 \int_{-\varepsilon}^{\varepsilon} \sin^2 \left\{ \frac{1}{2} q(x,t) \right\} f_1(x+t) \frac{\sin \lambda t}{t} dt$$

$$+ 2i \int_{-\varepsilon}^{\varepsilon} \sin \{q(x,t)\} f_2(x+t) \frac{\sin \lambda t}{t} dt,$$

from which (2.18) for the function $\alpha_1(x,\nu)$ follows on the basis of the Riemann-Lebesgue lemma. This proves the lemma.

Lemmas 2.1 – 2.4 easily imply the following, which is basic for this section.

THEOREM 2.1 (ON EQUICONVERGENCE). *Let $f(x) \in \mathcal{L}^2(0,\infty)$. If the coefficients $p(x)$ and $r(x)$ in equation* (1.1) *are measurable and locally bounded, then we have, uniformly in every finite interval,*

$$\lim_{\lambda \to \infty} \{ [S_k(x,\lambda) - S_k(x,-\lambda)] - [S_k^*(x,\lambda) - S_k^*(x,-\lambda)] \} = 0, \tag{2.20}$$

i.e. the difference between the expansions of the vector-function $f(x)$ as a generalized Fourier integral with respect to the eigenfunctions of the problems (1.1) – (1.2) *and* (2.1) *tends to zero uniformly in every finite interval.*

Proof. By virtue of the estimates (2.12) and the equality (2.10), Levitan's Tauberian theorem (cf. Theorem 4.1, Chapter 14) is applicable, on the basis of which we have for every fixed x

$$\lim_{\lambda\to\infty}\{R_k(x,\lambda) - R_k(x,-\lambda)\} = 0 \qquad (k=1,2),$$

i.e. according to (2.11)

$$\lim_{\lambda\to\infty}\{[S_k(x,\lambda) - S_k(x,-\lambda)] - [S_k^*(x,\lambda) - S_k^*(x,-\lambda)]\}$$

$$= \frac{1}{2\pi}\lim_{\lambda\to\infty}\int_{-\lambda}^{\lambda}\{\overline{\alpha(x,\nu)} - \overline{\beta(x,\nu)}\}\,d\nu.$$

Then (2.20) follows by virtue of Lemmas 2.3 and 2.4. This proves the theorem.

Theorem 2.1 gives a definitive solution to the question of the convergence of the expansion of a vector-function, which is square integrable over $[0,\infty)$, with respect to the eigen-vector-functions of a one-dimensional Dirac system. Indeed, if we put $h=0$ in the problem (2.1), then an expansion as a generalized Fourier integral with respect to the eigen-vector-functions of this problem is just an ordinary Fourier integral expansion since, as was noted at the beginning of this section, for $h=0$ the eigen-vector-functions of this problem are

$$\psi(x,\lambda) = \begin{pmatrix}\cos\lambda x\\ \sin\lambda x\end{pmatrix}.$$

Consequently, for $h=0$ Theorem 2.1 can be formulated as follows:

Theorem 2.2. *Suppose that $f(x)\in\mathscr{L}^2(0,\infty)$ and that the hypotheses of Theorem* 2.1 *are satisfied. Then the difference between the expansion of $f(x)$ with respect to the eigen-vector-functions of a one-dimensional Dirac system and its expansion as an ordinary Fourier integral tends to zero uniformly in every finite interval.*

In particular, there follows from Theorem 2.2:

Theorem 2.3. *Suppose that the hypotheses of Theorem* 2.1 *are satisfied. Then at every point x_0 at which local conditions for the expandibility of $f(x)$ as an ordinary Fourier integral are satisfied, we have*

$$\lim_{\lambda\to\infty}\{S(x_0,\lambda) - S(x_0,-\lambda)\} = f(x_0),$$

i.e. the expansion of $f(x)$ as a generalized Fourier integral with respect to *the eigen-vector-functions of a one-dimensional Dirac system tends at the point x_0 to the value $f(x_0)$.*

§3. Proof of a theorem on the equisummability of differentiated expansions

In this section, basing our argument on (1.25) and the aforementioned Tauberian theorem of Levitan, we will prove a theorem on the Riesz equisummability of the differentiated expansions with respect to the eigen-vector-functions of the problem (1.1) + (1.2) and of the simpler problem (2.1).

We will first take up the problem of transforming the right side of (1.24). Inserting the expression for the matrix $W(x, s; \epsilon)$ from (1.5)[1] in (1.24), first differentiating with respect to x, we obtain

$$\int_{-\infty}^{\infty} \psi_\varepsilon(\lambda)\, d_\lambda \left\{\frac{\partial S(x, \lambda)}{\partial x} - \frac{\partial S^*(x, \lambda)}{\partial x}\right\}$$

$$\begin{aligned} &= \int_{x-\varepsilon}^{x+\varepsilon} \frac{\partial A(x, s)}{\partial x} f(s)\, g_\varepsilon(s-x)\, ds \\ &+ \int_{x-\varepsilon}^{x+\varepsilon} [A(x, s) - I] f(s) \frac{\partial g_\varepsilon(s-x)}{\partial x}\, ds \\ &- \int_{x-\varepsilon}^{x+\varepsilon} \{H(x, t, s) - \tilde{H}(x, -t, s)\}\big|_{t=|x-s|} f(s)\, g_\varepsilon(s-x)\, ds \\ &+ \int_{x-\varepsilon}^{x+\varepsilon} \left\{\int_{|x-s|}^{\varepsilon} H'_x(x, t, s) + \tilde{H}'_x(x, -t, s)]\, g_\varepsilon(t)\, dt\right\} f(s)\, ds \\ &= I_1 + I_2 + I_3 + I_4, \end{aligned} \tag{3.1}$$

where

$$A(x, s) = \begin{pmatrix} \cos q(x, s), & -\sin q(x, s) \\ \sin q(x, s), & \cos q(x, s) \end{pmatrix}, \tag{3.1'}$$

and

$$q(x, s) = \frac{1}{2} \int_0^s \{p(x+\tau) + r(x+\tau)\}\, d\tau. \tag{3.1''}$$

---

[1] For our purposes we can restrict ourselves in the sequel to the case of an even function $g_\epsilon(t)$. For this case the expression for $W(x, s; \epsilon)$ simplifies somewhat.

Suppose that the vector-function $f(x)$ is differentiable in an ϵ-neighborhood of the point x. Then integrating by parts in I_2, we can bring it to the form

$$I_2 = \int_{x-\varepsilon}^{x+\varepsilon} \frac{\partial A(x, s)}{\partial s} f(s) g_\varepsilon(s-x)\, ds + \int_{x-\varepsilon}^{x+\varepsilon} [A(x, s) - I] f'(s) g_\varepsilon(s-x)\, ds.$$

Therefore

$$I_1 + I_2 = \int_{x-\varepsilon}^{x+\varepsilon} \left\{\left[\frac{\partial A(x, s)}{\partial x} + \frac{\partial A(x, s)}{\partial s}\right] f(s) + [A(x, s) - I] f'(s)\right\} g_\varepsilon(s-x)\, ds,$$

from which we obtain, by the change of variables $s - x = t$,

$$I_1 + I_2 = \int_{-\varepsilon}^{\varepsilon} \left\{\left[\frac{\partial A(x, s)}{\partial x} + \frac{\partial A(x, s)}{\partial s}\right]_{s=x+t} f(x+t) + [A(x, x+t) - I] f'(x+t)\right\} g_\varepsilon(t)\, dt. \tag{3.2}$$

If we now put

$$\alpha(x, \lambda) = \int_{-\varepsilon}^{\varepsilon} \left\{\left[\frac{\partial A(x, s)}{\partial x} + \frac{\partial A(x, s)}{\partial s}\right]_{s=x+t} f(x+t) + [A(x, x+t) - I] f'(x+t)\right\} \cos \lambda t\, dt, \tag{3.3}$$

then it follows from (2.5) by the Parseval equality for the Fourier transform that[2)]

$$I_1 + I_2 = \frac{1}{\pi} \int_{-\infty}^{\infty} \psi_\varepsilon(\lambda)\, \alpha(x, \lambda)\, d\lambda = \int_{-\infty}^{\infty} \psi_\varepsilon(\lambda)\, d_\lambda \left\{\frac{1}{\pi} \int_0^\lambda \alpha(x, \nu)\, d\nu\right\}. \tag{3.4}$$

Further, by a change of variables the integral I_3 can be represented in the form

$$I_3 = -\int_{-\varepsilon}^{\varepsilon} \{H(x, |t|, x+t) + \tilde{H}(x, -|t|, x+t)\} f(x+t) g_\varepsilon(t)\, dt$$

and therefore, putting

[2)] Note that for even $g_\epsilon(t)$ the relation (2.5) assumes the form

$$\psi_\varepsilon(\lambda) = \int_{-\varepsilon}^{\varepsilon} g_\varepsilon(t) \cos \lambda t\, dt.$$

$$\beta(x,\lambda)=-\int_{-\varepsilon}^{\varepsilon}\{H(x,|t|,x+t)+\tilde H(x,-|t|,x+t)\}f(x+t)\cos\lambda t\,dt \tag{3.5}$$

and again applying the Parseval equality, we obtain from (3.5) and (2.5)

$$I_3=\frac{1}{\pi}\int_{-\infty}^{\infty}\psi_\varepsilon(\lambda)\beta(x,\lambda)\,d\lambda=\int_{-\infty}^{\infty}\psi_\varepsilon(\lambda)\,d_\lambda\left\{\frac{1}{\pi}\int_0^\lambda\beta(x,\nu)\,d\nu\right\}. \tag{3.6}$$

Finally, we transform the integral I_4. Changing the order of integration, we find that

$$I_4=\int_{-\varepsilon}^{\varepsilon}\left\{\int_{x-t}^{x+t}\frac{\partial}{\partial x}[H(x,t,s)+\tilde H(x,-t,s)]f(s)\,ds\right\}g_\varepsilon(t)\,dt.$$

Now, putting

$$\gamma(x,\lambda)=\int_{-\varepsilon}^{\varepsilon}\left\{\int_{x-t}^{x+t}\frac{\partial}{\partial x}[H(x,t,s)+\tilde H(x,-t,s)]f(s)\,ds\right\}\cos\lambda t\,dt \tag{3.7}$$

and applying the Parseval equality, we obtain from (3.7) and (2.5)

$$I_4=\frac{1}{\pi}\int_{-\infty}^{\infty}\psi_\varepsilon(\lambda)\gamma(x,\lambda)\,d\lambda=\int_{-\infty}^{\infty}\psi_\varepsilon(\lambda)\,d_\lambda\left\{\frac{1}{\pi}\int_0^\lambda\gamma(x,\nu)\,d\nu\right\}. \tag{3.8}$$

Consequently, by virtue of (3.4), (3.6) and (3.8) we can rewrite (3.1) in the form

$$\int_{-\infty}^{\infty}\psi_\varepsilon(\lambda)\,d_\lambda\Phi(x,\lambda)=0, \tag{3.9}$$

where

$$\Phi(x,\lambda)=\frac{\partial S(x,\lambda)}{\partial x}-\frac{\partial S^*(x,\lambda)}{\partial x}-\frac{1}{\pi}\int_0^\lambda\alpha(x,\nu)\,d\nu-\frac{1}{\pi}\int_0^\lambda\beta(x,\nu)\,d\nu-\frac{1}{\pi}\int_0^\lambda\gamma(x,\nu)\,d\nu. \tag{3.10}$$

We will precede the proof of the basic theorem of this section with several lemmas.

Lemma 3.1. *If the coefficients $p(x)$ and $r(x)$ and the vector-function $f(x)$ have locally bounded derivatives in every finite interval, then for every fixed x we have, as $a\to\infty$,*

$$\bigvee_a^{a+1} \{\Phi_k(x,\lambda)\} = o(a) \qquad (k=1,2),$$

(3.11)

$$\Phi(x,\lambda) = \begin{pmatrix} \Phi_1(x,\lambda) \\ \Phi_2(x,\lambda) \end{pmatrix}.$$

These estimates hold uniformly in every finite interval.

Proof. It is obvious that to prove the estimates (3.11) it is sufficient to prove them for each term appearing in (3.10). For the components of the vector-functions $\partial S(x,\lambda)/\partial x$ and $\partial S^*(x,\lambda)/\partial x$ the estimates (3.11) follow from the estimates (1.25) and (1.28). Further, since the vector-function

$$\left[\frac{\partial A(x,s)}{\partial x} - \frac{\partial A(x,s)}{\partial s}\right]_{s=x+t} f(x+t) \tag{3.12}$$
$$+ [A(x,x+t) - I] f'(x+t) \equiv B(x,t)$$

is summable (in a neighborhood of the point $t=0$), the Riemann-Lebesgue lemma is applicable, according to which

$$\lim_{n\to\infty} \alpha_k(x,\lambda) = 0 \qquad (k=1,2), \quad \alpha(x,\lambda) = \begin{pmatrix} \alpha_1(x,\lambda) \\ \alpha_2(x,\lambda) \end{pmatrix}.$$

Therefore

$$\bigvee_a^{a+1} \left\{ \int_0^\lambda \alpha_k(x,\nu) \right\} = \int_a^{a+1} |\alpha_k(x,\lambda)|\, d\nu = o(1) \quad (k=1,2).$$

Further, proceeding similarly, we conclude from the definition of the vector-functions $\beta(x,\lambda)$ and $\gamma(x,\lambda)$ and the Riemann-Lebesgue lemma that

$$\bigvee_a^{a+1} \left\{ \int_0^\lambda \beta_k(x,\nu)\, d\nu \right\} = o(1), \quad \bigvee_a^{a+1} \left\{ \int_0^\lambda \gamma_k(x,\lambda)\, d\nu \right\} = o(1)$$
$$(k=1,2).$$

This proves the lemma.

Lemma 3.2. *Suppose that the vector-function $f(x)$ has a square-summable derivative in an ϵ-neighborhood of the point x, and that the coefficients $p(x)$ and $r(x)$ in equation* (1.1) *have locally bounded first derivatives. Then we have*

$$\lim_{\lambda\to\infty} \int_0^\lambda \left(1 - \frac{\nu^2}{\lambda^2}\right) \alpha_k(x,\nu)\, d\nu = 0 \qquad (k=1,2). \tag{3.13}$$

Proof. From the definition of the matrix $A(x,s)$, i.e. from (3.1′) and (3.1″), there follows

$$(3.14)\quad \left.\frac{\partial A(x, s)}{\partial x}\right|_{s=x+t} = \frac{1}{2} h(x) \begin{pmatrix} \sin q(x, x+t) & \cos q(x, x+t) \\ -\cos q(x, x+t) & \sin q(x, x+t) \end{pmatrix},$$

$$(3.15)\quad \left.\frac{\partial A(x, s)}{\partial s}\right|_{s=x+t} = -\frac{1}{2} h(x+h) \begin{pmatrix} \sin q(x, x+t) \cos q(x, x+t) \\ -\cos q(x, x+t) \sin q(x, x+t) \end{pmatrix},$$

where

$$(3.16)\quad q(x, x+t) = \frac{1}{2} \int_0^t h(x+\tau)\, d\tau, \quad h(x) = p(x) + r(x).$$

Further, by virtue again of (3.1′) we have

$$(3.17)\quad A(x, x+t) - I = \begin{pmatrix} 2\sin^2\left\{\frac{1}{2} q(x, x+t)\right\} & \sin q(x, x+t) \\ -\sin q(x, x+t) & 2\sin^2\left\{\frac{1}{2} q(x, x+t)\right\} \end{pmatrix}.$$

We will now prove (3.13) for the function $\alpha_1(x,\nu)$. The proof for $\alpha_2(x,\nu)$ is similar since, as is not hard to see, these functions contain terms of the same type. From the definition of $\alpha_1(x,\nu)$ (cf. (3.3)) it follows that it is the Fourier cosine transform of the first component of the vector-function (3.12), which by (3.14), (3.15) and (3.17) has the form

$$(3.18)\quad \begin{aligned} B_1(x, t) = &-\frac{1}{2}[h(x+t) - h(x)] f_1(x+t) \sin q(x, x+t) \\ &-\frac{1}{2}[h(x+t) - h(x)] f_2(x+t) \cos q(x, x+t) \\ &+ 2f_1'(x+t) \sin\left\{\frac{1}{2} q(x, x+t)\right\} + f_2'(x+t) \sin q(x, x+t) \\ &\equiv b_1 + b_2 + b_3 + b_4. \end{aligned}$$

Therefore $\alpha_1(x,\lambda)$ can be represented in the form

$$\alpha_1(x, \lambda) = \sum_{k=1}^{4} \int_{-\varepsilon}^{\varepsilon} b_k(x, t) \cos \lambda t\, dt.$$

Consequently, to prove the lemma we have to prove the relation

$$(3.19)\quad \lim_{\lambda\to\infty} \sum_{k=1}^{4} \int_{-\lambda}^{\lambda} \left(1 - \frac{\nu^2}{\lambda^2}\right) \left\{ \int_{-\varepsilon}^{\varepsilon} b_k(x, t) \cos \nu t\, dt \right\} d\nu = 0.$$

We will prove (3.19) for each term separately. We have

$$I_1=\int_{-\lambda}^{\lambda}\left(1-\frac{\nu^2}{\lambda^2}\right)\left\{\int_{-\varepsilon}^{\varepsilon} b_1(x,\ t)\cos\nu t\,dt\right\}d\nu$$
$$=2\int_{-\varepsilon}^{\varepsilon} b_1(x,\ t)\left\{\int_{0}^{\lambda}\left(1-\frac{\nu^2}{\lambda^2}\right)\cos\nu t\,d\nu\right\}dt.$$

Inserting the expression for $b_1(x,t)$ from (3.18) and taking account of (4.27), Chapter 10, we find that

$$I_1=-\sqrt{\frac{\pi}{2}}\frac{1}{\sqrt{\lambda}}\int_{-\varepsilon}^{\varepsilon}[h(x+t)-h(x)]$$
$$\times f_1(x+t)\sin q(x,\ x+t)\frac{I_{3/2}(\lambda t)}{t^{3/2}}dt.$$

Since by hypothesis the functions $p(x)$ and $r(x)$ have locally bounded derivatives, it follows from the definition of the function $h(x)$ that as $t\to 0$

(3.20) $$h(x+t)-h(x)=O(t).$$

Therefore, using also the boundedness of the function $f_1(x+t)$, we obtain

$$|I_1|\leqslant C_1\frac{1}{\sqrt{\lambda}}\int_{-\varepsilon}^{\varepsilon}|t|^{1/2}\left|\frac{I_{3/2}(\lambda t)}{t^{3/2}}\right|dt=\frac{C_1}{\sqrt{\lambda}}\int_{-\varepsilon\lambda}^{\varepsilon\lambda}\left|\frac{I_{3/2}(u)}{u}\right|du=\frac{\tilde{C}_1}{\sqrt{\lambda}},$$

from which it follows that as $\lambda\to\infty$

(3.21) $$I_1\to 0.$$

For the term in (3.19) which contains $b_2(x,t)$, the proof is completely analogous. The proofs for the last two terms are similar. We will carry out the proof for the last term.

By definition we have

$$I_4=\int_{-\lambda}^{\lambda}\left(1-\frac{\nu^2}{\lambda^2}\right)\left\{\int_{-\varepsilon}^{\varepsilon} b_4(x,\ t)\cos\nu t\,dt\right\}d\nu$$
$$=2\int_{-\varepsilon}^{\varepsilon} b_4(x,\ t)\sqrt{\frac{\pi}{2}}\frac{1}{\sqrt{\lambda}}\frac{I_{3/2}(\lambda t)}{t^{3/2}}dt.$$

Inserting the expression for $b_4(x,t)$ from (3.18), we obtain

$$I_4=2\sqrt{\frac{\pi}{2}}\cdot\frac{1}{\sqrt{\lambda}}\int_{-\varepsilon}^{\varepsilon} f_2'(x+t)\sin q(x,\ x+t)\frac{I_{3/2}(\lambda t)}{t^{3/2}}dt.$$

Since by (3.16) we have $\sin q(x, x+t) = O(t)$ for small t, it follows that

$$|I_4| \leqslant \frac{C_2}{\sqrt{\lambda}} \int_{-\varepsilon}^{\varepsilon} |f_2'(x+t)|\,|t| \left| \frac{I_{3/2}(\lambda t)}{t^{3/2}} \right| dt$$

$$\leqslant \frac{C_2}{\sqrt{\lambda}} \left(\int_{-\varepsilon}^{\varepsilon} |f'(x+t)|^2 dt \right)^{1/2} \left(\int_{-\varepsilon}^{\varepsilon} t^2 \frac{I_{3/2}^2(\lambda t)}{t^3} dt \right)^{1/2}$$

$$\leqslant \frac{C_2'}{\sqrt{\lambda}} \left(\int_{-\varepsilon\lambda}^{\varepsilon\lambda} \frac{I_{3/2}^2(u)}{u} du \right)^{1/2} \leqslant \frac{\tilde{C}_2}{\sqrt{\lambda}}.$$

Hence $I_4 \to 0$ as $\lambda \to \infty$, which completes the proof of the lemma.

Lemma 3.3. *Let $f(x) \in \mathscr{L}^2(0, \infty)$. If the coefficients $p(x)$ and $r(x)$ in equation* (1.1) *have locally bounded derivatives, then for every fixed x we have*

$$\lim_{\lambda \to \infty} \int_{-\lambda}^{\lambda} \left(1 - \frac{\nu^2}{\lambda^2}\right) \beta_k(x, \nu)\, d\nu = 0 \qquad (k = 1, 2). \tag{3.22}$$

These relations hold uniformly in every finite interval.

Proof. By the differentiability of the matrix

$$H(x, |t|, x+t) + \tilde{H}(x, -|t|, x+t)$$

with respect to t, as well as (2.16), (2.17) and (5.12), Chapter 9, we have (for small t)

$$H(x, |t|, x+t) + \tilde{H}(x, -|t|, x+t) = Ct.$$

Therefore the proof of (3.22) can be carried out in exactly the same way as was the proof of (3.21).

Lemma 3.4. *Let $f(x) \in \mathscr{L}^2(0, \infty)$. If the coefficients $p(x)$ and $r(x)$ in equation* (1.1) *have locally bounded derivatives, then for every fixed x we have*

$$\lim_{\lambda \to \infty} \int_{-\lambda}^{\lambda} \left(1 - \frac{\nu^2}{\lambda^2}\right) \gamma_k(x, \nu)\, d\nu = 0 \qquad (k = 1, 2). \tag{3.23}$$

These relations hold uniformly in every finite interval.

Proof. Proceeding as in the proof of Lemma 2.2, we establish that as $t \to 0$

$$|c_k(x, t)| = 2Ct \qquad (k = 1, 2), \tag{3.24}$$

where

$$C(x, t) = \begin{pmatrix} c_1(x, t) \\ c_2(x, t) \end{pmatrix} = \int_{x-t}^{x+t} \frac{\partial}{\partial x} [H(x, t, s) + \tilde{H}(x, t, s)]\, f(s)\, ds,$$

$$C = \max_{x-t \leqslant s \leqslant x+t} \left| \frac{\partial}{\partial x} \{H(x, t, s) + \tilde{H}(x, t, s)\} \right|.$$

Then using (3.24), the proof can be completed in the same way as that of (3.21).

From Lemmas 3.1 – 3.4 and (3.9) it easily follows that:

THEOREM 3.1 (ON EQUISUMMABILITY). *Suppose that the vector-function* $f(x)$ *belongs to* $\mathscr{L}^2(0, \infty)$ *and satisfies the following conditions:*

1. *the vector-function* $f'(x)$ *is square-summable in some neighborhood of the point* x_0,
2. *the coefficients* $p(x)$ *and* $r(x)$ *in equation* (1.1) *have locally bounded first derivatives.*

Then we have

$$\lim_{\lambda \to \infty} \int_{-\lambda}^{\lambda} \left(1 - \frac{\nu^2}{\lambda^2}\right) d_\nu \left\{ \frac{\partial S_k(x, \lambda)}{\partial x} - \frac{\partial S_k^*(x, \lambda)}{\partial x} \right\}_{x=x_0} = 0 \tag{3.25}$$

$$(k = 1, 2),$$

i.e. the difference between the Riesz means of order one of the derivatives of the expansions of $f(x)$ *with respect to the eigen-vector-functions of the problems* (1.1) – (1.2) *and* (2.1) *tends to zero at the point* x_0.

PROOF. By Lemma 3.1 and (3.9), Levitan's Tauberian theorem is applicable, according to which

$$\lim_{\lambda \to \infty} \int_{-\lambda}^{\lambda} \left(1 - \frac{\nu^2}{\lambda^2}\right) d_\nu \Phi_k(x, \nu) = 0 \qquad (k = 1, 2). \tag{3.26}$$

The relation (3.25) and thus the theorem follows from (3.26), by virtue of Lemmas 3.2, 3.3 and 3.4.

If we put $h = 0$ in the problem (2.1), then, as was noted in §2, an expansion with respect to the eigen-vector-functions of this problem coincides with an ordinary Fourier integral expansion. Therefore from Theorem 3.1 and Theorem 4.2, Chapter 8, follows:

THEOREM 3.2 (ON SUMMABILITY). *Suppose that the coefficients* $p(x)$ *and* $r(x)$ *in equation* (1.1) *have locally bounded first derivatives, and that the following conditions are satisfied:*

1. $f(x) \in \mathscr{L}^2(0, \infty)$.
2. *The vector-function* $f'(x)$ *is continuous at some point* x_0.

Then

$$\lim_{\lambda\to\infty}\int_{-\lambda}^{\lambda}\left(1-\frac{\nu^2}{\lambda^2}\right)d_\nu\frac{\partial S(x,\nu)}{\partial x}\bigg|_{x=x_0}=f'(x_0),$$

i.e. the first derivative at the point x_0 *of the expansion of* $f(x)$ *with respect to the eigen-vector-functions of a one-dimensional Dirac system is summable by the Riesz method of order one to the value* $f'(x_0)$.

Bibliographical references

The results of this chapter are due to Sargsjan [**6, 10, 12**].

Chapter 12
Asymptotic Behaviour of the Number of Eigenvalues of a Sturm-Liouville Operator

§1. An integral equation for the Green's function

1. From the results of Chapter 4 we know that if the function $q(x)$ in the Sturm-Liouville operator

$$Ly = -y'' + q(x)y \qquad (a < x < b) \tag{1.1}$$

is bounded from below and tends to $+\infty$ as $x \to a$ or as $x \to b$ (or both), then the spectrum of L is discrete.[1]

In this chapter we will give another proof of this fact for the cases: 1) $a = -\infty$, $b = +\infty$ and 2) $a = 0$, $b = +\infty$. Besides the condition $q(x) \to +\infty$ (for $x \to \pm\infty$) on the function $q(x)$ we will impose additional conditions which will enable us not only to prove the discreteness of the spectrum of the operator L, but also to obtain an asymptotic formula for the function $N(\lambda) = \sum_{\lambda_n < \lambda} 1$, i.e. for the number of eigenvalues.

Since by hypothesis the function $q(x)$ is bounded from below, we may assume without loss of generality that $q(x) > 1$ for all $x \in (a, b)$.

We will consider in detail the case of the entire line, i.e. the case $a = -\infty$, $b = +\infty$. The changes required for the case of the halfline are not complicated, and will be indicated briefly.

We put for fixed x and $\mu > 0$

$$\kappa = \{q(x) + \mu\}^{1/2}, \quad g(x, \eta; \mu) = \tfrac{1}{2}\kappa^{-1}\exp(-|x - \eta|\kappa) \tag{1.2}$$

and consider the integral equation

$$G(x, \eta; \mu) = g(x, \eta; \mu) - \int_{-\infty}^{\infty} g(x, \xi; \mu)[q(\xi) - q(x)]\, G(\xi, \eta; \mu)\, d\xi. \tag{1.3}$$

It will be shown below that for sufficiently large μ equation (1.3) can be solved by an iteration method and that the solution is the Green's

---

[1] Under the assumption that at least one of the boundary points is singular. In this connection, if one of the boundary points is regular, then we have to specify a boundary condition at this point.

function of the operator (1.1). (The uniqueness of the Green's function follows from Theorem 2.3, Chapter 2.)

2. Let us denote by X the Banach space[2] of numerically valued functions $A(x,\eta)$ $(-\infty < x, \eta < +\infty)$ with the norm

$$\|A(x,\ \eta)\|_X^2 = \int_{-\infty}^{\infty}\left\{\int_{-\infty}^{\infty} |A(x,\ \eta)|^2\, d\eta\right\} dx$$

and define an operator N in X by

$$NA(x,\ \eta) = \int_{-\infty}^{\infty} g(x,\ \xi;\ \mu)[q(\xi) - q(x)]\, A(\xi,\ \eta)\, d\xi, \tag{1.4}$$

where the function $g(x,\xi;\mu)$ is given by (1.2).

Further on an important role will be played by the following:

LEMMA 1.1. *Suppose that $q(x)$ satisfies the conditions:*

1. *For* $|x-\xi| \leqq 1$

$$|q(\xi) - q(x)| \leqq Kq^a(x)\,|\xi - x|, \tag{1.5}$$

where $K > 0$ *and* $0 < a < 3/2$ *are constants;*

2. *for* $|x - \xi| > 1$

$$q(\xi)\exp(-\tfrac{1}{2}|x-\xi|\sqrt{q(x)}) \leqq B, \tag{1.6}$$

where B is a constant.

Then for sufficiently large μ the operator N is a contraction (i.e. $\|N\| < 1$).

PROOF. Put

$$NA(x,\eta) \equiv M(x,\eta) = \int_{|x-\xi|\leqslant 1} g(x,\ \xi;\ \mu)[q(\xi) - q(x)]\, A(\xi,\ \eta)\, d\xi$$

$$+ \int_{|x-\xi|>1} g(x,\ \xi;\ \mu)[q(\xi) - q(x)]\, A(\xi,\ \eta)\, d\xi = \alpha(x,\ \eta;\ \mu) + \beta(x,\ \eta;\ \mu).$$

Then

$$\begin{aligned} &|M(x,\ \eta)| \leqslant |\alpha(x,\ \eta;\ \mu)| + |\beta(x,\ \eta;\ \mu)|,\\ &M^2(x,\ \eta) \leqslant 2\{\alpha^2(x,\ \eta;\ \mu) + \beta^2(x,\ \eta;\ \mu)\}. \end{aligned} \tag{1.7}$$

Further,

---

[2] For the definition of a Banach space cf. Chapter 13.

$$(1.8)\quad \alpha^2(x,\eta;\mu) \leqslant \left\{ \int\limits_{|x-\xi|\leqslant 1} g(x, \xi; \mu)\,|q(\xi)-q(x)|\cdot|A(\xi, \eta)|\,d\xi \right\}^2.$$

By virtue of the condition (1.5) we have for $|x-\xi| \leqq 1$

$$g(x,\xi;\mu)\,|q(\xi)-q(x)| \leqq Kg(x,\xi;\mu)q^a(x)\,|\xi - x|.$$

Therefore

$$\left(\int\limits_{|x-\xi|\leqslant 1} g(x, \xi; \mu)\,|q(\xi)-q(x)|\,|A(\xi, \eta)|\,d\xi \right)^2$$
$$\leqslant K^2 \left(\int\limits_{|x-\xi|\leqslant 1} q^a(x)\, g(x, \xi; \mu)\,|x-\xi|\,|A(\xi, \eta)|\,d\xi \right)^2.$$

Inserting the value of the function $g(x,\xi;\mu)$ from (1.2), we obtain

$$\int\limits_{|x-\xi|\leqslant 1} |x-\xi|\, q^a(x)\, g(x, \xi; \mu)\,|A(\xi, \eta)|\,d\xi$$
$$(1.9)\quad = \frac{1}{2} \int\limits_{|x-\xi|\leqslant 1} |x-\xi|^{1+\varepsilon}\,|x-\xi|^{-\varepsilon}\, q^a(x)\,\varkappa^{-2+\varepsilon}$$
$$\times \varkappa^{1-\varepsilon} \exp(-\varkappa|x-\xi|)\,|A(\xi, \eta)|\,d\xi,$$

Since $\kappa = [q(x)+\mu]^{1/2}$, we have as $\mu \to +\infty$

$$(1.10)\quad q^a(x)\kappa^{-2+\epsilon} = O(1/\mu^{\zeta}),$$

where $\zeta > 0$ since by virtue of $0 < a < 3/2$ one can find $\epsilon < 1$ such that $\zeta = 1 + \frac{1}{2}\epsilon - a$ will be a positive number.

Then it follows from the estimates (1.8), (1.9) and (1.10) that

$$\alpha^2(x, \eta; \mu) \leqslant \frac{K^2}{\mu^{2\zeta}} \left(\int\limits_{|x-\xi|\leqslant 1} |x-\xi|^{-\varepsilon}\,|A(\xi, \eta)|\,d\xi \right)^2$$
$$(1.11)\quad = \frac{K^2}{\mu^{2\zeta}} \int\limits_{|x-\xi|\leqslant 1} |x-\xi|^{-\varepsilon}\,|A(\xi, \eta)|\,d\xi \int\limits_{|x-\xi'|\leqslant 1} |x-\xi'|^{-\varepsilon}\,|A(\xi', \eta)|\,d\xi'.$$

Integrating the inequality (1.11) with respect to x from $-\infty$ to $+\infty$, we obtain, following an obvious change of variables and rearrangement of the limits of integration (using also the Cauchy-Bunjakovskiĭ inequality),

$$\int\limits_{-\infty}^{\infty} \alpha^2(x, \eta; \mu)\,dx \leqslant \frac{K^2}{\mu^{2\zeta}} \int\limits_{|u|\leqslant 1} |u|^{-\varepsilon}\,du \int\limits_{|u'|\leqslant 1} |u'|^{-\varepsilon}\,du'$$
$$\times \left\{ \int\limits_{-\infty}^{\infty} |A(x+u, \eta)|\,|A(x+u', \eta)|\,dx \right\}$$

$$\leqslant \frac{K^2}{\mu^{2\zeta}} \int\limits_{|u|\leqslant 1} |u^{-\varepsilon}|\, du \int\limits_{|u'|\leqslant 1} |u'|^{-\varepsilon}\, du' \left\{ \int\limits_{-\infty}^{\infty} |A(x, \eta)|^2\, dx \right\}$$

$$= \frac{K_1}{\mu^{2\zeta}} \int\limits_{-\infty}^{\infty} |A(x, \eta)|^2\, dx.$$

Integrating now with respect to η from $-\infty$ to $+\infty$, we obtain

$$\|\alpha(x, \eta; \mu)\|_X \leqslant \frac{C}{\mu^\zeta} \|A(x, \eta)\|_X, \tag{1.12}$$

where C is a constant.

Let us now estimate $\|\beta(x, \eta; \mu)\|_X$ as $\mu \to +\infty$. By the definition of $\beta(x, \eta; \mu)$, we have

$$\beta(x, \eta; \mu) = \int\limits_{|x-\xi|>1} g(x, \xi; \mu)[q(\xi) - q(x)] A(\xi, \eta)\, d\xi$$

$$= \int\limits_{|x-\xi|>1} g(x, \xi; \mu) q(\xi) A(\xi, \eta)\, d\xi$$

$$- q(x) \int\limits_{|x-\xi|>1} g(x, \xi; \mu) A(\xi, \eta)\, d\xi = \beta_1 + \beta_2.$$

From (1.6) and (1.2) follows

$$\beta_1^2 \leqslant \frac{B^2}{4\mu^2} \int\limits_{|x-\xi|>1} \exp\left(-\frac{1}{2}\sqrt{\mu}\,|x-\xi|\right) |A(\xi, \eta)|\, d\xi$$

$$\times \int\limits_{|x-\xi'|>1} \exp\left(-\frac{1}{2}\sqrt{\mu}\,|x-\xi'|\right) |A(\xi', \eta)|\, d\xi$$

$$= \frac{B^2}{4\mu^2} \int\limits_{|u|>1} \exp\left(-\frac{1}{2}\sqrt{\mu} u\right) |A(x+u, \eta)\, du$$

$$\times \int\limits_{|u'|>1} \exp\left(-\frac{1}{2}\sqrt{\mu} u'\right) |A(x+u', \eta)|\, du'.$$

Now, just as in the derivation of the estimate (1.12), integrating the preceding inequality with respect to x from $-\infty$ to $+\infty$, we obtain

$$\int\limits_{-\infty}^{\infty} \beta_1^2(x, \eta; \mu)\, dx \leqslant \frac{4B^2}{\mu^2} e^{-\frac{1}{2}\sqrt{\mu}} \int\limits_{-\infty}^{\infty} |A(x, \eta)|^2\, dx.$$

Therefore by integrating this inequality with respect to η from $-\infty$ to $+\infty$, we obtain the needed estimate for $\|\beta_1(x, \eta; \mu)\|_X$ as $\mu \to +\infty$, namely

$$\|\beta_1(x, \eta; \mu)\|_X \leqslant \frac{C}{\mu^r} \|A(x, \eta)\|_X, \tag{1.13}$$

where C is a constant, and r is an arbitrarily large positive number.

A similar estimate holds for $\|\beta_2(x,\eta;\mu)\|_X$.

Then by virtue of the estimates (1.12) and (1.13), we obtain from (1.7) the following estimate for $NA(x,\eta)$ as $\mu \to +\infty$:

$$\|NA(x,\eta)\|_X \leqslant \frac{C}{\mu^{\zeta}} \|A(x,\eta)\|_X,$$

which proves the lemma.

Remark. Let us introduce the Banach spaces $X_2^{(p)}$, $X^{(\tau)}$, $X_1^{(\tau)}$ ($p \geqq 1$; τ is an arbitrary real number), whose elements are numerically-valued functions $A(x,\eta)$ ($-\infty < x, \eta < \infty$), with the norms defined respectively by:

$$\|A(x,\eta)\|^p_{X_2^{(p)}} = \sup_{-\infty<x<\infty} \int_{-\infty}^{\infty} |A(x,\eta)|^p \, d\eta, \tag{1.14}$$

$$\|A(x,\eta)\|^2_{X^{(\tau)}} = \int_{-\infty}^{\infty} \left\{ \int_{-\infty}^{\infty} |A(x,\eta)|^2 q^{2\tau}(\eta) \, d\eta \right\} dx, \tag{1.15}$$

$$\|A(x,\eta)\|^2_{X_1^{(\tau)}} = \sup_{-\infty<x<\infty} \int_{-\infty}^{\infty} |A(x,\eta)|^2 q^{2\tau}(\eta) \, d\eta. \tag{1.16}$$

It is easily seen that a lemma analogous to Lemma 1.1 holds for all of these Banach spaces.

3. We will again consider the integral equation (1.3), which can be written, using the definition of the operator N (cf. (1.4)), in the form

$$G(x,\eta;\mu) = g(x,\eta;\mu) - NG(\xi,\eta;\mu). \tag{1.17}$$

Since for sufficiently large μ the operator N is a contraction in all of the Banach spaces considered above, it follows that if $g(x,\eta;\mu)$ belongs to one of these spaces, then equation (1.17) can be solved by the method of iteration and its solution belongs to that space.

In the following lemmas we will indicate sufficient conditions for the function $g(x,\eta;\mu)$ to belong, respectively, to the spaces $X_2^{(p)}$ ($p > 0$), X, $X_1^{(1/2)}$ and $X^{(-\tau)}$ ($\tau > 0$).

Lemma 1.2. *If $q(x) > 0$, then the functions $g(x,\eta;\mu)$ and $\partial g(x,\eta;\mu)/\partial\eta$ belong to the space $X_1^{(p)}$ ($p > 0$), i.e.*

$$\sup_{-\infty<x<\infty} \int_{-\infty}^{\infty} \{g(x,\eta;\mu)\}^p \, d\eta < \infty, \tag{1.18}$$

$$\sup_{-\infty<x<\infty} \int_{-\infty}^{\infty} \left\{\frac{\partial}{\partial\eta} g(x,\eta;\mu)\right\}^p d\eta < \infty. \tag{1.18$'$}$$

Proof. From the definition of the function $g(x, \eta; \mu)$, i.e. from (1.2), follows

$$\int_{-\infty}^{\infty} \{g(x, \eta; \mu)\}^p \, d\eta \leqslant \int_{-\infty}^{\infty} \exp(-p\mu^{1/2} |x-\eta|) \, d\eta$$
$$= \int_{-\infty}^{\infty} \exp(-p\mu^{1/2} |u|) \, du = \frac{2}{p\sqrt{\mu}},$$

whence (1.18) follows at once. The proof of (1.18′) is analogous.

Lemma 1.3. *If the condition* $\int_{-\infty}^{\infty} \{q(x)\}^{-3/2} dx < \infty$ *is satisfied, then* $g(x, \eta; \mu)$ *belongs to the space* X, *i.e.*

(1.19) $$\int_{-\infty}^{\infty} \left\{ \int_{-\infty}^{\infty} [g(x, \eta; \mu)]^2 \, d\eta \right\} dx < \infty.$$

Proof. In fact, for $\mu > 0$ we have

$$\int_{0}^{\infty} \{g(x, \eta; \mu)\}^2 \, d\eta = \int_{-\infty}^{\infty} \frac{1}{4\varkappa^2} \exp(-\varkappa |x-\eta|) \, d\eta$$
$$= \frac{1}{2\varkappa^3} = \frac{1}{2\{q(x)+\mu\}^{3/2}} \leqslant \frac{1}{2q^{3/2}(x)}.$$

Therefore (1.19) follows from the hypothesis of the lemma.

Lemma 1.4. *If for* $|x-\eta| \leqq 1$ *the function* $q(x)$ *satisfies the condition*

(1.20) $$q(\eta) q^{-1}(x) \leqq C,$$

where C *is a constant, then* $g(x, \eta; \mu)$ *belongs to the space* $X_1^{(1/2)}$, *i.e.*

(1.21) $$\sup_{-\infty < x < \infty} \int_{-\infty}^{\infty} \{g(x, \eta; \mu\}^2 q(\eta) \, d\eta < \infty.$$

Proof. By the definition of $g(x, \eta; \mu)$ we have

$$\int_{-\infty}^{\infty} \{g(x, \eta; \mu)\}^2 q(\eta) \, d\eta$$
$$= \int_{|x-\eta| \leqslant 1} \frac{q(\eta)}{4[q(x)+\mu]} \exp\{-2|x-\eta| [q(x)+\mu]^{1/2}\} \, d\eta$$
$$+ \int_{|x-\eta| > 1} \frac{q(\eta)}{4[q(x)+\mu]} \exp\{-2|x-\eta| [q(x)+\mu]^{1/2}\} \, d\eta$$
$$= a(x) + b(x).$$

From the condition (1.20) follows

$$\sup_x a(x) \leqslant \sup_x \int\limits_{|x-\eta|\leqslant 1} \exp(-2\varkappa|x-\eta|)\, d\eta$$
(1.22)
$$= \sup_x \int\limits_{|u|\leqslant 1} \exp(-2\varkappa|u|)\, du < \infty.$$

Further, by virtue of (1.6) we have for $|x-\eta|>1$

$$g(\eta)\{g(x, \eta; \mu)\}^2 \leqslant C\exp\left(-\frac{3}{2}|x-\eta|\sqrt{\mu}\right).$$

Therefore, proceeding as we did to obtain the bound (1.22), we obtain $\sup_x b(x) < \infty$, which together with (1.22) proves (1.21) and thus the lemma.

LEMMA 1.5. *Suppose that for* $|x-\eta| \leqq 1$ *we have*[3]

(1.20′) $$q(x)q^{-1}(\eta) < C$$

Then the function $\partial^2 g(x,\eta;\mu)/\partial\eta^2$ $(x \neq \eta)$ *belongs to the space* $X_1^{(-1/2)}$, *i.e.*[4]

(1.23) $$\sup_x \int_{-\infty}^{\infty} \left\{\frac{\partial^2}{\partial\eta^2} g(x, \eta; \mu)\right\}^2 q^{-1}(\eta)\, d\eta < \infty.$$

PROOF. If $\eta \neq x$, then

$$\partial^2 g/\partial\eta^2 = \kappa^2 g(x,\eta;\mu) = \tfrac{1}{2}\kappa \exp(-\kappa|x-\eta|),$$

where $\kappa^2 = q(x)+\mu$. Therefore (1.23) can be obtained in the same way as was (1.21).

LEMMA 1.6. *Suppose that* $q(x)$ *satisfies the condition* (1.20′) *and also the condition*

(1.24) $$\{q(x)\}^{-3/2-2\tau} \in \mathcal{L}_1(-\infty, \infty).$$

Then $g(x,\eta;\mu)$ *belongs to the space* $X^{(-\tau)}$, *i.e.*

$$\int_{-\infty}^{\infty}\left\{\int_{-\infty}^{\infty} [g(x, \eta; \mu)]^2 [q(\eta)]^{-2\tau}\, d\eta\right\} dx < \infty.$$

---

[3] If instead of the condition (1.20) one assumes that $q(\eta)q^{-1}(x)$ is bounded not only from above, but also from below by a positive number, then we obtain (1.20′) as a consequence.

[4] By definition

$$\int_{-\infty}^{\infty}\left\{\frac{\partial^2 g}{\partial\eta^2} q^{-1/2}(\eta)\right\}^2 d\eta = \lim_{\varepsilon\to 0}\int_{-\infty}^{x-\varepsilon}\left\{\frac{\partial^2 g}{\partial\eta^2} q^{-1/2}(\eta)\right\}^2 d\eta + \lim_{\varepsilon\to 0}\int_{x+\varepsilon}^{\infty}\left\{\frac{\partial^2 g}{\partial\eta^2} q^{-1/2}(\eta)\right\}^2 d\eta.$$

PROOF. Put

$$\int_{-\infty}^{\infty} \{g(x, \eta; \mu)\}^2 q^{-2\tau}(\eta)\, d\eta = \int_{|x-\eta|\leq 1} \{g(x, \eta; \mu)\}^2 q^{-2\tau}(\eta)\, d\eta$$

$$+ \int_{|x-\eta|>1} \{g(x, \eta; \mu)\}^2 q^{-2\tau}(\eta)\, d\eta = a(x) + b(x).$$

From the definition of $g(x, \eta; \mu)$ and the condition (1.20′) for $|x-\eta| \leqq 1$ it follows for $\mu > 0$ that

$$\{g(x, \eta; \mu)\}^2 q^{-2\tau}(\eta) \leqq Cq^{-(2\tau+1)}(x) \exp\{-2|x-\eta|\sqrt{q(x)}\}.$$

Therefore

$$a(x) \leqslant Cq^{-(2\tau+1)}(x) \int_{|x-\eta|\leqslant 1} \exp\{-2|x-\eta|\sqrt{q(x)}\}\, d\eta \leqslant C\{q(x)\}^{-\left(2\tau+\frac{3}{2}\right)},$$

from which on the basis of (1.24) follows $a(x) \in \mathscr{L}_1(-\infty, \infty)$.

Further, for $|x-\eta| > 1$

$$\{g(x, \eta; \mu)\}^2 q^{-\tau}(\eta) \leqq C\exp(-2|x-\eta|\kappa).$$

Therefore

$$b(x) \leqslant C\int_1^{\infty} \exp(-2u\kappa)\, du = C_1 e^{-2\kappa}$$

and so $b(x)$ also belongs to $\mathscr{L}_1(-\infty, \infty)$. Consequently $a(x) + b(x) \in \mathscr{L}_1(-\infty, \infty)$, which proves the lemma.

§2. The first derivative of the function $G(x, \eta; \mu)$

Differentiating the equation (1.3) formally with respect to η, we obtain

$$\frac{\partial G}{\partial \eta} = \frac{\partial g}{\partial \eta} - \int_{-\infty}^{\infty} g(x, \xi; \mu)[g(\xi) - q(x)]\frac{\partial G}{\partial \eta}\, d\xi. \tag{2.1}$$

Replacing $\partial G(x, \eta; \mu)/\partial \eta$ by $K(x, \eta; \mu)$ in equation (2.1), we obtain the integral equation

$$K(x, \eta; \mu) = \frac{\partial}{\partial \eta} g(x, \eta; \mu) - \int_{-\infty}^{\infty} g(x, \xi; \mu)[g(\xi) - q(x)] K(\xi, \eta; \mu)\, d\xi. \tag{2.2}$$

By Lemma 1.2, $\partial g(x, \eta; \mu)/\partial \eta \in X_2^{(p)}$. Therefore from Lemma 1.1 (cf. the Remark following this lemma, and §1.3) it follows that $K(x, \eta; \mu) \in X_2^{(p)}$. It is also not hard to verify that for $\eta \neq x$ the function $K(x, \eta; \mu)$ is continuous with respect to η.

Integrating (2.2) with respect to η from $-\infty$ to η, we obtain

$$\int_{-\infty}^{\eta} K(x, \eta; \mu)\, d\eta = g(x, \eta; \mu)$$
$$(2.2') \qquad - \int_{-\infty}^{\infty} g(x, \xi; \mu)[q(\xi) - q(x)] \left\{ \int_{-\infty}^{\eta} K(\xi, \eta; \mu)\, d\eta \right\} d\xi.$$

For $\eta = x$ the function $\partial g(x, \eta; \mu)/\partial \eta$, and consequently $K(x, \eta; \mu)$, has a discontinuity of the first kind, which does not prevent the application of the Newton-Leibnitz formula. The existence of the integral in the right side of equation (2.2′), and also the validity of interchanging the order of integration, follow from the fact that $K(x, \eta; \mu) \in X_2^{(1)}$, i.e. from the fact that

$$\sup_{-\infty < x < \infty} \int_{-\infty}^{\infty} |K(x, \eta; \mu)|\, d\eta < \infty.$$

Equation (2.2′) coincides with the integral equation (1.3). Therefore it follows from the uniqueness of the solution of (1.3) that

$$\int_{-\infty}^{\eta} K(x, \eta; \mu)\, d\eta = G(x, \eta; \mu).$$

Differentiating this relation with respect to η, we obtain for $\eta \neq x$

$$K(x, \eta; \mu) = \partial G(x, \eta; \mu)/\partial \eta.$$

Therefore the integral equation (2.2) can be written in the form

$$\frac{\partial G}{\partial \eta} - \frac{\partial g}{\partial \eta} = - \int_{-\infty}^{\infty} g(x, \xi; \mu)[q(\xi) - q(x)] \frac{\partial}{\partial \eta} g(\xi, \eta; \mu)\, d\xi$$
$$(2.3) \qquad - \int_{-\infty}^{\infty} g(x, \xi; \mu)[q(\xi) - q(x)] \left\{ \frac{\partial G}{\partial \eta} - \frac{\partial g}{\partial \eta} \right\} d\xi.$$

Applying estimates similar to those which we used in the proof of the basic Lemma 1.1, one can show that the function

$$l(x, \eta; \mu) \equiv - \int_{-\infty}^{\infty} g(x, \xi; \mu)[q(\xi) - q(x)] \frac{\partial}{\partial \eta} g(\xi, \eta, \mu)\, d\xi$$

is continuous with respect to η. To do this it is sufficient to verify that the improper integral which defines the function $l(x, \eta; \mu)$ converges uniformly with respect to η, which follows easily from (1.6).

Now, solving the equation (2.3) by iteration, it is not hard to show the continuity with respect to η of the function

$$\partial G(x,\eta;\mu)/\partial\eta - \partial g(x,\eta;\mu)/\partial\eta.$$

Therefore the function $\partial G(x,\eta;\mu)/\partial\eta$ has the same discontinuities as the function $\partial g(x,\eta;\mu)/\partial\eta$. The latter has a unique discontinuity for $\eta = x$. In fact,

$$\frac{\partial}{\partial\eta} g(x,\eta;\mu) = \begin{cases} -\frac{1}{2}\exp\{-(\eta - x)\varkappa\}, & \text{if } \eta > x, \\ \frac{1}{2}\exp\{-(x-\eta)\varkappa\}, & \text{if } \eta < x \end{cases} \qquad \varkappa = \{q(x)+\mu\}^{1/2}.$$

Therefore

$$\frac{\partial}{\partial\eta} g(x,\eta;\mu)\Big|_{\eta=x+0} - \frac{\partial}{\partial\eta} g(x,\eta;\mu)\Big|_{\eta=x-0} = -1. \tag{2.4}$$

Thus the function $\partial G(x,\eta;\mu)/\partial\eta$ is continuous with respect to η everywhere except for the point $\eta = x$, at which it has a discontinuity of the first kind, and

$$\frac{\partial}{\partial\eta} G(x,\eta;\mu)\Big|_{\eta=x+0} - \frac{\partial}{\partial\eta} G(x,\eta;\mu)\Big|_{\eta=x-0} = -1. \tag{2.5}$$

§3. The second derivative of the function $G(x,\eta;\mu)$

In the preceding section we proved that the first derivative of $G(x,\eta;\mu)$ exists as an element of the space $X_2^{(p)}$ for any $p \geqq 1$ (for the definition of $X_2^{(p)}$ cf. the Remark following the proof of Lemma 1.1 in §1), is continuous with respect to η for $\eta \neq x$, and satisfies the integral equation

$$\frac{\partial G}{\partial\eta} = \frac{\partial g}{\partial\eta} - \int_{-\infty}^{\infty} g(x,\xi;\mu)[q(\xi)-q(x)]\frac{\partial G(\xi,\eta;\mu)}{\partial\eta}\,d\xi.$$

This equation can be reduced to the form

$$\mathscr{L}(x,\eta;\mu) = l(x,\eta;\mu) - \int_{-\infty}^{\infty} g(x,\xi;\mu)[q(\xi)-q(x)]\mathscr{L}(\xi,\eta;\mu)\,d\xi, \tag{3.1}$$

where

$$\mathscr{L}(x,\eta;\mu) = \frac{\partial}{\partial\eta} G(x,\eta;\mu) - \frac{\partial}{\partial\eta} g(x,\eta;\mu),$$

$$l(x,\eta;\mu) = \int^{\infty} g(x,\xi;\mu)[q(\xi)-q(x)]\frac{\partial}{\partial\eta} g(\xi,\eta;\mu)\,d\xi. \tag{3.2}$$

Differentiating the equation (3.1) formally with respect to η, we obtain

$$\frac{\partial}{\partial\eta}\mathcal{L}(x, \eta; \mu) = \frac{\partial}{\partial\eta} l(x, \eta; \mu)$$

(3.3)
$$- \int_{-\infty}^{\infty} g(x, \xi; \mu)[q(\xi) - q(x)]\frac{\partial}{\partial\eta}\mathcal{L}(\xi, \eta; \mu)\, d\xi.$$

Further, differentiating (3.2), again formally for the moment, with respect to η and using (2.4) (cf. also the footnote on p. 000), we obtain

$$\frac{\partial}{\partial\eta} l(x, \eta, \mu) = g(x, \eta; \mu)[q(\eta) - q(x)]$$
$$- \int_{-\infty}^{\infty} g(x, \xi; \mu)[q(\xi) - q(x)]\frac{\partial^2}{\partial\eta^2} g(\xi, \eta; \mu)\, d\xi$$
$$\equiv l_1(x, \eta; \mu) \quad (\eta \neq x).$$

Let us now consider the integral equation

$$M(x, \eta; \mu) = l_1(x, \eta; \mu)$$

(3.4)
$$- \int_{-\infty}^{\infty} g(x, \xi; \mu)[q(\xi) - q(x)]\, M(\xi, \eta; \mu)\, d\xi.$$

We will show that $l_1(x, \eta; \mu)$ belongs to the space $X_1^{(-1/2)}$. It then follows from Lemma 1.1 that the solution of (3.4) also belongs to this Banach space. We have

$$l_1(x, \eta; \mu) = g(x, \eta; \mu)[q(\eta) - q(x)]$$
$$- \int_{|\xi-\eta|\leqslant 1} g(x, \xi; \mu)[g(\xi) - q(x)]\frac{\partial^2}{\partial\eta^2} g(\xi, \eta; \mu)\, d\xi$$
$$- \int_{|\xi-\eta|>1} g(x, \xi; \mu)[q(\xi) - q(x)]\frac{\partial^2}{\partial\eta^2}(\xi, \eta; \mu)\, d\xi = I_1 + I_2 + I_3.$$

From the conditions (1.5) and (1.6) it is not hard to deduce that I_1 belongs to the space $X_1^{(1)}$ and hence to $X_1^{(-1/2)}$.

Let us now consider I_2. Assuming that the condition (1.20′) is satisfied, we obtain

$$I_2 q^{-1/2}(\eta) \leqslant \int_{|\xi-\eta|\leqslant 1} g(x, \xi; \mu)\,|q(\xi) - q(x)|$$
$$\times \exp\{-|\xi - \eta| \cdot [q(\xi) + \mu]^{-1/2}\}\, q^{-1/2}(\eta)[q(\xi) + \mu]^{1/2} d\xi$$
$$< C(\mu) \int_{|\xi-\eta|\leqslant 1} g(x, \xi; \mu)\,|q(\xi) - q(x)| \exp\{-|\xi - \eta|[q(\xi) + \mu]^{1/2}\}\, d\xi,$$

where $C(\mu)$ is a constant, depending only upon μ.

Finally, using the condition (1.6), we obtain the following bound for I_3:

$$I_3 q^{-1/2}(\eta)$$

$$\leqslant C(\mu) \int_{|\xi-\eta|>1} g(x, \xi; \mu | q(\xi) - q(x) | \exp\left\{-\frac{1}{2}|\xi-\eta|[q(\xi)+\mu]^{1/2}\right\} d\xi.$$

Therefore

$$(I_2 + I_3) q^{-1/2}(\eta)$$

$$\leqslant C(\mu) \int_{-\infty}^{\infty} g(x, \xi; \mu) | q(\xi) - q(x) | \exp\left\{-\frac{1}{2}|\xi-\eta|[q(\xi)+\mu]^{1/2}\right\} d\xi.$$

The integral on the right can be bounded as in the proof of Lemma 1.1. As a consequence it is easily proved that $l_1(x, \eta; \mu)$ belongs to the space $X_1^{(-1/2)}$, i.e.

$$\sup_{-\infty<x<\infty} \int_{-\infty}^{\infty} \{l_1(x, \eta; \mu)\}^2 q^{-1}(\eta)\, d\eta < \infty.$$

We will now show that $M(x, \eta; \mu)$, i.e. the solution of the integral equation (3.4) (which, as was proved, belongs to the space $X_1^{(-1/2)}$) is equal for $\eta \neq x$ to

$$\frac{\partial^2}{\partial \eta^2}[G(x, \eta; \mu) - g(x, \eta; \mu)].$$

Integrating the equation (3.4) with respect to η from η_0 to η (η_0 is a fixed number), we obtain

$$\begin{aligned} \int_{\eta_0}^{\eta} M(x, \eta; \mu)\, d\eta &= \int_{\eta_0}^{\eta} l_1(x, \eta; \mu)\, d\eta \\ &- \int_{-\infty}^{\infty} g(x, \xi; \mu)[q(\xi) - q(x)] \left\{ \int_{\eta_0}^{\eta} M(\xi, \eta; \mu)\, d\eta \right\} d\xi. \end{aligned} \tag{3.5}$$

Further, from equation (3.1) follows

$$\begin{aligned} \mathscr{L}(x, \eta; \mu) - \mathscr{L}(x, \eta_0; \mu) &= l(x, \eta; \mu) - l(x, \eta_0; \mu) \\ &- \int_{-\infty}^{\infty} g(x, \xi; \mu)[q(\xi) - q(x)][\mathscr{L}(\xi, \eta; \mu) - \mathscr{L}(\xi, \eta_0; \mu)]\, d\xi. \end{aligned} \tag{3.6}$$

If we show that

$$\int_{\eta_0}^{\eta} l_1(x, \eta; \mu)\, d\eta = l(x, \eta; \mu) - l(x, \eta_0; \mu), \tag{3.7}$$

then from the uniqueness of the solution of equations (3.5) and (3.6) it would follow that

$$\int_{\eta_0}^{\eta} M(x, \eta; \mu)\, d\eta = \mathscr{L}(x, \eta; \mu) - \mathscr{L}(x, \eta_0; \mu),$$

i.e.

$$\begin{aligned} M(x, \eta; \mu) &= \frac{\partial}{\partial \eta} \mathscr{L}(x, \eta; \mu) \\ &= \frac{\partial}{\partial \eta}\left[\frac{\partial}{\partial \eta} G(x, \eta; \mu) - \frac{\partial}{\partial \eta} g(x, \eta; \mu)\right]. \end{aligned} \tag{3.8}$$

Now since $\partial^2 g(x,\eta;\mu)/\partial\eta^2$ exists for $\eta \neq x$, it follows from (3.8) that $\partial^2 G(x,\eta;\mu)/\partial\eta^2$ exists for $\eta \neq x$ and is an element of the Banach space $X_1^{(-1/2)}$.

Thus it remains for us to prove (3.7).

It is easy to show from the condition (1.6) that the function

$$l(x, \eta; \mu) = -\int_{-\infty}^{\infty} g(x, \xi; \mu)[q(\xi) - q(x)] \frac{\partial}{\partial \eta} g(\xi, \eta; \mu)\, d\xi \tag{3.9}$$

can be differentiated under the integral sign. In fact, we choose N so large that the point η lies in the interval $(-N, N)$, and represent the function $l(x, \eta; \mu)$ in the form

$$\begin{aligned} l(x, \eta; \mu) = &- \int_{-\infty}^{-N} g(x, \xi; \mu)[q(\xi) - q(x)] \frac{\partial}{\partial \eta} g(\xi, \eta; \mu)\, d\xi \\ &- \int_{-N}^{N} g(x, \xi; \mu)[q(\xi) - q(x)] \frac{\partial}{\partial \eta} g(\xi, \eta; \mu)\, d\xi \\ &- \int_{N}^{\infty} g(x, \xi; \mu)[q(\xi) - q(x)] \frac{\partial}{\partial \eta} g(\xi, \eta; \mu)\, d\xi \\ &= \alpha_1(x, \eta; \mu) + \alpha_2(x, \eta; \mu) + \alpha_3(x, \eta; \mu). \end{aligned}$$

The functions $\alpha_1(x,\eta;\mu)$ and $\alpha_3(x,\eta;\mu)$ can be differentiated under the integral sign, since following differentiation we obtain absolutely (and uniformly) convergent integrals, which easily follows from the condition (1.6). The function $\alpha_2(x,\eta;\mu)$ is an integral with finite limits, and its differentiation does not occasion any difficulties.

Differentiating (3.9) with respect to η and using (2.4) (cf. also the footnote on p. 439), we obtain

$$\frac{\partial}{\partial\eta} l(x, \eta; \mu) = g(x, \eta; \mu)[q(\eta) - q(x)]$$

$$\int_{-\infty}^{\infty} g(x, \xi; \mu)[q(\xi) - q(x)] \frac{\partial^2}{\partial\eta^2} g(\xi, \eta; \mu)\, d\xi \equiv l_1(x, \eta; \mu).$$

Therefore

$$\int_{\eta_0}^{\eta} l_1(x, \eta; \mu)\, d\eta = l(x, \eta; \mu) - l(x, \eta_0; \mu),$$

which coincides with (3.7), and the assertion is proved.

§4. Further properties of the function $G(x, \eta; \mu)$

1. Besides the properties which were established above, the function $G(x, \eta; \mu)$ has still other important properties, which will be derived in this section.

Lemma 4.1. *For $\eta \neq x$ the function $G(x, \eta; \mu)$ satisfies the differential equation*

$$\partial^2 G/\partial\eta^2 = \{ q(\eta) + \mu \} G. \tag{4.1}$$

Proof. By (3.8) the equation (3.4) can be written in the form

$$\frac{\partial^2}{\partial\eta^2} G(x, \eta; \mu) - \{q(x) + \mu\} g(x, \eta; \mu)$$
$$= \{q(\eta) - q(x)\} g(x, \eta; \mu) - \int_{-\infty}^{\infty} g(x, \xi; \mu) \{q(\xi) - q(x)\} \frac{\partial^2}{\partial\eta^2} G(\xi, \eta; \mu)\, d\xi,$$

from which it follows that

$$\frac{\partial^2 G}{\partial\eta^2} = \{q(\eta) + \mu\} g - \int_{-\infty}^{\infty} g(x, \xi; \mu) \{q(\xi) - q(x)\} \frac{\partial^2 G}{\partial\eta^2}\, d\xi$$

or

$$\frac{\partial^2 G}{\partial\eta^2} \{q(\eta) + \mu\}^{-1} = g - \int_{-\infty}^{\infty} g(x, \xi; \mu) \{q(\xi) - q(x)\} \frac{\partial^2 G}{\partial\eta^2} \{q(\eta) + \mu\}^{-1}\, d\xi.$$

Comparing this equation with equation (1.3), we conclude on the basis of uniqueness of solutions that

$$\frac{\partial^2 G}{\partial\eta^2} \{q(\eta) + \mu\}^{-1} = G(x, \eta; \mu),$$

which coincides with (4.1). This proves the lemma.

LEMMA 4.2. *The function* $G(x,y;\mu)$ *is symmetric in the variables* x *and* y, *i.e.*

$$G(x,y;\mu) = G(y,x;\mu). \tag{4.2}$$

PROOF. Let x and y be two distinct real numbers, and suppose for definiteness that $x<y$. Let us consider the integral

$$I = \int_{-\infty}^{\infty} \left\{ G(y, \xi; \mu) \frac{\partial^2}{\partial \xi^2} G(x, \xi; \mu) - G(x, \xi; \mu) \frac{\partial^2}{\partial \xi^2} G(y, \xi; \mu) \right\} d\xi,$$

whose existence follows from the fact that $G(x,\eta;\mu) \in X_1^{(1/2)}$ and $\partial^2 G(x,\eta;\mu)/\partial\eta^2 \in X_1^{(-1/2)}$, which we mentioned in the preceding sections.

Put

$$I = \left\{ \int_{-\infty}^{x} + \int_{x}^{y} + \int_{y}^{\infty} \right\} \left\{ G(y, \xi; \eta) \frac{\partial^2}{\partial \xi^2} G(x, \xi; \mu) - G(x, \xi; \mu) \frac{\partial^2}{\partial \xi^2} G(y, \xi; \mu) \right\} d\xi$$
$$= I_1 + I_2 + I_3.$$

Since

$$G(y, \xi; \mu) \frac{\partial^2}{\partial \xi^2} G(x, \xi; \mu) - G(x, \xi; \mu) \frac{\partial^2}{\partial \xi^2} G(y, \xi; \mu)$$
$$= \frac{\partial}{\partial \xi} \left[G(y, \xi; \mu) \frac{\partial}{\partial \xi} G(x, \xi; \mu) - G(x, \xi; \mu) \frac{\partial}{\partial \xi} G(y, \xi; \mu) \right],$$

we have

$$I_1 = G(y, x; \mu) \frac{\partial}{\partial \xi} G(x, \xi; \mu)\big|_{\xi=x-0} - G(x, x; \mu) \frac{\partial}{\partial \xi} G(y, \xi; \mu)\big|_{\xi=x},$$

$$I_2 = G(y, y; \mu) \frac{\partial}{\partial \xi} G(x, \xi; \mu)\big|_{\xi=y} - G(x, y; \mu) \frac{\partial}{\partial \xi} G(y, \xi; \mu)\big|_{\xi=y-0}$$
$$- G(y, x; \mu) \frac{\partial}{\partial \xi} G(x, \xi; \mu)\big|_{\xi=x+0} + G(x, x; \mu) \frac{\partial}{\partial \xi} G(y, \xi; \mu)\big|_{\xi=x},$$

$$I_3 = -G(y, y; \mu) \frac{\partial}{\partial \xi} G(x, \xi; \mu)\big|_{\xi=y} + G(x, y, \mu) \frac{\partial}{\partial \xi} G(y, \xi; \mu)\big|_{\xi=y+0}.$$

Adding these equalities and taking (2.5) into account, we obtain $I = G(y,x;\mu) - G(x,y;\mu)$, which proves (4.2) since it follows from (4.1) that $I=0$.

2. If the function $q(x)$ is bounded from below and $\lim_{|x|\to+\infty} q(x) = +\infty$, then the Sturm-Liouville operator has a discrete spectrum and a unique Green's function for all μ outside the spectrum (cf. Chapter 4, §1).

It was shown above that the function $G(x, \eta; \mu)$ satisfies all the conditions of the Green's function of the Sturm-Liouville operator. Consequently, by the uniqueness of the Green's function, $G(x, \eta; \mu)$ is the Green's function. We know from Chapter 2 that the Green's function can be expressed in terms of the solutions $\psi_1(x, \mu) \in \mathscr{L}^2(-\infty, 0)$ and $\psi_2(x, \mu) \in \mathscr{L}^2(0, \infty)$ which for the case $q(x) \to +\infty$ as $|x| \to \infty$ are essentially uniquely determined.

From the integral representation of the resolvent (cf. Chapter 2, §3) it follows that for any function $f(x) \in \mathscr{L}^2(-\infty, \infty)$

$$\int_{-\infty}^{\infty} G(x, y; \mu) f(y)\, dy = \sum_{n=1}^{\infty} \frac{a_n \varphi_n(x)}{\lambda_n + \mu}, \tag{4.3}$$

where $a_n = \int_{-\infty}^{\infty} f(x) \varphi_n(x)\, dx$.

Here the λ_n are the eigenvalues, and the $\varphi_n(x)$ corresponding eigenfunctions of the Sturm-Liouville operator, i.e.

$$-\varphi_n''(x) + q(x)\varphi_n(x) = \lambda_n \varphi_n(x).$$

§5. Differentiation of the Green's function with respect to μ

1. It follows from Lemmas 1.1 and 1.3 that if $q^{-3/2}(x) \in \mathscr{L}^1(-\infty, \infty)$, then the Green's function $G(x, \eta; \mu)$ belongs to the space X, i.e.

$$\int_{-\infty}^{\infty} \left\{ \int_{-\infty}^{\infty} G^2(x, \eta; \mu)\, d\eta \right\} dx < \infty. \tag{5.1}$$

If $q^{-3/2}(x) \notin \mathscr{L}^1(-\infty, \infty)$, then (5.1) may fail to hold.

Let us assume that for some integer $\tau > 0$

$$q^{-(3/2+2\tau)}(x) \in \mathscr{L}^1(-\infty, \infty). \tag{5.2}$$

In this case it is natural to consider the iteration $G^{(\tau)}(x, \eta; \mu)$ of the Green's functions, and one may hope that this iterate will be a Hilbert-Schmidt kernel, i.e. will satisfy the condition (5.1).

From the integral representation of the Green's function (cf. Chapter 2, §3) it follows that the iterated Green's functions coincide up to sign with the corresponding derivatives with respect to μ.

We will study in detail $\partial G(x, \eta; \mu)/\partial\mu$, which corresponds to the first iterate. The case of higher derivatives (iterates) can be studied similarly, and we will restrict ourselves in this case to simply stating the final results.

From equation (1.3) follows

$$\frac{\partial G}{\partial \mu} = \frac{\partial g}{\partial \mu} - \int_{-\infty}^{\infty} \frac{\partial}{\partial \mu} \{g(x, \xi; \mu)\} [q(\xi) - q(x)] G(\xi, \eta; \mu) \, d\xi$$
(5.3)
$$- \int_{-\infty}^{\infty} g(x, \xi; \mu) [q(\xi) - q(x)] \frac{\partial}{\partial \mu} G(\xi, \eta; \mu) \, d\xi.$$

If $q^{-7/2}(x) \in \mathscr{L}^1(-\infty, \infty)$, then it follows from Lemma 1.6 that $g(x, \eta; \mu)$, and consequently by Lemma 1.1 $G(x, \eta; \mu)$, belongs to the space $X^{(-1)}$, i.e.

$$\int_{-\infty}^{\infty} \left\{ \int_{-\infty}^{\infty} G^2(x, \eta; \mu) q^{-2}(\eta) \, d\eta \right\} dx < \infty. \tag{5.4}$$

Since the function $G(x, \eta; \mu)$ is symmetric, the condition (5.4) can be written in the form

$$\int_{-\infty}^{\infty} \left\{ \int_{-\infty}^{\infty} G^2(\eta, x; \mu) q^{-2}(\eta) \, d\eta \right\} dx < \infty. \tag{5.5}$$

Moreover, it follows from Lemma 1.1 (and subsequent approximations) that as $\mu \to \infty$

$$\left\{ \int_{-\infty}^{\infty} \left[\int_{-\infty}^{\infty} G^2(\eta, x; \mu) \{q(\eta) + \mu\}^{-2} d\eta \right] dx \right\}^{1/2}$$
(5.6)
$$= O\left(\int_{-\infty}^{\infty} \left\{ \int_{-\infty}^{\infty} g^2(x, \eta; \mu) [q(\eta) + \mu]^{-2} d\eta \right\} dx \right).$$

Put

$$\alpha(x, \eta; \mu) = \int_{-\infty}^{\infty} \frac{\partial g(x, \xi; \mu)}{\partial \mu} [q(\xi) - q(x)] G(\xi, \eta; \mu) \, d\xi. \tag{5.7}$$

LEMMA 5.1. *Let the function $q(x)$ satisfy the conditions*:

1. *for* $|x - \xi| \leq 1$

$$|q(\xi) - q(x)| \leq A q^a(x) |\xi - x|,$$

where $A > 0$ *and* $0 < a < 3/2$ *are constants*;

2. *for* $|x - \xi| \leq 1$

$$q(\xi) q^{-1}(x) \leq B,$$

where B *is a constant*;

3. *for* $|x - \xi| > 1$

$$q^2(\xi) \exp\{-\tfrac{1}{2} |x - \xi| \sqrt{q(x)}\} < C,$$

where C is a constant;

4. $q^{-7/2}(x) \in \mathscr{L}^1(-\infty, \infty)$.

Then the function $\alpha(x, \eta; \mu)$, *defined by* (5.7), *belongs to the space X, i.e.*

$$\int_{-\infty}^{\infty}\left\{\int_{-\infty}^{\infty} \alpha^2(x, \eta; \mu)\, d\eta\right\} dx < \infty. \tag{5.7'}$$

PROOF. Let us write the function $\alpha(x, \eta; \mu)$ in the form

$$\alpha(x, \eta; \mu) = \int_{-\infty}^{\infty} \frac{\partial g(x, \xi; \mu)}{\partial \mu} [q(\xi) + \mu][q(\xi) - q(x)][q(\xi) + \mu]^{-1} G(\xi, \eta; \mu)\, d\xi.$$

Estimating this integral analogously to what was done in the proof of Lemma 1.1, we can prove (5.7′). Moreover, arguing as in the proof of Lemma 1.1 and using the estimate (5.6), it is not hard to obtain the estimate

$$\begin{aligned}\alpha(x, \eta; \mu)\|_X &= o(\|G(x, \eta; \mu)[q(\eta) + \mu]^{-1}\|_X) \\ &= o(\|g(x, \eta; \mu)[q(\eta) + \mu]^{-1}\|_X).\end{aligned} \tag{5.8}$$

2. It can be verified directly that if $q^{-7/2}(x) \in \mathscr{L}^1(-\infty, \infty)$, then the function $\partial g(x, \eta; \mu)/\partial\mu$ belongs to the space X, i.e.

$$\int_{-\infty}^{\infty}\left\{\int_{-\infty}^{\infty}\left[\frac{\partial}{\partial\mu} g(x, \eta; \mu)\right]^2 d\eta\right\} dx < \infty.$$

Taking account of (5.7), the integral equation (5.3) can be rewritten in the form

$$\frac{\partial G}{\partial \mu} = \frac{\partial g}{\partial \mu} - \alpha(x, \eta; \mu) - \int_{-\infty}^{\infty} g(x, \xi; \mu)[q(\xi) - q(x)] \frac{\partial}{\partial \mu} G(\xi, \eta; \mu)\, d\xi. \tag{5.3'}$$

Solving this equation by iteration, we obtain

$$\frac{\partial G}{\partial \mu} = \frac{\partial g}{\partial \mu} - \alpha(x, \eta; \mu) + \beta(x, \eta; \mu), \tag{5.9}$$

where it follows from Lemma 1.1 that as $\mu \to \infty$

$$\|\beta(x, \eta; \mu)\|_X = o\left(\left\|\frac{\partial g}{\partial \mu} - \alpha\right\|_X\right). \tag{5.10}$$

Since

$$\left\|\frac{\partial g}{\partial \mu} - \alpha\right\|_X \leqslant \left\|\frac{\partial g}{\partial \mu}\right\|_X + \|\alpha\|_X,$$

(5.10) can be written in the form

$$\|\beta(x, \eta; \mu)\|_X = o\left(\left\|\frac{\partial g}{\partial \mu}\right\|_X + \|\alpha\|_X\right). \tag{5.11}$$

We will show that

$$\|\alpha(x, \eta; \mu)\|_X = o\left(\left\|\frac{\partial}{\partial \mu} g(x, \eta; \mu)\right\|_X\right). \tag{5.12}$$

This estimate will follow from the estimate (5.8), if we can show that as $\mu \to \infty$

$$\|g(x, \eta; \mu)[g(\eta) + \mu]^{-1}\|_X = o\left(\left\|\frac{\partial}{\partial \mu} g(x, \eta; \mu)\right\|_X\right). \tag{5.13}$$

A direct computation yields

$$\left\|\frac{\partial}{\partial \mu} g(x, \eta; \mu)\right\|_X^2 = C \int_{-\infty}^{\infty} \frac{dx}{\{q(x) + \mu\}^{7/2}}, \tag{5.14}$$

where C is a constant.

Further, using the condition (1.20) we obtain

$$\|g(x, \eta; \mu)[q(\eta) + \mu]^{-1}\|_X =$$

$$\begin{aligned}
&= \int_{-\infty}^{\infty}\left\{\int_{-\infty}^{\infty} g^2(x, \eta; \mu)[q(\eta) + \mu]^{-2}\, d\eta\right\} dx \\
&= \int_{-\infty}^{\infty}\left\{\int_{|x-\eta|\leqslant 1} g^2(x, \eta; \mu)[q(\eta) + \mu]^{-2} d\eta\right. \\
&\qquad \left. + \int_{|x-\eta|>1} g^2(x, \eta; \mu)[q(\eta) + \mu]^{-2}\, d\eta\right\} dx \\
&\leqslant C_1 \int_{-\infty}^{\infty}\left\{\int_{-\infty}^{\infty} g^2(x, \eta; \mu)[q(x) + \mu]^{-2}\, d\eta\right\} dx \\
&\qquad + C_2 \int_{-\infty}^{\infty}\left\{\int_1^{\infty} \exp(-u\sqrt{q(x) + \mu})\, du\right\} dx \\
&\leqslant C_3 \int_{-\infty}^{\infty} \frac{dx}{\{q(x) + \mu\}^{7/2}} + C_4 \int_{-\infty}^{\infty} \exp(-\sqrt{q(x) + \mu})\, dx \\
&\leqslant C_5 \int_{-\infty}^{\infty} \frac{dx}{\{q(x) + \mu\}^{7/2}}.
\end{aligned} \tag{5.15}$$

Obviously (5.13) follows from (5.14) and (5.15). Consequently we have proved (5.12). Now from the estimates (5.11) and (5.12) we obtain

$$\|\beta(x, \eta; \mu)\|_X = o\left(\left\|\frac{\partial}{\partial\mu} g(x, \eta; \mu)\right\|_X\right). \tag{5.16}$$

From equation (5.9) there follows, by virtue of (5.12) and (5.16),

$$\left\|\frac{\partial}{\partial\mu} G(x, \eta; \mu)\right\|_X \leqslant \left\|\frac{\partial}{\partial\mu} g(x, \eta; \mu)\right\|_X (1 + o(1)). \tag{5.17}$$

We now return to equation (5.3′), which we rewrite in the form

$$\frac{\partial g}{\partial\mu} = \frac{\partial G}{\partial\mu} + \alpha + \int_{-\infty}^{\infty} g(x, \xi; \mu)[q(\xi) - q(x)] \frac{\partial}{\partial\mu} G(\xi, \eta; \mu)\, d\xi.$$

Taking the norm of both sides and estimating the norm of the integral by means of Lemma 1.1, we obtain

$$\left\|\frac{\partial g}{\partial\mu}\right\|_X \leqslant \left\|\frac{\partial G}{\partial\mu}\right\|_X + o\left(\left\|\frac{\partial G}{\partial\mu}\right\|_X\right) + o\left(\left\|\frac{\partial g}{\partial\mu}\right\|_X\right),$$

from which

$$\left\|\frac{\partial}{\partial\mu} g(x, \eta; \mu)\right\|_X \leqslant \left\|\frac{\partial}{\partial\mu} G(x, \eta; \mu)\right\|_X (1 + o(1)). \tag{5.18}$$

From the estimates (5.17) and (5.18) we conclude that

$$\left\|\frac{\partial}{\partial\mu} G(x, \eta; \mu)\right\|_X = \left\|\frac{\partial}{\partial\mu} g(x, \eta; \mu)\right\|_X (1 + o(1)), \tag{5.18′}$$

i.e.

$$\lim_{\mu\to\infty} \frac{\left\|\frac{\partial}{\partial\mu} G(x, \eta; \mu)\right\|_X}{\left\|\frac{\partial}{\partial\mu} g(x, \eta; \mu)\right\|_X} = 1.$$

3. We will now formulate without proof the results for the general case, which covers iterates of higher order.

THEOREM 5.1. *Suppose that $q(x)$ satisfies conditions* 1 *and* 2 *of Lemma* 5.1 *and, moreover, the conditions*:

3. *for some integer* $\tau \geqq 0$

$$q^{-3/2-2\tau}(x) \in \mathscr{L}^1(-\infty, \infty);$$

4. *for* $|x - \xi| \geqq 1$

$$q^{\tau+1}(\xi)\exp\{-\tfrac{1}{2}|x-\xi|\sqrt{q(x)}\} < C,$$

where C is a constant.

Then the function $\partial^\tau G(x,\eta;\mu)/\partial\mu^\tau$ belongs to the space X, i.e.

$$\int_{-\infty}^{\infty}\left\{\int_{-\infty}^{\infty}\left[\frac{\partial^\tau}{\partial\mu^\tau}G(x,\eta;\mu)\right]^2 d\eta\right\}dx<\infty$$

and as $\mu\to\infty$

$$\left\|\frac{\partial^\tau}{\partial\mu^\tau}G(x,\eta;\mu)\right\|_X=\left\|\frac{\partial^\tau}{\partial\mu^\tau}g(x,\eta;\mu)\right\|_X(1+o(1)). \tag{5.19}$$

Since by (4.3)

$$\int_{-\infty}^{\infty}\frac{\partial^\tau}{\partial\mu^\tau}\{G(x,\eta;\mu)\}\varphi_n(\eta)\,d\eta=(-1)^\tau\tau!\,\frac{\varphi_n(x)}{(\lambda_n+\mu)^{\tau+1}},$$

it follows from the Parseval equality that

$$\int_{-\infty}^{\infty}\left\{\frac{\partial^\tau}{\partial\mu^\tau}G(x,\eta;\mu)\right\}^2 d\eta=(\tau!)^2\sum_{n=1}^{\infty}\frac{\varphi_n^2(x)}{(\lambda_n+\mu)^{2\tau+2}}, \tag{5.20}$$

from which

$$\int_{-\infty}^{\infty}\left\{\int_{-\infty}^{\infty}\left[\frac{\partial^\tau}{\partial\mu^\tau}G(x,\eta;\mu)\right]^2 d\eta\right\}dx=\sum_{n=1}^{\infty}\frac{(\tau!)^2}{(\lambda_n+\mu)^{2\tau+2}}\equiv(\tau!)^2S_\tau.$$

Therefore the asymptotic formula (5.19) can be written in the form

$$(\tau!)^2S_\tau\sim\left\|\frac{\partial^\tau}{\partial\mu^\tau}g(x,\eta;\mu)\right\|_X^2=\int_{-\infty}^{\infty}\left\{\int_{-\infty}^{\infty}\left[\frac{\partial^\tau}{\partial\mu^\tau}g(x,\eta;\mu)\right]^2 d\eta\right\}dx. \tag{5.21}$$

Let us calculate the integral in the right side of (5.21).
Putting $\beta=\kappa=[q(x)+\mu]^{1/2}$, $\tau=|x-\eta|$ in the well-known equality

$$\frac{1}{\pi}\int_{-\infty}^{\infty}\frac{e^{i\alpha t}}{\alpha^2+\beta^2}\,d\alpha=\frac{e^{-\beta t}}{\beta}\qquad(t\geqslant 0),$$

we obtain (cf. (1.2))

$$g(x,\eta;\mu)=\frac{1}{2\kappa}\exp(-|x-\eta|\kappa)=\frac{1}{2\pi}\int_{-\infty}^{\infty}\frac{e^{i\alpha|x-\eta|}}{\alpha^2+q(x)+\mu}\,d\alpha.$$

Differentiating this relation τ times with respect to μ, we obtain

$$\frac{\partial^\tau}{\partial\mu^\tau}g(x,\eta;\mu)=\frac{(-1)^\tau\tau!}{2\pi}\int_{-\infty}^{\infty}\frac{e^{i\alpha|x-\eta|}}{\{\alpha^2+q(x)+\mu\}^{\tau+1}}\,d\alpha.$$

Therefore it follows from the Parseval equality that

$$\int_{-\infty}^{\infty} \left\{\frac{\partial^\tau}{\partial\mu^\tau} g(x, \eta; \mu)\right\}^2 d\eta = \frac{(\tau!)^2}{2\pi} \int_{-\infty}^{\infty} \frac{d\alpha}{\{\alpha^2 + q(x) + \mu\}^{2\tau+2}},$$

from which, on the basis of the preceding equality and (2.1), follows

$$\int_{-\infty}^{\infty} \left\{\frac{\partial^\tau}{\partial\mu^\tau} g(x, \eta; \mu)\right\}^2 d\eta = \frac{(-1)^{2\tau+1}(\tau!)^2}{(2\tau+1)!} \frac{\partial^{2\tau+1}}{\partial\mu^{2\tau+1}} g(x, \eta; \mu)\Big|_{\eta=x}$$

(5.22)
$$= (-1)^{2\tau+1} \frac{(\tau!)^2}{(2\tau+1)!} \frac{\partial^{2\tau+1}}{\partial\mu^{2\tau+1}} \left(\frac{1}{2[q(x)+\mu]^{1/2}}\right)$$

$$= \frac{(\tau!)^2 \cdot 1 \cdot 3 \cdot 5 \ldots (4\tau+1)}{2^{2\tau+2}(2\tau+1)!} \frac{1}{[q(x)+\mu]^{2\tau+3/2}}.$$

From the asymptotic formula (5.21) and the above equality we finally obtain, for $\mu \to \infty$,

(5.20′)
$$S_\tau = \sum_{n=1}^{\infty} \frac{1}{(\lambda_n + \mu)^{2\tau+2}} \sim \frac{(4\tau+1)!!}{2^{2\tau+2}(2\tau+1)!} \int_{-\infty}^{\infty} \frac{dx}{\{q(x)+\mu\}^{2\tau+3/2}}.$$

§6. Asymptotic distribution of the eigenvalues

1. In this section, using a Tauberian theorem of Keldyš (cf. Chapter 14, Theorem 4.3), we will obtain from the asymptotic formula (5.20′) an asymptotic formula for the distribution of the eigenvalues. This formula leads to the following theorem.

THEOREM 6.1. *Let $q(x)$ satisfy the conditions of Theorem 5.1. We introduce the monotonic function $\sigma(\lambda) = \mathrm{mes}\{q(x) < \lambda\}$ and put*

$$\psi(\lambda) = \int_0^\lambda (\lambda - \nu)^{1/2} d\sigma(\lambda).$$

Let us assume that there exist positive constants α and β such that for sufficiently large λ we have the inequality

(6.1)
$$\alpha\psi(\lambda) < \lambda\psi'(\lambda) < \beta\psi(\lambda).$$

Then for $\lambda \to \infty$ we have the asymptotic formula

(6.2)
$$N(\lambda) = \sum_{\lambda_n < \lambda} 1 \sim \frac{1}{\pi} \int_0^\lambda (\lambda - \nu)^{1/2} d\sigma(\lambda) = \frac{1}{\pi} \int_{q(x)<\lambda} \{\lambda - q(x)\}^{1/2} dx.$$

PROOF. Let $\sigma(\lambda) = \mathrm{mes}\{q(x) < \lambda\}$. Then

(6.3)
$$\int_{-\infty}^{\infty} \frac{dx}{\{q(x)+\mu\}^{2\tau+1+1/2}} = \int_0^{\infty} \frac{d\sigma(\lambda)}{(\lambda+\mu)^{2\tau+1+1/2}}.$$

We will show that

(6.4)
$$\int_0^{\infty} \frac{d\sigma(\lambda)}{(\lambda+\mu)^{2\tau+3/2}} = \frac{\Gamma(2\tau+2)}{\Gamma(2\tau+3/2)\,\Gamma(1/2)} \int_0^{\infty} \frac{d\lambda}{(\lambda+\mu)^{2\tau+2}} \left\{ \int_0^{\lambda} (\lambda-\nu)^{-1/2} d\sigma(\nu) \right\}.$$

In fact, changing the order of integration, we obtain

(6.5)
$$\int_0^{\infty} \frac{d\lambda}{(\lambda+\mu)^{2\tau+2}} \left\{ \int_0^{\lambda} \frac{d\sigma(\nu)}{(\lambda-\nu)^{1/2}} \right\} = \int_0^{\infty} d\sigma(\nu) \left\{ \int_{\nu}^{\infty} \frac{(\lambda-\nu)^{-1/2}}{(\lambda+\mu)^{2\tau+2}} d\lambda \right\}.$$

Further,

$$\int_{\nu}^{\infty} \frac{(\lambda-\nu)^{-1/2}}{(\lambda+\mu)^{2\tau+2}} d\lambda = \int_0^{\infty} \frac{t^{-1/2}}{(t+\nu+\mu)^{2\tau+2}} dt = \frac{1}{(\mu+\nu)^{2\tau+3/2}} \int_0^{\infty} \frac{z^{-1/2}}{(1+z)^{2\tau+2}} dz.$$

Hence, by well-known formulas for Euler integrals we obtain

$$\int^{\infty} \frac{(\lambda-\nu)^{-1/2}}{(\lambda+\mu)^{2\tau+2}} d\lambda = \frac{\Gamma(2\tau+3/2)\,\Gamma(1/2)}{\Gamma(2\tau+2)\,(\nu+\mu)^{2\tau+3/2}},$$

which together with (6.5) proves (6.4).

Let us put $\psi(\lambda) = \int_0^{\lambda} (\lambda-\nu)^{1/2} d\sigma(\nu)$. Then

$$d\psi(\lambda) = \frac{1}{2} \int_0^{\lambda} (\lambda-\nu)^{-1/2} d\sigma(\nu)\, d\lambda.$$

Therefore (6.3) can be written in the form

(6.6)
$$\int_{\infty}^{\infty} \frac{d\sigma(\lambda)}{(\lambda+\mu)^{2\tau+3/2}} = \frac{2\Gamma(2\tau+2)}{\Gamma(2\tau+3/2)\,\Gamma(1/2)} \int_0^{\infty} \frac{d\psi(\lambda)}{(\lambda+\mu)^{2\tau+2}}.$$

From the asymptotic formula (5.20′) and from (6.3) it follows, by virtue of (6.6), that

$$S_{\tau} \sim \frac{(4\tau+1)!!}{2^{2\tau+2}(2\tau+1)!} \frac{2\Gamma(2\tau+2)}{\Gamma(2\tau+3/2)\,\Gamma(1/2)} \int_0^{\infty} \frac{d\psi(\lambda)}{(\lambda+\mu)^{2\tau+2}}.$$

Inserting the values of the Euler functions $\Gamma(p)$, we obtain

(6.7)
$$S_{\tau} \sim \frac{1}{\pi} \int_0^{\infty} \frac{d\psi(\lambda)}{(\lambda+\mu)^{2\tau+2}}.$$

By definition

$$S_\tau = \sum_{n=1}^{\infty} \frac{1}{(\lambda_n + \mu)^{2\tau+2}}.$$

Putting $N(\lambda) = \sum_{\lambda_n < \lambda} 1$, we can write S_τ in the form

$$S_\tau = \int_0^\infty \frac{dN(\lambda)}{(\lambda + \mu)^{2\tau+2}}.$$

Therefore from (6.7) we finally obtain the asymptotic formula

$$\int_0^\infty \frac{dN(\lambda)}{(\lambda + \mu)^{2\tau+2}} \sim \frac{1}{\pi} \int_0^\infty \frac{d\psi(\lambda)}{(\lambda + \mu)^{2\tau+2}}. \tag{6.8}$$

By virtue of the condition (6.1) the above-mentioned Tauberian theorem of Keldyš is applicable to (6.8), from which follows the asymptotic formula $(\lambda \to \infty)$ $N(\lambda) \sim \psi(\lambda)/\pi$ or in expanded form

$$N(\lambda) = \sum_{\lambda_n < \lambda} 1 \sim \frac{1}{\pi} \int_0^\lambda (\lambda - \nu)^{1/2} d\sigma(\nu) = \frac{1}{\pi} \int_{q(x) < \lambda} \{\lambda - q(x)\}^{1/2} dx,$$

and the theorem is proved.

2. In this section we will obtain a formula from which, on the one hand, (6.2) will follow as a special case, and on the other hand, using this formula we will prove in the next section a theorem on the convergence and summability of expansions and differential expansions, in the eigenfunctions of Sturm-Liouville operators, of functions which grow at infinity like polynomials.

THEOREM 6.2. *Let τ and p be positive integers. Assume that*

4′. for $|x - \xi| \geqq 1$

$$q^{\tau+1}(\xi) q^p(x) \exp\{-\tfrac{1}{2} |x - \xi| \sqrt{q(x)}\} < B,$$

and suppose that the other conditions of Theorem 5.1 are satisfied in their original form. Then together with the asymptotic formula (5.19) one has the asymptotic formula

$$\left\| \frac{\partial^{\tau+p}}{\partial \mu^{\tau+p}} G(x, \eta; \mu) \right\|_{X^{(p)}} = \left\| \frac{\partial^{\tau+p}}{\partial \mu^{\tau+p}} g(x, \eta; \mu) \right\|_{X^{(p)}} (1 + o(1)). \tag{6.9}$$

PROOF. We will prove the asymptotic formula (6.9) for the simple case $\tau = 0$, $p = 1$. The proof for the general case is completely analogous, but involves somewhat more cumbersome calculations.

From the integral equation (5.3) follows (multiplying by $q(x)$)

$$
\begin{aligned}
q(x)\frac{\partial G}{\partial \mu} = q(x)\frac{\partial g}{\partial \mu} &- \int_{-\infty}^{\infty} q(x)\frac{\partial}{\partial \mu}\{g(x, \xi; \mu)\}\,[q(\xi) - q(x)]\,G(\xi, \eta; \mu)\,d\xi \\
&- \int_{-\infty}^{\infty} q(x)\,g(x, \xi; \mu)\,[q(\xi) - q(x)]\,q^{-1}(\xi)\left[q(\xi)\frac{\partial}{\partial \mu}G(\xi, \eta; \mu)\right]d\xi.
\end{aligned}
\tag{6.10}
$$

Put

$$\alpha(x, \eta; \mu) = -\int_{-\infty}^{\infty} q(x)\frac{\partial}{\partial \mu}\{g(x, \xi; \mu)\}\,[q(\xi) - q(x)]\,G(\xi, \eta; \mu)\,d\xi.$$

Arguing as in the proof of Lemma 1.1, one can show that

$$\| \alpha(x, \eta; \mu) \|_X = o(\| G(x, \eta; \mu) \|_X).$$

Further, from this same lemma and the integral equation (1.3) follows

$$\| G(x, \eta; \mu) \|_X = O(\| g(x, \eta; \mu) \|_X).$$

Therefore

$$\| \alpha(x, \eta; \mu) \|_X = o(\| g(x, \eta; \mu) \|_X).$$

By a direct calculation it is not hard to establish the estimate

$$\| g(x, \eta; \mu) \|_X = O\left(\left\| q(x)\frac{\partial}{\partial \mu} g(x, \eta; \mu)\right\|_X\right),$$

which with the preceding estimates leads to the estimate

$$\| \alpha(x, \eta; \mu) \|_X = o\left(\left\| q(x)\frac{\partial}{\partial \mu} g(x, \eta; \mu)\right\|_X\right).$$

From this estimate and the integral equation (6.10) there follows, analogously to the derivation of the asymptotic equality (5.18),

$$\left\| q(x)\frac{\partial}{\partial \mu} G(x, \eta; \mu)\right\|_X = \left\| q(x)\frac{\partial}{\partial \mu} g(x, \eta; \mu)\right\|_X (1 + o(1)),$$

which proves the assertion of the theorem for $\tau = 0$ and $p = 1$, since

$$\left\| q^p(x)\frac{\partial}{\partial \mu} G(x, \eta; \mu)\right\|_X = \left\| \frac{\partial}{\partial \mu} G(x, \eta; \mu)\right\|_{X^{(p)}}.$$

THEOREM 6.3. *Suppose that the function $q(x)$ satisfies the conditions of Theorem 6.2. We introduce the monotonic function $\sigma(\lambda) = \mathrm{mes}\{q(x) < \lambda\}$ and put*

$$\psi_p(\lambda) = \int_0^\lambda (\lambda - \nu)^{1/2} \nu^{2p} d\sigma(\nu) \tag{6.11}$$

and assume that the function $\psi_p(\lambda)$ *satisfies the inequalities* (6.1).

Then as $\lambda \to \infty$

$$\sum_{\lambda_n < \lambda} a_n^{(p)} \sim \frac{1}{\pi} \int_0^\lambda (\lambda - \nu)^{1/2} \nu^{2p} d\sigma(\nu) = \frac{1}{\pi} \int_{q(x)<\lambda} q^{2p}(x) \{\lambda - q(x)\}^{1/2} dx, \tag{6.12}$$

where

$$a_n^{(p)} = \int_{-\infty}^{\infty} q^{2p}(x) \varphi_n^2(x)\, dx, \tag{6.13}$$

and $\varphi_n(x)$ *is an eigenfunction of the operator* (1.1) *corresponding to the eigenvalue* λ_n.

PROOF. Replacing the number τ in (5.20) by $\tau + p$, we obtain

$$\int_{-\infty}^{\infty} \left\{ \frac{\partial^{\tau+p}}{\partial \mu^{\tau+p}} G(x, \eta; \mu) \right\}^2 d\eta = (\tau + p)! \sum_{n=1}^{\infty} \frac{\varphi_n^2(x)}{(\lambda_n + \mu)^{2(\tau+p)+2}}.$$

If we now multiply both sides of this equality by $q^{2p}(x)$ and integrate with respect to x from $-\infty$ to $+\infty$ and then use the asymptotic formula (6.9) and equality (6.12), we obtain for $\mu \to \infty$

$$\{(\tau + p)!\}^2 \sum_{n=1}^{\infty} \frac{a_n^{(p)}}{(\lambda_n + \mu)^{2(\tau+p)+2}} \sim \int_{-\infty}^{\infty} q^{2p}(x) \left\{ \int_{-\infty}^{\infty} \left[\frac{\partial^{\tau+p}}{\partial \mu^{\tau+p}} g(x, \eta; \mu) \right]^2 d\eta \right\} dx.$$

From this asymptotic equality, by virtue of (5.22) (putting there $\tau + p$ in place τ) we arrive at the equality

$$\sum_{n=1}^{\infty} \frac{a_n^{(p)}}{(\lambda_n + \mu)^{2(\tau+p)+2}} \sim \frac{(4\tau + 4p + 1)!!}{2^{2(\tau+p)+2}(2\tau + 2p + 1)!} \int_{-\infty}^{\infty} \frac{q^{2p}(x)}{\{q(x) + \mu\}^{2(\tau+p)+3/2}} dx. \tag{6.14}$$

Proceeding in the same way as we did to obtain the asymptotic equality (6.8) from (5.20′), for the case under consideration we arrive from (6.14) at the equality

$$\int_{-\infty}^{\infty} \frac{d\chi_p(\lambda)}{(\lambda + \mu)^{2(\tau+p)+2}} \sim \frac{1}{\pi} \int_{-\infty}^{\infty} \frac{d\psi_p(\lambda)}{(\lambda + \mu)^{2(\tau+p)+2}}, \tag{6.15}$$

where $\chi_p(\lambda) = \sum_{\lambda_n < \lambda} a_n^{(p)}$, and the function $\psi_p(\lambda)$ is defined by (6.11). Then the asymptotic formula (6.12) follows from (6.15) by Keldyš's theorem. This proves the theorem.

Formula (6.2) is obviously a special case of (6.12) (corresponding to the case $p = 0$).

THE CASE OF THE HALFLINE. Similar results can be obtained for the case of the halfline $[0, \infty)$. Suppose that $q(x)$ is defined on the halfline $[0, \infty)$ and satisfies the conditions (1.5), (1.6) and (1.20).

We will consider the operator

$$l(y) = -y'' + q(x)y \quad (x \geqq 0), \tag{6.16}$$

$$y'(0) - hy(0) = 0, \tag{6.17}$$

where h is an arbitrary fixed number.

Let us first clarify what the analog of the function $g(x, \eta; \mu)$ is for this case. Let $\kappa^2 = q(x) + \mu$ $(\mu > 0)$,

$$X(x) = \operatorname{ch} \kappa x + h\kappa^{-1} \operatorname{sh} \kappa x.$$

We consider the function

$$g^+(x, \eta; \mu) = \begin{cases} (\kappa + h)^{-1} X(\eta) e^{-\kappa x} & (\eta \leqslant x), \\ (\kappa + h)^{-1} X(x) e^{-\kappa \eta} & (\eta > x). \end{cases} \tag{6.18}$$

It is not hard to verify that the function $g^+(x, \eta; \mu)$ satisfies the equation $\partial^2 g^+ / \partial \eta^2 = \kappa^2 g^+$ and the conditions

$$\left(\frac{\partial g^+}{\partial \eta} - hg^+\right)\Big|_{\eta=0} = 0, \quad \frac{\partial g^+}{\partial \eta}\Big|_{\eta=x+0} - \frac{\partial g^+}{\partial \eta}\Big|_{\eta=x-0} = -1.$$

With the help of the function $g^+(x, \eta; \mu)$ we can construct an integral equation analogous to equation (1.3), and then prove the existence of a Green's function $G^+(x, \eta; \mu)$ for the operator (6.16)–(6.17).

To ultimately see this, we only have to see how to estimate the function $g^+(x, \eta; \mu)$. From (6.18) follows

$$g^+(x, \eta; \mu) = \begin{cases} \frac{1}{2} \kappa^{-1} (1 + h\kappa^{-1})^{-1} e^{-(x-\eta)\kappa} \{1 + O(\kappa^{-1})\} & (\eta \leqslant x), \\ \frac{1}{2} \kappa^{-1} (1 + h\kappa^{-1}) e^{-(\eta - x)\kappa} \{1 + O(\kappa^{-1})\} & (\eta > 0), \end{cases} \tag{6.19}$$

Since $\kappa^{-1} = o(1)$ as $\mu \to +\infty$, it follows from (6.19) that

$$g^+(x, \eta; \mu) = \tfrac{1}{2} \kappa^{-1} e^{-|x-\eta|\kappa} \{1 + o(1)\} = g(x, \eta; \mu) \{1 + o(1)\}. \tag{6.20}$$

The representation (6.20) makes it easy to extend the results obtained earlier to the case of the halfline.

We remark only that for the case of the halfline the coefficient in front of the integral in the asymptotic formulas (6.2) and (6.12) has to be replaced by $1/2\pi$

To conclude this section we will indicate various conditions which imply the inequality (6.1).

Suppose that $q(x)$ is monotone increasing, has a first and second derivative and is concave. Then $q'(x) > 0$, $q''(x) > 0$. Further, suppose that the condition

$$q(x)q''(x) < C[q'(x)]^2 \tag{6.21}$$

is satisfied, where C is some finite nonzero constant.

Let $x = p(\lambda)$ denote the function inverse to $q(x) = \lambda$. From consideration of the graph of $q(x)$ it is evident that $p(\lambda)$ is monotone increasing, has a first and second derivative and is convex, i.e. $p'(\lambda) > 0$, $p''(\lambda) < 0$.

It is easily seen that from (6.21) follows the inequality

$$p'(\lambda) > -C^{-1}\lambda p''(\lambda). \tag{6.22}$$

In fact, by differentiating $q(x) = \lambda$ we obtain

$$q'(x) = \frac{d\lambda}{dx} = \frac{1}{\frac{dx}{d\lambda}} = \frac{1}{p'(\lambda)},$$

$$q''(x) = \frac{d}{dx}\left(\frac{d\lambda}{dx}\right) = \frac{d}{dx}\left(\frac{1}{\frac{dx}{d\lambda}}\right) = \frac{d}{d\lambda}\left(\frac{1}{p'(\lambda)}\right)\frac{d\lambda}{dx} = -\frac{p''(\lambda)}{\{p'(\lambda)\}^3}.$$

Inserting these values of $q(x)$, $q'(x)$ and $q''(x)$ in (6.21), we obtain

$$-\lambda \frac{p''(\lambda)}{\{p'(\lambda)\}^3} < C\left\{\frac{1}{p'(\lambda)}\right\}^2, \quad p'(\lambda) > -C^{-1}\lambda p''(\lambda),$$

i.e. the condition (6.22).

Let us further express the function $\sigma(\lambda) = \operatorname{mes}\{q(x) < \lambda\}$ in terms of the function $p(\lambda)$. It is obvious that

$$\sigma(\lambda) = \operatorname{mes}\{q(x) < \lambda\} = \operatorname{mes}\{x < p(\lambda)\} = p(\lambda).$$

Therefore from the definition of the function $\psi_p(\lambda)$ follows ($\psi(\lambda) = \psi_0(\lambda)$)

$$\psi_p'(\lambda) = \frac{1}{2}\int_0^\lambda \frac{\nu^p d\sigma(\nu)}{(\lambda-\nu)^{1/2}} = \frac{1}{2}\int_0^\lambda \frac{\nu^p p'(\nu)\,d\nu}{(\lambda-\nu)^{1/2}}.$$

Integrating by parts and taking into account that $p'(\lambda) > 0$ and $p''(\lambda) < 0$, we obtain

(6.23)
$$\psi_p'(\lambda) = -\tau \int_0^\lambda (\lambda - \nu)^{1/2} \nu^p p'(\nu)\, d\nu + \int_0^\lambda (\lambda - \nu)^{1/2} \nu^p [-p''(\nu)]\, d\nu$$
$$< \int_0^\lambda \nu^p (\lambda - \nu)^{1/2} [-p''(\nu)]\, d\nu.$$

On the other hand, by (6.22) we have

$$\psi_p(\lambda) = \int_0^\lambda \nu^p (\lambda - \nu)^{1/2} p'(\nu)\, d\nu > C^{-1}\lambda \int_0^\lambda \nu^p (\lambda - \nu)^{1/2} [-p''(\nu)]\, d\nu.$$

From this inequality there follows, by (6.23), $\lambda\psi_p'(\lambda) < C\psi_p(\lambda)$, i.e. the second inequality in (6.1). The first inequality in (6.1) is trivial.

§7. Eigenfunction expansions for an unbounded increasing potential

We will consider the following eigenvalue problem:

(7.1) $$y'' + \{\lambda - q(x)\} y = 0 \qquad (0 \leqq x < \infty),$$

(7.2) $$y(0) \cos\alpha + y'(0) \sin\alpha = 0.$$

Suppose that $q(x)$ satisfies the conditions of Theorem 6.2, and moreover that for large x it satisfies the inequality

(7.3) $$q(x) \geqq ax^\delta,$$

where α and δ are certain positive constants.

It is known that if the condition (7.3) is satisfied (more precisely, if $q(x) \to +\infty$ as $x \to +\infty$), then the eigenfunctions $\varphi_n(x)$ of the problem (7.1) + (7.2) decrease rapidly. Thus, for example, if the function $q(x)$ has polynomial growth at infinity, then the eigenfunction $\varphi_n(x)$ decreases more rapidly than $e^{-C|x|}$, where $C > 0$ is any constant. Therefore for any $s > 0$ the integral

(7.4) $$a_n^{(s)} = \int_0^\infty q^s(x)\, \varphi_n^2(x)\, dx < +\infty$$

exists, and to construct the Fourier series of a function we do not need to require that it be square integrable.

In this section, using the asymptotic formula (6.14), we will prove several theorems concerning the convergence and summability of expansions and differentiated expansions, in the eigenfunctions of the problem (7.1) + (7.2), of functions which have polynomial growth at infinity.

We will first prove several lemmas.

Lemma 7.1. *Suppose that the function $q(x)$ satisfies the conditions of Theorem 6.2 and the condition (7.3). Then for $\mu\to\infty$ we have the estimate*

$$\sum_{\mu\leqslant\mu_n\leqslant\mu+1} a_n^{(s)} = O\left(\mu^{2s+\frac{2}{\delta}+1}\right), \tag{7.5}$$

where we have put $\mu_n^2=\lambda_n$ (the λ_n are the eigenvalues of the problem (7.1) – (7.2)), and the numbers $a_n^{(s)}$ are defined by (7.4).

Proof. By virtue of (7.3) we have

$$\sigma(\lambda) = \operatorname{mes}\{q(x)<\lambda\} \leqq \operatorname{mes}\{ax^\delta<\lambda\} = (\lambda/a)^{1/\delta} = C\lambda^{1/\delta}. \tag{7.6}$$

Put $\sigma_s(\lambda) = \int_0^\lambda \nu^s d\sigma(\nu)$, where s is some fixed number. Then, using the notation of the preceding section, and replacing $\tau+p+1$ in (6.14) by p, and $2p$ by s, we can write (6.14) in the form[5]

$$\int_0^\infty \frac{d\chi_s(\lambda)}{(\lambda+\mu)^{2p}} < C\int_0^\infty \frac{d\sigma_s(\lambda)}{(\lambda+\mu)^{2p-1/2}} = C\int_0^\infty \frac{\sigma_s(\lambda)\,d\lambda}{(\lambda+\mu)^{2p+1/2}}. \tag{7.7}$$

From (7.6) we obtain the following bound for $\sigma_s(\lambda)$: $\sigma_s(\lambda)<C\lambda^{s+1/\delta}$. Therefore from (7.7) follows the inequality

$$\int_0^\infty \frac{d\chi_s(\lambda)}{(\lambda+\mu)^{2p}} < C\int_0^\infty \frac{\lambda^{s+1/\delta}d\lambda}{(\lambda+\mu)^{2p+1/2}} = \frac{C}{\mu^{2p-s-1/\delta-1/2}}\int_0^\infty \frac{t^{s+1}}{(1+t)^{2p+1/2}}\,dt = \frac{C}{\mu^{2p-s-1/\delta-1/2}},$$

and so a fortiori

$$\int_{\mu^2}^{(\mu+1)^2} \frac{d\chi_s(\lambda)}{(\lambda+\mu^2)^{2p}} < C\mu^{4p-2s-2/\delta-1} \tag{7.8}$$

On the other hand,

$$\int_{\mu^2}^{(\mu+1)^2} \frac{d\chi_s(\lambda)}{(\lambda+\mu^2)^{2p}} > \frac{C}{\mu^{4p}}\{\chi_s[(\mu+1)^2]-\chi_s(\mu^2)\}. \tag{7.9}$$

From the bounds (7.8) and (7.9) follows the bound ($\mu_n=\sqrt{\lambda_n}$):

---

[5] To be correct, we will use a formula analogous to (6.14) for the case of the halfline $[0,\infty)$. These formulas differ only by a constant factor.

$$\chi_s[(\mu+1)^2]-\chi_s(\mu^2)=\sum_{\mu\leqslant\mu_n\leqslant\mu+1} a_n^{(s)} < C\mu^{2s+2/\delta+1},$$

as was to be proved.

Suppose that the function $f(x)$ is summable over every finite interval, and suppose that there exists a positive number s such that

$$\int_0^\infty f^2(x)\,q^{-s}(x)\,dx<\infty. \tag{7.10}$$

Put

$$S(x,\mu)=\sum_{\mu_n<\mu} c_n\varphi_n(x),$$

where

$$c_n=\int_0^\infty f(x)\,\varphi_n(x)\,dx.$$

Lemma 7.2. *If the function $f(x)$ satisfies the condition* (7.10), *and $q(x)$ satisfies the hypothesis of Lemma* 7.1, *then for $\mu\to\infty$ we have the estimate*

$$\bigvee_{\mu}^{\mu+1}\{S(x,\mu)\}=O(\mu^{s+1/\delta+1/2}). \tag{7.11}$$

Proof. By the definition of the function $S(x,\mu)$ we have

$$\begin{aligned}\bigvee_{\mu}^{\mu+1}\{S(x,\mu)\}&=\sum_{\mu\leqslant\mu_n\leqslant\mu+1}|c_n|\,|\varphi_n(x)|\\ &\leqslant\Big(\sum_{\mu\leqslant\mu_n\leqslant\mu+1} c_n^2\Big)^{1/2}\Big(\sum_{\mu\leqslant\mu_n\leqslant\mu+1}\varphi_n^2(x)\Big)^{1/2}.\end{aligned} \tag{7.12}$$

Further, applying the Cauchy-Bunjakovskiĭ inequality, by virtue of (7.10) we obtain

$$c_n=\int_0^\infty f(x)\,\varphi_n(x)\,dx=\int_0^\infty \frac{f(x)}{q^{s/2}(x)}\,\varphi_n(x)\,q^{s/2}(x)\,dx$$

$$\leqslant\left(\int_0^\infty f^2(x)\,q^{-s}(x)\,ds\right)^{1/2}\left(\int_0^\infty \varphi_n^2(x)\,q^s(x)\,dx\right)^{1/2}<C\{a_n^{(s)}\}^{1/2}.$$

Therefore

$$\bigvee_{\mu}^{\mu+1}\{S(x,\mu)\}\leqslant C\Big(\sum_{\mu\leqslant\mu_n\leqslant\mu+1} a_n^{(s)}\Big)^{1/2}\Big(\sum_{\mu\leqslant\mu_n\leqslant\mu+1}\varphi_n^2(x)\Big)^{1/2}.$$

The lemma follows from this inequality by virtue of (7.5) and (2.12), Chapter 7.

In the very same way, there follows from (7.5) and (1.21), Chapter 8:

Lemma 7.3. *Suppose that the hypothesis of Lemma 7.2 is satisfied, and moreover that $q(x)$ has a kth derivative which is summable over every finite interval. Then for $\mu \to \infty$ we have the estimate*

$$\bigvee_{\mu}^{\mu+1} \left\{ \frac{\partial^k S(x, \mu)}{\partial x^k} \right\} = O\left(\mu^{s+1/\delta+1/2+k}\right). \tag{7.13}$$

Using the estimates (7.11) and (7.13) and the methods of Chapters 7 and 8, it is easy to prove the following theorems.

Theorem 7.1. *Suppose that the function $q(x)$ satisfies the hypotheses of Theorem 6.2 as well as the condition (7.3), and suppose that $f(x)$ is summable over every finite interval and there exists a positive number s such that*

$$\int_0^\infty f^2(x)\, q^{-s}(x)\, dx < \infty.$$

Then for $\tau > s + 1/\delta + \frac{1}{2}$ we have, at every point of continuity of $f(x)$,

$$\lim_{\mu \to \infty} \sum_{\mu_n < \mu} \left(1 - \frac{\mu_n^2}{\mu^2}\right)^\tau c_n \varphi_n(x) = f(x),$$

where $c_n = \int_0^\infty f(x) \varphi_n(x) dx$, i.e. the Riesz mean (of order $> s + 1/\delta + \frac{1}{2}$) of the expansion of $f(x)$ in the eigenfunctions of the Sturm-Liouville operator tends to $f(x)$ at every point of continuity of $f(x)$.

Theorem 7.2. *Let $f(x)$ and $q(x)$ satisfy all the hypotheses of the preceding theorem, and suppose moreover that $q(x)$ has a $(k-1)$th derivative, and $f(x)$ a kth derivative, which derivatives are summable over every finite interval. Then for $\tau > s + 1/\delta + \frac{1}{2} + k$ we have, at every point of continuity of $f^{(k)}(x)$,*

$$\lim_{\mu \to \infty} \sum_{\mu_n < \mu} \left(1 - \frac{\mu_n^2}{\mu^2}\right)^\tau c_n \varphi^{(k)}(x) = f^{(k)}(x),$$

i.e. the Riesz mean of order $> s + 1/\delta + \frac{1}{2} + k$ of the kth derivative of the expansion of $f(x)$ in the eigenfunctions of the Sturm-Liouville operator tends at every point of continuity of $f^{(k)}(x)$ to the value $f^{(k)}(x)$.

Theorem 7.3. *Suppose that $q(x)$ satisfies the hypothesis of Theorem 7.1. Suppose further that there exist operators $\mathscr{L}^0 f = f(x)$, $\mathscr{L} f = f'' - q(x) f$, $\mathscr{L}^2 f, \cdots, \mathscr{L}^k f$ on the interval $[0, \infty)$ which satisfy the following conditions:*

1. $\{\mathscr{L}^i f(x)\}_{x=0} \cos\alpha + \{\mathscr{L}^i f(x)\}'_{x=0} \sin\alpha = 0$ $(i = 1, 2, \cdots, k-1)$.

2. $\mathcal{L}^i f(x) = o\left\{\exp\left[\int_0^x \sqrt{q(x)}\,dx\right]\right\}$ for $x \to \infty$ $(i = 0, 1, 2, \cdots, k-1)$,

3. $\{\mathcal{L}^i f(x)\}'_x = o\left\{\exp\left[\int_0^x \sqrt{q(x)}\,dx\right]\right\}$ for $x \to \infty$ $(i = 0, 1, 2, \cdots, k-1)$,

4. *There exists a positive number s such that*

$$\int_0^\infty \{\mathcal{L}^k f(x)\}^2 q^{-s}(x)\,dx < \infty.$$

Then if $k > [s/2 + 1/2\delta + 1/4]$, *the equality*

$$f(x) = \sum_{n=1}^{\infty} c_n \varphi_n(x)$$

holds uniformly in every finite interval.

THEOREM 7.4. *Suppose that all the conditions of the preceding theorem are satisfied, and moreover that* $f(x)$ *and* $q(x)$ *have mth derivatives which are summable over every finite interval.*

Then if $k > [s/2 + 1/2\delta + 1/4 + m/2]$, *we have at every point of continuity of* $f^{(m)}(x)$

$$f^{(m)}(x) = \sum_{n=1}^{\infty} c_n \varphi_n^{(m)}(x).$$

§8. Asymptotic behaviour of the number of eigenvalues of an operator of order $2n$

We will consider the selfadjoint ordinary differential operator L on the entire line $-\infty < x < \infty$

$$Ly = (-1)^n y^{(2n)} + p_{2n-2}(x) y^{(2n-2)} + \cdots + p_0(x) y + L_1(x, d/dx) y = L_0 + L_1,$$

where $L_1(x, d/dx)$ is an operator of order not greater than $2n-2$. We will assume that the characteristic polynomial

$$P(s, x; \lambda) = s^{2n} + p_{2n-2}(x)(is)^{2n-2} + \cdots + p_0(x) + \lambda$$

is positive for all sufficiently large x and $\lambda > 0$ and all real s. Further, we will assume that the coefficients $p_i(x)$ of L and the roots $s_i(x, \lambda)$ of $P(s, x; \lambda)$ are such that:

1. $$B \leqslant \left|\frac{s_k(x, \lambda)}{s_j(x, \lambda)}\right| \leqslant A, \quad B \leqslant \left|\frac{\operatorname{Im} s_k(x, \lambda)}{s_j(x, \lambda)}\right| \leqslant A.$$

2. $$|p_j(x)-p_j(\xi)|\leqslant c|x-\xi|p_j^{\alpha_j}(x),\quad |x-\xi|\leqslant 1,$$
$$\alpha_j<1+\frac{1}{2n-j}\quad (j=0,1,\ldots,2n-2).$$

3. $|p_j(x)|\leqq c\exp[c_0|x-\xi|p_0^{1/2n}(\xi)]$ for $|x-\xi|\geqq 1$ and for some constant $c_0>0$ which depends upon the characteristic polynomial $P(s,x;\lambda)$.

4. $p_0(x)\leqq cp_0(\xi)$ for $|x-\xi|\leqq 1$ and some $c>0$.

5. $\int_{-\infty}^{\infty}p_0^{-l}(x)\,dx<\infty$ for some $l>0$.

6. $$\frac{1}{P'[s_k(x,\lambda)]}\leqslant\frac{c}{|\lambda+p_0(x)|^{\frac{2n-1}{2n}}}$$

for all x and $\lambda>0$, $k=1,2,\cdots,2n$.

We now introduce the function

$$\rho(\lambda)=i\sum_{k=1}^{n}\int_{-\infty}^{\infty}\frac{ax}{P'[s_k(x,\lambda)]},$$

where the $s_k(x,\lambda)$ are the characteristic roots for which $\operatorname{Im}s_k(x,\lambda)>0$. As is evident, $\rho(\lambda)$ is an analytic function in the $\lambda=\xi+i\tau$ plane slit along the positive semiaxis.

Now let $\sigma(t)$ be defined by

$$\sigma(t)=\frac{1}{\pi}\lim_{\tau\to 0}\int_0^t\operatorname{Im}\rho(-\xi+i\tau)\,d\xi\quad(\tau>0,\ \xi>0).$$

The following holds.

THEOREM 8.1. *Suppose that the operator L satisfies the conditions* 1–6, *and that (for each i) the coefficient $q_i(x)$ of the ith derivative in the operator $L_1(x,d/dx)$ is subordinate to $p_0(x)$ in the sense that*

$$|q_i(x)|\leqq c_1\{p_0(x)\}^{(2n-1)/2n-\epsilon}$$

for some $\epsilon>0$. Further, let the function $\sigma(t)$ defined above satisfy the Tauberian condition $t\sigma'(t)\leqq\alpha\sigma(t)$ $(\alpha>0)$.

Then for $\lambda\to\infty$ we have the asymptotic formula $\sum_{\lambda_n<\lambda}1\equiv N(\lambda)\sim\sigma(\lambda)$, where the λ_n are the eigenvalues of the operator L.

This theorem can be proved in accordance with the classical scheme of T. Carleman: first one finds the asymptotic behaviour of the Green's function $K(x,\xi;\lambda)$, the kernel of the resolvent of L, for large $\lambda<0$. Then one applies the Tauberian theorem of Keldyš. We note that for our purposes we need the asymptotic behaviour of $K(x, \xi; \lambda)$ uniformly with respect to x over the entire space. The asymptotic behaviour is sought by the "parametrics" method, i.e. one writes down in explicit form the kernel of the resolvent of the principal part of the operator with "frozen" coefficients, i.e. of the operator

$$\tilde{L}_0 = (-1)^n \frac{d^{2n}}{dx^{2n}} + p_{2n-2}(\xi) \frac{d^{2n-2}}{dx^{2n-2}} + \ldots + p_0(\xi).$$

This kernel can be found by means of the Fourier transform. One then constructs an integral equation for the kernel of the resolvent of the original operator L in the usual way. One can prove that it can be solved by the method of successive approximations, if the hypotheses of the theorem are satisfied.

We note two special cases.

I. Let $p_k(x) \equiv 0$ $(k = 1, 2, \cdots, 2n-2)$. Then the calculation of the function $\sigma(t)$ can easily be carried through. In this case

$$N(\lambda) \sim \frac{1}{\pi} \int\limits_{q(x)<\lambda} \{\lambda - q(x)\}^{\frac{1}{2n}} dx.$$

II. $2n = 4$. We write the operator L in the form

$$Ly = y^{\mathrm{IV}} - 2\{p_2(x)y\}' + p_0(x)y + L_1(x, d/dx)y.$$

It is not hard to verify that the conditions 1, 3 and 6 will be satisfied if $p_2^2(x) \leq c[p_0(x) + \lambda_0]$ for some λ_0 and $c < 1$. In this case also the calculation of the function $\sigma(t)$ can be carried through. Namely

$$N(\lambda) \sim \frac{1}{\pi} \int\limits_{p_0(x)<\lambda} \sqrt{-p_2(x) + \sqrt{p_2^2(x) + \{\lambda - p_0(x)\}}}\, dx.$$

This same result can be obtained by studying the Green's function $G(x, \xi; t)$ of a parabolic equation for small t. By this route one can obtain results concerning the asymptotic behaviour of $N(\lambda)$ for systems of ordinary differential equations, and also for singular elliptic operators with a discrete spectrum.

Bibliographical references

The first to obtain the asymptotic behaviour of $N(\lambda)$ were de Wet and Mandl [**1**], who used a variational theorem of Courant. Titchmarsh [**2**] applied Carleman's method to this problem and obtained an asymptotic formula for $N(\lambda)$ under less restrictive conditions on the potential function $q(x)$ (in the book cited the case of two variables was considered; however the method of this book is also applicable to the case of a single variable).

The complete series for the Green's function was used by Levitan [21]. The application of estimates for the weighted trace of the Green's function to questions concerning eigenfunction expansions with unbounded increasing potentials is due to Levitan (cf. also the authors' joint paper [3]).

The formula for $N(\lambda)$ for the case of an nth order operator, given in §8, is due to Kostjučenko [3,4].

CHAPTER 13

ELEMENTS OF THE SPECTRAL THEORY OF LINEAR OPERATORS IN HILBERT SPACE. RELATION TO DIFFERENTIAL OPERATORS

For a deeper understanding of many questions which have been discussed in the previous chapters of this book, it is useful to consider these questions from the point of view of the general spectral theory of linear selfadjoint operators in Hilbert space.

At the present time there are many excellent texts on the abstract theory of linear operators in Hilbert space; however in applying the general theory to differential operators a number of analytic difficulties arise and, unfortunately, not everything as yet appears as smooth and definitive in this topic as in the general theory. Possibly this stems from an incompleteness of the theory, in particular when speaking about partial differential operators.

In this chapter we give a brief presentation of some basic topics from the spectral theory of linear selfadjoint operators in an abstract Hilbert space. As we have already mentioned, at the present time this theory has been brought to a high degree of completenes, and can be studied from many texts. Therefore this part of the chapter will in general be presented without proofs, and can serve as a first introduction to abstract spectral theory in Hilbert space (the reader who is acquainted with this theory to the extent, for example, of the books by Ahiezer and Glazman [1], Ljusternik and Sobolev [1], or Smirnov [1] can omit the corresponding sections of this chapter).

The end of this chapter (§§6 and 7) is devoted to various applications of the abstract theory to differential operators.

§1. Abstract Hilbert space

1. Definition of an abstract Hilbert space. Let H be a complex linear space with elements $x, y, z, \cdots$. We assume that with every ordered pair $x, y \in H$ there is associated a complex number denoted by (x, y), called the *scalar product* of these elements and satisfying the following conditions (axioms):

a) $(x, y) = \overline{(y, x)}$,
b) $(x_1 + x_2, y) = (x_1, y) + (x_2, y)$,
c) $(\lambda x, y) = \lambda(x, y)$ for any complex number λ,
d) $(x, x) \geqq 0$, and $(x, x) = 0$ if and only if $x = \theta$ (the zero element of the space H).

The number $\|x\| = (x, x)^{1/2}$ *is called the norm of the element* x. From the axioms of the scalar product one can deduce that this quantity satisfies all the conditions of the norm in a linear normed space.

If H *is complete relative to the metric* $\rho(x, y) = \|x - y\|$ *and is infinite dimensional, then it is called a complex (unitary) Hilbert space.*

Remark. A real Hilbert space is defined analogously. For this case the scalar product is assumed to be a real number.

We give two important examples of Hilbert space.

1°. *The space* l_2. The elements of this space are infinite sequences $x = (\xi_1, \xi_2, \cdots)$, $y = (\eta_1, \eta_2, \cdots)$ of complex numbers such that

$$\sum_{k=1}^{\infty} |\xi_k|^2 < \infty, \quad \sum_{k=1}^{\infty} |\eta_k|^2 < \infty.$$

The scalar product is defined by

$$(x, y) = \sum_{k=1}^{\infty} \xi_k \cdot \bar{\eta}_k.$$

One can show that l_2 satisfies all the conditions of a unitary Hilbert space.

2°. *The space* $\mathscr{L}^2(a, b)$. The elements of this space are measurable complex-valued functions $f(x), g(x), \cdots$ defined on the interval $[a, b]$ and having square-integrable modulus. The scalar product is defined by

$$(f(x), g(x)) = \int_a^b f(x) \overline{g(x)}\, dx.$$

We have essentially been dealing with this space throughout the book. Namely, we have studied the spectral theory of a Stúrm-Liouville differential operator in the spaces $\mathscr{L}^2(0, \pi)$, $\mathscr{L}^2(0, \infty)$ and $\mathscr{L}^2(-\infty, \infty)$. In connection with Parseval's formula we were led to consider the space $\mathscr{L}^2_{\{\rho(\lambda)\}}(a, b)$, which also is a Hilbert space.

Finally, in connection with the spectral theory of Dirac-type systems of equations we considered a Hilbert space whose elements are vector-functions.

2. Orthogonality. *Two elements* $x, y \in H$ *are said to be orthogonal, if* $(x, y) = 0$. In this case one writes $x \perp y$. *An element* x *is said to be orthogonal to a subspace* $\mathscr{L} \in H$ *if* x *is orthogonal to every* $y \in \mathscr{L}$. In this case we write $x \perp \mathscr{L}$.

One has the following important result.

THEOREM 1.1. *Let $\mathscr{L}$ be a closed subspace of the space H. Then every $x \in H$ has a unique decomposition*

$$x = y + z, \tag{1.1}$$

where $y \in \mathscr{L}$ and $z \perp \mathscr{L}$.

The element y in the decomposition (1.1) is called the *projection* of the element x on $\mathscr{L}$. We remark that the set M of all elements in H which are orthogonal to $\mathscr{L}$ is itself a closed subspace. Therefore the element z from (1.1) is the projection of x on M. The subspace M is called the *orthogonal complement* of $\mathscr{L}$ and is denoted by $H \dot{-} \mathscr{L}$. One also says that H is the orthogonal sum of $\mathscr{L}$ and M, written as $H = \mathscr{L} \dot{+} M$.

3. Orthogonal systems. *A system $e_1, e_2, \cdots$ of elements of the space H is said to be orthonormal if*

$$(e_i, e_j) = \begin{cases} 0, & \text{if } i \neq j, \\ 1, & \text{if } i = j. \end{cases}$$

Let $\mathscr{L}$ be the closed subspace generated by an orthonormal system $e_1, e_2, \cdots$ (i.e., the closure in the norm of H of the set of all linear combinations of the form $\sum_{k=1}^N \xi_k e_k$, where the ξ_k are complex or real numbers according as H is a complex or real Hilbert space). Obviously for any element $x \in \mathscr{L}$ and any $\epsilon > 0$ there exists a linear combination $\sum_{k=1}^N \xi_k e_k$ such that

$$\left\| x - \sum_{k=1}^N \xi_k e_k \right\| < \varepsilon. \tag{1.2}$$

A simple computation shows that

$$\left\| x - \sum_{k=1}^N \xi_k e_k \right\|^2 = \left(x - \sum_{k=1}^N \xi_k e_k,\ x - \sum_{k=1}^N \xi_k e_k \right)$$

$$= (x, x) - \sum_{k=1}^N \xi_k (x, e_k) - \sum_{k=1}^N \bar{\xi}_k (e_k, x) + \sum_{k=1}^N \sum_{i=1}^N \xi_k \bar{\xi}_i (e_k, e_i) \tag{1.3}$$

$$= \|x\|^2 - \sum_{k=1}^N \xi_k c_k - \sum_{k=1}^N \bar{\xi}_k \bar{c}_k + \sum_{k=1}^N |\xi_k|^2 = \|x\|^2 + \sum_{k=1}^N |\xi_k - c_k|^2 - \sum_{k=1}^N |c_k|^2,$$

where $c_k = (x, e_k)$. The numbers c_k are called the Fourier coefficients of x (relative to the system $e_1, e_2, \cdots$).

It follows from the identity (1.3) that the difference $x - \sum_{k=1}^N \xi_k e_k$

has minimum norm when the numbers ξ_k equal the corresponding Fourier coefficients: $\xi_k = c_k$. In this case we obtain from (1.2) and (1.3)

$$0 \leqslant \left\| x - \sum_{k=1}^{N} c_k e_k \right\|^2 = \| x \|^2 - \sum_{i=1}^{N} | c_i |^2 < \varepsilon^2, \tag{1.4}$$

and since ϵ can be chosen arbitrarily small, it follows from (1.4) that

$$x = \lim_{N \to \infty} \sum_{k=1}^{N} c_k e_k = \sum_{k=1}^{\infty} c_k e_k.$$

From (1.4) also follows the convergence of the series $\sum_{k=1}^{\infty} |c_k|^2$, and that $\|x\|^2 = \sum_{k=1}^{\infty} |c_k|^2$ $(x \in \mathscr{L})$.

If now x is an arbitrary element of H, we can put $x = y + z$, where $y \in \mathscr{L}$, $z \perp \mathscr{L}$. Then

$$y = \sum_{k=1}^{\infty} c_k e_k, \quad c_k = (y, e_k) = (y, e_k) + (z, e_k) = (x, e_k), \tag{1.5}$$
$$\| y \|^2 = \sum_{k=1}^{\infty} | c_k |^2.$$

Since $\|x\|^2 = \|y + z\|^2 = \|y^2\| + \|z\|^2$, it follows that

$$\|x\|^2 \geqq \|y\|^2. \tag{1.6}$$

From (1.5) and (1.6) follows

$$\sum_{k=1}^{\infty} | c_k |^2 \leqslant \| x \|^2, \qquad c_k = (x, e_k). \tag{1.7}$$

The inequality (1.7) is called Bessel's inequality.

4. Closed orthonormal systems. Let $\{e_i\}$ be an orthonormal system in a Hilbert space H.

DEFINITION 1.1. *The orthonormal system $\{e_i\}$ is said to be complete, if there exists no vector $x \in H$ other than zero which is orthogonal to all the elements of the system $\{e_i\}$.*

DEFINITION 1.2. *The orthonormal system $\{e_i\}$ is said to be closed if the closed subspace $\mathscr{L}$ generated by this system coincides with H.*

Obviously the Fourier series of an element x, corresponding to a closed orthonormal system, converges to x for every $x \in H$, and one has the Parseval equality $\sum_{k=1}^{\infty} |c_k|^2 = \|x\|^2$.

A closed orthonormal system is also called an orthonormal basis of the space H.

THEOREM 1.2. *The concepts of a complete and a closed orthonormal system coincide.*

Proof. If an orthonormal system is complete, then there exists no nonzero element which is orthogonal to the closed subspace $\mathscr{L}$ generated by this system. But then $z=\theta$ in the decomposition (1.1), and consequently $x=y$.

Conversely, if the orthonormal system $\{e_i\}$ is closed, then $\mathscr{L}=H$ and so for every element $x\in H$ we have $\|x\|^2=\sum_{k=1}^{\infty}|c_k|^2$. Therefore if $x\perp e_i$ $(i=1,2,\cdots)$, then $\|x\|=0$, i.e. $x=\theta$, which says that the system $\{e_i\}$ is complete.

Remark. The considerations of §1.4 are concerned with so-called *separable Hilbert spaces*. These spaces are characterized by the fact that they contain a countable complete orthonormal system.

There exist, however, nonseparable Hilbert spaces, for example the space of almost periodic functions. In these spaces every complete orthonormal system is uncountable. In this book we do not encounter nonseparable Hilbert spaces.

§2. Linear functionals and bounded linear operators

1. Linear functionals and linear operators. Let D be some set of elements of a Hilbert space H.

Definition 2.1. *If with each element $f\in D$ we associate some number $\Phi(f)$, then we say that we have defined a functional Φ in H with domain D.*

If with each element $f\in D$ we associate some element $Tf\in H$, then T is called an operator in H with domain D. The set $\Delta=\{Tf;f\in D\}$ is called the range of T.

The notation D_Φ, D_T and Δ_T is also used for the domain of Φ and for the domain and range of T.

From now on we will in this section consider only so-called linear functionals and linear operators, and we will assume that $D=H$.

Definition 2.2. *A functional Φ (operator T) is said to be additive and homogeneous if for any complex numbers α, β and any elements $f,g\in H$ we have*

$$\Phi(\alpha g+\beta g)=\alpha\Phi(f)+\beta\Phi(g)\quad (T(\alpha f+\beta g)=\alpha Tf+\beta Tg).$$

Definition 2.3. *A functional Φ (operator T) is said to bounded if*

$$\sup_{\|f\|\leqslant 1}|\Phi(f)|<\infty,\qquad \Big(\sup_{\|f\|\leqslant 1}\|Tf\|<\infty\Big).$$

Definition 2.4. *A linear functional (operator) is a functional (operator) which is homogeneous, additive and bounded.*

The norm $\|T\|$ of a linear operator T is defined by

$$\|T\|=\sup_{\|f\|\leqslant 1}\|Tf\|.$$

The norm can also be defined by

$$\| T \| = \sup_{f \in H} \frac{\| Tf \|}{\| f \|}.$$

A bounded linear operator is continuous.

If a homogeneous and additive operator is continuous at one point, then it is bounded. The analogous property holds for linear functionals.

2. Bilinear functionals. *If with every ordered pair of elements $f, g \in H$ we associate some (generally speaking, complex) number $\Omega(f,g)$ in such a way that*

$$\begin{aligned} &\text{a)}\ \Omega(\alpha_1 f_1 + \alpha_2 f_2,\ g) = \alpha_1 \Omega(f_1,\ g) + \alpha_2 \Omega(f_2,\ g), \\ (2.1)\qquad &\text{b)}\ \Omega(f,\ \beta_1 g_1 + \beta_2 g_2) = \bar{\beta}_1 \Omega(f,\ g_1) + \bar{\beta}_2 \Omega(f,\ g_2), \\ &\text{c)}\ \sup_{\|f\| \leq 1,\ \|g\| \leq 1} | \Omega(f,\ g) | < \infty, \end{aligned}$$

then we say that we have defined a bilinear functional in H.

The quantity in c) *is called the norm of the bilinear functional Ω and is denoted by* $\| \Omega \|$.

THEOREM 2.1. *Every bilinear functional Ω in H can be represented in the form $\Omega(f,g) = (Af,g)$, where A is a bounded operator in H which is uniquely determined by the functional Ω; furthermore $\| A \| = \| \Omega \|$.*

3. The adjoint operator. Let A be a bounded linear operator defined everywhere in H. The expression (f, Ag) is obviously a bilinear functional in H with norm $\| A \|$. Therefore it follows from the preceding theorem that there exists a unique (bounded) linear operator A^* for which

$$(f, Ag) = (A^* f, g), \tag{2.2}$$

and that $\| A^* \| = \| A \|$.

The operator A^*, satisfying (2.2) *for all $f, g \in H$, is called the adjoint of A.* It is easily seen that $(A^*)^* = A$.

If $A^* = A$, then A is said to be a *selfadjoint* operator.

A bounded selfadjoint operator A is said to be *positive* if $(Af, f) \geqq 0$ for every $f \in H$.

4. Projection operators. Let H be a Hilbert space and G a closed subspace of H. *The operator P which associates with every element $f \in H$ its projection g on G is called the projection operator or projector* (*on* G), *and is denoted by P_G.*

A projection operator is obviously additive and homogeneous. Moreover, it is bounded, and its norm equals unity. We give without proofs the basic properties of projection operators.

a) *In order that an operator* P, *defined everywhere in* H, *be a projector, it is necessary and sufficient that* 1) $P^2 = P$, 2) $P^* = P$.

b) *The product of two projectors* P_{G_1}, P_{G_2} *is a projector if and only if they commute, i.e.* $P_{G_1}P_{G_2} = P_{G_2}P_{G_1}$, *and if this condition is satisfied, then* $P_{G_1}P_{G_2} = P_G$, *where* $G = G_1 \cap G_2$.

c) *A sum of projectors,* $P_{G_1} + P_{G_2} + \cdots + P_{G_n} = Q$ $(n < \infty)$ *is a projector if and only if* $P_{G_k}P_{G_j} = 0$ $(j \neq k)$, *i.e. if the subspaces* G_i *are pairwise orthogonal. In this case* $Q = P_G$, *where* $G = G_1 \dotplus G_2 \dotplus \cdots \dotplus G_n$.

d) *The difference of two projectors,* $P_{G_1} - P_{G_2}$, *is a projector if and only if* $G_2 \subseteq G_1$.

e) *If* $\{P_k\}$ $(k = 1, 2, \cdots)$ *is an infinite monotonic sequence of projectors, than as* $k \to \infty$ *the projector* P_k *converges strongly to a projector* P, *i.e. for every* $f \in H$ *we have* $\lim_{k\to\infty} P_k f = Pf$.

5. Unitary operators. *A linear operator* U, *defined on all of the space* H, *and having as its range all of* H, *is called a unitary operator if*

$$(2.3) \qquad (Uf, Ug) = (f, g)$$

for all $f, g \in H$.

We state, again without proofs, the following simple properties of unitary operators.

a) *If* U *is a unitary operator, then its inverse* U^{-1} *exists and is also unitary.*

b) $U^* = U^{-1}$, *i.e.* $U^*U = UU^* = I$ *(the identity operator).*

6. Isometric mappings. Suppose that we are given two Hilbert spaces H_1 and H_2; let us agree to differentiate between the scalar products in these spaces by the subscripts 1 and 2.

DEFINITION 2.5. *A linear operator* V, *defined on all of* H_1 *and mapping* H_1 *onto all of* H_2, *is said to be an isometric mapping (of* H_1 *onto* H_2) *if*

$$(2.4) \qquad (Vf, Vg)_2 = (f, g)_1$$

for all $f, g \in H_1$.

If the range Δ_V of V does not coincide with all of H_2 (but (2.4) still holds), then V is said to be *partially isometric*.

By virtue of these definitions Theorem 1.1 of Chapter 2 can be regarded as proving that the indicated mapping of the space $\mathscr{L}^2(0, \infty)$ into the space $\mathscr{L}^2_{\{\rho(\lambda)\}}(-\infty, \infty)$ is partially isometric, and Theorem 4.2 of Chapter 2, that the indicated mapping between these spaces is isometric.

7. Isometric operators.

DEFINITION 2.6. *A linear operator* V, *defined on a Hilbert space* H, *is said to be isometric if it preserves the norm of every element, i.e. if* $\|Vx\| = \|x\|$ *for all* $x \in H$.

Like any linear operator, V is defined on all of H, but it is not required that V shall map H onto all of H. Therefore an isometric operator need not be unitary, which is evident from the following simple example. Let $e_1, e_2, \cdots$ be an orthonormal basis in H, and let x be an arbitrary element of H. Then $x = \sum_{k=1}^{\infty} x_k e_k$, where $x_k = (x, e_k)$.

The operator V defined by $Vx = \sum_{k=1}^{\infty} x_k e_{k+1}$ is isometric and maps all of H onto the subspace of H which is orthogonal to e_1.

§3. Some general concepts of the theory of linear operators

In this section we will consider some general concepts of the theory of linear but, as a rule, unbounded operators. Such operators are generally defined not on the entire space, but only on some dense subset of it. Thus, for example, differential operators are defined on differentiable functions which form a dense set in the space $\mathscr{L}^2(a,b)$.

Two operators A and B are regarded as equal or as coinciding if $D_A = D_B$ *and* $Ax = Bx$ *for every* $x \in D_A$. *If* $D_A \subseteq D_B$ *and* $Ax = Bx$ *for every* $x \in D_A$, *then B is said to be an extension of A, and A is said to be a restriction of B. In this case we write* $A \subseteq B$.

1. The concept of a closed operator.

DEFINITION 3.1. *An operator T (not necessarily linear) is said to be closed, if from the existence of the limits*

$$\lim_{n\to\infty} f_n = f, \quad \lim_{n\to\infty} Tf_n = g, \quad \text{with } f_n \in D_T,$$

it follows that $f \in D_T$ *and* $g = Tf$.

Thus every continuous operator (with closed domain) is obviously closed; but the converse is false, since for a discontinuous operator the existence of $\lim_{n\to\infty} f_n$ does not necessarily imply that of $\lim_{n\to\infty} Tf_n$.

If an operator T is not closed, but whenever two sequences $\{f_n\}$ and $\{g_n\}$ in D_T converge to the same limit, then the sequences $\{Tf_n\}$ and $\{Tg_n\}$ either both diverge or both converge to the same limit, then T has closed extensions. Among these extensions we distinguish the minimal closed extension, which is a restriction of every other closed extension of T. This operator is usually denoted by $\overline{T}$ and is called the closure of T. To obtain $\overline{T}$ we adjoin to D_T all those elements $f \in \overline{D}_T$ (the closure of D_T) for which there exists at least one sequence $f_1, f_2, \cdots \in D_T$ which converges to f, and for which the sequence $\{Tf_n\}$ converges; we then put $\overline{T}f = \lim_{n\to\infty} Tf_n$.

2. The general definition of the adjoint operator. Let T be an arbitrary linear operator[1] whose domain D_T is dense in H. We consider the scalar product (Tf,g), $f \in D_T$.

There exist pairs of elements g, g^* for which

$$(Tf,g) = (f,g^*) \tag{3.1}$$

holds for every $f \in D_T$. For example, (3.1) holds for $g = g^* = \theta$.

If D_T is dense in H (which we are assuming), then the element g^* is uniquely determined by the element g.

We put $g^* = T^*g$. By definition, T^* is called the adjoint of T. Its domain D_{T^*} is the set of all those g for which there exists a g^* satisfying (3.1) for every $f \in D_T$. The following assertions follow easily from the definition of the adjoint operator.

a) *T^* is linear.*

b) *If $S \subseteq T$, then $S^* \supseteq T^*$.*

c) *The operator T^* is closed even when T is not closed.*

d) *If T has a closure $\overline{T}$, then $(\overline{T})^* = T^*$.*

e) *If the operator T^{**} exists, then $T \subseteq T^{**}$.*

3. Symmetric and selfadjoint operators. *A linear operator A is said to be symmetric if*

a) *its domain D_A is dense in H,*

b) *for every pair of elements $f,g \in D_A$ one has $(Af,g) = (f,Ag)$.*

If there exists a number $\gamma > -\infty$ such that $(Af,f) \geqq \gamma(f,f)$, $f \in D_A$, then A is said to be lower semibounded.

Obviously for a symmetric operator A we have $A^* \supseteq A$, from which it follows that a symmetric operator always has a closure.

If B is a symmetric extension of A, then $B \subseteq A^*$.

An operator A is said to be selfadjoint if it coincides with its adjoint, i.e. $A^* = A$.

We present without proofs the following properties of selfadjoint operators.

a) *If a selfadjoint operator has an inverse, then this inverse is selfadjoint (bounded or unbounded).*

b) *The eigenvalues of a selfadjoint operator are real.*

c) *Two eigenvectors f_1, f_2 of a selfadjoint operator corresponding to distinct eigenvalues λ_1, λ_2 are orthogonal.*

---

[1] In the theory of unbounded operators the term "linear" stands for "homogeneous and additive".

4. The spectrum of a linear operator. Let T be a closed linear operator defined on a dense subset D_T of H. We consider the equation $Tf - \lambda f = g$, $f \in D_T$, $g \in H$, with a complex parameter λ. Let us denote by $\Delta_T(\lambda)$ the range of the operator $T - \lambda I$. If $T - \lambda I$ establishes a one-to-one correspondence between D_T and $\Delta_T(\lambda)$, then $T - \lambda I$ has an inverse $(T - \lambda I)^{-1}$ with domain $\Delta_T(\lambda)$ and range D_T.

DEFINITION 3.2. *The values of the parameter λ for which the inverse operator $(T - \lambda I)^{-1}$ exists, is defined everywhere in H (i.e. $\Delta_T(\lambda) = H$) and is bounded, is called a regular value (or regular point) of the operator T. The remaining points of the complex λ-plane form the spectrum of T.*

One can show that the spectrum of a selfadjoint operator lies entirely on the real axis.

5. The resolvent. Let T be a closed linear operator with domain which is dense in H. *The resolvent of T is defined as the operator-valued function of λ,*

$$R_\lambda = (T - \lambda I)^{-1},$$

defined on the set of all those values of λ for which it exists and for which its domain, i.e. $\Delta_T(\lambda)$, is dense in H. At every regular point of T the resolvent R_λ is a bounded operator defined on all of H.

One can show that for a selfadjoint operator A we have $R_\lambda^* = R_{\bar\lambda}$ if λ is a regular point of A.

§4. Spectral analysis of selfadjoint operators

1. DEFINITION 4.1. *By a spectral function is meant an operator-valued function E_λ of a real parameter λ satisfying the following conditions:*

a) *E_λ is a projector for every value of λ,*

b) *$E_\lambda \leqq E_\mu$ for $\lambda < \mu$ (i.e. $E_\mu - E_\lambda$ is a nonnegative operator),*

c) *$E_{-\infty} = 0$, $E_{+\infty} = I$, i.e. for every $x \in H$ we have*

$$\lim_{\lambda \to -\infty} \| E_\lambda x \| = 0, \qquad \lim_{\lambda \to +\infty} \| E_\lambda x - x \| = 0,$$

d) *for every x the vector-function $E_\lambda x$ is continuous from the right, i.e.*

$$\lim_{\epsilon \to +0} \| E_{\lambda+\epsilon} x - E_\lambda x \| = 0.$$

It easily follows from property b) that the limits

$$\lim_{\epsilon \to 0} E_{\mu-\epsilon} x = E_{\mu-0} x, \quad \lim_{\epsilon \to 0} E_{\mu+\epsilon} x = E_{\mu+0} x \quad (\epsilon > 0)$$

exist for every $x \in H$. Condition d) says that $E_{\mu+0} = E_\mu$. This condition is not essential and represents only a specific normalization of the spectral function.

We will use the following notation: if Δ denotes one of the intervals (α,β), $[\alpha,\beta)$, $(\alpha,\beta]$ or $[\alpha,\beta]$, then E_Δ will denote, respectively,

$$E_{\beta-0}-E_{\alpha+0}=E_{\beta-0}-E_{\alpha};\ E_{\beta-0}-E_{\alpha-0},\ E_{\beta+0}-E_{\alpha+0}=E_{\beta}-E_{\alpha};$$
$$E_{\beta+0}-E_{\alpha-0}=E_{\beta}-E_{\alpha-0}.$$

It easily follows from property b) that E_Δ is a projector for any Δ. It is also easy to verify that if the intervals $\Delta_1,\Delta_2,\cdots,\Delta_n$ are pairwise disjoint, then $E_{\Delta_i}E_{\Delta_j}=0$ for $i\neq j$.

2. The integral with respect to a spectral function. By means of an arbitrary spectral function one can construct an integral analogous to the Stieltjes integral.

Let f be an arbitrary numerical function, defined and continuous on an interval $[a,b]$. We consider an arbitrary partition of $[a,b]$ by means of certain points $a=\lambda_0<\lambda_1<\cdots<\lambda_{n-1}<\lambda_n=b$ into parts $\Delta_1=[a,\lambda_1]$, $\Delta_2=(\lambda_1,\lambda_2],\cdots,\Delta_{n-1}=(\lambda_{n-1},\lambda_{n-2}],\Delta_n=(\lambda_{n-1},b]$ and associate with this partition the sum

$$S=\sum_{i=1}^{n} f(\xi_i)E_{\Delta_i},$$

where ξ_i is an arbitrary point of the interval Δ_i.

The sum S is a bounded linear operator in H.

One can show that as the length of the largest of the intervals Δ_i tends to zero, the sum S tends in the operator (uniform) norm to a definite limit J, i.e. for any $\epsilon>0$ there exists a $\delta>0$ such that if the length of every Δ_i is less than δ, then any corresponding sum S satisfies the inequality $\|S-J\|<\delta$.

The operator J is called the integral of the function f with respect to E_λ, and is denoted by

$$\int_a^b f(\lambda)\,dE_\lambda. \tag{4.1}$$

From the existence of the integral (4.1) there obviously follows the existence of the integral

$$\int_a^b f(\lambda)\,dE_\lambda x,\quad x\in H, \tag{4.2}$$

which is defined as the limit of the sums $\sum_{i=1}^n f(\xi_i)E_{\Delta_i}x=Sx$.

If in the equality

$$\|Sx\|^2=\left\|\sum_{i=1}^{n} f(\xi_i)E_{\Delta_i}x\right\|^2=\sum_{i=1}^{n}|f(\xi_i)|^2\|E_{\Delta_i}x\|^2=\sum_{i=1}^{n}|f(\xi_i)|^2(E_{\Delta_i}x,\ x) \tag{4.3}$$

we pass to the limit, we obtain

$$\left\| \int_a^b f(\lambda)\, dE_\lambda x \right\|^2 = \int_a^b |f(\lambda)|^2 |dE_\lambda x|^2 = \int_a^b |f(\lambda)|^2 d(E_\lambda x, x). \tag{4.4}$$

The right side of this equality is an ordinary Stieltjes integral. If the function f is continuous for all finite λ, then one can define the integral

$$\int_{-\infty}^{\infty} f(\lambda)\, dE_\lambda x \tag{4.5}$$

as the limit of the integral (4.2) as $a \to -\infty$, $b \to +\infty$. By virtue of the Cauchy criterion and formula (4.4), the integral (4.5) exists if and only if the ordinary Stieltjes integral

$$\int_{-\infty}^{\infty} |f(\lambda)|^2 d|E_\lambda x|^2 = \int_{-\infty}^{\infty} |f(\lambda)|^2 d(E_\lambda x, x)$$

exists. We then have

$$\left\| \int_{-\infty}^{\infty} f(\lambda)\, dE_\lambda x \right\|^2 = \int_{-\infty}^{\infty} |f(\lambda)|^2 d(E_\lambda x, x),$$

which is obtained from (4.4) by passing to the limit.

3. The basic spectral theorem.

THEOREM 3.1. *For every selfadjoint operator A there exists a unique spectral function E_λ having the following properties*:

a) *a given vector x belongs to D_A if and only if* $\int_{-\infty}^{\infty} |\lambda|^2 d(E_\lambda x, x) < \infty$;

b) *when this condition is satisfied, then*

$$Ax = \int_{-\infty}^{\infty} \lambda\, dE_\lambda x, \quad \|Ax\|^2 = \int_{-\infty}^{\infty} \lambda^2 d(E_\lambda x, x). \tag{4.6}$$

Conversely, every operator defined by a) *and* b) *by means of some spectral function E_λ is a selfadjoint operator.*

For a proof of this theorem see, for example, Ahiezer and Glazman [1].

Theorem 1.1 of Chapter 2 is easily seen to be a special case of Theorem 4.1, for which

$$E_\lambda f = \int_{-\infty}^{\lambda} F(\lambda) \bar{\varphi}(x, \lambda)\, d\rho(\lambda), \quad F(\lambda) = \int_0^{\infty} f(t) \varphi(t, \lambda)\, dt.$$

From this same theorem it follows that to every point of Weyl's limit circle there corresponds a selfadjoint extension of the Sturm-

Liouville operator (consequently, in the case of a limit point there exists only one selfadjoint extension. See also the following section of this chapter).

4. Reducibility. Let M be a closed subspace of H, and P the projector on M.

DEFINITION 4.2. *The subspace M reduces an operator A, if $x \in D_A$ implies that $Px \in D_A$ and $APx = PAx$.*

THEOREM 4.2. *A closed subspace M reduces a selfadjoint operator A if and only if the projector P_M commutes with the spectral function E_λ of A for every value of λ.*

For the proof see Ahiezer and Glazman [**1**].

5. Description of the spectrum of a selfadjoint operator by means of its spectral decomposition. Recall that a number λ is called a regular point of an operator A if the operator $R_\lambda = (A - \lambda I)^{-1}$ exists, is defined on the entire space H and is bounded. In this case *the operator R_λ is called the resolvent of the operator A. The set of all nonregular points of A is called its spectrum.* Obviously every eigenvalue λ of A belongs to its spectrum, since in this case the operator $(A - \lambda I)^{-1}$ does not exist. *The set of all eigenvalues of A is called its discrete spectrum. All other points of the spectrum, if such exist, are called points of the continuous spectrum, and the set of these points is called the continuous spectrum of the operator.* If A is selfadjoint, then eigenvectors corresponding to distinct eigenvalues of A are mutually orthogonal, and therefore A can have only a finite or countable set of eigenvalues (in a separable space there cannot exist an uncountable set of pairwise orthogonal vectors). Thus *the discrete spectrum of a selfadjoint operator* (in a separable Hilbert space) *is either a finite or countable set of real numbers.*

We will show that every nonreal number λ is a regular point of any selfadjoint operator A, and that

$$R_\lambda = (A - \lambda I)^{-1} = \int_{-\infty}^{\infty} \frac{dE_\mu}{\mu - \lambda}, \tag{4.7}$$

where E_μ is the spectral function of A (compare with Theorems 3.1 and 7.1 of Chapter 2).

Indeed, the integral in the right side of (4.7) exists (see subsection 2 of this section). Furthermore, for any $x \in H$ we have

$$\left\| \int_{-\infty}^{\infty} \frac{dE_\mu x}{\mu - \lambda} \right\|^2 = \int_{-\infty}^{\infty} \frac{dE_\mu x}{|\mu - \lambda|^2} \leqslant \frac{1}{|\mathrm{Im}\,\lambda|^2} \int_{-\infty}^{\infty} \| dE_\mu x \|^2 = \frac{\|x\|^2}{|\mathrm{Im}\,\lambda|^2}. \tag{4.8}$$

Hence this integral is a bounded operator, and it is easy to verify that it is equal to R_λ.

It follows from the inequality (4.8) that $\|R_\lambda\| \leqq 1/|\mathrm{Im}\,\lambda|$. Moreover, it follows from (4.7) that $(R_\lambda)^* = R_{\bar{\lambda}}$.

The following theorem gives a certain characterization of the spectrum of a selfadjoint operator A.

THEOREM 4.3. *Let A be a selfadjoint operator in H, and let E_λ be its spectral function. Then*

a) *A real number λ_0 is a regular point of A if and only if the function E_λ is constant in some neighborhood of this point.*

b) *A real number λ_0 is an eigenvalue of A if and only if $E_{\lambda_0} - E_{\lambda_0 - 0} \neq 0$; in this case $E_{\lambda_0} - E_{\lambda_0 - 0}$ is the projector on the eigenspace of A corresponding to the eigenvalue λ_0.*

For a proof see Naimark [1], §12.5 or Ahiezer and Glazman [1], §82.

§5. The extension of symmetric operators

1. The deficiency subspaces and deficiency indices. One of the basic problems of the abstract theory of symmetric operators is the description of all symmetric extensions of a given symmetric operator A. A particularly important special case of this problem is that of clarifying conditions under which a given symmetric operator has selfadjoint extensions, and describing in this case all such extensions.

If B is a symmetric extension of a symmetric operator A, i.e. $A \subseteq B$, then $B^* \subseteq A^*$. But B is a symmetric operator; therefore $B \subseteq B^*$ and hence

$$A \subseteq B \subseteq B^* \subseteq A^*. \tag{5.1}$$

It follows from these inclusion relations that every symmetric extension of A is a restriction of the operator A^*.

DEFINITION 5.1. *A symmetric operator A is said to be maximal if it has no symmetric extensions.*

It follows from this definition that a selfadjoint operator A is maximal. In fact, in this case $A^* = A$ and (5.1) shows that $B = A$, i.e. every symmetric extension B of A coincides with A.

Let us denote by $\mathscr{L}_i$ the subspace consisting of all elements of the form $y = (A + iI)x$, $x \in D_A$, where $i = \sqrt{-1}$, and by $\mathscr{L}_{-i}$ the subspace of all elements of the form $(A - iI)x$, $x \in D_A$.

Lemma 5.1. *$\mathscr{L}_i$ and $\mathscr{L}_{-i}$ are closed subspaces, if A is closed.*

Proof. Simple calculations yield $\|(A \pm iI)x\|^2 = \|Ax\|^2 + \|x\|^2$, and therefore

$$\|(A \pm iI)x\| \geqq \|x\|. \tag{5.2}$$

Let us now assume that $y_n = (A \pm iI)x_n$ and that $y_n \to y_0$. Then $\|y_n - y_m\| \to 0$, and consequently by (5.2) $\|x_n - x_m\| \to 0$. Since H is complete, it follows that $x_n \to x_0 \in H$. Thus $x_n \in D_A$, $x_n \to x_0$, $Ax_n = y_n \mp ix_n \to y_0 \mp ix_0$. Since A is a closed operator, we have $x_0 \in D_A$ and $Ax_0 = y_0 \mp ix_0$. It follows that $y_0 \in \mathscr{L}_{\pm i}$, as was to be proved.

We denote by $N_{\pm i}$ the orthogonal complement of $\mathscr{L}_{\pm i}$.

Definition 5.2. *The subspaces $N_{\pm i}$ are called the deficiency subspaces of the operator A. Their dimensions are called the deficiency indices of A.*

Remark 1. Let λ denote an arbitrary nonreal number. Similarly to the foregoing one can introduce the two subspaces $\mathscr{L}_\lambda$ and $\mathscr{L}_{\bar{\lambda}}$, prove that they are closed, define the subspaces N_λ and $N_{\bar{\lambda}}$ and define the deficiency indices as the dimensions of the subspaces N_λ and $N_{\bar{\lambda}}$. The subspaces N_λ and $N_{\bar{\lambda}}$ depend upon λ, but one can prove that their dimensions in each of the halfspaces $\operatorname{Im}\lambda > 0$ and $\operatorname{Im}\lambda < 0$ do not depend upon λ (for a proof see, for example, Ahiezer and Glazman [**1**]). Therefore the deficiency indices of a symmetric operator do not depend upon the choice of the (nonreal) number λ.

Remark 2. If for a Sturm-Liouville operator

$$l(y) \equiv -y'' + q(x)y, \qquad 0 \leqq x < \infty$$

with the boundary condition

$$y'(0)\cos\alpha + y(0)\sin\alpha = 0$$

the limit-point case holds, then the deficiency indices are $(0,0)$, while in the limit-circle case they are $(1,1)$.

2. Description of the domain of the adjoint operator. An element $z \in H$ is orthogonal to the subspace $\mathscr{L}_i$ if and only if, for every $x \in D_A$, one has $(z, Ax - ix) = 0$ or $(Ax, z) = (x, iz)$, i.e. when $z \in D_{A^*}$ and $A^*z = iz$. Thus the orthogonal complement of $\mathscr{L}_i$, which we introduced above and denoted by N_i, is the subspace of eigenvectors of A^* corresponding to the eigenvalue i. A similar result holds for $\mathscr{L}_{-i}$ and N_{-i}.

THEOREM 5.1. *The domain D_{A^*} of A^* (the adjoint of a closed symmetric operator) is the direct sum*[2] *of D_A, N_i and N_{-i}, i.e.*

(5.3) $$D_{A^*} = D_A \oplus N_i \oplus N_{-i}.$$

PROOF. Let y be an arbitrary element in D_{A^*} and consider the element $A^*y - iy$. Since $H = \mathscr{L}_{-i} \oplus N_{-i}$, we have

(5.4) $$A^*y - iy = (Ax - ix) + \tilde{z}_0, \qquad \tilde{z}_0 \in N_{-i}.$$

Further, $Ax = A^*x$ and $A^*\tilde{z}_0 = -i\tilde{z}_0$. Therefore (5.4) implies that

$$A^*\left(y - x - \frac{1}{2} i\tilde{z}_0\right) = i(y-x) + \tilde{z}_0 + A^*\left(-\frac{1}{2} i\tilde{z}_0\right)$$
$$= i(y-x) + \tilde{z}_0 - \frac{1}{2} i(-i)\tilde{z}_0 = i(y-x)$$
$$+ \frac{1}{2}\tilde{z}_0 = i(y-x) + \frac{1}{2} i(-i)\tilde{z}_0 = i\left(y - x - \frac{1}{2} i\tilde{z}_0\right).$$

Hence $y - x - \frac{1}{2} i\tilde{z}_0 = z \in N_i$. Putting $\frac{1}{2} i\tilde{z}_0 = \tilde{z}$, we obtain

(5.5) $$y = x + z + \tilde{z},$$

which yields the desired decomposition (5.3). We shall show that this decomposition is unique. Suppose that along with (5.5) we have another decomposition of the element y: $y = x_1 + z_1 + \tilde{z}_1$. Then

(5.6) $$(x - x_1) + (z - z_1) + (\tilde{z} - \tilde{z}_1) = 0.$$

Applying A^* to both sides of (5.6), we obtain

(5.7) $$A(x - x_1) + i(z - z_1) - i(\tilde{z} - \tilde{z}_1) = 0.$$

Multiplying (5.6) by i and subtracting from (5.7), we have

(5.8) $$A(x - x_1) - i(x - x_1) - 2i(\tilde{z} - \tilde{z}_1) = 0.$$

Since $A(x - x_1) - i(x - x_1) \perp (\tilde{z} - \tilde{z}_1)$, it follows from (5.8) that

$$A(x - x_1) - i(x - x_1) = 0, \qquad 2i(\tilde{z} - \tilde{z}_1) = 0.$$

Therefore $\tilde{z}_1 = \tilde{z}$. Similarly $z_1 = z$, and therefore $x_1 = x$.

If instead of i we choose an arbitrary nonreal number λ, we obtain the decomposition

(5.9) $$D_{A^*} = D_A \oplus N_\lambda \oplus N_{\bar{\lambda}}.$$

---

[2] Let E be a linear space and let $\mathscr{L}_1, \cdots, \mathscr{L}_n$ be subspaces of E. If every element $x \in E$ can be represented in a unique way as a sum $x = x_1 + \cdots + x_n$, $x_k \in \mathscr{L}_k$, $k = 1, \cdots, n$, then one says that E is the direct sum of the $\mathscr{L}_k$, and writes $E = \sum_{k=1}^{n} \oplus \mathscr{L}_k$.

If the deficiency indices of A are $(0,0)$, i.e. the corresponding deficiency subspaces N_i and N_{-i} are the zero subspace, it follows from (5.5) that $D_{A^*}=D_A$, and so A is selfadjoint.

3. Let us denote by Γ the set of those elements $y\in D_{A^*}$ for which the quadratic form (A^*y,y) assumes real values.

THEOREM 5.2. *In order that an element $y\in D_{A^*}$ belong to the set Γ, it is necessary and sufficient that in* (5.5) *one have $\|z\|=\|\tilde{z}\|$.*

PROOF. If $y=x+z+\tilde{z}$, then

$$(5.10)\quad \begin{aligned}(A^*y,y)&=(Ax+A^*(z+\tilde{z}),x+(z+\tilde{z}))\\&=(Ax,x)+(Ax,z+\tilde{z})+(A^*(z+\tilde{z}),x)+(A^*(z+\tilde{z}),z+\tilde{z})\\&=(Ax,x)+(x,A^*(z+\tilde{z}))+(A^*(z+\tilde{z}),x)+(A^*(z+\tilde{z}),z+\tilde{z})\\&=A(x,x)+2\operatorname{Re}(x,A^*(z+\tilde{z}))+(iz-i\tilde{z},z+\tilde{z})\\&=A(x,x)+2\operatorname{Re}(x,A^*(z+\tilde{z}))+i\|z\|^2+i(z,\tilde{z})-i(\tilde{z},z)-i\|\tilde{z}\|^2.\end{aligned}$$

Since (Ax,x) and $i(z,\tilde{z})-i(\tilde{z},z)$ are real numbers, it follows from (5.10) that $\operatorname{Im}(A^*y,y)=\|z\|^2-\|\tilde{z}\|^2$, from which the theorem follows.

4. THEOREM 5.3. *To every symmetric extension B of a closed symmetric operator A there correspond two subspaces $T_i\subseteq N_i$ and $T_{-i}\subseteq N_{-i}$ and an isometric operator U mapping T_i onto T_{-i}, such that*

a) *the domain D_B consists of all elements of the form*

$$(5.11)\quad y=x+z+Uz,$$

where $x\in D_A$ and $z\in T_i$,

b) *for every y of the form* (5.11)

$$(5.12)\quad By=Ax+iz-iUz.$$

Conversely, given two subspaces $T_i\subseteq N_i$ and $T_{-i}\subseteq N_{-i}$ and an isometric operator U which maps T_i onto T_{-i}, then the operator B defined on the set of elements of the form (5.11) *by formula* (5.12) *is a symmetric extension of A.*

The operator B is closed if and only if the subspaces T_i and T_{-i} are closed.

PROOF. Let B be a symmetric extension of A and let $y\in D_B$. As we have seen, $D_B\subseteq D_{A^*}$ and therefore according to (5.5) $y=x+z+\tilde{z}$ where, since $y\in\Gamma$, we have by Theorem 5.2

$$(5.13)\quad \|\tilde{z}\|=\|z\|.$$

If y runs through all of D_B, then z runs through some subspace T_i, and $\tilde{z}$ runs through some subspace T_{-i}. We shall show that with each element $z \in T_i$ one can associate only one element $\tilde{z} \in T_{-i}$. In fact, if

$$y_1 = x_1 + z_1 + \tilde{z}_1, \quad y_2 = x_2 + z_1 + \tilde{z}_2, \quad y_1, y_2 \in D_B,$$

then

$$y_1 - y_2 = x_1 - x_2 + 0 + (\tilde{z}_1 - \tilde{z}_2) \in D_B \subseteq \Gamma.$$

Therefore from (5.13) follows $\|\tilde{z}_1 - \tilde{z}_2\| = 0$, i.e. $\tilde{z}_1 = \tilde{z}_2$. Associating with each element $z \in T_i$ the unique element $\tilde{z}$ corresponding to it from (5.5), we obtain an isomorphic and isometric mapping of T_i onto T_{-i}. Denoting this mapping by U, we arrive at (5.11), and then at

$$By = A^*y = A^*(x + z + Uz) = Ax + iz - iUz,$$

which is (5.12).

Conversely, let $T_i \subseteq N_i$ and $T_{-i} \subseteq N_{-i}$ be two subspaces and let U be an isometric operator mapping T_i onto T_{-i}. The operator B defined on the elements of the form (5.11) by the formula (5.12) is a symmetric extension of A, since the subspace D_B of element of the form (5.11) satisfies the condition $D_B \subseteq \Gamma \cap D_{A^*}$, and $By = A^*y$, $y \in D_B$.

It remains to prove the last part of the theorem. To this end we note that for B to be closed it is necessary and sufficient that the subspace $\mathscr{L}_i'$ of elements of the form $(B + iI)y$, $y \in D_B$, be closed. The necessity was proved in Lemma 5.1 of this section. We prove its sufficiency. Let us assume that $\mathscr{L}_i'$ is closed, but B is not closed. Taking the closure[3] of B, we would have to add new limit elements to $\mathscr{L}_i'$, which would contradict $\mathscr{L}_i'$ being closed.

Further, for any $y \in D_B$

$$(B + iI)y = (B + iI)(x + z + \tilde{z}) = (A + iI)x + 2iz.$$

Therefore

$$\mathscr{L}_i' = \mathscr{L}_i \oplus T_i, \tag{5.14}$$

where $\mathscr{L}_i$ is the set of elements of the form $(A + iI)x$, $x \in D_A$. Since $\mathscr{L}_i$ is closed, $\mathscr{L}_i'$ is closed if and only if T_i is. This proves sufficiency.

5. Let us assume that we have extended a symmetric operator A, by the method indicated in Theorem 5.3, to a symmetric operator B. The question arises, what are the deficiency subspaces and the deficiency indices of this extension? The answer is given by the following theorem.

---

[3] Every symmetric operator has a closure; cf. §3.

THEOREM 5.4. *Let B be a closed symmetric extension of a closed symmetric operator A, with domain $D_B = D_A + (I+U)T_i$.*

Let (m_1, m_2) be the deficiency indices of A: $m_1 = \dim N_i$, $m_2 = \dim N_{-i}$, and (m_1', m_2') the deficiency indices of B: $m_1' = \dim N_i'$, $m_2' = \dim N_{-i}'$.

Then $N_i = N_i' \oplus T_i$, $N_{-i} = N_{-i}' \oplus T_{-i}$.

Therefore if $\dim T_i = \dim T_{-i} = l$, we have $m_1 = m_1' + l$, $m_2 = m_2' + l$.

PROOF. Since $H = \mathscr{L}_i' \oplus N_i'$, we have by (5.14) $H = \mathscr{L}_i \oplus N_i' \oplus T_i$. On the other hand, $H = \mathscr{L}_i \oplus N_i$. Therefore $N_i = N_i' \oplus T_i$. Similarly one proves that $N_{-i} = N_{-i}' \oplus T_{-i}$. The relations between the deficiency indices follow at once from these relations.

By means of Theorem 5.3 and 5.4 one can give a complete description of all closed symmetric extensions of a given closed symmetric operator A. One can also state conditions for the existence of selfadjoint extensions and describe all of these extensions. For this and other topics in abstract spectral theory the reader can turn to the monographs of Plesner [1] and Ahiezer and Glazman [1].

§6. A sufficient condition for the selfadjointness of a Sturm-Liouville operator on the real line or halfline

1. It follows directly from the definition of the adjoint and of a selfadjoint operator that to prove the selfadjointness of a symmetric operator A it is sufficient to show that if some element x is in D_{A^*}, then $x \in D_A$, i.e. that

$$D_{A^*} = D_A. \tag{6.1}$$

In order to prove the equality (6.1), we must have a detailed description of D_{A^*}. For a Sturm-Liouville operator

$$l(y) = -y'' + q(x)y \tag{6.2}$$

this can be accomplished, and on the basis of such a description one can give a sufficient condition for the selfadjointness of the operator $\bar{l}$ (the closure of the operator l) in terms of the function $q(x)$. The same can be accomplished for first-order Dirac-type systems, and will be done in the following section.

In the subsequent discussion an important role is played by the following result.

LEMMA 6.1. *Let $f(x)$ $(a \leq x \leq b)$ satisfy the following conditions:*

a) *it has an absolutely continuous first derivative in the interval (a, b),*

b) $f''(x) \in \mathscr{L}^2(a,b)$,
c) $f(a) = f'(a) = f(b) = f'(b) = 0$.
Let $h(x)$ *be an arbitrary fixed function from* $\mathscr{L}^2(a,b)$.
If for every function $f(x)$ *satisfying conditions* a), b) *and* c) *we have*

$$\int_a^b f''(x)h(x)\,dx = 0, \tag{6.3}$$

then $h(x) = \alpha + \beta x$ *almost everywhere, where* α *and* β *are constants.*

Proof. First we give a characterization of the class of functions satisfying conditions a), b) and c). Let $f''(t) = \lambda(t)$. Then

$$f'(x) = \int_a^x \lambda(t)\,dt, \qquad f(x) = \int_a^x (x-t)\lambda(t)\,dt.$$

It therefore follows from condition c) that

$$\int_a^b \lambda(t)\,dt = 0, \qquad \int_a^b t\lambda(t)\,dt = 0. \tag{6.4}$$

Conversely, if a function $\lambda(t)$ from $\mathscr{L}^2(a,b)$ satisfies the conditions (6.4), then putting

$$f(x) = \int_a^x (x-t)\lambda(t)\,dt,$$

we obtain a function $f(x)$ satisfying conditions a), b) and c).

If now a function $h(t) \in \mathscr{L}^2(a,b)$ does not satisfy the conditions (6.4), we can choose constants α and β such that the function

$$h_1(t) = h(t) - \alpha - \beta t \tag{6.5}$$

will satisfy these conditions. Indeed, to determine α and β we have two equations

$$\begin{aligned} \int_a^b h_1(t)\,dt &= \int_a^b h(t)\,dt - \alpha(b-a) - \frac{\beta}{2}(b-a)^2 = 0, \\ \int_a^b th_1(t)\,dt &= \int_a^b th(t)\,dt - \frac{\alpha}{2}(b-a)^2 - \frac{\beta}{3}(b-a)^3 = 0, \end{aligned} \tag{6.6}$$

from which α and β are uniquely determined.

Let us now turn to equation (6.3). It can be written in the form

$$\int_a^b \lambda(x) h(x)\,dx = 0. \tag{6.7}$$

Using (6.4), we rewrite this last equation in the form

$$\int_a^b \lambda(x) h_1(x)\,dx = 0, \tag{6.8}$$

where $h_1(x)$ has the form (6.5), with constants α and β satisfying the equations (6.6). Putting, in (6.8), $\lambda(x) = h_1(x)$ (which we can do since $h_1(x)$ satisfies the conditions (6.4)), we obtain $\int_a^b h_1^2(x)\,dx = 0$, from which it follows that $h(x) = \alpha + \beta x$ almost everywhere, as was to be proved.

2. The case of the entire line. Let $l(y) = -y'' + q(x)y$ $(-\infty < x < \infty)$, and let $q(x)$ be everywhere continuous. We denote by $D(l)$ the set of all twice continuously differentiable functions having compact supports. First of all we shall give a description of $D(l^*)$. According to the definition of the adjoint of an operator, if $y \in D(l)$ and $z \in D(l^*)$, then there exists a function $z^* \in \mathcal{L}^2(-\infty, \infty)$ such that

$$\int_{-\infty}^{\infty} (-y'' + q(x) y) z\,dx = \int_{-\infty}^{\infty} y \cdot z^*\,dx. \tag{6.9}$$

Integrating by parts, we obtain

$$\int_{-\infty}^{\infty} y \cdot z^*\,dx = \int_{-\infty}^{\infty} y'' \cdot Z^*(x)\,dx,$$

$$\int_{-\infty}^{\infty} y \cdot q(x) \cdot z\,dx = \int_{-\infty}^{\infty} y'' \cdot Y(x)\,dx,$$

where

$$Z^*(x) = \int_0^x (x-t) z^*(t)\,dt, \qquad Y(x) = \int_0^x (x-t) q(t) z(t)\,dt.$$

Therefore (6.9) can be rewritten in the form

$$\int_{-\infty}^{\infty} y'' [-z(x) + Y(x) - Z^*(x)]\,dx = 0. \tag{6.10}$$

On the basis of Lemma 6.1 we obtain

$$z(x) = Y(x) - Z^*(x) + \alpha + \beta x = \int_0^x (x-t)[q(t) z(t) - z^*(t)]\,dt + \alpha + \beta x.$$

From this equality there follows a complete description of the functions $z(x)$ belonging to $D(l^*)$, which is given in the following lemma.

LEMMA 6.2. *The set* $D(l^*)$ *consists of all functions* $z(x)$ *satisfying the following conditions*:

a) $z(x) \in \mathscr{L}^2(-\infty, \infty)$,

b) $z(x)$ *is continuous and has a first derivative which is absolutely continuous in every finite interval,*

c) $-z'' + q(x)z = z^*(x) \in \mathscr{L}^2(-\infty, \infty)$.

With the help of Lemma 6.2 we can prove the following theorem.

THEOREM 6.1. *If the function* $q(x)$ *is bounded from below, then the operator* $\bar{l}$ *(the closure of* l*) is selfadjoint.*

We will first prove another lemma, which contains the main part of the proof of Theorem 6.1.

LEMMA 6.3. *If the function* $q(x)$ *is bounded from below, then from*

$$y \in \mathscr{L}^2(-\infty, \infty), \quad l(y) = -y'' + q(x)y \in \mathscr{L}^2(-\infty, \infty)$$

follows

$$y' \in \mathscr{L}^2(-\infty, \infty). \tag{6.11}$$

PROOF. Integrating by parts, we have

$$\int_{-a}^{a} y'^2 dx + \int_{-a}^{a} yy'' dx = y'(a)\, y(a) - y'(-a)\, y(-a). \tag{6.12}$$

From $y \in \mathscr{L}^2(-\infty, \infty)$ and $-y'' + q(x)y \in \mathscr{L}^2(-\infty, \infty)$ follows the existence, as $a \to +\infty$, of a finite limit for the sum

$$-\int_{-a}^{a} yy''\, dx + \int_{-a}^{a} q(x)\, y^2\, dx. \tag{6.13}$$

Since by hypothesis the function $q(x)$ is bounded from below, it follows from the boundedness of the sum (6.13) that the first term of this sum either tends to a finite limit or tends to $-\infty$. But then the left side of (6.12), and therefore the quantity $y'(a)y(a) - y'(-a)y(-a)$, either tends, as $a \to +\infty$, to a finite limit or to $+\infty$. If we can show that the case

$$\lim_{a \to +\infty} [y'(a)y(a) - y'(-a)y(-a)] = +\infty \tag{6.14}$$

is not possible, then from (6.12) we would obtain

$$\lim_{a\to\infty} \int_{-a}^{a} y'^2(x)\,dx < \infty,$$

which would prove the lemma. To prove that (6.14) is not possible, consider the function $I(a) = y^2(a) + y^2(-a)$ and its derivative $I'(a) = 2y'(a)y(a) - 2y'(-a)y(-a)$.

If (6.14) holds, then $I'(a) \to \infty$ and a fortiori $I(a) \to \infty$, which contradicts the fact that $y(x) \in \mathscr{L}^2(-\infty, \infty)$.

PROOF OF THEOREM 6.1. We will show that the operator l^* is symmetric on its domain. It will follow that

$$l^* \subseteq (l^*)^* = l^*. \tag{6.15}$$

On the other hand, for a symmetric operator l we always have the inclusion

$$l^{**} \subseteq l^*. \tag{6.16}$$

It follows from (6.15) and (6.16) that $l^{**} = l^*$, i.e., the operator l^* is selfadjoint. Further, for any symmetric operator A (with a dense domain) $A^{**} = \bar{A}$, where $\bar{A}$ is the closure of A (cf. Ahiezer and Glazman [1], §§43 and 51). Therefore $\bar{l} = l^*$, and hence $\bar{l}$ is selfadjoint. Thus it remains for us to prove that l^* is symmetric. Let $y, z \in D(l^*)$. It follows from Lemma 6.3 that $y', z' \in \mathscr{L}^2(-\infty, \infty)$ and therefore $y'z, yz' \in \mathscr{L}(-\infty, \infty)$. Consequently there exist sequences $\{a_n\} \to -\infty$, $\{b_n\} \to +\infty$, such that

$$\begin{aligned} &\lim_{n\to\infty} y'(a_n)\,z(a_n) = 0, \qquad \lim_{n\to\infty} y(a_n)\,z'(a_n) = 0, \\ &\lim_{n\to\infty} y'(b_n)\,z(b_n) = 0, \qquad \lim_{n\to\infty} y(b_n)\,z'(b_n) = 0. \end{aligned} \tag{6.17}$$

Passing to the limit in Green's identity

$$\int_{a_n}^{b_n} [yl^*(z) - zl^*(y)]\,dx = (zy' - yz')_{a_n}^{b_n}$$

and using (6.17), we obtain

$$\int_{-\infty}^{\infty} [yl^*(z) - zl^*(y)]\,dx = 0,$$

which proves that l^* is symmetric on its domain $D(l^*)$.

3. The case of a halfline. We now consider the Sturm-Liouville operator $l(y) = -y'' + q(x)y$ on the halfline $(0, \infty)$, assuming as before that $q(x)$ is a continuous function. We take for $D(l)$ the set of all twice continuously differentiable functions with compact supports satisfying the boundary condition

$$y'(0)\cos\alpha + y(0)\sin\alpha = 0. \tag{6.18}$$

The following lemma gives a description of $D(l^*)$.

Lemma 6.4. *The domain $D(l^*)$ of l^* consists of those functions $z(x)$ satisfying the following conditions*:

a) $z(x) \in \mathscr{L}^2(0, \infty)$,

b) $z(x)$ *is continuous and has a first derivative which is absolutely continuous in every finite interval,*

c) $-z'' + q(x)z \in \mathscr{L}^2(0, \infty)$,

d) $z'(0)\cos\alpha + z(0)\sin\alpha = 0$.

The proof is based upon Lemma 6.1. Assume first that $y(x)$ satisfies the conditions $y(0) = y'(0) = 0$. Applying Lemma 6.1, we obtain

$$z(x) = \alpha + \beta x + \int_0^x (x - t)\,[q(t)\,z(t) - z^*(t)]\,dt.$$

It follows from this representation that $z(x)$ satisfies conditions b) and c), and also that the limits $z(+0) = \alpha$, $z'(+0) = \beta$ exist.

We now consider an arbitrary function satisfying the condition (6.18) and write down the equality which defines the adjoint operator:

$$\int_0^\infty \{-y'' + q(x)\,y\}\,z\,dx = \int_0^\infty y \cdot z^*\,dx. \tag{6.19}$$

Integrating by parts, we obtain

$$\int_0^\infty -y''z\,dx = -y'(0)\,z(0) + y(0)\,z'(0) - \int_0^\infty y\,(-z'')\,dx.$$

Therefore (6.19) can be written in the form

$$\int_0^\infty \{-y'' + q(x)\,y\}\,z\,dx$$

$$= \int_0^\infty y\,\{-z'' + q(x)\,z\}\,dx - y'(0)\,z(0) + y(0)\,z'(0) = \int_0^\infty yz^*dx. \tag{6.20}$$

Since, as was proved, $l^*(z) = -z'' + q(x)z$, it follows from (6.20) that

$$-y'(0)z(0) + y(0)z'(0) = 0. \tag{6.21}$$

By (6.18), $y'(0)/y(0) = -\operatorname{tg}\alpha$. Inserting this expression in (6.21), we obtain condition d) for $z(x)$.

Using Lemma 6.4, we can obtain the following theorem, which is analogous to Theorem 6.1.

THEOREM 6.2. *If* $q(x)$ $(0 \leq x < \infty)$ *is continuous and bounded from below, then the operator* $\bar{l}$ *(the closure of the operator* l*, whose definition was given at the beginning of this subsection) is selfadjoint.*

REMARK. Another proof of Theorem 6.2 was given in Chapter 2. A substantially more general theorem (Titchmarsh's theorem) was also proved in Chapter 2.

§7. The selfadjointness of a first-order system

We consider a first-order operator, defined on vector-functions $y(x) = \begin{pmatrix} y_1(x) \\ y_2(x) \end{pmatrix}$ $(0 \leq x < \infty)$,

$$\mathscr{L}(y) = \begin{cases} \dfrac{dy_2}{dx} + a(x)\,y_1 + b(x)\,y_2, \\ -\dfrac{dy_1}{dx} + b(x)\,y_1 + c(x)\,y_2, \end{cases} \tag{7.1}$$

$$y_1(x)\cos\alpha + y_2(0)\sin\alpha = 0. \tag{7.2}$$

We assume the coefficients a, b and c to be continuous functions, and take $D(\mathscr{L})$ to be the set of all continuous vector-functions with compact supports having continuous first derivative, and satisfying the condition (7.2).

Let us find the domain $D(\mathscr{L}^*)$ of the adjoint of $\mathscr{L}$. Instead of Lemma 6.1 we will use the following lemma, whose proof is completely analogous to that of Lemma 6.1.

LEMMA 7.1. *Let* $f(x)$ *be a continuous function satisfying the conditions*:
a) $f(x)$ *is absolutely continuous in the interval* (a, b),
b) $f'(x) \in \mathscr{L}^2(a, b)$,
c) $f(a) = f(b) = 0$.
If

$$\int_a^b f'(x)\,h(x)\,dx = 0, \quad h(x) \in \mathscr{L}^2(a, b),$$

then $h(x) = \text{const}$ *almost everywhere.*

Let

$$z = \begin{pmatrix} z_1 \\ z_2 \end{pmatrix} \in D(\mathscr{L}^*).$$

From the definition of the adjoint operator we obtain

$$\int_0^\infty \left\{\left[\frac{dy_2}{dx} + a(x)\,y_1 + b(x)\,y_2\right] z_1 + \left[-\frac{dy_1}{dx} + b(x)\,y_1 + c(x)\,y_2\right] z_2\right\} dx$$

$$= \int_0^\infty (y_1 z_1^* + y_2 z_2^*)\, dx, \tag{7.3}$$

where

$$z^* = \begin{pmatrix} z_1^* \\ z_2^* \end{pmatrix} \in \mathscr{L}^2(0, \infty).$$

Putting successively $y_2 = 0$, $y_1 = 0$ and using Lemma 7.1, it is not hard to obtain from (7.3) the following lemma (cf. the proof of Lemma 6.4).

LEMMA 7.2. *The domain $D(\mathscr{L}^*)$ consists of all vector-functions*

$$z(x) = \begin{pmatrix} z_1(x) \\ z_2(x) \end{pmatrix}$$

satisfying the following conditions:

a) $z_1(x), z_2(x) \in \mathscr{L}^2(0, \infty)$,

b) $z_1(x)$ *and* $z_2(x)$ *are absolutely continuous in every finite interval,*

c) $z_2' + a(x)z_1 + b(x)z_2 \in \mathscr{L}^2(0, \infty)$ *and* $-z_1' + b(x)z_1 + c(x)z_2 \in \mathscr{L}^2(0, \infty)$, *i.e.* $\mathscr{L}(z) \in \mathscr{L}^2(0, \infty)$.

d) $z_1(0) \cos \alpha + z_2(0) \sin \alpha = 0$.

From Lemma 7.2 we deduce the following remarkable theorem.

THEOREM 7.1. *If the coefficients of the system* (7.1) *are continuous, then the operator $\bar{\mathscr{L}}$ (the closure of the operator $\mathscr{L}$) is selfadjoint.*

PROOF. It is sufficient to prove the symmetry of the operator $\mathscr{L}^*$ (cf. Theorem 6.1). Let

$$y = \begin{pmatrix} y_1 \\ y_2 \end{pmatrix} \quad \text{and} \quad z = \begin{pmatrix} z_1 \\ z_2 \end{pmatrix}$$

belong to $D(\mathscr{L}^*)$. By Lemma 7.2 we have $y_1 z_2, y_2 z_1 \in \mathscr{L}^2(0, \infty)$. Therefore there exists a sequence $\{a_n\} \to \infty$ such that

$$\lim_{n\to\infty} y_1(a_n) z_2(a_n) = 0, \quad \lim_{n\to\infty} y_2(a_n) z_1(a_n) = 0. \tag{7.4}$$

Further, on the basis of Green's identity we have (again using the result of Lemma 7.2)

$$(y, \mathscr{L}^*(z))_{a_n} - (\mathscr{L}^*(y), z)_{a_n}$$
$$= \int_0^{a_n} \left[\frac{d}{dx}(y_1, z_2) - \frac{d}{dx}(z_1, y_2)\right] dx$$
$$= y_1(a_n) z_2(a_n) - z_1(a_n) y_2(a_n).$$

Passing to the limit and using (7.4), we obtain $(y, \mathscr{L}^*(z)) = (\mathscr{L}^*(y), z)$ for any $y, z \in D(\mathscr{L}^*)$, as was to be proved.

Theorem 7.1 shows that for a first-order system whose coefficients are continuous, the limit circle case cannot occur (i.e. the deficiency indices are $(0, 0)$).

Bibliographical references

§§1–5. In these sections we have given a brief discussion of the foundations of the theory of symmetric and selfadjoint operators in Hilbert space. The reader can find a detailed discussion of this theory in the books by Ahiezer and Glazman [1] and Riesz and Sz.-Nagy [1], for example.

§6. Theorem 6.1 is due to Weyl [1]. In presenting the results of this section we have used the book by Glazman [1].

§7. Theorem 7.1 is due to Levitan. A similar theorem was obtained independently by Martynov [1].

CHAPTER 14

SOME THEOREMS OF ANALYSIS

§1. The Riemann-Lebesgue lemmas

Throughout this book we have frequently referred to lemmas on the behaviour, as $|\lambda| \to \infty$, of trigonometric integrals of the form

$$(1.1) \qquad \int_a^b f(x) \sin \lambda x \, dx, \qquad \int_a^b f(x) \cos \lambda x \, dx,$$

where (a, b) is a finite or infinite interval, and $f(x) \in \mathscr{L}(a, b)$. These lemmas are well known and are usually called the Riemann-Lebesgue lemmas. Less known are lemmas on the behaviour of the integrals (1.1) when the function $f(x)$ depends upon a parameter.

LEMMA 1.1. *Let (a, b) be a finite interval and $f(x) \in \mathscr{L}(a, b)$. Then*

$$(1.2) \qquad \lim_{|\lambda| \to \infty} \int_a^b f(x) \sin \lambda x \, dx = 0.$$

PROOF. Let us first assume that $f(x)$ is the characteristic function of an interval $(\alpha, \beta) \subseteq (a, b)$, i.e. $f(x) = 1$ for $x \in (\alpha, \beta)$ and $f(x) = 0$ for $x \notin (\alpha, \beta)$. Then

$$\int_a^b f(x) \sin \lambda x \, dx = \int_\alpha^\beta \sin \lambda x \, dx = \frac{\cos \lambda\alpha - \cos \lambda\beta}{\lambda},$$

from which (1.1) follows immediately.

Now assume that $f(x)$ is a step function, i.e. the interval (a, b) can be subdivided into a finite number of subintervals (α_{k-1}, α_k), $k = 1, \cdots, n$, such that $f(x)$ is equal to a constant d_k throughout the interval (α_{k-1}, α_k), $k = 1, \cdots, n$. Then

$$\lim_{|\lambda| \to \infty} \int_a^b f(x) \sin \lambda x \, dx = \sum_{k=1}^n d_k \lim_{|\lambda| \to \infty} \int_{\alpha_{k-1}}^{\alpha_k} \sin \lambda x \, dx = 0.$$

If $f(x)$ is an arbitrary function from $\mathscr{L}(a, b)$, then it is known[1)] that

1) Lusin's theorem; see, for example, Šilov [1].

for any $\epsilon > 0$ one can find a step function $\varphi_\epsilon(x)$ such that

$$\int_a^b |f(x) - \varphi_\varepsilon(x)|\, dx < \varepsilon. \tag{1.3}$$

It follows from the foregoing that for fixed ϵ we can choose $\lambda_0 = \lambda_0(\epsilon)$ so large that for $|\lambda| > \lambda_0$ we have

$$\left| \int_a^b \varphi_\varepsilon(x) \sin \lambda x\, dx \right| < \varepsilon. \tag{1.4}$$

It therefore follows from (1.3) and (1.4) that for $|\lambda| > \lambda_0$

$$\left| \int_a^b f(x) \sin \lambda x\, dx \right| \leqslant \left| \int_a^b [f(x) - \varphi_\varepsilon(x)] \sin \lambda x\, dx \right| + \left| \int_a^b \varphi_\varepsilon(x) \sin \lambda x\, dx \right|$$

$$\leqslant \int_a^b |f(x) - \varphi_\varepsilon(x)|\, dx + \left| \int_a^b \varphi_\varepsilon(x) \sin \lambda x\, dx \right| < 2\varepsilon,$$

whence the lemma follows by the arbitrariness of ϵ.

REMARK 1. One can similarly prove that

$$\lim_{|\lambda| \to \infty} \int_a^b f(x) \cos \lambda x\, dx = 0.$$

The same remark holds for the subsequent lemmas of this section. We will therefore state them only for the first of the integrals (1.1).

LEMMA 1.2. *Let* (a, b) *be an infinite (on one or both sides) interval and let* $f(x) \in \mathscr{L}(a, b)$. *Then* (1.2) *holds.*

PROOF. For definiteness let us consider an interval of the form $(a, +\infty)$, where a is finite. For arbitrary $\epsilon > 0$ one can find $A = A(\epsilon)$ such that

$$\int_A^\infty |f(x)|\, dx < \varepsilon. \tag{1.5}$$

By Lemma 1.1, for fixed A we can choose $\lambda_0 = \lambda_0(\epsilon)$ so large that

(1.6)

for $|\lambda| > \lambda_0$. It follows from (1.5) and (1.6) that for $|\lambda| > \lambda_0$ one has

$$\left|\int_a^\infty f(x)\sin\lambda x\,dx\right| \leqslant \left|\int_a^A f(x)\sin\lambda x\,dx\right| + \left|\int_A^\infty f(x)\sin\lambda x\,dx\right|$$

$$\leqslant \left|\int_a^A f(x)\sin\lambda x\,dx\right| + \int_A^\infty |f(x)|\,dx < 2\varepsilon,$$

from which the lemma follows.

The following lemma gives us information on the behaviour of integrals of the form (1.1) when the function $f(x)$ depends upon a parameter (or parameters).

Lemma 1.3. *Suppose that we have an infinite set $E = \{f(x)\}$ of functions which satisfy the following conditions:*

1°. *$f(x) \in \mathscr{L}(a,b)$, where the interval (a,b) can be finite as well as infinite.*

2°. *The set E is relatively compact in the metric space $\mathscr{L}(a,b)$.*

Then the relation (1.2) *holds uniformly, i.e. for any $\epsilon > 0$ one can find $\lambda_0 = \lambda_0(\epsilon)$ such that for $|\lambda| > \lambda_0$ the inequality*

$$\left|\int_a^b f(x)\sin\lambda x\,dx\right| < \varepsilon \tag{1.7}$$

holds for all $f(x) \in E$.

Proof. Let $\epsilon > 0$. From a well-known criterion for the relative compactness of a set in a metric space it follows that one can find a finite ϵ-set for E, i.e. there exists a finite set of functions $E_\epsilon = \{f_1(x), \cdots, f_n(x)\} \subseteq E$ such that for each function $f(x) \in E$ there exists a function $f_k(x) \in E_\epsilon$ such that

$$\int_a^b |f(x) - f_k(x)|\,dx < \varepsilon. \tag{1.8}$$

For fixed ϵ we can obviously find $\lambda_0 = \lambda_0(\epsilon) > 0$ such that the inequality

$$\left|\int_a^b f_k(x)\sin\lambda x\,dx\right| < \varepsilon \tag{1.9}$$

holds for $|\lambda| > \lambda_0$ and for all $k = 1, \cdots, n$. The lemma follows from (1.8) and (1.9).

Remark 2. If the interval (a,b) is finite and the functions $f(x)$ of the family E are continuous, then for the relative compactness of E in the metric of $\mathscr{L}(a,b)$ it is sufficient that E be relatively compact in the metric of the space $C(a,b)$, and for this it is sufficient that E be uniformly bounded and equicontinuous.[2)]

If the interval (a,b) is infinite, then for the relative compactness of E it is sufficient to require that for any $\epsilon > 0$ there exist a finite interval $(a',b') \subset (a,b)$ such that

$$\int_a^{a'} |f(x)|\,dx + \int_{b'}^{b} |f(x)|\,dx < \varepsilon$$

for every function $f(x) \in E$, and that the restrictions of the functions in E to the interval (a',b') form a relatively compact family in the metric of $\mathscr{L}(a',b')$. In fact, denoting the set of all restrictions mentioned above by E', it is clear that if the restrictions to (a',b') of functions $f_1(x), \cdots, f_n(x) \in E$ form an ϵ-set in E', then these functions themselves will form a 3ϵ-set in E; hence E is relatively compact since ϵ is arbitrary.

A case of interest is that in which the family E is generated by one summable function $f(x): E = \{f_t(x)\} = \{f(x+t)\}$, $x \in (a,b)$, (a,b) is a finite interval and t belongs to some closed bounded set M on the line.

Let $\{x+t\} \in (A,B)$ and $f(x) \in \mathscr{L}(A,B)$. Then the set E is compact. In fact, let $\{t_n\}$ be an infinite sequence of numbers belonging to the set M. If we choose a subsequence $\{t_n'\}$ of $\{t_n\}$ which converges to some number t_0, then

$$\int_a^b |f(x+t_n') - f(x+t_0)|\,dx$$

$$= \int_{a+t_0}^{b+t_0} |f(x+t_n'-t_0) - f(x)|\,dx \leqslant \int_A^B |f(x+t_n'-t_0) - f(x)|\,dx \to 0.$$

§2. Helly's theorems

1. Let $\sigma_1(\lambda), \sigma_2(\lambda), \cdots$ be an infinite sequence of nondecreasing bounded functions defined on some finite closed interval $[a,b]$. For definiteness let us assume that all the functions of this sequence are continuous from the left: $\sigma_n(\lambda - 0) = \sigma_n(\lambda)$.

HELLY'S FIRST THEOREM. *If the functions $\sigma_n(\lambda)$ are uniformly bounded, then we can find a monotone function $\sigma(\lambda)$ and a subsequence $\{\sigma_{n_j}(\lambda)\}$ which converges to $\sigma(\lambda)$ at every point of continuity of $\sigma(\lambda)$.*

2) This is the so-called Arzelà-Ascoli lemma; see, for example, Ljusternik and Sobolev [1].

Proof. Let $\lambda_1, \lambda_2, \cdots$ be a countable everywhere dense subset of the interval $[a, b]$. By means of the diagonal procedure (see, for example, Smirnov [1], §12) one can find a subsequence $\sigma_{n_1}, \sigma_{n_2}, \cdots$ which converges at each of the points $\lambda_1, \lambda_2, \cdots$. Put $\lim_{j\to\infty} \sigma_{n_j}(\lambda_\nu) = \tau_\nu$. If $\lambda_\mu < \lambda_\nu$, then $\tau_\mu \leqq \tau_\nu$. We now define the function σ by means of the numbers τ_ν, putting

$$\sigma(\lambda + 0) = \inf_{\lambda_\nu > \lambda} \tau_\nu,$$

$$\sigma(\lambda - 0) = \sup_{\lambda_\nu < \lambda} \tau_\nu,$$

$$\sigma(\lambda) = \frac{1}{2}\{\sigma(\lambda + 0) + \sigma(\lambda - 0)\}.$$

Obviously σ is a nondecreasing function.

We will now show that at every point of continuity of σ one has

$$\lim_{j\to\infty} \sigma_{n_j}(\lambda) = \sigma(\lambda). \tag{2.1}$$

Let λ be any point from the interval $[a, b]$. Since for $\lambda < \lambda_\nu$ we have $\sigma_{n_j}(\lambda) \leqq \sigma_{n_j}(\lambda_\nu)$, it follows that

$$\overline{\lim_{j\to\infty}}\, \sigma_{n_j}(\lambda) \leqslant \overline{\lim_{j\to\infty}}\, \sigma_{n_j}(\lambda_\nu) = \lim_{j\to\infty} \sigma_{n_j}(\lambda_\nu) = \tau_\nu.$$

The left side of this inequality does not depend upon ν, while in the right side there appears an arbitrary τ_ν for which the corresponding λ_ν is greater than λ. Therefore

$$\overline{\lim_{j\to\infty}}\, \sigma_{n_j}(\lambda) \leqslant \inf_{\lambda_\nu > \lambda} \tau_\nu = \sigma(\lambda + 0).$$

In the same way one can prove that

$$\sigma(\lambda - 0) \leqslant \underline{\lim_{j\to\infty}}\, \sigma_{n_j}(\lambda).$$

Thus

$$\sigma(\lambda - 0) \leqslant \underline{\lim_{j\to\infty}}\, \sigma_{n_j}(\lambda) \leqslant \overline{\lim_{j\to\infty}}\, \sigma_{n_j}(\lambda) \leqslant \sigma(\lambda + 0).$$

At every point of continuity of σ we have $\sigma(\lambda - 0) = \sigma(\lambda + 0)$. Therefore the preceding inequalities imply that $\lim_{j\to\infty} \sigma_{n_j}(\lambda) = \sigma(\lambda)$, as was to be shown.

Helly's Second Theorem. *Let $[a, b]$ be a finite interval and let f be a continuous function on $[a, b]$. Suppose that a sequence $\sigma_1, \sigma_2, \cdots$ of nondecreasing functions defined on $[a, b]$ converges to a nondecreasing function σ, defined on $[a, b]$, at all points of continuity of σ. If, moreover,*

$$\lim_{n\to\infty} \sigma_n(a) = \sigma(a), \qquad \lim_{n\to\infty} \sigma_n(b) = \sigma(b), \tag{2.2}$$

then

$$\lim_{n\to\infty}\int_a^b f(\lambda)\,d\sigma_n(\lambda)=\int_a^b f(\lambda)\,d\sigma(\lambda). \tag{2.3}$$

PROOF. Let $\epsilon > 0$ be arbitrary. We choose a positive number $\delta = \delta(\epsilon)$ such that $|\lambda' - \lambda''| < \delta$ implies $|f(\lambda') - f(\lambda'')| < \epsilon$.

Let $a = \lambda_0 < \lambda_1 < \cdots < \lambda_{p-1} < \lambda_p = b$ be a partition of $[a,b]$ such that $\lambda_i - \lambda_{i-1} < \delta$. Choosing an arbitrary point μ_i in each interval $[\lambda_{i-1}, \lambda_i]$, we obtain

$$\begin{aligned}
&\left|\int_a^b f(\lambda)\,d\sigma_n(\lambda)-\sum_{i=1}^{p} f(\mu_i)\,[\sigma_n(\lambda_i)-\sigma_n(\lambda_{i-1})]\right| \\
&\quad=\left|\sum_{i=1}^{p}\int_{\lambda_{i-1}}^{\lambda_i} f(\lambda)\,d\sigma_n(\lambda)-f(\mu_i)\int_{\lambda_{i-1}}^{\lambda_i} d\sigma_n(\lambda)\right| \\
&\quad\leqslant\sum_{i=1}^{p}\int_{\lambda_{i-1}}^{\lambda_i}|f(\lambda)-f(\mu_i)|\,d\sigma_n(\lambda)<\varepsilon\int_a^b d\sigma_n(\lambda) \\
&\quad=\varepsilon\,\{\sigma_n(b)-\sigma_n(a)\}<2M\varepsilon,
\end{aligned} \tag{2.4}$$

where M is chosen to satisfy the condition $|\sigma_n(\lambda)| < M$. In the same way we can prove the inequality

$$\left|\int_a^b f(\lambda)\,d\sigma(\lambda)-\sum_{i=1}^{p} f(\mu_i)\,[\sigma(\lambda_i)-\sigma(\lambda_{i-1})]\right|<2M\varepsilon. \tag{2.5}$$

We now choose the points $\lambda_1, \cdots, \lambda_{p-1}$ to be points of continuity of σ. It is obvious that, holding these points fixed, we can choose n so large that the inequality

$$\left|\sum_{i=1}^{p} f(\mu_i)\,[\sigma_n(\lambda_i)-\sigma_n(\lambda_{i-1})]-\sum_{i=1}^{p} f(\mu_i)\,[\sigma(\lambda_i)-\sigma(\lambda_{i-1})]\right|<\varepsilon \tag{2.6}$$

will hold. Since ϵ is arbitrary, (2.4), (2.5) and (2.6) imply (2.3).

2. Helly's second theorem admits the following generalization to an infinite interval: *Suppose that a sequence $\{\sigma_n\}$ of nondecreasing functions defined on $(-\infty, \infty)$ converges to a nondecreasing function σ at all points of continuity of σ, and suppose that f is a continuous function defined on $(-\infty, \infty)$. Assume that for any $\epsilon > 0$ there exists a number $A = A(\epsilon)$ such that for all $a, b > A$ and all n we have*

$$\int_{-\infty}^{-a}|f(\lambda)|\,d\sigma_n(\lambda)<\varepsilon,\qquad\int_b^{\infty}|f(\lambda)|\,d\sigma_n(\lambda)<\varepsilon. \tag{2.7}$$

Then

$$\lim_{n\to\infty}\int_{-\infty}^{\infty} f(\lambda)\,d\sigma_n(\lambda)=\int_{-\infty}^{\infty} f(\lambda)\,d\sigma(\lambda).$$

PROOF. We will first show that f is absolutely integrable with respect to the function σ. Indeed, if $b, c > A$, then by hypothesis $\int_b^c |f(\lambda)|\,d\sigma_n(\lambda) < \epsilon$.

Holding b and c fixed (we can assume that b and c are points of continuity of σ) and letting $n\to\infty$, we obtain $\int_b^c |f(\lambda)|\,d\sigma(\lambda) < \epsilon$.

Letting $c\to\infty$, we obtain

$$\int_b^{\infty} |f(\lambda)|\,d\sigma(\lambda) < \varepsilon. \tag{2.8}$$

In the same way we can prove that

$$\int_{-\infty}^{-a} |f(\lambda)|\,d\sigma(\lambda) < \varepsilon. \tag{2.9}$$

These inequalities imply the absolute integrability of f. By the previous theorem, if $-a$ and b are points of continuity of σ, then

$$\lim_{n\to\infty}\int_{-a}^{b} f(\lambda)\,d\sigma_n(\lambda)=\int_{-a}^{b} f(\lambda)\,d\sigma(\lambda). \tag{2.10}$$

Further, by virtue of the inequalities (2.7) we have

$$\overline{\lim_{n\to\infty}}\int_{-\infty}^{-a} |f(\lambda)|\,d\sigma_n(\lambda) < \varepsilon, \qquad \overline{\lim_{n\to\infty}}\int_b^{\infty} |f(\lambda)|\,d\sigma_n(\lambda) < \varepsilon. \tag{2.11}$$

It follows from (2.8) – (2.11) that

$$\begin{aligned}
&\overline{\lim_{n\to\infty}}\left|\int_{-\infty}^{\infty} f(\lambda)\,d\sigma_n(\lambda)-\int_{-\infty}^{\infty} f(\lambda)\,d\sigma(\lambda)\right| \\
&\quad\leqslant \overline{\lim_{n\to\infty}}\int_{-\infty}^{-a} |f(\lambda)|\,d\sigma_n(\lambda)+\int_{-\infty}^{-a} |f(\lambda)|\,d\sigma(\lambda) \\
&\quad\quad+\overline{\lim_{n\to\infty}}\left|\int_{-a}^{b} f(\lambda)\,d\sigma_n(\lambda)-\int_{-a}^{b} f(\lambda)\,d\sigma(\lambda)\right| \\
&\quad\quad\quad+\overline{\lim_{n\to\infty}}\int_b^{\infty} |f(\lambda)|\,d\sigma_n(\lambda)+\int_b^{\infty} |f(\lambda)|\,d\sigma(\lambda) < 4\varepsilon.
\end{aligned}$$

Since ϵ is arbitrary, the theorem is proved.

§3. The Stieltjes inversion formula

Let $\sigma(\lambda) = \sigma_1(\lambda) + i\sigma_2(\lambda)$ be a complex function of bounded variation on the entire real line.[3)]

We put

$$\varphi(z) = \int_{-\infty}^{\infty} \frac{d\sigma(\lambda)}{z - \lambda}.$$

The Stieltjes inversion formula gives us a means of expressing $\sigma(\lambda)$ in terms of $\varphi(z)$.

We put (with $z = \sigma + i\tau$)

$$\psi(\sigma, \tau) = \frac{\operatorname{sign} \tau}{\pi} \frac{\varphi(z) - \varphi(\bar{z})}{2i} = -\frac{1}{\pi} \int_{-\infty}^{\infty} \frac{|\tau|\, d\sigma(\lambda)}{(\lambda - \sigma)^2 + \tau^2}. \tag{3.1}$$

Theorem 3.1. *If the points a, b are points of continuity of $\sigma(\lambda)$, then*

$$\sigma(b) - \sigma(a) = \lim_{\tau \to 0} \int_a^b -\psi(\sigma, \tau)\, d\sigma. \tag{3.2}$$

Proof. We may clearly assume without loss of generality that $\sigma(-\infty) = 0$. Integrating (3.1) by parts, we obtain

$$-\psi(\sigma, \tau) = -\frac{1}{\pi} \int_{-\infty}^{\infty} \frac{\partial}{\partial\lambda} \left(\frac{|\tau|}{(\lambda - \sigma)^2 + \tau^2} \right) \sigma(\lambda)\, d\lambda$$

$$= \frac{1}{\pi} \int_{-\infty}^{\infty} \frac{\partial}{\partial\sigma} \left(\frac{|\tau|}{(\lambda - \sigma)^2 + \tau^2} \right) \sigma(\lambda)\, d\lambda.$$

Integrating this equality with respect to σ from a to b, we obtain

$$\begin{aligned} \int_a^b -\psi(\sigma, \tau)\, d\sigma &= \frac{1}{\pi} \int_{-\infty}^{\infty} \frac{|\tau|}{(\lambda - b)^2 + \tau^2} \sigma(\lambda)\, d\lambda \\ &- \frac{1}{\pi} \int_{-\infty}^{\infty} \frac{|\tau|}{(\lambda - a)^2 + \tau^2} \sigma(\lambda)\, d\lambda = I(\tau, b) - I(\tau, a). \end{aligned} \tag{3.3}$$

We begin by considering the first integral. We have

$$I(\tau, b) = \frac{1}{\pi} \int_{-\infty}^{\infty} \frac{|\tau|}{(\lambda - b)^2 + \tau^2} \sigma(\lambda)\, d\lambda = \frac{1}{\pi} \int_{-\infty}^{\infty} \frac{|\lambda|}{\lambda^2 + \sigma^2} \sigma(\lambda + b)\, d\lambda.$$

[3)] That is, the real functions σ_1 and σ_2 are of bounded variation on the real line.

Since

$$\frac{1}{\pi}\int_{-\infty}^{\infty}\frac{|\tau|}{\lambda^2-\tau^2}\,d\lambda=1,$$

it follows that

$$I(\tau,b)-\sigma(b)=\frac{1}{\pi}\int_{-\infty}^{\infty}\frac{|\tau|}{\lambda^2+\tau^2}\{\sigma(\lambda+b)-\sigma(b)\}\,d\lambda$$

$$=\frac{1}{\pi}\int_{|\lambda|\leqslant\delta}\frac{|\tau|}{\lambda^2+\tau^2}\{\sigma(\lambda+b)-\sigma(b)\}\,d\lambda$$

$$+\frac{1}{\pi}\int_{|\lambda|>\delta}\frac{|\tau|}{\lambda^2+\tau^2}\{\sigma(\lambda+b)-\sigma(b)\}\,d\lambda=I_1+I_2.$$

We pick $\epsilon>0$ and then choose δ so that the inequality $|\sigma(\lambda+b)-\sigma(b)|<\epsilon/2$ holds for $|\lambda|\leqq\delta$. This is possible, since by assumption b is a point of continuity of $\sigma(\lambda)$. Having chosen δ in this way, we obtain

$$|I_1|<\frac{\varepsilon}{2\pi}\int_{-\infty}^{\infty}\frac{|\tau|}{\lambda^2+\tau^2}\,d\lambda=\frac{\varepsilon}{2}.$$

Let $M>0$ be such that $|\sigma(\lambda)|<M$. Then

$$|I_2|\leqslant\frac{2M}{\pi}\left\{\int_{-\infty}^{-\delta}+\int_{\delta}^{\infty}\right\}\frac{|\tau|}{\lambda^2+\tau^2}\,d\lambda=\frac{4M}{\pi}\int_{\delta}^{\infty}\frac{|\tau|}{\lambda^2+\tau^2}\,d\lambda=\frac{4M}{\pi}\cdot\left(\frac{\pi}{2}-\operatorname{arctg}\frac{\delta}{\tau}\right).$$

For fixed δ we can choose τ so small that the last expression will be $<\epsilon/2$. Since ϵ is arbitrary, we have proved that $\lim_{\tau\to 0}I(\tau,b)=\sigma(b)$. In the same way we can prove that $\lim_{\tau\to 0}I(\tau,a)=\sigma(a)$. Therefore (3.2) follows from (3.3).

Theorem 3.2 (uniqueness theorem). *If $\sigma_1(\lambda)$ and $\sigma_2(\lambda)$ are two complex-valued functions of bounded variation on $(-\infty,\infty)$, and*

$$\int_{-\infty}^{\infty}\frac{d\sigma_1(\lambda)}{z-\lambda}=\int_{-\infty}^{\infty}\frac{d\sigma_2(\lambda)}{z-\lambda}$$

for all nonreal z, then $\sigma_1-\sigma_2=\text{const}$.

Proof. By Theorem 3.1, $\sigma_1(b)-\sigma_1(a)$ and $\sigma_2(b)-\sigma_2(a)$ coincide whenever a, b are points of continuity of both σ_1 and σ_2. Hence (assuming that all functions of bounded variation are taken to be continuous from the left) it follows that σ_1 and σ_2 can only differ by a constant.

Remark. If $\sigma(\lambda)$ is a real function, then $\varphi(\bar z)=\overline{\varphi(z)}$. Therefore in this case

$$\psi(\sigma, \tau) = \frac{\operatorname{sign} \tau}{\pi} \frac{\varphi(z) - \overline{\varphi(z)}}{2i} = \frac{\operatorname{sign} \tau}{\pi} \operatorname{Im} \{\varphi(z)\}.$$

For $\tau > 0$ the Stieltjes inversion formula assumes the form

$$\sigma(b) - \sigma(a) = \lim_{\tau \to 0} \frac{1}{\pi} \int_a^b - \operatorname{Im} \{\varphi(z)\} \, d\sigma.$$

§4. Tauberian theorems

1. Levitan's Tauberian theorems for Fourier integrals.

THEOREM 4.1. *Let the function $\sigma(\nu)$ satisfy the following conditions:*

a) *$\sigma(\nu)$ is of bounded variation in every finite interval.*

b) *For $\mu \to \infty$*

$$\bigvee_{\mu}^{\mu+1} \{\sigma(\nu)\} = o(|\mu|^r) \quad (r \geqslant 0). \tag{4.1'}$$

c) *For every function $g_\Lambda(t)$ which has a bounded derivative of order $r+2$ and which vanishes outside the interval $(-\Lambda, \Lambda)$,*

$$\int_{-\infty}^{\infty} h(\nu) \, d\sigma(\nu) = 0, \quad \textit{where} \quad h(\nu) = \frac{1}{2\pi} \int_{-\Lambda}^{\Lambda} g_\Lambda(t) e^{-i\nu t} dt.$$

Then for $\mu \to \infty$ we have the estimate

$$\int_{-\mu}^{\mu} \left(1 - \frac{\nu^2}{\mu^2}\right)^s d\sigma(\nu) = o(\mu^{r-s}) \quad (s \geqslant 0). \tag{4.1}$$

PROOF. Let us consider the integral

$$J(\mu, s) = \int_{-\mu}^{\mu} (\mu^2 - \nu^2)^s \, d\sigma(\nu). \tag{4.2}$$

The proof will consist in giving another representation for this integral, from which the estimate (4.1) follows easily. We put

$$\varphi(\mu, \nu, s) = \begin{cases} (\mu^2 - \nu^2)^s, & |\nu| \leqslant \mu \quad (s > 0), \\ 0, & |\nu| \geqslant \mu. \end{cases}$$

Let

$$h(t, \mu, s) = \int_{-\mu}^{\mu} \varphi(\mu, \nu, s) e^{i\nu t} d\nu. \tag{4.3}$$

By the Fourier inversion formula we have

$$\varphi(\mu, \nu, s) = \frac{1}{2\pi} \int_{-\infty}^{\infty} h(t, \mu, s) e^{-i\nu t} dt.$$

Inserting this expression in (4.2), we obtain

$$J(\mu, s) = \frac{1}{2\pi} \int_{-\infty}^{\infty} d\sigma(\nu) \left\{ \int_{-\infty}^{\infty} h(t, \mu, s) e^{-i\nu t} dt \right\}. \tag{4.2'}$$

Let n denote the integer part of $s : n = [s]$. Integrating the integral in (4.3) by parts n times, we obtain

$$h(t, \mu, s) = \frac{(-1)^n}{(it)^n} \int_{-\mu}^{\mu} \varphi_\nu^{(n)}(\mu, \nu, s) e^{i\nu t} d\nu. \tag{4.4}$$

We will consider the case of integer and noninteger s separately. Suppose first that s is an integer: $s = n$, with $n \geqq 1$ (the case $s = 0$ will be considered separately). Integrating the right side of (4.4) by parts once more, we obtain

$$\begin{aligned} h(t, \mu, s) = \frac{(-1)^n}{(it)^n} \{\varphi_\nu^{(n)}(\mu, \nu, s)\big|_{\nu=\mu} e^{i\mu t} - \varphi_\nu^{(n)}(\mu, \nu, s)\big|_{\nu=-\mu} e^{-i\mu t}\} \\ + \frac{(-1)^{n+1}}{(it)^{n+1}} \int_{-\mu}^{\mu} \varphi_\nu^{(n+1)}(\mu, \nu, s) e^{i\nu t} d\nu. \end{aligned} \tag{4.5}$$

Let $\psi(t)$ be a function which has a bounded derivative of order $n+2$, and which equals 1 for $-\frac{1}{2} \leqq t \leqq \frac{1}{2}$ and zero outside the interval $(-1, 1)$. Put $\psi_\Lambda(t) = \psi(t/\Lambda)$. The function $\psi_\Lambda(t)$ equals zero outside the interval $(-\Lambda, \Lambda)$. Therefore condition c) of the theorem implies that

$$\int_{-\infty}^{\infty} d\sigma(\nu) \left\{ \int_{-\infty}^{\infty} \psi_\Lambda(t) h(t, \mu, s) e^{-i\nu t} dt \right\} = 0.$$

Subtracting this relation from (4.2), we obtain

$$J(\mu, s) = \frac{1}{2\pi} \int_{-\infty}^{\infty} d\sigma(\nu) \left\{ \int_{-\infty}^{\infty} [1 - \psi_\Lambda(t)] h(t, \mu, s) e^{-i\nu t} dt \right\}.$$

If we replace here $h(t, \mu, s)$ by the expression (4.5), we obtain

$$J(\mu, s) = J^{(1)}(\mu, s) + J^{(2)}(\mu, s) + J^{(3)}(\mu, s),$$

where

$$J^{(1)}(\mu,s) = \frac{(-1)^n\varphi_\nu^{(n)}(\mu,\nu,s)\,|_{\nu=\mu}}{2\pi}\int_{-\infty}^{\infty} d\sigma(\nu)\left\{\int_{-\infty}^{\infty}\frac{1-\psi_\Lambda(t)}{(it)^n}e^{-i(\nu-\mu)t}dt\right\},$$

$$J^{(2)}(\mu,s) = \frac{(-1)^n\varphi_\nu^{(n)}(\mu,\nu,s)\,|_{\nu=-\mu}}{2\pi}\int_{-\infty}^{\infty} d\sigma(\nu)\left\{\int_{-\infty}^{\infty}\frac{1-\psi_\Lambda(t)}{(it)^n}e^{-i(\nu+\mu)t}dt\right\},$$

$J^{(3)}(\mu,s)$

$$=\frac{(-1)^{n+1}}{2\pi}\int_{-\infty}^{\infty} d\sigma(\nu)\left\{\int_{-\infty}^{\infty}\frac{1-\psi_\Lambda(t)}{(it)^{n+1}}\left[\int_{-\mu}^{\mu}\varphi_\xi^{(n+1)}(\mu,\xi,s)\,e^{-i(\nu-\xi)t}\,d\xi\right]dt\right\}.$$

We now put

$$A_s(\nu)=\frac{(-1)^s}{2\pi}\int_{-\infty}^{\infty}\frac{1-\psi_\Lambda(t)}{(it)^s}e^{-i\nu t}dt. \tag{4.6}$$

It is easily seen that this integral converges absolutely for $s > 1$, and converges for $s = 1$. Moreover, integrating by parts, it is not hard to show that for $\nu \to \infty$ one has the estimate

$$A_s(\nu) = O(1/\nu^{n+2}). \tag{4.7}$$

Interchanging the order of integration in $J^{(3)}(\mu,s)$, which is permissible by virtue of the convergence of the integral (4.6), we obtain

$$J^{(3)}(\mu,s)=\int_{-\mu}^{\mu}\varphi_\xi^{(s+1)}(\mu,\xi,s)\left[\int_{-\infty}^{\infty}A_{s+1}(\nu)\,d_\nu\,\sigma(\nu+\xi)\right]d\xi. \tag{4.8}$$

Further, using the notation (4.6), we obtain

$$J^{(1)}(\mu,s)=\frac{\varphi_\nu^{(s)}(\mu,\nu,s)\,|_{\nu=\mu}}{2\pi}\int_{-\infty}^{\infty}A_s(\nu)\,d_\nu\sigma(\nu+\mu), \tag{4.9}$$

$$J^{(2)}(\mu,s)=\frac{\varphi_\nu^{(s)}(\mu,\nu,s)\,|_{\nu=-\mu}}{2\pi}\int_{-\infty}^{\infty}A_s(\nu)\,d_\nu\,\sigma(\nu-\mu). \tag{4.10}$$

It is easily seen that as $\mu \to \infty$

$$\varphi_\nu^{(s)}(\mu,\nu,s)\,|_{\nu=\pm\mu} = O(\mu^s). \tag{4.11}$$

Let us now estimate the integrals (4.9) and (4.10). We will consider (4.9); the estimation of (4.10) is similar.

Let N be a positive integer and put

$$i=\int_{-\infty}^{\infty} A_s(\nu)\,d_\nu\sigma(\nu+\mu)=\left\{\int_{-\infty}^{-N}+\int_{-N}^{N}+\int_{N}^{\infty}\right\} A_s(\nu)\,d_\nu\sigma(\nu+\mu)=i_1+i_2+i_3.$$

Given an arbitrary $\epsilon > 0$, for $\mu > \mu(\epsilon)$ and $|\nu| \leqq N$ we will have

$$\bigvee_{j}^{j+1}\{\sigma(\nu+\mu)\}\leqslant\varepsilon\,|\mu+j\,|^r\leqslant\varepsilon\,|\,\mu\,|^r\,|\,j\,|^r\left(\frac{1}{\mu}+\frac{1}{j}\right)^r.$$

Therefore

$$(4.12)\qquad |\,i_2|\leqslant C\varepsilon\,|\,\mu\,|^r\sum_{j=-N}^{N}\frac{|\,j\,|^r}{|\,j\,|^{r+2}+1}=C\varepsilon\,\mu^r.$$

Further,

$$(4.13)\qquad |\,i_1\,|\leqslant C\mu^r\sum_{j=N+1}^{\infty}\frac{1}{j^2}=\frac{C}{N+1}\,\mu^r$$

and a similar estimate holds for i_3:

$$(4.13')\qquad i_3<\frac{C}{N+1}\,\mu^r.$$

If we first choose N, and then $\mu(\epsilon)$, then from the estimates (4.12), (4.13) and (4.13′) we obtain for $\mu > \mu(\epsilon)$

$$(4.14)\qquad |\,i\,| < C\epsilon|\,\mu\,|^r.$$

It follows from (4.11) that for $\mu \to \infty$

$$(4.15)\qquad J^{(1)}(\mu, s) = o(\mu^{r+s}).$$

A similar estimate holds for $J^{(2)}(\mu, s)$. Therefore for $\mu \to \infty$

$$(4.16)\qquad J^{(1)}(\mu, s) + J^{(2)}(\mu, s) = o(\mu^{s+r}).$$

We must now estimate $J^{(3)}(\mu, s)$ for large μ. As before, let N be a positive integer. We put

$$\int_{-\infty}^{\infty} A_s(\nu)\,d_\nu\sigma(\nu+\xi)=\left\{\int_{-\infty}^{-N}+\int_{-N}^{N}+\int_{N}^{\infty}\right\} A_s(\nu)\,d_\nu\sigma(\nu+\xi)=j_1+j_2+j_3.$$

Just as before,

$$(4.17)\qquad |\,j_1\,|+|\,j_2\,|\leqslant\frac{C}{N+1}(|\,\xi\,|^r+1).$$

Further,

$$J^{(3)}(\mu,\ s)=\int_{-\mu}^{\mu}\varphi_\xi^{(s+1)}(\mu,\ \xi,\ s)\,\{j_1+j_2+j_3\}\,d\xi=\ j_1'+j_2'+j_3'.$$

It follows from (4.17) that

$$|j_1'|+|j_2'|\leqslant \frac{C}{N+1}\int_{-\mu}^{\mu} \varphi_\xi^{(s+1)}(\mu,\ \xi,\ s)(|\xi|^r+1)\,d\xi \leqslant \frac{C}{N+1}\mu^{s+r}, \tag{4.18}$$

since $\varphi_\xi^{(s+1)}(\mu,\xi,s)=O(\mu^{s-1})$.

Let us now consider j_3'. We have

$$j_3'=\int_{-\mu}^{\mu}\varphi_\xi^{(s+1)}(\mu,\ \xi,\ s)\left[\int_{-N}^{N}A_s(\nu)\,d_\nu\sigma(\nu+\xi)\right]d\xi$$

$$=\left\{\int_{-\mu}^{-a}+\int_{-a}^{a}+\int_{a}^{\mu}\right\}\varphi_\xi^{(s+1)}(\mu,\ \xi,\ s)\left[\int_{-N}^{N}A_s(\nu)\,d_\nu\sigma(\nu+\xi)\right]d\xi=e_1+e_2+e_3.$$

Let η be an arbitrary positive number and choose a so large that for $|\nu|<N$ and $|\xi|>a$ one has

$$\bigvee_j^{j+1}\{\sigma(\nu+\xi)\}<\eta|\nu+\xi|^r<C\eta|j|^r|\xi|^r.$$

We then obtain

$$\left|\int_{-N}^{N}A_s(\nu)\,d_\nu\sigma(\nu+\xi)\right|\leqslant C\eta\sum_{j=-N}^{N}\frac{|\xi|^r|j|^r}{|j|^{r+2}+1}<C\eta|\xi|^r.$$

Therefore

$$|e_1|+|e_3|<C\eta\int_{-\mu}^{\mu}|\varphi_\xi^{(s+1)}(\mu,\ \xi,\ s)||\xi|^r\,d\xi\leqslant 2C\eta|\mu|^{r+s}.$$

Further,

$$|e_2|\leqslant C\int_{-a}^{a}|\varphi_\xi^{(s+1)}(\mu,\ \xi,\ s)||\xi|^r\,d\xi\leqslant C\mu^{s-1}a^{r+1},$$

and therefore $|j_3'|\leqq C(2\eta\mu^{r+s}+\mu^{s-1}a^{r+1})$. Hence

$$|J^{(3)}(\mu,\ s)|\leqslant C\mu^{s+r}\left[\frac{2}{N+1}+2\eta+\left(\frac{a}{\mu}\right)^{r+1}\right]. \tag{4.19}$$

Let ϵ be an arbitrary positive number. We first choose N so that $2/(N+1)<\epsilon/3$. We then choose a so large that $2C\eta<\epsilon/3$. Finally, we choose μ so large that $(a/\mu)^{r+1}<\epsilon/3$. Then from (4.19) we obtain the estimate

$$J^{(3)}(\mu,s)=o(\mu^{r+s}). \tag{4.19'}$$

From the estimates (4.16) and (4.19′) it follows that

(4.20) $$J(\mu, s) = o(\mu^{r+s}).$$

Dividing by μ^{2s}, we obtain

$$\frac{1}{\mu^{2s}} J(\mu, s) = \int_{-\mu}^{\mu} \left(1 - \frac{\nu^2}{\mu^2}\right)^s d\sigma(\nu) = o(\mu^{r-s}),$$

as was to be proved.

Let us now consider the case $s = 0$. In this case (cf. the notation (4.3))

$$h(t, \mu, 0) = \int_{-\mu}^{\mu} e^{i\nu t} d\nu = \frac{e^{i\nu t}}{it}\Big|_{-\mu}^{\mu} = \frac{e^{i\mu t} - e^{-i\mu t}}{it}.$$

Therefore

$$J(\mu, 0) = \frac{1}{2\pi} \int_{-\infty}^{\infty} d\sigma(\nu) \left\{ \int_{-\infty}^{\infty} [1 - \psi_\Lambda(t)] \frac{e^{i\mu t} - e^{-i\mu t}}{it} e^{-i\nu t} dt \right\}$$

$$= \int_{-\infty}^{\infty} d\sigma(\nu) \{A_0(\nu - \mu) - A_0(\nu + \mu)\}$$

$$= \int_{-\infty}^{\infty} A_0(\nu)\, d_\nu \sigma(\nu + \mu) - \int_{-\infty}^{\infty} A_0(\nu)\, d_\nu \sigma(\nu - \mu).$$

Since $A_0(\nu)$ has all the properties of the function $A_s(\nu)$, we can carry out the previous estimates, as a consequence of which we obtain

$$J(\mu, 0) = \int_{-\mu}^{\mu} d\sigma(\nu) = \sigma(\mu) - \sigma(-\mu) = o(|\mu|^r),$$

which proves the theorem for $s = 0$.

Now suppose that the number $s = n + \alpha$ $(\alpha \neq 0)$ is noninteger. In this case we have to estimate the integral

$$J(\mu, s) = \int_{-\mu}^{\mu} (\mu^2 - \nu^2)^{n+\alpha} d\sigma(\nu).$$

We will transform this integral by means of integration by parts, and reduce its estimation to the case of integer powers of s. Integrating by parts $n + 1$ times, we obtain

$$J(\mu, n + \alpha) = C \int_0^{\mu} (\mu^2 - \nu^2)^{\alpha - 1} \nu \left[\int_{-\nu}^{\nu} (\nu^2 - t^2)^n d\sigma(t) \right] d\nu =$$

$$= C \int_0^{\mu} (\mu^2 - \nu^2)^{\alpha-1} \nu J(\nu, n)\, d\nu,$$

where C is some constant. Further, put

$$J(\mu, n+\alpha) = C \left\{ \int_0^{\mu-1} + \int_{\mu-1}^{\mu} \right\} (\mu^2 - \nu^2)^{\alpha-1} \nu J(\nu, n)\, d\nu = J' + J''.$$

Then the estimate (4.20) implies that

$$(4.21) \qquad |J''| = o\left(\mu^{n+r} \int_{\mu-1}^{\mu} \nu (\mu^2 - \nu^2)^{\alpha-1} d\nu \right) = o(\mu^{n+r+\alpha}).$$

Further, integrating by parts once more, we obtain

$$J' = -C_1 [\mu^2 - (\mu-1)^2]^{\alpha-1} J(\mu-1, n+1)$$
$$+ C_2 \int_0^{\mu-1} (\mu^2 - \nu^2)^{\alpha-2} \nu J(\nu, n+1)\, d\nu.$$

Therefore, applying the estimate (4.20) again, we obtain

$$(4.22) \qquad J' = o(\mu^{n+r+\alpha}) + o\left(\mu^{n+1+r} \int_0^{\mu-1} \nu (\mu^2 - \nu^2)^{\alpha-2} d\nu \right) = o(\mu^{n+r+\alpha}).$$

It follows from (4.21) and (4.22) that $J(\mu, s) = o(\mu^{r+s})$ for $\mu \to \infty$. Thus the proof of the theorem is complete.

REMARK. One can similarly prove that in the estimates (4.1) and (4.1′) one can replace o by O. For the case of an odd $\sigma(\nu)$ (precisely this case is applied in this book) the functions $A_s(\nu)$ and $h(t, \mu, s)$ can be assumed even, i.e. one can consider the cosine Fourier transform of the even function $g_\Lambda(t)$, which equals zero outside the interval $(-\Lambda, \Lambda)$. For this case the basic estimate assumes the form

$$\int_0^{\mu} (\mu^2 - \nu^2)^s\, d\sigma(\nu) = o(\mu^{s+r})$$

or

$$\int_0^{\mu} \left(1 - \frac{\nu^2}{\mu^2}\right)^s d\sigma(\nu) = o(\mu^{r-s}).$$

Theorem 4.1 in fact gives us the asymptotic behaviour of the odd component of the function $\sigma(\nu)$. To obtain the asymptotic behaviour of $\sigma(\nu)$ itself, one has to consider a kernel of a more general kind than

$\varphi(\mu, \nu; s)$ (cf. Marčenko's theorems in this section) or else consider, together with the kernel $\varphi(\mu, \nu; s)$, some odd kernel as well, which would enable us to clarify the asymptotic behaviour of the even component of σ.

In the following theorem this is done in one of the simplest cases.[4)]

THEOREM 4.1′. *Let us assume that the function $\sigma(\nu)$ satisfies the following conditions*:

a) *It has bounded variation in every finite interval.*

b) *For $|a| \to \infty$*

$$\bigvee_a^{a+1} \{\sigma(\nu)\} = O(1).$$

c) *For every function $g_\Lambda(t)$ which has a continuous second derivative and is equal to zero outside the interval $(-\Lambda, \Lambda)$,*

$$\int_{-\infty}^{\infty} h(\nu)\, d\sigma(\nu) = 0,. \quad h(\nu) = \frac{1}{2\pi} \int_{-\Lambda}^{\Lambda} g_\Lambda(t)\, e^{-i\nu t} dt.$$

d) *The point $\nu = 0$ is a point of continuity of $\sigma(\nu)$.*

Then for $\lambda \to \infty$

$$\sigma(\lambda) + \sigma(-\lambda) = O(1). \tag{4.23}$$

REMARK. We note that the condition d) is not a restriction, as there always exists a number ν_0 such that the function $\sigma_1(\nu) = \sigma(\nu + \nu_0)$ will be continuous at $\nu = 0$. At the same time, it is easy to see that σ_1 will satisfy the conditions a), b) and c) if σ does, and that σ will satisfy (4.23) if σ_1 does (in view of σ satisfying b)).

We assume in (4.23) that the points λ and $-\lambda$ are points of continuity of $\sigma(\nu)$.

PROOF OF THEOREM 4.1′. For $\lambda > 0$ we put

$$\chi(\lambda, \nu) = \begin{cases} 0 & \text{for} \quad \nu < -\lambda, \\ -\frac{1}{2} & \text{for} \quad \nu = -\lambda, \\ -1 & \text{for} \quad -\lambda < \nu < 0, \\ 0 & \text{for} \quad \nu = 0, \\ 1 & \text{for} \quad 0 < \nu < \lambda, \\ \frac{1}{2} & \text{for} \quad \nu = \lambda, \\ 0 & \text{for} \quad \nu > \lambda, \end{cases}$$

[4)] This theorem is due to Levitan and is published here for the first time.

$$K(t, \lambda) = \int_{-\lambda}^{\lambda} \chi(\lambda, \nu) e^{i\nu t} d\nu = 2i \frac{1 - \cos \lambda t}{t}.$$

It follows from the Fourier inversion formula that

$$\chi(\lambda, \nu) = \frac{i}{\pi} \int_{-\infty}^{\infty} \frac{1 - \cos \lambda t}{t} e^{-i\nu t} dt. \tag{4.24}$$

We put, by definition

$$\begin{gathered} \int_{-\lambda}^{-0} f(\nu)\, d\sigma(\nu) = \lim_{\varepsilon \to 0} \int_{-\lambda}^{-\varepsilon} f(\nu)\, d\sigma(\nu), \quad \int_{+0}^{\lambda} f(\nu)\, d\sigma(\nu) = \lim_{\varepsilon \to 0} \int_{\varepsilon}^{\lambda} f(\nu)\, d\sigma(\nu), \\ \int_{-\lambda}^{\lambda} f(\nu)\, d\sigma(\nu) = \int_{-\lambda}^{-0} f(\nu)\, d\sigma(\nu) + \int_{+0}^{\lambda} f(\nu)\, d\sigma(\nu). \end{gathered} \tag{4.25}$$

It is not hard to see that

$$\sigma(\lambda) + \sigma(-\lambda) = 2\sigma(0) + \int_{-\lambda}^{\lambda} \chi(\lambda, \nu)\, d\sigma(\nu),$$

or, inserting for $\chi(\lambda, \nu)$ the expression (4.24),

$$\sigma(\lambda) + \sigma(-\lambda) = 2\sigma(0) + \frac{i}{\pi} \int_{-\lambda}^{\lambda} d\sigma(\nu) \left\{ \int_{-\infty}^{\infty} \frac{1 - \cos \lambda t}{t} e^{-i\nu t} dt \right\}. \tag{4.26}$$

Let $\psi_\Lambda(t)$ be the same as in Theorem 4.1. It follows from condition c) of the theorem that

$$\frac{i}{\pi} \int_{-\infty}^{\infty} d\sigma(\nu) \left\{ \int_{-\infty}^{\infty} \psi_\Lambda(t) \frac{1 - \cos \lambda t}{t} e^{-i\nu t} dt \right\} = 0. \tag{4.27}$$

Since $\sigma(\nu)$ and

$$\int_{-\infty}^{\infty} \psi_\Lambda(t) \frac{1 - \cos \lambda t}{t} e^{-i\nu t} dt$$

are continuous at the point $\nu = 0$, the outer integral in (4.27) can be understood in the sense of (4.25). Subtracting (4.27) from (4.26), we obtain

$$
\begin{aligned}
\sigma(\lambda)+\sigma(-\lambda) &= 2\sigma(0)+\frac{i}{\pi}\int_{-\infty}^{\infty} d\sigma(\nu)\left\{\int_{-\infty}^{\infty}\frac{1-\psi_\Lambda(t)}{t}(1-\cos\lambda t)\,e^{-i\nu t}dt\right\} \\
&= 2\sigma(0)+\frac{i}{\pi}\int_{-\infty}^{\infty} d\sigma(\nu)\left\{\int_{-\infty}^{\infty}\frac{1-\psi_\Lambda(t)}{t}e^{-i\nu t}dt\right\} \\
&\quad -\frac{i}{2\pi}\int_{-\infty}^{\infty} d\sigma(\nu)\left\{\int_{-\infty}^{\infty}\frac{1-\psi_\Lambda(t)}{t}e^{-i(\nu-\lambda)t}dt\right\} \\
&\quad -\frac{i}{2\pi}\int_{-\infty}^{\infty} d\sigma(\nu)\left\{\int_{-\infty}^{\infty}\frac{1-\psi_\Lambda(t)}{t}e^{-i(\nu+\lambda)t}dt\right\} \\
&= 2\sigma(0)+2\int_{-\infty}^{\infty} A_1(\nu)\,d\sigma(\nu)-\int_{-\infty}^{\infty} A_1(\nu-\lambda)\,d\sigma(\nu) \\
&\quad -\int_{-\infty}^{\infty} A_1(\nu+\lambda)\,d\sigma(\nu),
\end{aligned} \tag{4.28}
$$

where $A_1(\nu)$ is the same as in Theorem 4.1.

Since $|A_1(\nu)| = O(1/\nu^2)$ for large ν, the estimate (4.23) can be obtained from the integral representation (4.28) by means of estimates similar to those which were carried out in the proof of Theorem 4.1.

2. Marčenko's Tauberian theorems for Fourier integrals.

THEOREM 4.2. *Suppose that for some integer $n \geq 0$ the nondecreasing and left-continuous functions $\rho(\lambda)$ and $\sigma(\lambda)$ $(-\infty<\lambda<\infty)$ satisfy the following conditions:*

I. *On the set of all infinitely differentiable functions $f(x)$ which vanish outside a given interval $(-h,h)$ one has the identity*

$$\int_{-\infty}^{\infty} \mathcal{E}_f(\lambda)\,d\{\rho(\lambda)-\sigma(\lambda)\} = \int_{-\infty}^{\infty} f^{(n)}(x)\,G(x)\,dx,$$

where $G(x)$ is some function, defined and summable over $(-h,h)$, and

$$\mathcal{E}_f(\lambda) = \int_{-\infty}^{\infty} f(x)e^{-i\lambda x}dx.$$

II. *One of the functions* $\rho(\lambda)$ *and* $\sigma(\lambda)$, *say* $\sigma(\lambda)$, *is such that*

$$\overline{\lim_{|a|\to\infty}} \frac{\sigma(a+\lambda_0)-\sigma(a)}{|a|^n} = \sigma^*(\lambda_0) < \infty$$

for some $\lambda_0 > 0$.

Further, suppose that the kernel $T(N,\lambda)$ $(0 \leqq N < \infty$; N *is a parameter*) *satisfies the following conditions*:

A. *For every* N *the function* $T(N,\lambda)$ *is of bounded variation in* λ $(-\infty < \lambda < \infty)$ *and*

$$\int_{-\infty}^{\infty} (1+|\lambda|^{n+2})\,|T(N,\lambda)|\,d\lambda < \infty.$$

B. *For every* N *the function* $T(N,\lambda)$ *has* $k \geqq 0$ *derivatives with respect to* λ, *and* $|T^{(m)}(N,\lambda)| = o(|\lambda|^{-n-2})$ *for* $|\lambda| \to \infty$ $(m = 0, 1, \cdots, k)$. *The* k*th derivative* $T^{(k)}(N,\lambda)$ *is of bounded variation (as a function of* λ) *and*

$$\int_{-\infty}^{\infty} (1+|\lambda|^n)\,|d_\lambda T^{(k)}(N,\lambda)| < \infty.$$

C. *For every finite* a

$$\lim_{N\to\infty} \frac{\int_{-a}^{a} (1+|\lambda|^n)\,|d_\lambda T^{(k)}(N,\lambda)|}{\int_{-\infty}^{\infty} (1+|\lambda|^n)\,|d_\lambda T^{(k)}(N,\lambda)|} = 0$$

(*the number* n *in* A, B *and* C *is the same as in conditions* I *and* II).

Let us denote by $\varphi(x)$ *an even infinitely differentiable function such that*

$$\varphi(x) = 1, \quad |x| < \frac{2}{3},$$

$$0 \leqslant \varphi(x) \leqslant 1, \quad \frac{2}{3} \leqslant |x| \leqslant 1,$$

$$\varphi(x) = 0, \quad |x| \geqslant 1$$

and put

$$f_1(N,x) = \frac{1}{2\pi}\varphi\left(\frac{x}{h}\right)\int_{-\infty}^{\infty} T(N,\lambda)\,e^{i\lambda x}\,d\lambda.$$

Then

$$\overline{\lim_{N\to\infty}}\left|\frac{\int_{-\infty}^{\infty} T(N,\lambda)\,d_\lambda\{\rho(\lambda)-\sigma(\lambda)\} - \int_{-\infty}^{\infty}[f_1(N,x)]_x^{(n)}G(x)\,dx}{\int_{-\infty}^{\infty}(1+|\lambda|^n)\,|d_\lambda T^{(k)}(N,\lambda)|}\right| \leqslant C2^k h^{-k}\sigma^*\left(\frac{1}{h}\right),$$

e the constant C, which depends only upon $\varphi(x)$, can be chosen to fy $C < 8\cdot 10^7$.

pecial case of Theorem 4.2. A special case of Theorem 4.2 has been ied in this book, namely:

$$T(N,\lambda) = \begin{cases} 1, & |\lambda| < N,\\ \frac{1}{2}, & |\lambda| = N,\\ 0, & |\lambda| > N,\end{cases}$$

$k = n = 0$,

the function $G(x)$ is of bounded variation in some neighborhood ro.

en if the hypotheses of Theorem 4.2 are satisfied for any h, and $(+0) = 0$ (for example, if $\sigma(\lambda) = \lambda$), then for $N \to \infty$

$$N) - \rho(-N) = \sigma(N) - \sigma(-N) + \frac{1}{2}\{G(+0) + G(-0)\} + o(1).$$

EOREM 4.2′.[5] *Suppose that the functions $\rho(\lambda)$ and $\sigma(\lambda)$ satisfy ions I and II of Theorem 4.2 with $n = 0$. Let $g(x)$ be an arbitrary on with compact support which coincides with $G(x)$ in some neigh- d of zero, and let*

$$\tilde{g}(\mu) = \int_{-\infty}^{\infty} g(x)e^{-i\mu x}dx,$$

$$S_g(\lambda) = \frac{1}{2\pi}\int_{-\lambda}^{\lambda}\tilde{g}(\mu)\,d\mu;\quad S_g^*(\lambda) = \frac{1}{2\pi}\int_{-\lambda}^{\lambda}\frac{\mu}{|\mu|}\tilde{g}(\mu)\,d\mu.$$

en there exists an absolute constant C such that

$$\overline{\lim_{\lambda\to+\infty}}\,|\{\rho(\lambda)-\rho(-\lambda)\} - \{\sigma(\lambda)-\sigma(-\lambda)\} - S_g(\lambda)| \leq C\sigma^*(1/h),$$

$$\overline{\lim_{\lambda\to+\infty}}\,|\{\rho(\lambda)-\rho(-\lambda)\} - \{\sigma(\lambda)-\sigma(-\lambda)\} - S_g^*(\lambda) - A| \leq C\sigma^*(1/h),$$

e A is some constant.

[5] Theorem 4.2′ is due to Marčenko and is published here for the first time.

Let us remark, first of all, that if $g_1(x)$ is another function with compact support which coincides with $G(x)$ in a neighborhood of zero, then by a well-known localization theorem

$$\lim_{\lambda\to\infty}\{S_g(\lambda)-S_{g_1}(\lambda)\}=0,$$

and

$$\lim_{\lambda\to\infty}\{S_g^*(\lambda)-S_{g_1}^*(\lambda)\}$$

exists. Therefore it suffices to prove the theorem for one function $g(x)$ with compact support which coincides with $G(x)$ in a neighborhood of zero.

We also remark that the function $\rho(\lambda)$ is defined up to an additive constant; therefore we can always assume it to be normalized in such a way that $A=0$.

PROOF. The assertion I) is a corollary of Theorem 4.2, if we take the kernel $T(N,\lambda)=1$ for $|\lambda|<N$, and $=0$ for $|\lambda|>N$.

We precede the proof of the assertion II) by the following remark: if the function $a(\lambda)$, $\lambda\in(0,+\infty)$, is such that for some $\delta\geqq 0$ we have, for any sequences $\{d_N\}$, $\{b_N\}$, $b_N>d_N$, $\lim_{N\to\infty}d_N=\infty$, the inequality

$$\overline{\lim_{N\to\infty}}\,|a(d_N)-a(b_N)|\leqq\delta, \tag{4.29}$$

then there exists a number A such that

$$\overline{\lim_{\lambda\to\infty}}\,|a(\lambda)-A|\leqslant\frac{\delta}{\sqrt{2}}. \tag{4.30}$$

Indeed, it is obvious from (4.29) that

$$\overline{\lim_{\lambda\to\infty}}\,|a(\lambda)|<\infty,\qquad \overline{\lim_{\lambda\to\infty}}\,\mathrm{Re}\,a(\lambda)-\underline{\lim_{\lambda\to\infty}}\,\mathrm{Re}\,a(\lambda)\leqslant\delta,$$

$$\overline{\lim_{\lambda\to\infty}}\,\mathrm{Im}\,a(\lambda)-\underline{\lim_{\lambda\to\infty}}\,\mathrm{Im}\,a(\lambda)\leqslant\delta.$$

Therefore all the limit values of $a(\lambda)$, $\lambda\to+\infty$, lie in the rectangle

$$\overline{\lim}\,\mathrm{Re}\,a(\lambda)\geqq\mathrm{Re}\,z\geqq\underline{\lim}\,\mathrm{Re}\,a(\lambda),$$

$$\overline{\lim}\,\mathrm{Im}\,a(\lambda)\geqq\mathrm{Im}\,z\geqq\underline{\lim}\,\mathrm{Im}\,a(\lambda),$$

whose sides are of length $\leqq\delta$. It is clear that if the complex number A is taken as the center of this rectangle, then the inequality (4.30) will hold.

Let us now verify that the function

(4.31) $$a(\lambda)=\{\rho(\lambda)+\rho(-\lambda)\}-\{\sigma(\lambda)+\sigma(-\lambda)\}-S_g^*(\lambda)$$

satisfies the condition (4.29), if $g(x)=\varphi(x/h)G(x)$ and $\delta=C\sigma^*(1/h)$. For this we put

$$T(N,\lambda)=\begin{cases}\lambda/|\lambda| & \text{for } |\lambda|\in(d_N,b_N),\\ 0 & \text{for } |\lambda|\notin(d_N,b_N).\end{cases}$$

This kernel satisfies the conditions A, B and C (for $k=0$) of Theorem 4.2, and therefore Theorem 4.2 is applicable to it. Taking note of the inequality proved in that theorem, we see that the function $a(\lambda)$ defined by (4.31) satisfies the condition (4.29) with $\delta=C\sigma^*(1/h)$. There follows the existence of A for which (4.30), and hence condition II), holds, which was to be proved.

COROLLARY. *If the conditions of Theorem 4.2 are satisfied for any h, and $\sigma^*(+0)=0$, while $G(x)$ satisfies conditions in the neighborhood of zero which guarantee the convergence of its Fourier series and conjugate series, then for $\lambda\to\pm\infty$ one has the following asymptotic formula*:

$$\rho(\lambda)=\sigma(\lambda)+\frac{1}{2}\cdot\frac{\lambda}{|\lambda|}G(0)+A+o(1).$$

3. A Tauberian theorem of Keldyš.

THEOREM 4.3. *Let $\varphi(x)$ and $\psi(x)$ be positive increasing functions, defined for $x>0$, where $\varphi(x)$ is differentiable and satisfies the conditions $\lim_{x\to\infty}\varphi(x)=\infty$ and $\alpha\varphi(x)<x\varphi'(x)<\beta\varphi(x)$ for $x>a>0$, where α and β are positive numbers,*[6] *while $\varphi(x)=\psi(x)=0$ for $x<a/2$, and let*

$$f(x)=\int_0^\infty\frac{d\varphi(\xi)}{(\xi-x)^{m+1}},\qquad g(x)=\int_0^\infty\frac{d\psi(\xi)}{(\xi-x)^{m+1}},$$

where m is the integer part of β. If $f(x)/g(x)\to 1$ for $x\to-\infty$, then

$$\lim_{x\to\infty}\frac{\varphi(x)}{\psi(x)}=1.$$

---

[6] Keldyš proved his theorem under the assumption that the numbers α and β satisfy the inequality $0<\beta<\alpha+1$. Later Korenbljum [1] showed that the theorem is true without this condition.

BIBLIOGRAPHY

E. Abdukadyrov

1. *Computation of the regularized trace for a Dirac system*, Vestnik Moskov. Univ. Ser. I Mat. Meh. **22** (1967), no. 4, 17-24. (Russian) MR **36** #2862.

N. I. Ahiezer

1. *Lectures on the theory of approximation*, 2nd rev. ed., "Nauka", Moscow, 1965; English transl. of 1st ed., Ungar, New York, 1956. MR **20** #1872; **32** #6108.

N. I. Ahiezer and I. M. Glazman

1. *Theory of linear operators in Hilbert space*, 2nd rev. ed., "Nauka", Moscow, 1966; English transl. of 1st ed., Ungar, New York, 1961. MR **34** #6527.

F. V. Atkinson

1. *Discrete and continuous boundary problems*, Academic Press, New York, 1964; Russian transl., "Mir", Moscow, 1968. MR **31** #416; **39** #4473.

G. D. Birkhoff

1. *On the asymptotic character of the solutions of certain linear differential equations containing a parameter*, Trans. Amer. Math. Soc. **9** (1908), 219-231.
2. *Boundary value and expansion problems of ordinary linear differential equations*, Trans. Amer. Math. Soc. **9** (1908), 373-395.
3. *Existence and oscillation theorems for a certain boundary value problem*, Trans. Amer. Math. Soc. **10** (1909), 259-270.

G. D. Birkhoff and R. E. Langer

1. *The boundary problems and developments associated with a system of ordinary linear differential equations of the first order*, Proc. Amer. Acad. Arts and Sci. **58** (1923), 51-128.

D. I. Blohincev

1. *Foundations of quantum mechanics*, GITTL, Moscow, 1949; 3rd ed., Vyšs. Škola, Kiev, 1961; English transl., Reidel, Dordrecht, 1964. MR **24** #B1896; **29** #6806.

E. A. Coddington

1. *On the spectral representation of ordinary self-adjoint differential operators*, Proc. Nat. Acad. Sci. U.S.A. **38** (1952), 732-737. MR **14**, 278.

E. A. Coddington and N. Levinson

1. *On the nature of the spectrum of singular second-order linear differential equations*, Canad. J. Math. **3** (1951), 335-338. MR **13**, 133.
2. *Theory of ordinary differential equations*, McGraw-Hill, New York, 1955; Russian transl., IL, Moscow, 1958. MR **16**, 1022.

S. D. Conte and W. C. Sangren

1. *An asymptotic solution for a pair of first order equations*, Proc. Amer. Math. Soc. **4** (1953), 696-702. MR **15**, 314.
2. *An expansion theorem for a pair of singular first order equations*, Canad. J. Math. **6** (1954), 554-560. MR **17**, 159.

R. Courant and D. Hilbert

1. *Methoden der Mathematischen Physik*. Vol. I, Springer, Berlin, 1931; English transl., Interscience, New York, 1953; Russian transl., GITTL, Moscow, 1951. MR **5**, 97; **13**, 800; **16**, 426.

L. A. DIKIĬ

1. *The zeta function of an ordinary differential equation on a finite interval,* Izv. Akad. Nauk SSSR Ser. Mat. **19** (1955), 187-200. (Russian) MR **17**, 370.

A. A. DORODNICYN

1. *Asymptotic laws of distribution of the characteristic values for certain special forms of differential equations of the second order,* Uspehi Mat. Nauk **7** (1952), no. 6 (52), 3-96; English transl., Amer. Math. Soc. Transl. (2) **16** (1960), 1-101. MR **14**, 876; **22** # 8161.

V. A. FOK

1. *Beginnings of quantum mechanics,* Izdat. Leningrad. Gos. Univ., Leningrad, 1932. (Russian)

N. G. GASYMOV

1. *The inverse scattering problem for a system of Dirac equations of order 2n,* Trudy Moskov. Mat. Obšč. **19** (1968), 41-112 = Trans. Moscow Math. Soc. **1968**, 41-120. MR **39** # 2418.

I. M. GEL'FAND

1. *On identities for eigenvalues of a differential operator of second order,* Uspehi Mat. Nauk **11** (1956), no. 1 (67), 191-198. (Russian) MR **18**, 129.

I. M. GEL'FAND AND B. M. LEVITAN

1. *On the determination of a differential equation from its spectral function,* Izv. Akad. Nauk SSSR Ser. Mat. **15** (1951), 309-360; English transl., Amer. Math. Soc. Transl. (2) **1** (1955), 253-304.
2. *On a simple identity for the eigenvalues of a second-order differential operator,* Dokl. Akad. Nauk SSSR **88** (1953), 593-596. (Russian) MR **15**, 33.

I. M. GLAZMAN

1. *Direct methods of the qualitative spectral analysis of singular differential operators,* Fizmatgiz, Moscow, 1963; English transl., Israel Program for Scientific Translations, Jerusalem, 1965; Davey, New York, 1966. MR **32** # 2938; # 8210.

A. HAAR

1. *Zur Theorie der Orthogonalen Funktional-systeme.* I, Math. Ann. **69** (1910), 331-371.
2. *Zur Theorie der Orthogonalen Funktional-systeme.* II, Math. Ann. **71** (1911), 38-53.

G. HAMEL

1. *Über die lineare Differentialgleichungen zweiter Ordnung mit periodischen Koeffizienten,* Math. Ann. **73** (1913), 371-412.

O. HAUPT

1. *Über lineare homogene Differentialgleichungen zweiter Ordnung mit periodischen Koeffizienten,* Math. Ann. **79** (1919), 278-285.

D. HILBERT

1. *Grundzuge einer allgemeinen Theorie der linearen Integralgleichungen.* I – VI, Nachr. Akad. Wiss. Göttingen Math. Phys. Kl. **1904**, 49-91; **1905**, 213-259, 307-338; **1906**, 157-227, 439-480; **1910**, 355-477.

E. W. HOBSON

1. *On a general convergence theorem, and the theory of the representation of a function by series of normal functions,* Proc. London Math. Soc. (2) **6** (1908), 349-395.

G. HOHEISEL

1. *Gewöhnliche Differentialgleichungen. Sammlung Göschen,* Band 920, Fünfte, durchgesehene Auflage, De Gruyter, Berlin, 1956; Russian transl. of an earlier ed., GITTL, Moscow, 1937. MR **18**, 127.

W. A. HURWITZ

1. *An expansion theorem for a system of linear differential equations of the first order,* Trans. Amer. Math. Soc. **22** (1921), 526-543.

I. S. Kac

1. *On Hilbert spaces generated by monotone Hermitian matrix-functions,* Zap. Mat. Otd. Fiz.-Mat. Fak. i Har'kov. Mat. Obšč. (4) **22** (1950), 95-113. (Russian) MR **18**, 222.

M. V. Keldyš

1. *On a Tauberian theorem,* Trudy Mat. Inst. Steklov. **38** (1951), 77-86; English transl., Amer. Math. Soc. Transl. (2) **102** (1973), 133-143. MR **13**, 738.

K. Kodaira

1. *On ordinary differential equations of any even order and the corresponding eigenfunction expansions,* Amer. J. Math. **72** (1950), 502-544. MR **12**, 103.

B. I. Korenbljum

1. *A general Tauberian theorem for the ratio of functions,* Dokl. Akad. Nauk SSSR **88** (1953), 745-748. (Russian) MR **14**, 866.

A. G. Kostjučenko

1. *Asymptotic behavior of the spectral function of selfadjoint elliptic operators,* Fourth Math. Summer School (Kaciveli, 1966), "Naukova Dumka", Kiev, 1968, 42-117. (Russian) MR **41** #8852.
2. *Asymptotic behavior of the spectral functions of a singular differential operator of order 2m,* Dokl. Akad. Nauk SSSR **168** (1966), 276-279 = Soviet Math. Dokl. **7** (1966), 632-635. MR **34** #1884.
3. *Asymptotic distribution of the eigenvalues of elliptic operators,* Dokl. Akad. Nauk SSSR **158** (1964), 41-44 = Soviet Math. Dokl. **5** (1964), 1171-1174. MR **29** #4985.
4. *Distributions of eigenvalues for singular differential operators,* Dokl. Akad. Nauk SSSR **168** (1966), 21-24 = Soviet Math. Dokl. **7** (1966), 584-587. MR **33** #2947.

A. G. Kostjučenko and B. M. Levitan

1. *Asymptotic behavior of eigenvalues of the operator Sturm-Liouville problem,* Funkcional. Anal. i Priložen. **1** (1967), no. 1, 86-96 = Functional Anal. Appl. **1** (1967), 75-83. MR **34** #8229.

M. G. Kreĭn

1. *On Hermitian operators and directing functionals,* Akad. Nauk Ukrain. RSR. Zbirnik Prac. Mat. Inst. **10** (1948), 83-106. (Ukrainian)
2. *On a one-dimensional singular boundary problem of even order in the interval* $(0, \infty)$, Dokl. Akad. Nauk SSSR **74** (1950), 9-12. (Russian) MR **12**, 502.
3. *On the indeterminate case of the Sturm-Liouville boundary problem in the interval* $(0, \infty)$, Izv. Akad. Nauk SSSR Ser. Mat. **16** (1952), 293-324. (Russian) MR **14**, 558.

B. Ja. Levin

1. *Distribution of zeros of entire functions,* GITTL, Moscow, 1956; English transl., Transl. Math. Monographs, vol. 5, Amer. Math. Soc., Providence, R. I., 1964. MR **19**, 402; **28** #217.

N. Levinson

1. *Criteria for the limit-point case for second order linear differential operators,* Časopis. Pěst. Mat. Fys. **74** (1949), 17-20. MR **11**, 248.
2. *The inverse Sturm-Liouville problem,* Mat. Tidsskr. B. **1949**, 25-30. MR **11**, 248.
3. *A simplified proof of the expansion theorem for singular second order linear differential equations,* Duke Math. J. **18** (1951), 57-71; Addendum, 719-722. MR **12**, 828; **13**, 463.
4. *The expansion theorem for singular self-adjoint linear differential operators,* Ann. of Math. (2) **59** (1954), 300-315. MR **15**, 709.

B. M. Levitan

1. *Expansion in characteristic functions of differential equations of the second order,* GITTL, Moscow, 1950. (Russian) MR **12**, 183.

2. *On a decomposition theorem for characteristic functions of differential equations of the second order,* Dokl. Akad. Nauk SSSR **71** (1950), 605-608. (Russian) MR **11**, 720.
3. *Representation by Bessel functions in a Fourier-Stieltjes integral,* Dokl. Akad. Nauk SSSR **73** (1950), 453-456. (Russian) MR **12**, 95.
4. *Proof of the theorem on the expansion in eigenfunctions of self-adjoint differential equations,* Dokl. Akad. Nauk SSSR **73** (1950), 651-654. (Russian) MR **12**, 502.
5. *Expansion in Fourier series and integrals with Bessel functions,* Uspehi Mat. Nauk **6** (1951), no. 2 (42), 102-143. (Russian) MR **14**, 163.
6. *Remark on a theorem of V. A. Marčenko,* Trudy Moskov. Mat. Obšč. **1** (1952), 421-422; English transl., Amer. Math. Soc. Transl. (2) **101** (1973), 105-106. MR **15**, 315.
7. *On the asymptotic behavior of the spectral function of a self-adjoint differential equation of the second order,* Izv. Akad. Nauk SSSR Ser. Mat. **16** (1952), 325-352; English transl., Amer. Math. Soc. Transl. (2) **101** (1973), 192-221. MR **15**, 315.
8. *On the asymptotic behavior of the spectral function of a selfadjoint differential equation of the second order,* Uspehi Mat. Nauk **7** (1952), no. 3 (49), 108-109. (Russian)
9. *On the completeness of the squares of characteristic functions,* Dokl. Akad. Nauk SSSR **83** (1952), 349-352. (Russian) MR **14**, 171.
10. *On the spectral function of the equation* $y'' + \{\lambda - q(x)\} y = 0$, Izv. Akad. Nauk SSSR Ser. Mat. **17** (1953), 473-484; English transl., Amer. Math. Soc. Transl. (2) **102** (1973), 231-243. MR **15**, 316.
11. *On the expansion in characteristic functions of differential equations of the second order,* Uspehi Mat. Nauk **8** (1953), no. 3 (55), 156-157. (Russian)
12. *On expansion in characteristic functions of the equation* $y'' + \{\lambda - q(x)\} y = 0$, Dokl. Akad. Nauk SSSR **90** (1953), 17-20. (Russian) MR **15**, 316.
13. *On a special Tauberian theorem,* Izv. Akad. Nauk SSSR Ser. Mat. **17** (1953), 269-284; English transl., Amer. Math. Soc. Trânsl. (2) **101** (1973), 222-238. MR **15**, 316.
14. *On the asymptotic behavior of the spectral function of a self-adjoint differential equation of the second order and on eigenfunction expansions,* Izv. Akad. Nauk SSSR Ser. Mat. **17** (1953), 331-364; English transl., Amer. Math. Soc. Transl. (2) **102** (1973), 191-229. MR **15**, 316.
15. *On the asymptotic behavior of the spectral function of a self-adjoint differential equation of the second order and on eigenfunction expansions.* II, Izv. Akad. Nauk SSSR Ser. Mat. **19** (1955), 33-58; English transl., Amer. Math. Soc. Transl. (2) (to appear). MR **16**, 1027.
16. *Certain questions in the spectral theory of self-adjoint differential operators,* Uspehi Mat. Nauk **11** (1956), no. 6 (72), 117-144; English transl., Amer. Math. Soc. Transl. (2) **18** (1961), 49-79. MR **21** #2805; **23** #A1876.
17. *On the justification of Fourier's method in the one-dimensional case,* Naučn. Zap. Moskov. Zaočn. Ped. Inst. **1958**, 12-16. (Russian)
18. Appendix (pp. 217-277) to the Russian transl. of Titchmarsh [1].
19. *On a theorem of Titchmarsh and Sears,* Uspehi Mat. Nauk **16** (1961), no. 4 (100), 175-178. (Russian) MR **24** #A2133.
20. *Calculation of the regularized trace for the Sturm-Liouville operators,* Uspehi Mat. Nauk **19** (1964), no. 1 (115), 161-165. (Russian) MR **28** #3195.
21. *Investigation of the Green's function of a Sturm-Liouville equation with an operator coefficient,* Mat. Sb. **76** (118) (1968), 239-270 = Math. USSR Sb. **5** (1968), 229-260. MR **37** #2039.

B. M. LEVITAN AND I. S. SARGSJAN

1. *Theorem on convergence of twice differential expansion according to eigenfunctions of a Sturm-Liouville operator,* Akad. Nauk Armjan. SSR. Izv. Fiz.-Mat. Estest. Tehn. Nauki **9** (1956), no. 3, 3-15. (Russian) MR **18**, 310.
2. Addendum to [1], ibid. **9** (1956), no. 7, 105-106. (Russian)
3. *Some problems in the theory of the Sturm-Liouville equation,* Uspehi Mat. Nauk **15** (1960), no. 1 (91), 3-98 = Russian Math. Surveys **15** (1960), no. 1, 1-95. MR **22** # 5776.
4. *An extension of the solutions of a one-dimensional Dirac system,* Dokl. Akad. Nauk SSSR **165** (1965), 1241-1244 = Soviet Math. Dokl. **6** (1965), 1588-1591. MR **33** # 4362.

V. B. LIDSKIĬ AND V. A. SADOVNIČIĬ

1. *Regularized sums of roots of a class of entire functions,* Funkcional Anal. i Priložen. **1** (1967), no. 2, 52-59 = Functional Anal. Appl. **1** (1967), 133-140. MR **35** # 4380.
2. *Asymptotic formulas for the roots of a class of entire functions,* Mat. Sb. **75 (117)** (1968), 558-566 = Math. USSR Sb. **4** (1968), 519-528. MR **36** # 6621.
3. *Trace formulas in the case of the Orr-Sommerfeld equation,* Izv. Akad. Nauk SSSR Ser. Mat. **32** (1968), 633-648 = Math. USSR Izv. **2** (1968), 585-600. MR **37** # 3089.

L. A. LJUSTERNIK AND V. I. SOBOLEV

1. *Elements of functional analysis,* 2nd rev. ed., "Nauka", Moscow, 1965; English transl. of 1st ed., Hindustan, Delhi; Gordon and Breach, New York; Ungar, New York, 1961. MR **25** # 5361; # 5362; **35** # 698.

R. MACLEOD

1. *Some problems in the theory of eigenfunctions,* Thesis, Oxford University, Oxford, 1966.

V.A.MARČENKO

1. *Some questions of the theory of one-dimensional linear differential operators of the second order.* I, Trudy Moskov. Mat. Obšč. **1** (1952), 327-420; English transl., Amer. Math. Soc. Transl. (2) 101 (1973), 1-104. MR **15**, 315.
2. *Some questions of the theory of one-dimensional linear differential operators.* II, Trudy Moskov. Mat. Obšč. **2** (1953), 3-83; English transl., Amer. Math. Soc. Transl. (2) **101** (1973), 107-191. MR **15**, 315.
3. *Theorems of Tauberian type in spectral analysis of differential operators,* Izv. Akad. Nauk SSSR Ser. Mat. **19** (1955), 381-422; English transl., Amer. Math. Soc. Transl. (2) **102** (1973), 145-190. MR **17**, 852.

V. V. MARTYNOV

1. *Conditions for discreteness and continuity of the spectrum of a selfadjoint system of first-order differential equations,* Dokl. Akad. Nauk SSSR **165** (1965), 996-999 = Soviet Math. Dokl. **6** (1965), 1546-1550. MR **33** # 1762.

A. M. MOLČANOV

1. *On conditions for discreteness of the spectrum of self-adjoint differential equations of the second order,* Trudy Moskov. Mat. Obšč. **2** (1953), 169-199. (Russian) MR **15**, 224.

M. A. NAĬMARK

1. *Linear differential operators,* GITTL, Moscow, 1954; 2nd rev. ed., "Nauka", Moscow, 1969; English transl., Ungar, New York, 1967. MR **16**, 702; **35** # 6885.
2. *Investigation of the spectrum and the expansion in eigenfunctions of a non-selfadjoint differential operator of the second order on a semi-axis,* Trudy Moskov. Mat. Obšč. **3** (1954), 181-270; English transl., Amer. Math. Soc. Transl. (2) **16** (1960), 103-193. MR **15**, 959; **22** # 8162.

I. P. NATANSON

1. *Theory of functions of a real variable,* 2nd rev. ed., GITTL, Moscow, 1957; English transl., Ungar, New York, 1961. MR **16**, 120; **26** # 6309.

N. Nilsson

1. *Some estimates for eigenfunction expansions and spectral functions corresponding to elliptic differential operators,* Math. Scand. **9** (1961), 107-121. MR **23** # A1924.

I. G. Petrovskiĭ

1. *Lectures on the theory of integral equations,* GITTL, Moscow, 1948; English transl., Graylock Press, Rochester, N. Y., 1957. MR **19**, 42.

M. Plancherel

1. *Le passage à la limite des équations aux différences aux équations différentielles dans les problèmes aux limites,* Bull. Sci. Math. **47** (1923), 153-160, 170-177.

A. I. Plesner

1. *Spectral theory of linear operators,* "Nauka", Moscow, 1965. (Russian) MR **33** # 3106.

A. Ja. Povzner

1. *On differential equations of Sturm-Liouville type on a half-axis,* Mat. Sb. **23** (**65**) (1948), 3-52; English transl., Amer. Math. Soc. Transl. (1) **4** (1962), 24-101. MR **10**, 299.

F. Prats and J. S. Toll

1. *Construction of the Dirac equation central potential from phase shifts and bound states,* Phys. Rev. (2) **113** (1959), 363-370. MR **21** # 2488.

F. Riesz and B. Sz.-Nagy

1. *Leçons d'analyse fonctionnelle,* Akad. Kiadó, Budapest, 1952; English transl., Ungar, New York, 1955. MR **14**, 286; **17**, 175.

B. W. Roos and W. C. Sangren

1. *Spectra for a pair of singular first order differential equations,* Proc. Amer. Math. Soc. **12** (1961), 468-476. MR **23** # A2576.

V. A. Sadovničiĭ

1. *On the trace of the difference of two ordinary differential operators of higher orders,* Differencial'nye Uravnenija **2** (1966), 1611-1624 = Differential Equations **2** (1966), 836-842. MR **34** # 7866.
2. *On the traces of ordinary differential operators of higher orders,* Mat. Sb. **72** (**114**) (1967), 293-317 = Math. USSR Sb. **1** (1967), 263-288. MR **35** # 450.

I. S. Sargsjan

1. *Summability of the derivatives of an expansion in eigenfunctions of a Sturm-Liouville operator,* Dokl. Akad. Nauk SSSR **104** (1955), 821-824. (Russian) MR **17**, 853.
2. *Theorems concerning summation of derivatives of expansions into the standard and the generalized Fourier integral,* Akad. Nauk Armjan. SSR. Dokl. **23** (1956), 3-10. (Russian) MR **18**, 392.
3. *On differentiation of an expansion in eigenfunctions of a Sturm-Liouville operator,* Izv. Akad. Nauk SSSR Ser. Mat. **21** (1957), 263-282. (Russian) MR **19**, 961.
4. *Asymptotic behavior of the derivatives of a spectral function of a Sturm-Liouville operator,* Izv. Akad. Nauk Armjan. SSR. Ser. Fiz.-Mat. Nauk **10** (1957), no. 3, 3-16. (Russian) MR **19**, 746.
5. *Asymptotic behavior of the spectral matrix of the one-dimensional Dirac system,* Dokl. Akad. Nauk SSSR **166** (1966), 1058-1061 = Soviet Math. Dokl. **7** (1966), 245-248. MR **33** # 4366.
6. *Eigenfunction expansions for the one-dimensional Dirac system,* Dokl. Akad. Nauk SSSR **166** (1966), 1292-1295 = Soviet Math. Dokl. **7** (1966), 282-286. MR **33** # 2862.
7. *A theorem on the completeness of the eigenfunctions of the generalized Dirac system,* Akad. Nauk Armjan. SSR Dokl. **42** (1966), no. 2, 77-82. (Russian) MR **33** # 4724.
8. *Solution of the Cauchy problem for a one-dimensional nonstationary Dirac system,* Akad. Nauk Armjan. SSR Dokl. **42** (1966), 129-133. (Russian)

9. *Summation of derivatives of the spectral matrix of a one dimensional Dirac system,* Akad. Nauk Armjan. SSR Dokl. **42** (1966), no. 4, 193-197. (Russian) MR **34** # 4954.

10. *A theorem on equisummability of differentiated expansions in eigenfunctions of a one-dimensional Dirac system,* Akad. Nauk Armjan. SSR Dokl. **42** (1966), no. 5, 262-267. (Russian) MR **34** # 412.

11. *Solution of the Cauchy problem for a one-dimensional Dirac system,* Izv. Akad. Nauk Armjan. SSR. Ser. Mat. **1** (1966), no. 6, 392-436. (Russian) MR **34** # 7944.

12. *Asymptotic behavior of the spectral function and expansion in eigenfunctions of a one-dimensional Dirac system,* Izv. Akad. Nauk Armjan. SSR Ser. Mat. **2** (1967), no. 3, 155-194. (Russian) MR **35** # 6901.

E. Schrödinger

1. *Quantisierung als Eingenwertproblem,* Ann. Physik (4) **80** (1926), 437-490.

D. B. Sears

1. *Note on the uniqueness of the Green's functions associated with certain differential equations,* Canad. J. Math. **2** (1950), 314-325. MR **12**, 102.

D. B. Sears and E. C. Titchmarsh

1. *Some eigenfunction formulae,* Quart. J. Math. Oxford Ser. (2) **1** (1950), 165-175. MR **12**, 261.

R. F. Ševčenko

1. *On the trace of a differential operator,* Dokl. Akad. Nauk SSSR **164** (1965), 62-65 = Soviet Math. Dokl. **6** (1965), 1183-1186. MR **32** # 4313.

G. E. Šilov

1. *Mathematical analysis: Special course,* 2nd ed., Fizmatgiz, Moscow, 1961; English transl., Pergamon Press, New York, 1965. MR **24** # A1180; **32** # 2519.

V. I. Smirnov

1. *A course in higher mathematics.* Vol. V, rev. ed., Fizmatgiz, Moscow, 1959; English transl., Pergamon Press, Oxford; Addison-Wesley, Reading, Mass., 1964. MR **29** # 5964.

V. A. Steklov

1. *The problem of cooling a nonhomogeneous rigid bar,* Soobšč. Har'kov Mat. Obšč. (2) **5** (1896), 136-181. (Russian). Cf. *Problème de la refroidissement d'une barre hétérogène,* Ann. Fac. Sci. Univ. Toulouse (2) **3** (1901), 281-313.

V. V. Stepanov

1. *Course of differential equations,* Fizmatgiz, Moscow, 1958; German transl., Hochschulbücher für Mathematik, Band 20, VEB Deutscher Verlag, Berlin, 1963. MR **31** # 1418.

G. Szegö

1. *Orthogonal polynomials,* Amer. Math. Soc. Colloq. Publ., vol. 23, Amer. Math. Soc., Providence, R. I., 1959; Russian transl., Fizmatgiz, Moscow, 1962. MR **21** # 5029.

J. D. Tamarkin

1. *General problems in the theory of ordinary linear differential equations, and series expansion of arbitrary functions,* Dissertation, Petrograd, 1917. (Russian)

2. *Some general problems of the theory of ordinary linear differential equations and expansion of an arbitrary functions in series of fundamental functions,* Math. Z. **27** (1927), 1-54.

E. C. Titchmarsh

1. *Eigenfunction expansions associated with second-order differential equations.* Part I, Clarendon Press, Oxford, 1946; Russian transl., IL, Moscow, 1960. MR **8**, 458; **22** # 9662.

2. *Eigenfunction expansions associated with second-order differential equations.* Part II, Clarendon Press, Oxford, 1958; Russian transl., IL, Moscow, 1961. MR **20** # 1065.

3. *Introduction to the theory of Fourier integrals,* Clarendon Press, Oxford, 1937; Russian transl., GITTL, Moscow, 1948.

4. *On the uniqueness of the Green's function associated with a second-order differential equation,* Canad. J. Math. 1 (1949), 191-198. MR **10**, 537.

5. *Some eigenfunction expansion formulae,* Proc. London Math. Soc. (3) **11** (1961), 159-168. MR **23** # A384.

6. *Eigenfunction problems with periodic potentials,* Proc. Roy. Soc. London Ser. A. **203** (1950), 501-514. MR **12**, 502.

G. N. Watson

1. *A treatise on the theory of Bessel functions,* Cambridge Univ. Press. Cambridge; Macmillan, New York, 1922; Russian transl., GITTL, Moscow, 1959.

J. S. de Wet and F. Mandl

1. *On the asymptotic distribution of eigenvalues,* Proc. Roy. Soc. London Ser. A. **200** (1950), 572-580. MR **11**, 723.

H. Weyl

1. *Über gewöhnliche lineare Differentialgleichungen mit singulären Stellen und ihre Eigenfunktionen,* Nachr. Akad. Wiss. Göttingen **1909**, 37-64.

2. *Über gewöhnliche Differentialgleichungen mit Singularitaten und die zugehörigen Entwicklungen wilkürlicher Funktionen,* Math. Ann. **68** (1910), 220-269.

3. *Über gewöhnliche Differentialgleichungen mit singularen Stellen und ihre Eigenfunktionen,* Nachr. Akad. Wiss. Göttingen **1910**, 442-467.

A. Wintner

1. *On Dirac's theory of continuous spectra,* Physical Rev. (2) **73** (1948), 781-785. MR **9**, 595.

K. Yosida

1. *On Titchmarsh-Kodaira's formula concerning Weyl-Stone's eigenfunction expansion,* Nagoya Math. J. 1 (1950), 49-58. MR **13**, 39.